88 Springer Series in Solid-State Sciences

Edited by Peter Fulde

Springer

Berlin
Heidelberg
New York
Barcelona
Budapest
Hong Kong
London
Milan
Paris
Singapore
Tokyo

Springer Series in Solid-State Sciences

Editors: M. Cardona   P. Fulde   K. von Klitzing   H.-J. Queisser

Managing Editor: H. K. V. Lotsch   Volumes 1–89 are listed at the end of the book

T. Ishiguro   K. Yamaji   G. Saito

Organic Superconductors

Second Edition
With 295 Figures and 32 Tables

 Springer

Professor Dr. Takehiko Ishiguro
Department of Physics, Graduate School of Science
Kyoto University, Kyoto, 606-8502 Japan

Professor Dr. Kunihiko Yamaji
Electrotechnical Laboratory, 1-1-4 Umezono
Tsukuba, 305-8568 Japan
and
Institute of Materials Science, University of Tsukuba
Tsukuba, 305-8573 Japan

Professor Dr. Gunzi Saito
Department of Chemistry, Graduate School of Science
Kyoto University, Kyoto, 606-8502 Japan

Series Editors:

Professor Dr., Dres. h. c. Manuel Cardona
Professor Dr., Dres. h. c. Peter Fulde*
Professor Dr., Dres. h. c. Klaus von Klitzing
Professor Dr., Dres. h. c. Hans-Joachim Queisser

Max-Planck-Institut für Festkörperforschung, Heisenbergstrasse 1, D-70569 Stuttgart, Germany
* Max-Planck-Institut für Physik komplexer Systeme, Nöthnitzer Strasse 38
  D-01187 Dresden, Germany

Managing Editor:

Dr.-Ing. Helmut K. V. Lotsch

Springer-Verlag, Tiergartenstrasse 17, D-69121 Heidelberg, Germany

ISSN 0171-1873
ISBN 3-540-63025-2 Springer-Verlag Berlin Heidelberg New York
ISBN 3-540-51321-3 1st Edition Springer-Verlag Berlin Heidelberg New York

Library of Congress Cataloging-in-Publication Data
Ishiguro, Takehiko, 1930– Organic superconductors / T. Ishiguro, K. Yamaji, G. Saito. – 2nd ed. p. cm. – (Springer series in solid-state sciences; 88) Includes bibliographical references and index. ISBN 3-540-63025-2 (alk. paper) 1. Organic superconductors. 2. Organic conductors. I. Yamaji, K. (Kunihiko), 1942– . II. Saito, G. (Gunzi), 1945– . III. Title. IV. Series. QC611.98.O74I74 1998 537.6'23–dc21 98-8807 CIP

Typesetting: PS^{TM} Technical Word Processor
Cover concept: eStudio Calamar Steinen
Cover production: *design & production* GmbH, Heidelberg

SPIN: 10629903   54/3144 – 5 4 3 2 1 0 – Printed on acid-free paper

Preface

On publication of the first edition, the field of organic superconductors was already rich in material data, but understandings based upon appropriate theories and serious measurements to determine specific features were not mature. We had obtained intriguing data on the superconductivity that needed to be clarified, particularly with regard to the restricted dimensionality. Investigations to obtain knowledge on the electronic structure were still at an early stage.

In recent years, investigations to clarify those issues have been clearly promoted. To understand superconductivity and the associated nature, interactions with other fields, such as the study of the cuprate two-dimensional superconductor, have been quite effective. The analysis of the electronic structure has been substantially promoted through the observation of the de Haas effect and the angle-dependent magnetoresistance oscillation by utilizing high-magnetic-field facilities. It is noteworthy that new organic superconductors with a high T_c exceeding 10 K and fullerence superconductors with T_c exceeding 30 K have been reported.

In this second edition we describe organic and molecule-based superconductors, taking into account the advances. The material data are put in a more organized way based on the present-day understanding. In addition, we introduce a new chapter on material chemistry to describe the design, synthesis, and crystal growth of the salts; this was prepared by G. Saito as a new co-author. We believe that this is useful in helping physics and chemistry to coalesce so as to promote advances to new frontieres.

We are deeply indebted to Helmut K.V. Lotsch for his kind effort to make this monograph well readable. Finally, we express our sincere gratitude and deep appreciation for the permission to reproduce figures from authors and publishers of the following journals:

Acta Crystallographica
Advanced Materials
Advances in Physics
Angewandte Chemie
Applied Physics Letters
Bulletin of the Chemical Society of Japan
Chemistry Letters

Chemical Physics Letters
Europhysics Letters
Inorganic Chemistry
International Journal of Quantum Chemistry
Japanese Journal of Applied Physics
JETP Letters
Journal de Physique
Journal de Physique Letters
Journal of Magnetism and Magnetic Materials
Journal of Materials Chemistry
Journal of Molecular Electronics
Journal of Physics, Condensed Matter
Journal of Physics and Chemistry of Solids
Journal of Superconductivity
Journal of the American Chemical Society
Journal of the Chemical Society, Chemical Communications
Kagaku to Kogyo (Chemistry and Chemical Industry)
Kotai Butsuri (Solid State Physics)
Journal of the Physical Society of Japan
Molecular Crystals and Liquid Crystals
Physica
Physical Review
Physical Review Letters
Physics Letters
Physics Today
Science
Solid State Communications
Soviet Physics - JETP
Synthetic Metals
Zeitschrift für Physik

Kyoto, Ibaraki*
November 1997

Takehiko Ishiguro
*Kunihiko Yamaji**
Gunzi Saito

Preface to the First Edition

The initial impetus for the search for an organic superconductor was the proposal of the existence of a polymer superconductor with a high critical temperature (T_c). This spurred on activities having the aim of synthesizing and characterizing organic conductors, which had already been going on for two decades. These efforts have resulted in the thriving field of low-dimensional conductors and superconductors.

This monograph is intended to be an introduction as well as a review of the study of organic conductors and superconductors. The investigations are sufficiently rich to warrant a treatise of some length. At the same time they have produced a few active subfields, each containing exciting topics. This situation seems to necessitate a monograph describing the current status of the field for both researchers and newcomers to the field. Such a need may also be felt by scientists engaged in the study of the high-T_c oxide superconductors for comparison of the two kinds of new superconductors, which share some important aspects, for example, the low-dimensionality and the competition or coexistence of superconductivity and magnetism.

However, available experimental and theoretical results are sometimes conflicting and have not yet been arranged into a coherent standard picture of the whole field. Further developments are continually being reported and therefore it is still premature to write a textbook about some of the topics. However, we have tried to include discussions of recent topics in this volume.

The material is divided into ten chapters. Following a historical introduction in Chap. 1, we describe the organic conductors in Chap. 2 as background for the superconductors. The superconductors, categorized according to the major types of organic molecular structures, are discussed in Chaps. 3 and 5-7. Theoretical arguments about the possible mechanism of superconductivity are presented in Chap. 8. Spin Density Waves (SDWs) are dealt with in Chap. 4, where a picture is provided of how the SDW is suppressed and superconductivity appears in organic materials. Chapter 9 shows that the same picture gives a coherent description of a new effect, the field-induced SDW. The book concludes with remarks on possible future developments. Each chapter may be read separately, although it is desirable to read through the whole. Because we are not expert in the areas, chemical synthesis and crystal growth are not discussed. Chapters 4, 8 and

9 have been written by K.Y. and the rest by T.I. It should be noted that the references listed are not intended as a complete bibliography.

We are indebted to our colleagues at the Electrotechnical Laboratory whose research efforts have made it possible for us to be deeply involved in this attractive field: Dr. S. Abe, Dr. H. Anzai, Dr. H. Bando, Dr. K. Kajimura, Mr. N. Kinoshita, Dr. K. Murata, and Dr. M. Tokumoto. We also wish to acknowledge the research cooperation of Prof. G. Saito (Institute for Solid State Physics, University of Tokyo), Prof. H. Kobayashi (Toho University), Dr. K. Kikuchi (Tokyo Metropolitan University), and we are grateful to Prof. J. Kondo (Electrotechnical Laboratory) and Prof. H. Inokuchi (Institute for Molecular Science) for their informative conversations and encouragement. Thanks are also extended to Ms. Makiko Isomata and Ms. Junko Kusumoto for typing manuscripts, and to Dr. H. Lotsch and collaborators at Springer-Verlag for their efforts to make the text read fluently. It is also a pleasure to acknowledge many unnamed scientists with whom we exchanged scientific information to promote our mutual research. Finally, thanks are due to the authors and publishers of the distinguished papers cited here for allowing us to quote their persuasive data.

Kyoto, Ibaraki   *Takehiko Ishiguro*
December 1989   *Kunihiko Yamaji*

Contents

1. The Evolution of Organic Superconductors

This introductory chapter briefly reviews the development of organic materials exhibiting conducting and superconducting properties. It also outlines the build-up of the book.

1.1 Emergence of Organic Conductors

Organic materials have primarily been regarded as electrical insulators. This appears to be true even today when we observe man-made materials such as plastics in our surroundings. In the 1940s, however, the electrical conduction in organic crystals began to draw the attention of scientists: The mobility of electrons and its underlying mechanism, as well as the synthesis of materials of higher conductivity, became "hot" subjects of investiagtion. These research efforts have brought about the field of organic semiconductors.

In 1954 the perylene bromine complex [1.1] was found to display a marked increase in conductivity over other materials. This has led to the development of highly-conducting organics, such as the organic metals. A high concentration of current carriers, which according to the principles of solid-state physics is regarded as a prerequisite to metallic conduction, was found in TCNQ (7, 7, 8, 8-tetracyano-p-quinodimethane) salts. These compounds have a large electrical conductivity, a virtually zero activation energy, and a temperature-independent paramagnetism [1.2]. Subsequently, the number of organic metals expanded explosively. These are the richly varied organic charge-transfer salts, the intermolecular compounds stabilized by the partial transfer of electrons between constituent molecules. In Fig. 1.1 the chemical structures of the principal molecules discussed in this monograph are illustrated.

The extraordinarily high electrical conductivity found in **TTF·TCNQ** (tetrathiafulvalene tetracyanoquinodimethane) in 1973 [1.3] accelerated interest in organic conductors, not only because of their huge electrical conductivity at 60 K, but also by the possibility of fluctuating superconductivity at a high-critical temperature (T_c). Although the interpretation of fluctu-

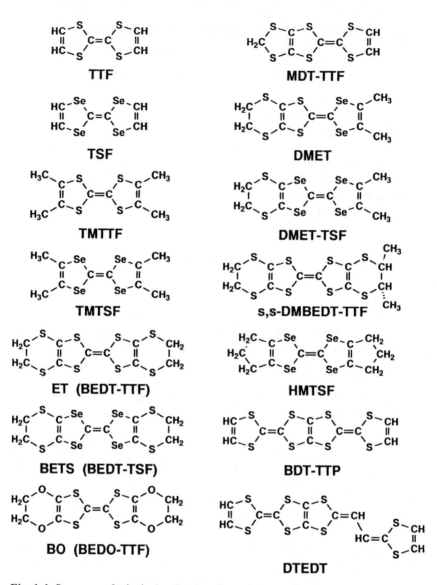

Fig. 1.1. Structures of principal molecules of organic conductors and superconductors

ating superconductivity was refuted by later investigations, intensive studies on the remarkably anisotropic high conductivity, e.g., quasi one-dimensional conductivity, opened the door to the active field of restricted-dimensional conductors [1.4].

The TTF·TCNQ salt and its relatives such as TSF·TCNQ (tetraselena fulvalene TCNQ) have primarily been regarded as a convenient prototype

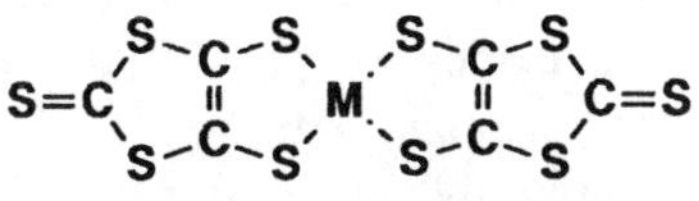

M(dmit)$_2$ (M=Ni,Pd)

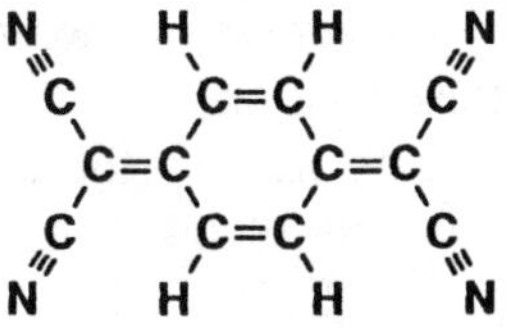

TCNQ

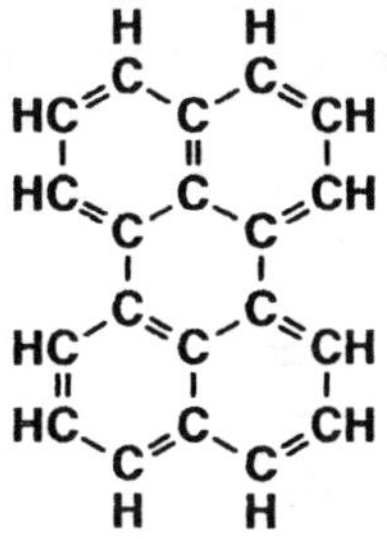

Perylene

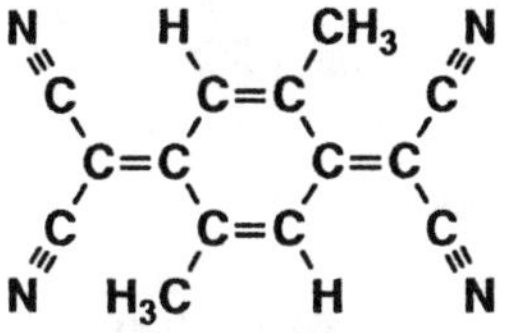

Me$_2$TCNQ

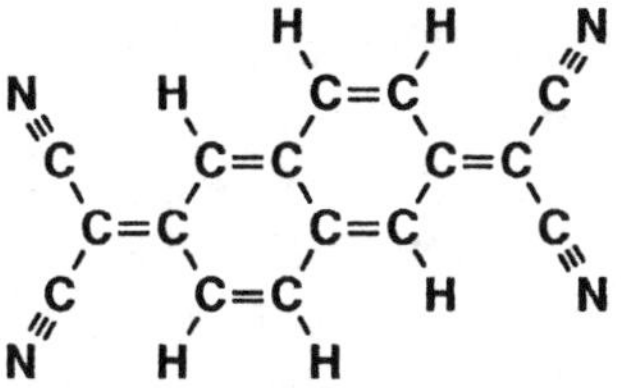

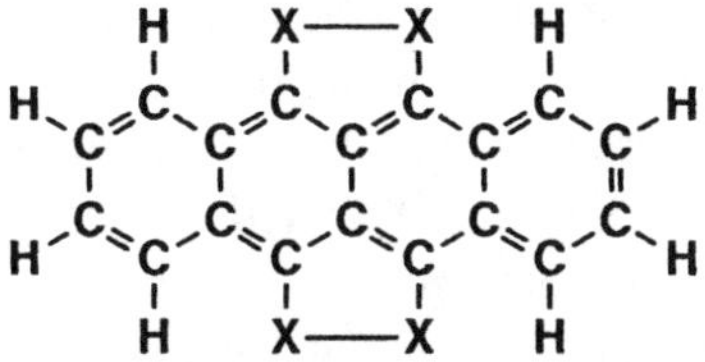

TTT(X=S)
TST(X=Se)
TTeT(X=Te)

TNAP

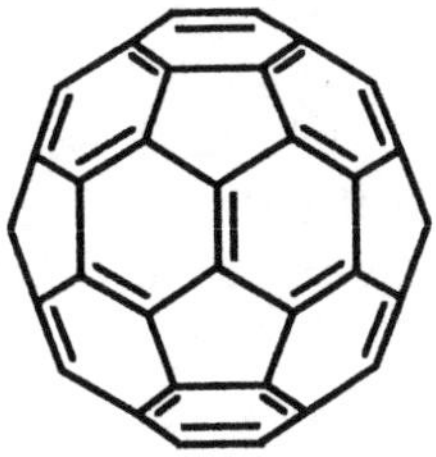

C$_{60}$

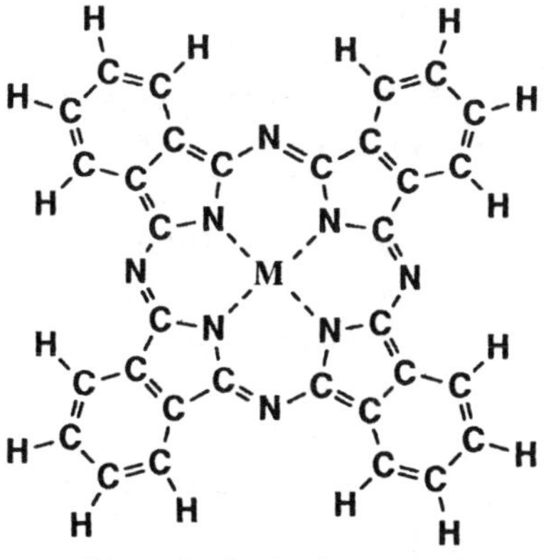

Metal phthalocyanine

for testing the theories of quasi one-dimensional conductors, where four types of interesting ground states are expected to occur: Charge Density Wave (CDW), Spin Density Wave (SDW), Singlet Superconductivity (SS), and Triplet Superconductivity (TS) states. From the viewpoint of materials science, the influence of molecular and crystalline structures on the observed phenomena are particularly interesting. This can be investigated by noting the consequences of modifying the structures either chemically, through substitution of the constituent atoms/molecules, or physically, by applying pressure to modify the intermolecular spacings. The effect of a chemical modification of the molecule, for example, by substitution of selenium for sulfur in TTF on the conductivity of the resultant crystals, enables one to investigate the roles of constituting atoms and molecules played in the conductance. The application of pressure leads to the information on the effect of intermolecular spacings.

1.2 Development of Organic Superconductors

The search for a high critical temperature (high-T_c) superconductor has been one of the principal goals in the course of investigation on organic conductors. This activity was spurred by *Little* [1.5] following the Bardeen, Cooper, and Schrieffer (BCS) theory of superconductivity which is based on the coherent motion of paired electrons with a phonon-mediated attractive interaction. *Little* extended the electron pairing mechanism to describe the electrons moving along an organic polymer with highly polarizable side chains. He also proposed a design to promote a high polarizability of the part attached to the conducting polymers. His idea stimulated much activity in the synthesis of superconductor. Although these superconducting polymers have not yet been successfully synthesized, *Little's* idea made great impact on the development of the field of organic conductors. The highly conducting organic materials that have been synthesized, through their novel quasi one- or two-dimensional behavior, have brought about new frontiers in condensed-matter physics. It is noteworthy that the synthesis of these new materials has generated a new area of research at the interface between chemistry and physics [1.6].

Until 1978 it was unclear whether attempts to synthesize organic superconductors would eventually be successful. The first superconductivity in an organic material was found in pressurized $(TMTSF)_2 PF_6$ (bis-tetramethyl-tetraselenafulvalene-hexafluorophosphate) in 1979 [1.7]. By replacing PF_6 with AsF_6, SbF_6, ClO_4, etc., a series of organic superconductors were discovered, as listed in Table 1.1. In these $(TMTSF)_2 X$ species, where X stands for an electron acceptor molecule such as PF_6 or ClO_4, the TMTSF molecules are stacked in columns along which the highest conductivity occurs. It also turns out that a considerable transverse coupling between the columns is crucial, firstly, in realizing good metallic conductivity down to low temperatures and, secondly, in impeding the appearance of an insulating phase, which has been recognized as being an SDW phase (Chaps. 3 and

Table 1.1. Development of molecule-based superconductors in which symmetric donors are dominant. P_c denotes the critical pressure and T_c the critical temperature

Material	Symmetry of counter molecule	P_c [kbar]	T_c [K]	Year of report
$(TMTSF)_2PF_6$	Octahedral	12	0.9[a]	1980
$(TMTSF)_2SbF_6$	Octahedral	10.5	0.38	1980
$(TMTSF)_2TaF_6$	Octahedral	11	1.35	1981
$(TMTSF)_2ClO_4$	Tetrahedral	0	1.4	1981
$(TMTSF)_2AsF_6$	Octahedral	9.5	1.1	1982
$(TMTSF)_2ReO_4$	Tetrahedral	9.5	1.2	1982
$(TMTSF)_2FSO_3$	Tetrahedral-like	5	≈ 3	1983
$(TMTTF)_2Br$	Spherical	26	0.8	1994
$(ET)_2ReO_4$	Tetrahedral	4.0	2.0	1983
$\beta_L\text{-}(ET)_2I_3$	Linear	0	1.5	1984
$\beta\text{-}(ET)_2IBr_2$	Linear	0	2.7	1984
$\beta_H\text{-}(ET)_2I_3$	Linear	0	8.1	1985
$\beta\text{-}(ET)_2AuI_2$	Linear	0	4.9	1985
$\gamma\text{-}(ET)_3I_{2.5}$	Linear	0	2.5	1985
$\kappa\text{-}(ET)_4Hg_{2.89}Cl_8$	Polymeric	12	1.8	1985
$\theta\text{-}(ET)_2I_3$	Linear	0	3.6	1986
$\kappa\text{-}(ET)_2I_3$	Linear	0	3.6	1987
$\kappa\text{-}(ET)_4Hg_{2.78}Br_8$	Polymeric	0	4.3	1987
$(ET)_3Cl_2\cdot(H_2O)_2$	Cluster	16	2	1987
$\kappa\text{-}(ET)_2Cu(NCS)_2$	Polymeric	0	10.4 (8.7)[b]	1988
$\kappa\text{-}(ET)_2Cu(NCS)_2$ deuterated	Polymeric	0	11.2 (9.0)[b]	1988
$\alpha\text{-}(ET)_2NH_4Hg(SCN)_4$	Polymeric	0	0.8÷1.7[1]	1990
$\kappa\text{-}(ET)_2Cu[N(CN)_2]Br$	Polymeric	0	11.8 (10.9)[b]	1990
$\kappa\text{-}(ET)_2Ag(CN)_2H_2O$	Cluster	0	5.0	1990
$\kappa\text{-}(ET)_2Cu[N(CN)_2]Cl$	Polymeric	0.3	12.8	1990
$\kappa\text{-}(ET)_2Cu[N(CN)_2]Cl$ deuterated	Polymeric	0.3	13.1	1991
$\kappa\text{-}(ET)_2Cu[N(CN)_2]Br$ deuterated	Polymeric	0	11.2 (10.6)[b]	1991
$\kappa\text{-}(ET)_2Cu_2(CN)_3$	Polymeric	1.5	2.8	1991
$\kappa'\text{-}(ET)_2Cu_2(CN)_3$	Polymeric	0	4.1	1991
$(ET)_4Pt(CN)_4H_2O$	Cluster	6.5	2	1991
$\kappa\text{-}(ET)_2Cu(CN)[N(CN)_2]$	Polymeric	0	11.2	1991
$\kappa\text{-}(ET)_2Cu(CN)[N(CN)_2]$ deuterated	Polymeric	0	12.3	1992
$(ET)_4Pd(CN)_4H_2O$	Cluster	7	1.2	1992
$\kappa\text{-}(ET)_2Cu[N(CN)_2]Cl_{0.5}Br_{0.5}$	Ploymeric	0	11.3	1993
$\alpha\text{-}(ET)_2KHg(SCN)_4$	Polmeric	0	0.3	1993
		1.2[d]	1.2	1994
$\alpha\text{-}(ET)_2RbHg(SCN)_4$	Polymeric	0	0.5	1994
$\alpha\text{-}(ET)_2TlHg(SCN)_4$	Polymeric	0	0.1	1994
$\kappa\text{-}(ET)_2Cu[N(CN)_2]Cl_{0.25}Br_{0.75}$	Polymeric	0	11.5	1994
$\kappa\text{-}(ET)_2Cu[NC(N)_2]Cl_{0.15}Br_{0.85}$	Polymeric	0	10	1994

5

Table 1.1 (Continued)

Material	Symmetry of counter molecule	P_c [kbar]	T_c [K]	Year of report
κ_L-$(ET)_2 Cu(CF_3)_4 \cdot TCE$	Planar	0	4.0	1994
κ_H-$(ET)_2 Cu(CF_3)_4 \cdot TCE$	Planar	0	9.2	1994
κ_H-$(ET)_2 Ag(CF_3)_4 \cdot TCE$	Planar	0	11.1	1994
κ-$(ET)_2 Cu[N(CN)_2]Br_{0.9}I_{0.1}$	Polymeric	3	5.9	1995
κ_H-$(ET)_2 Ag(CF_3)_4 \cdot TBE$	Planar	0	7.2	1995
κ_L-$(ET)_2 Cu(CF_3)_4 \cdot TBE$	Planar	0	5.2	1995
κ_L-$(ET)_2 Ag(CF_3)_4 \cdot 121DBCE$	Planar	0	4.5	1995
κ_L-$(ET)_2 Ag(CF_3)_4 \cdot 121DCBE$	Planar	0	3.8	1995
κ_H-$(ET)_2 Ag(CF_3)_4 \cdot 112DCBE$	Planar	0	10.2	1995
κ_L-$(ET)_2 Cu(CF_3)_4 \cdot 112DCBE$	Planar	0	4.9	1995
κ_L-$(ET)_2 Ag(CF_3)_4 \cdot 112DCBE$	Planar	0	4.1	1995
β''-$(ET)_4 Fe(C_2 O_4)_3 H_2 O \cdot PhCN$	Octahedral	0	$6.5 \div 7.7$[1]	1995
κ_H-$(ET)_2 Au(CF_3)_4 \cdot TCE$	Planar	0	10.5	1996
κ_H-$(ET)_2 Ag(CF_3)_4 \cdot 121DCBE$	Planar	0	7.3	1996
κ_L-$(ET)_2 Au(CF_3)_4 \cdot TBE$	Planar	0	5.8	1996
κ_L-$(ET)_2 Cu(CF_3)_4 \cdot 121DBCE$	Planar	0	5.5	1996
κ_L-$(ET)_2 Au(CF_3)_4 \cdot 112DCBE$	Planar	0	5.0	1996
κ_L-$(ET)_2 Au(CF_3)_4 \cdot 121DBCE$	Planar	0	5.0	1996
κ_L-$(ET)_2 Ag(CF_3)_4 \cdot TBE$	Planar	0	4.8	1996
κ_L-$(ET)_2 Ag(CF_3)_4 \cdot 121DBCE$	Planar	0	4.5	1996
κ_L-$(ET)_2 Cu(CF_3)_4 \cdot 121DCBE$	Planar	0	3.5	1996
κ_L-$(ET)_2 Au(CF_3)_4 \cdot 121DCBE$	Planar	0	3.2	1996
κ_L-$(ET)_2 Ag(CF_3)_4 \cdot TCE$	Planar	0	2.4	1996
κ_L-$(ET)_2 Au(CF_3)_4 \cdot TCE$	Planar	0	2.1	1996
β''-$(ET)_2 SF_5 CH_2 CF_2 SO_3$		0	5.3	1996
$(BO)_3 Cu_2 (NCS)_3$	Polymeric	0	1.06	1990
$(BO)_2 ReO_4 \cdot H_2 O$	Tetrahedral	0	1.5	1991
λ-$(BETS)_2 GaCl_4$	Tetrahedral	0	8	1993
λ-$(BETS)_2 GaBr_x Cl_y$	Tetrahedral	0	$7 \div 8$[1]	1995
λ-$(BETS)_2 GaCl_3 F$	Tetrahedral	0	3.5	1995

[a] The value was given in the first report of *Jérome* et al. [1.7], while later work provided $T_c = 1.1$ K at 6.5 kbar [1.8]

[b] The parenthesized values are given through an orthodox way of analysis, taking accout of the dimensionality and thermal fluctuations (Sect. 5.3, Appendix A)

[c] TCE : 1, 1, 2-trichloroethane. [d] Uniaxial
 TBE : 1, 1, 2-tribromoethane
 121DBCE : 1, 2-dibromo-1-chloroethane
 121DCBE : 1, 2-dichloro-1-bromoethane
 112DCBE : 1, 1-dichloro-2-bromoethane
 PhCN : benzonitrile

---

[1] The symbol $\div$ is used throughout the text as a shorthand for "from - to" or "between".

4). This type of insulating state is suppressed by breaking the Fermi-surface nesting by increasing or modulating the transverse intercolumnar coupling. Then, the superconducting phase appears at low temperatures. The Fermi-surface nesting is a typical property of low-dimensional metals which will be described in Chaps. 2 and 4. Chapter 4 discusses how the Fermi-surface nesting is broken in $(TMTSF)_2 X$ with a realistic band model.

To understand the basic mechanism of the superconductivity, the electronic states and the dynamics of the electrons must be known. During measurements taken for this purpose under a magnetic flux density of up to 10 T or more, an oscillatory behavior in the conductivity was found. In ordinary metals, the electrons have a circular motion in the plane normal to the magnetic field direction, whereas in a quasi one-dimensional conductor, the circular motions are impeded when an electron moves to the less conductive direction. The study of $(TMTSF)_2 X$ led to the discovery of the field-induced SDW states. This effect not only promotes further understanding of the relation between the SDW and superconducting phases but also is an amazing phenomenon discovered in low-dimensional synthetic metals in itself. The SDW and the field-induced SDW states will be described at length in Chaps. 4 and 9, respectively.

The second molecule which became the building block of organic superconductors so far discovered is **BEDT-TTF** (bisethylenedithio-tetrathiafulvalene, or simply **ET** as we call it hereafter). The first superconductor consisting of this molecule was $(ET)_4 (ReO_4)_2$ (Table 1.1). The chemical formula of ET is depicted in Fig. 1.1. It is based on sulfur atoms. In contrast to the TMTSF molecule which forms isomorphous crystal structures with different X, the ET molecule forms different types of compounds of varying composition ratio and crystal structure. Even for the same composition ratio, different types of crystal structure are possible. Among them a certain crystal called the **β-type of $(ET)_2 I_3$** exhibits superconductivity at ambient pressure [1.10]. Furthermore, it was found that by applying moderate pressure on the order of 1 kbar to this salt, T_c was raised as high as 8 K [1.11, 12]. The nature of its superconductivity will be described in Chap. 5.

Typical ET salts contrast with the TMTSF salts, in that they have quasi two-dimensional electronic states. In this case circular motions of the electrons are possible if the direction of the magnetic field is applied normal to the two-dimensional conducting plane. Through the dynamics of the electrons under the influence of the magnetic field, the size of the Fermi surface and the parameters of the dynamical aspects of the electrons are being elucidated for ET salts, as described in Sect. 5.1.

In 1987, the critical temperature T_c of the organic superconductor with κ-$(ET)_2 Cu(NCS)_2$ was raised to 10.4 K [1.13]. Successively, T_c of an ambient-pressure organic superconductor with κ-$(ET)_2 Cu[N(CN)_2]Br$ was further raised to 11.8 K [1.14]. In addition, its isostructural salt, κ-$(ET)_2 Cu \cdot$

[N(CN)$_2$]Cl, displays nonmetallic behavior at ambient pressure, but by applying a moderate pressure of 30 MPa (0.3 kbar) it was found to exhibit superconductivity at 12.8 K [1.15]. These **κ-type ET salts** with T$_c$ higher than 10 K exhibit a pronounced two-dimensionality where the coherence length of the superconductivity in the direction perpendicular to the conducting plane is shorter than the interlayer spacing. The high-temperature environment promotes thermal fluctuation [1.16]. These high-T$_c$ organic superconductors will be described in Sect. 5.3.

The salt α-(ET)$_2$NH$_4$Hg(SCN)$_4$, with an enhanced two-dimensionality, exhibits superconductivity near 1 K [1.17]. Its isostructural salts, such as α-(ET)$_2$KHg(SCN)$_4$ and α-(ET)$_2$RbHg(SCN)$_4$, on the other hand, do display a resistive decrease below 0.3 and 0.5 K, respectively, but down to 10 mK, they do not exhibit zero resistance. As a result, T$_c$ cannot be determined uniquely. This behavior is ascribed to the superconductivity localization effect specific to two-dimensional material with a high sheet resistance [1.18]. However, the antiferromagnetism appearing near 8 K in these salts is believed to suppress the superconductivity [1.19]. The interplay between superconductivity and magnetism has also been an intriguing subject not only for the TMTSF salts [1.20] but for κ-type ET salts in which antiferromagnetism exists adjacent to superconductivity. This interplay is more evident in **BETS** (bisethylenedithio-tetraselenafulvalene): λ-(BETS)$_2$GaCl$_4$ undergoes a superconductivity transition at 8 K, while the isostructural salt with Fe replacing Ga exhibits a non-metal transition at 8 K [1.21]. It is also worthy to note that the non-metallic state is suppressed by applying a magnetic flux density stronger than 9 T, presumably due to a ferromagnetic spin ordering [1.22], as will be described in Sect. 5.5.

It is interesting that the non-symmetrical donor **DMET** (dimethylethylenedithio-diselenadithiafulvalene, Fig. 1.1) which is a hybrid of TMTSF and ET, also yields a superconductor (Table 1.2) [1.23]. Interestingly, another hybrid molecule **MDT-TTF** (methylenedithio-tetrathiafulvalene, Fig. 1.1) which is composed of two parent molecules, TTF and BMDT-TTF (bismethylenedithio-tetrathiafulvalene), neither of which has yet been used successfully to produce superconductors, is itself a constituent of known superconductors [1.24]. In these cases, the non-symmetrical molecules are stacked so as to construct a symmetrical crystal by alternatingly changing the direction of the molecules. On the other hand, **DTEDT** (2-(1, 3-dithiol-2-ylidene)-5-(2-ethanediylidene-1, 3-dithiole)-1, 3, 4, 6-tetrathiapentalene, Fig. 1.1) is stacked by aligning the molecular direction uniformly, resulting in a non-symmetrical crystal with respect to the donor molecular arrangements [1.25], as described in Sect. 6.3. All the constituent molecules of molecule-based superconductors above mentioned are donor molecules, so they are superconductors of the cation-radical salts.

Table 1.2. Development of molecule-based superconductors in which hybrid donors are dominant. P_c denotes the critical pressure and T_c the critical temperature

Material	Symmetry of counter molecule	P_c [kbar]	T_c [K]	Year of report
$(DMET)_2 Au(CN)_2$	Linear	3.5	0.8	1987
$(DMET)_2 AuCl_2$	Linear	0	0.83	1987
$(DMET)_2 AuBr_2$	Linear	1.5	1.6	1987
$(DMET)_2 AuI_2$	Linear	5	0.55	1987
$(DMET)_2 I_3$	Linear	0	0.47	1987
$(DMET)_2 IBr_2$	Linear	0	0,58	1987
κ-$(DMET)_2 AuBr_2$	Linear	0	1.9	1988
κ-$(MDT\text{-}TTF)_2 AuI_2$	Linear	0	4.1	1988
κ-$(S, S\text{-}DMBEDT\text{-}TTF)_2 ClO_4$	Tetrahedral	5.8	2.6	1991
$(DMET\text{-}TSF)_2 AuI_2$	Linear	0	0.58	1993
$(DMET\text{-}TSF)_2 I_3$	Linear	0	0.4	1994
$(DTEDT)_3 Au(CN)_2$	Linear	0	4	1995
$(DTEDT)_3 SbF_6$	Octahedral	0	0.3	1996
$(TMET\text{-}STF)_2 BF_4$	Tetrahedral	0	3.8	1997

An acceptor molecule **Ni(dmit)$_2$** (nickel-bis-4, 5-dimercapto-3-dithiole-2-thione, Fig. 1.1) has been developed as a constituent molecule of superconducting materials [1.26], as will be described in Chap. 7. Even though the M(dmit)$_2$ (M: Ni, Pd) includes superconductors of TTF[M(dmit)$_2$] (M: Ni, Pd) and EDT-TTF[Ni(dmit)$_2$] (Table 1.3), which are of the ionic donor-acceptor type, they are classified into anion-radical-salt superconductors in this monograph since TTF and EDT-TTF molecules do not directly contribute to the superconductivity.

In 1991, a new class of molecule-based superconductors was opened up by the discovery of superconductivity in alkali-metal-doped fullerenes in $K_3 C_{60}$ and $Rb_3 C_{60}$ with T_c of 18 and 29 K, respectively (Table 1.4) [1.27, 28]. It is not customary to regard the fullerene molecules consisting solely of carbon atoms (Fig. 1.1) as organic molecules. The type of molecular structure of the fullerene molecules belong to a category different from those of TMTSF and ET, and the dimensionality of the crystal is three, in contrast to one and two for (TMTSF)$_2$X and (ET)$_2$X, respectively. In spite of these differences, the fullerene superconductors, which are anion-radical salts, display a number of features also exhibited by the organic supercon-

Table 1.3. Development of molecule-based superconductors in which metal-(dmit)$_2$ are dominant. P_c denotes the critical pressure and T_c the critical temperature

Material	Symmetry of counter molecule	P_c [kbar]	T_c [K]	Year of report
α-(TTF)[Ni(dmt)$_2$]$_2$	Planar	7	1.6	1986
β-N(CH$_3$)$_4$[Ni(dmit)$_2$]$_2$	Tetrahedral	7	5.0	1987
α-(EDT-TTF)[Ni(dmit)$_2$]	Planar	0	1.3	1993
α-(TTF)[Pd(dmit)$_2$]$_2$	Planar	22	1.7	1989
α'-(TTF)[Pd(dmit)$_2$]$_2$	Planar	20	6.5	1989
β-N(CH$_3$)$_4$[Pd(dmit)$_2$]$_2$	Tehrahedral	6.5	6.2	1991
α-N(CH$_3$)$_2$(C$_2$H$_5$)$_2$[Pd(dmit)$_2$]$_2$	Tetrahedral-like	4	2.4	1992
β'-P(CH$_3$)$_2$(C$_2$H$_5$)$_2$[Pd(dmit)$_2$]$_2$	Tetrahedral like	7.0	4	1997

ductors such as the dominant role played by the π-electrons in conduction and the relatively narrow bands with low electron density. Chapter 10 is therefore devoted to the **fullerene superconductorts**.

The organic superconductors described above are the charge-transfer compounds which are formed by stacking molecules with relatively low molecular weights. This is not in accordance with *Little*'s original proposal which is based on a polymer such as a polyacetylene-type material decorated with polarizable side groups [1.5]. The first polymer superconductor was realized as (SN)$_x$ with $T_c = 0.28$ K [1.29]. However, the material proposed by *Little* has been found to be very difficult to synthesize. Although the electrical conductivity of these conjugated polymers has reached as much as 10^5 S/cm [1.30], superconductivity has not yet been achieved. One of the principal reasons for this is presumably the inherent disorder in these polymers which may suppress the superconductivity in low-dimensional systems.

Mechanisms of organic superconductors will be reviewed from a theoretical viewpoint in Chap. 8. The structure and synthesis will be described, placing emphasis upon the designing of organic superconductors in Chap. 11. For elaborate descriptions of crystallographic structures, the reader is referred to the detailed work of *Williams* and co-workers [1.31], and others [1.32].

Table 1.4. Development of molecule-based superconductors in which fullerenes are dominant (fcc: face-centered cubic, sc: simple cubic, bcc: body-centered cubic, bct: body-centered tetragonal, OMTTF: octamethylenetetrathiafulvalene). P_c denotes the critical pressure and T_c the critical temperature

Material	Symmetry of the salts	P_c [kbar]	T_c [K]	Year of report
K_3C_{60}	fcc	0	19.3	1991
Rb_3C_{60}	fcc	0	29.6	1991
Cs_2RbC_{60}	fcc	0	33	1991
$CsRb_2C_{60}$	fcc	0	31	1991
Rb_2KC_{60}	fcc	0	27	1991
RbK_2C_{60}	fcc	0	23	1991
CsK_2C_{60}	fcc	0	24	1992
$RbNa_2C_{60}$	sc	0	3.5	1992
Ca_5C_{60}	sc	0	8.4	1992
Na_2CsC_{60}	sc	0	10.5	1992
$RbTl_{1.5}C_{60}$		0	27.5	1992
$Na_3N_3C_{60}$	sc	0	15	1993
$(NH_3)_4Na_2CsC_{60}$	fcc	0	29.6	1993
Sr_6C_{60}	bcc	0	4	1994
Cs_3C_{60}	bct/bcc	14.3	40	1995
$Na_2Rb_{0.25}Cs_{0.75}C_{60}$		0	9.4	1995
$Na_2Rb_{0.5}Cs_{0.5}C_{60}$		0	8.4	1995
$Na_2Rb_{0.75}Cs_{0.25}C_{60}$		0	5.2	1995
$NH_3K_3C_{60}$	Orthorhombic	14.8	28	1995
$Yb_{2.75}C_{60}$	Orthorhombic	0	6	1995
Sm_xC_{60} $(x \simeq 3)$		0	8	1995
$K_3Ba_3C_{60}$	bcc	0	5.6	1996
$(NH_3)_{0.5-1}NaRb_2C_{60}$	fcc	0	$8 \div 12$	1996
$(NH_3)_{0.5-1}NaK_2C_{60}$	fcc	0	$8.5 \div 17$	1996
$K_x(OMTTF)C_{60}$ (benzene)		0	18.8	1996
$Rb_x(OMTTF)C_{60}$ (benzene)		0	26	1996
Ba_4C_{60}			7.0	1997
$Rb_3Ba_3C_{60}$	bcc	0	2.0	1997

1.3 Exotic Features

The structure of organic conductors and superconductors are not simple, in the sense that a number of light atoms form basic molecules, and the solids are formed as combined structures of these molecules. However, the inter-molecular couplings in the conductors and superconductors are much weaker than the intramolecular couplings in these constituent molecular materials. This hierarchy in coupling strengths makes possible the reduction in complexity of the analysis.

Molecular-orbital theories have been employed to describe the electronic structure of the molecules, thus relating the crystal structures to the conducting and other properties. Calculations based on, for example, the extended Hückel method have provided a fairly accurate description of the electronic band structure and, hence, facilitated to link the chemical and physical concepts, or, in other words, to relate the molecular and crystalline structures of the materials to their observed properties. Needless to say, further progress is needed to describe actual materials satisfactorily. However, it is worth noting that the calculated results based on the extended Hückel approximation are quite useful in explaining the experimental observations with regard to the electronic structure near the Fermi level. The calculated electronic structures, thus derived, are characterized in terms of the low-dimensionality of $(TMTSF)_2 X$ and $(ET)_2 X$ reflecting the stacking structure: The quasi one-dimensional structure of $(TMTSF)_2 X$ and the quasi two-dimensional structure of $(ET)_2 X$ will be described in Sects. 3.1 and 5.1. With respect to the quasi one-dimensionality of $(TMTSF)_2 X$, it has been conjectured that a fluctuating superconductivity might emerge for temperatures much higher than T_c, where one would actually expect zero resistance. This idea had been supported by tunnel-junction spectroscopy and infrared-absorption experiments, showing a large superconducting energy gap. The thermal conductivity also seemed to exhibit a corresponding anomaly. Furthermore, in the metallic state, in contrast to ordinary metals, the resistivity was seen to decrease monotonically down to 1 K without showing residual resistivity, and was seen to increase significantly under the influence of a magnetic field. These facts could not be explained in terms of the conventional theory of metals.

The tunnel-junction data and the observed anomaly in the thermal conductivity in $(TMTSF)_2 X$ have not been reproduced in later investigations. Instead, the subsequent tunnel-junction experiment yielded a gap parameter in agreement with the prediction of the conventional BCS model. The infrared data have been found to be non-specific to the superconducting material, but rather characteristic to vibrational modes specific to $(TMTSF)_2 X$, irrespective of the superconducting gap. In addition, the specific heat increment related to the superconductivity is consitent with the BCS theory. Thus, the principal experimental bases for the postulated fluctuating superconductivity have been eliminated. The suppression of the fluctuation can be explained in terms of the coherence length reaching 2.2 nm even in the shortest direction, which is longer than the intercolumn spacing of 1.35 nm. This contrasts with the situation in $(ET)_2 X$ (with T_c exceeding 10 K), whose coherence length in the interlayer direction reaches only 0.3 nm, significantly shorter than the interlayer distance of 1.5 nm. The superconductivity-transition behaviors of the high-T_c two-dimensional conductors exhibit significant fluctuations. It should be noted, however, that the extraordinary magnetoresistance is open to further investigation.

It has been determined that superconductivity in these materials is very sensitive to non-magnetic defects. This has caused suspicion that these superconductors cannot be ordinary singlet superconductors which are usually insensitive to such defects. To interpret this phenomenon, the possible existence of a triplet superconductor has been argued. Currently, however, the accepted explanation involves a localization effect which dominates in a restricted-dimensional system. The temperature dependence of the Nuclear Magnetic Relaxation (NMR) relaxation in $(TMTSF)_2 ClO_4$ and κ-$(ET)_2$ $Cu[N(CN)_2]Br$ is also not consistent with the BCS description. This dependence suggests the presence of a zero-gap line on the Fermi surface, as in both CuO_2 superconductors and heavy-Fermion superconductors. These features are described in Sects. 3.2 and 5.3.

The fact that the superconducting temperature region lies close to that in which the antiferromagnetic ordering is displayed for TMTSF salts and κ-$(ET)_2 Cu[N(CN)_2]X$ (X: Br and Cl) has suggested that the electron pairing mechanism is mediated by a fluctuating SDW, as discussed in Sect. 8.2. On the other hand, the TTF-type molecule exhibits strong coupling between its Highest Occupied Molecular Orbital (HOMO) level and intermolecular vibrations, which is called the **Electron-Molecular-Vibration** (EMV) **coupling**. This interaction can contribute greatly to the attractive BCS-like interaction between electrons. In Sect. 8.3 we shall formulate the idea and discuss the applicability to TTF-derivative complexes. Experimental data suggesting a contribution coming from intermolecular phonons are presented in Sect. 8.4.

The **electron-electron interaction** is believed to be important in the molecular conductors, due to the low-dimensionality and the **low electron density**. Since the electron bandwidth is rather narrow, as a result of the weak intermolecular transfer integrals, the on-site Coulomb energy becomes comparable to, or larger than, the bandwidth corresponding to the kinetic energy of the conduction electrons. Consequently, the electron correlation can work effectively and drives the half-filled electron band systems into the **Mott-Hubbard insulating state**, as it is often found in the charge-transfer salts of interest. This has been argued for the two-dimensional half-filled metal κ-$(ET)_2 X$. We also note that the temperature dependence of the resistivity (represented by the T^2 curve and ascribed to the electron-electron interaction) has been found in some $(ET)_2 X$ salts [1.33].

It is noteworthy that a temperature-independent residual resistance has not been observed down to 1 K (or, in some sense, at lower temperatures) in not a few organic conductors. The magneto-quantum oscillations related to Fermi-surface topology such as the Shubnikov-de Haas and de Haas-van

Alphen effects, are seen quite clearly in these cases [1.34]. This implies that the sample may clean due to a reduction in the number of scattering centers formed by defects. In other words, organic conductors and superconductors can provide systems ideally suited for the investigation of the electron dynamics in crystalline solids.

2. Organic Conductors

Organic materials, such as organic polymers, are generally regarded as good insulators. Recent developments in materials science and technology, however, have brought about the discovery of organic conductors such as organic metals and even superconductors. The properties of rather simple materials such as ordinary metals and superconductors have well been described by theories of solid-state physics. To treat the more complicated organic conductors on the same basis, terms of solid-state physics and organic molecular materials need to be bridged.

One of the purposes of the present chapter is to provide this kind of bridge, derived mostly from chemical notions. For example, in describing the electrical properties of organic conductors, one uses the approximation of the π-electron model, which focuses on delocalized molecular orbitals. The band structure of the electronic system has been calculated on the basis of this molecular-orbital picture. Through such approaches the individual molecular properties and the electronic properties of the bulk systems are interconnected.

The molecular properties specify the electronic properties which, in turn, have revealed a few previously unseen physical phenomena in solid-state physics. The molecules which are the building blocks (bricks) in constructing organic conductors are sometimes arranged in linear columns or planar sheets. Thus, the resultant electrical properties are highly anisotropic. In some cases electrical conduction takes place principally along a unique crystalline direction. Such compounds are called **quasi one-dimensional conductors**. Quasi two-dimensional conductors and intermediate conductors show up as well. The discovery of the organic conductor first allowed the embodiment of the quasi one-dimensional conductor. The physics of such anisotropic compounds has been developed through the study of the former. Brief descriptions of physical aspects are given in this chapter, since they are indispensable to an understanding of the nature of organic superconductors.

2.1 Electronic Conduction in Organic Materials

Starting from organic semiconductors, the family of organic conductors has widened and spread to include organic metals and superconductors. In this respect the organic semiconductors are the direct predecessors of the organic conductors. In this monograph, however, when speaking of *organic conductors* we include only the organic metals and superconductors but not the organic semiconductors. In organic conductors, the concentration of charged carriers is rather independent of the temperature and is almost equal to the density of the constituent molecules similarly to ordinary metals. On the other hand, in organic semiconductors, the carriers are created through thermal excitation across an energy gap. Their carrier concentration is rather dilute and changes strongly with temperature.

In an alkali metal such as Na, one free valence electron is provided by each Na atom, leaving the other electrons in a closed inner shell within the atomic structure. For an organic molecular compound such as TTF·TCNQ which is formed by transferring electrons from TTF molecules (donors) to TCNQ molecules (acceptors), the transferred electrons or holes act as free carriers which cause electrical conduction. The carrier generation corresponds to the formation of free radicals from molecules in the terminology of chemistry. The high electrical conductivity originates from intermolecular charge-transfer interactions between electron donor and acceptor molecules [2.1]. Through charge transfer, which is dominated mainly by the strength of the molecules' ionization energy, and also by the Madelung energy, the number of free carriers is determined. In this case, it is also required to form unfilled electron bands so as to produce conducting carriers. That is, as long as the electrons can move throughout the compound, it may exhibit metallic conduction. For TTF·TCNQ, the charge-transfer ratio is given by $\rho = 0.59$ which means that on the average $-0.59e$ is transferred from TTF to TCNQ molecules, $-e$ being the electron charge. In the TTF·TCNQ crystal the TTF and TCNQ molecules are arranged to form two separate electronically conducting columns: the charge transfer produces 0.59 electrons per molecule on the TCNQ column and 0.59 holes per molecule on the TTF column. These carrier concentrations are nearly constant in the temperature range down to the metal-insulator transition appearing at 54 K. Further examples of charge-transfer compounds are $(TMTSF)_2 PF_6$ and $(ET)_2 I_3$ with $\rho = 0.5$. The properties of these superconducting charge-transfer compounds are the subjects of Chaps. 3 and 5.

Another way of creating free carriers is to dope the compounds by adding either molecular or atomic impurities. These dopants should have electronegativities (or affinities) quite different from the host materials. The number of free electrons produced in the charge transfer corresponds to the amount of dopant. In ordinary inorganic semiconductors such as Si and Ge, although the doping is crucial in determining their electronic properties, the amount of the dopant is very small, usually much less than 1% of the host atoms, to make the material semiconducting. In the terminology of semi-conductor physics, the dopants produce impurity levels within the electronic band structure constructed by the host material [2.2]. On the other hand, in making metallic conductors, host materials are heavily doped and the amount reaches even a few tens of percent. With this heavy doping a high carrier density is produced in the conduction or valence bands, and can be described by metallic electronic states. The carriers occupy electron levels to a certain extent into the conduction band called the **Fermi level**, and they obey the Fermi-Dirac statistics. This is one of the prerequisites for a system to be called **metallic**.

Polyacetylene $(CH)_x$, is a conjugated polymer, which has an alternating arrangement of single and double bonds, as shown in Fig. 2.1a. In this system the bond ordering can be changed (shifted) without loss of energy and hence the system with a changed bond order is energetically degenerate. When both kinds of bond ordering exist within the polymer, at some point they must connect (Fig. 2.1b). At the point of connection an unpaired electron (symbolized by the dot) is produced. Although the unpaired electron can move within the conjugated polymer freely, it cannot be a carrier for electricity. By combining with an electron acceptor, such as AsF_6, an electron is removed, leaving a positively charged defect which can carry electricity as a hole. This defect is called a **charged soliton** because its physical characteristics resemble that of an actual soliton in nonlinear dynamics. When their concentration reaches a certain level, the carriers produced in this way can also satisfy the conditions required by metallic conductors.

Another example of a molecular conductor is intercalated graphite. Although layered graphite is conducting by itself, strictly speaking it is not a metal but rather a semimetal in which the numbers of electrons and holes in

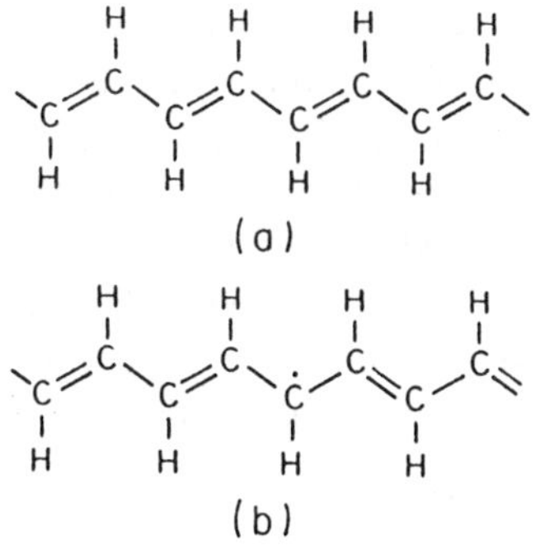

Fig. 2.1. Normal conjugated polyacetylene polymer (**a**). After a bond order shift (**b**) an unpaired electron remains at the point of contact between two phases of bond alternation

different bands are equal. By inserting electron acceptor or donor species into the interlayer spacings, the equilibrium in the number of electrons and holes is eliminated through charge transfer to or from the dopants and the system becomes metallic.

In addition to the high carrier density, the mobility of the carriers must be high enough to ensure the metallic behavior. That is, the free electrons or radicals should be able to move to carry electricity; the higher the mobility the higher the conductivity. In terms of a single-particle theory the mobility increases with the broadness of the electronic energy bandwidth. For charge-transfer compounds, this band is formed through the overlap of the molecular orbitals, and its width increases with the degree of overlap. For organic molecular crystals composed of stacks of planar molecules with orbitals normal to the plane, the resultant electron transfer and, of course, the electronic conduction become anisotropic, namely, the electrical conductivity along a certain direction in a crystal is much higher than that in another direction. This is referred to as **low-dimensional conduction**. If the conductivity is notably higher in one particular direction, the material is called a **quasi one-dimensional** (1-D) **conductor**, while if it is large in a given plane of a crystal, the material is a **quasi two-dimensional** (2-D) **conductor**. In charge transfer compounds one can find examples of both types, such as TTF·TCNQ as quasi one-dimensional and $(ET)_2I_3$ as quasi two-dimensional. Polymers like polyacetylene are expected to be quasi one-dimensional from their inherent linear structure. Intercalated graphite is a typical quasi two-dimensional conductor reflecting the layered structure of its host graphite.

Figure 2.2 shows the temperature dependence of the conductivity for typical organic conductors compared with ordinary metals such as Ag, Cu, and Fe. For TTF·TCNQ, the conductivity increases extraordinarily near 60 K, but then the compound undergoes a phase transition to an insulating state. On the other hand, $(TMTSF)_2ClO_4$ and $(ET)_2I_3$ are well behaved, exhibiting increasing conductivity towards lower temperatures and finally turn to be superconductors. Polyacetylene doped with 10% AsF_6 is denoted by $[CH(AsF_6)_{0.1}]_x$, whose conductivity is almost constant with a change of temperature. The formula $C_8(SbF_6)$ represents the first-stage intercalated graphite with an SbF_6 ion as intercalant, and exhibits metallic conductivity.

The characteristic increase in the conductivity of TTF·TCNQ near 60 K was originally claimed to be due to the emergence of fluctuating superconductivity. However, this was refuted by later studies, and a substantial part of the conductivity increase is ascribed to a sliding motion of Charge Density Waves (CDWs), a kind of collective mode of the electrons in a quasi one-dimensional system. This behavior is a typical feature of the low-dimensional conductor.

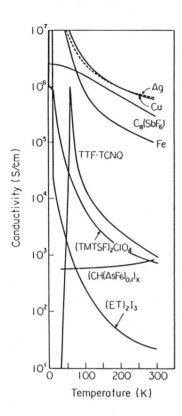

Fig. 2.2. Temperature dependence of the conductivity of typical organic conductors and ordinary metals

In such organic conductors the charge carriers can interact, at lower temperatures, with each other through Coulombic repulsion. This is because the screening due to the electron charges is much lower than in ordinary metals, and the electron-electron correlation effect dominates. As a result, even with a half-filled conduction band, a low-dimensional organic material may become an insulator, called a **Mott-Hubbard insulator**, due to strong correlations among charged carriers. To make this insulator metallic, doping the crystal to induce charge transfer thereby violating the condition of a half-filled band can be used.

2.2 π-Electron and Molecular Orbital Methods

2.2.1 π-Electron Approximation

The framework of an organic molecule such as benzene consists of carbon atoms. The electron configuration of a carbon atom in its ground state is $(1s)^2 (2s)^2 (2p)^2$. To form a hexagonal ring by bonding six atoms in one

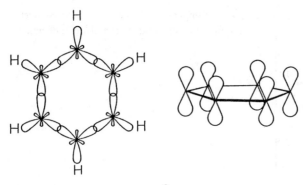

Fig. 2.3. σ-bonding through sp² hybrid orbitals (*left*) and p_z orbitals to form π-bonding (*right*) for benzene

plane, hybridization of one 2s- and two 2p-orbitals is required to form an sp²-orbital. The resultant hybrid orbital has three lobes symmetrically spaced with 120° angles between them. These orbitals are combined in bonds between the carbon atoms which are symmetrically distributed with respect to rotation about the bonding axis; in other words, whose electronic angular momenta along the C-C axes are zero. Thus, the framework of a benzene ring is formed by electrons occupying the σ-orbitals and hence by σ-bonds (left side of Fig. 2.3).

In addition to the σ-orbitals, orbitals with nonzero momentum along C-C (bonding axes) called **π-orbitals** are formed. In the conjugated hydrocarbon such as a benzene ring or a polyacetylene chain (right-hand sides of Figs. 2.3, 4), π-molecular orbitals are formed from p_z-orbitals, z denoting the direction normal to the molecular plane. Thus, the π-orbitals are also perpendicular to the bonding plane.

One of the unique features of these π-electrons is their delocalization. This property, i.e., the ability to move from one atom to another arises

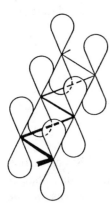

Fig. 2.4. p_z-orbitals to form π-bonding along a linear conjugated polymer such as polyacetylene

because the bonding energy of the π-electrons is much lower than in the σ-orbital, and they are easily excited. Low-energy excitation, orbital dia-magnetism, and electric conduction of molecules are most likely to occur with π-electrons. In simple quantum-mechanical calculations of these pro-cesses only the π-electrons are included to make the problem tractable. This procedure is called the **π-electron approximation**. In this method the σ-electrons, together with the nuclei and inner-shell electrons, determine the potential fields for the π-electrons. Most electronic and chemical properties of organic conductors can be described on the basis of the π-electron approximation [2.3].

For charge transfer compounds such as TTF·TCNQ and $(TMTSF)_2X$, the probability of electron transfer between adjacent planar molecules is calculated from the overlap of π-electron orbitals which extend normal to the molecular planes. In these compounds the stacking structures of planar molecules provide strong overlap and, hence, large electron-transfer proba-bilities along the stacking axis.

However, to calculate the electronic properties of realistic molecular salts, even the π-electron approximation is still too complex and requires further simplification. The easiest is the free electron approximation, which, however, overestimates the delocalization of the π-electron. The Hückel approximation, which is a typical molecular-orbital method and is described in the following section at some length, is used, too. In this meth-od one simplifies the calculated electronic structure semiquantitatively by including empirical molecular parameters. The Hückel method can further be improved [2.3]; of the many semi-empirical methods we would like to mention the extended Hückel method and the anti-symmetrized molecular-orbital method in which the electron-electron correlations are taken into account. Non-empirical methods have also been developed, in which elec-tronic states are calculated more elaborately. They need cumbersome calcu-lations but the issue is being resolved with the aid of supercomputers. Thus, many methods have been developed to attack the problem of electronic structure calculations, and the method of choice depends on the size of the calculation and available experimental data.

2.2.2 Molecular-Orbital Method

To calculate the electronic states of molecular systems with many electrons and nuclei, it is impossible to directly solve the Schrödinger equation. Thus, various approximations have been used to make the problem soluble. A-mong them, the Molecular Orbital (MO) method provides one of the most useful ways to evaluate the electronic state.

In the **MO method** one assumes that the available valence electrons are spread over the whole molecule, being released from each atom. In other words, we assume that each electron moves in a potential field produced by the nuclei and other electrons within the molecule. Then, the molecular orbitals are approximated by linear combinations of the constituent atomic orbitals. The number of MOs is equal to the number of adopted atomic orbitals, and each MO can contain only two electrons with antiparallel spins.

In the following we briefly describe the Hückel MO method which is one of the most elementary but instructive concepts of the MO methods [2.3]. We adopt some simplifications: first, only the π-electrons are taken into account, i.e., direct effects of the σ-electrons are neglected. Second, the total Hamiltonian is approximated by the sum of one-electron Hamiltonians as

$$\mathcal{H}(\mathbf{r}_1,\mathbf{r}_2,...,\mathbf{r}_N) = \sum_{i=1}^{N} h(\mathbf{r}_i) , \qquad (2.1)$$

where N is the number of π-electrons, $\mathbf{r}_i$ is the position of the ith electron, $h(\mathbf{r})$ is a one-electron Hamiltonian, with the eigenvalue ϵ_m and the eigenfunction $\psi_m(\mathbf{r})$,

$$h(\mathbf{r})\psi_m(\mathbf{r}) = \epsilon_m \psi_m(\mathbf{r}) . \qquad (2.2)$$

The π-electrons occupy orbitals starting from lower ones, each of which can contain at most two electrons. The ground-state wave function and its energy are represented by

$$\psi(\mathbf{r}_1,\mathbf{r}_2,...,\mathbf{r}_N) = \psi_1(\mathbf{r}_1)\alpha(\sigma_1)\psi_1(\mathbf{r}_2)\beta(\sigma_1)...$$
$$...\psi_{N/2}(\mathbf{r}_{N-1})\alpha(\sigma_{N/2})\psi_{N/2}(\mathbf{r}_N)\beta(\sigma_{N/2}) , \qquad (2.3)$$

$$E = 2 \sum_{m=1}^{N/2} \epsilon_m , \qquad (2.4)$$

where $\epsilon_1 \leq \epsilon_2 \leq ...$, and we assume that N is an even number. The functions $\alpha(\sigma_i)$ and $\beta(\sigma_i)$ represent the spin parts of the wave function. Now to solve (2.2), we approximate ψ_m by a linear combination of atomic orbitals $\chi_p(\mathbf{r})$ for the atomic position specified by p, i.e.,

$$\psi_m(\mathbf{r}) = \sum_p C_{pm} \chi_p(\mathbf{r}) . \tag{2.5}$$

The coefficients C_{pm} are determined from

$$\sum_p (h_{qp} - \epsilon_m S_{qp})C_{pm} = 0 , \tag{2.6}$$

where

$$h_{qp} = \int \chi_q^*(\mathbf{r}) \hbar(\mathbf{r}) \chi_p(\mathbf{r}) dv \tag{2.7}$$

$$S_{qp} = \int \chi_q^*(\mathbf{r}) \chi_p(\mathbf{r}) dv . \tag{2.8}$$

To make the problem simpler, we adopt the further approximation that

$$S_{qp} = \begin{cases} 1 & \text{for } q = p \\ 0 & \text{for } q \neq p \end{cases} \tag{2.9}$$

$$h_{qp} = \begin{cases} \alpha_p & \text{for } q = p \\ \beta_{qp} & \text{when q and p are adjacent} \\ 0 & \text{otherwise} \end{cases} \tag{2.10}$$

where S_{qp}, α_p, and β_{qp} are the overlap integral, the Coulomb integral, and the resonance integral or **transfer energy**, respectively; their values are estimated semi-empirically.

As a simple example, we apply the method to polyacetylene [2.4]. Here we assume that all the C atoms are equivalent, that each contributes one valence electron and that this electron spreads over the whole molecule. When 2n C atoms are involved, the eigenvalues are

$$\epsilon_j = \alpha + 2\beta \cos\left(\frac{j\pi}{2n+1}\right) \quad (j = 1, 2, ..., 2n) \tag{2.11}$$

where $\alpha_p = \alpha$, $\beta_{qp} = \beta$. Then, the energy difference ΔE between the Highest Occupied Molecular Orbital (HOMO), and the Lowest Unoccupied Molecular Orbital (LUMO) becomes

$$\Delta E = \epsilon_{n+1} - \epsilon_n = 4|\beta|\sin\left[\frac{\pi}{2(2n+1)}\right]. \tag{2.12}$$

For polyacetylene, n reaches tens of thousands, thus one can assume $n \to \infty$. Consequently, $\Delta E \to 0$, which means that there is no gap between HOMO and LUMO, and the calculation predicts that polyacetylene is metallic.

However, real polyacetylene is not metallic but insulating ($\Delta E \neq 0$) unless doped. The discrepancy comes from the assumption that the π-electrons are distributed uniformly. In reality the polyactetylene bonds are alternating, as depicted in Fig.2.1a. Therefore, we require two β values, β_1 and β_2 ($\beta_1 \neq \beta_2$), for the different bond lengths, and under this condition

$$\Delta E = 2\sqrt{(\beta_1-\beta_2)^2 + 4\beta_1\beta_2\sin^2[\pi/(4n+2)]} \tag{2.13}$$

and for $n \to \infty$,

$$\Delta E = |2(\beta_1 - \beta_2)| .$$

To treat this case of alternating bonds systematically, we may utilize the extended Hückel method and include the σ-electrons and equilibrium positions of the C atoms. That is, we take the total energy V as the sum of the σ-bonding energy F and the π-bonding energy E. The shape of the molecule is determined when $V = F + E$ is minimized. In reality, the lengths for the single and the double bonds are 1.42 Å and 1.39 Å, respectively. This has been confirmed both experimentally and theoretically.

In conventional solid-state physics, the approximation that is generally used is that electrons move in a mean-field formed by ions and other electrons, and it is periodic. The time-independent wave functions of electrons in such a periodic field are given by the Bloch function which has the following periodic property

$$\psi_{n\mathbf{k}}(\mathbf{r}+\mathbf{R}) = \exp(i\mathbf{k}\cdot\mathbf{R})\psi_{n\mathbf{k}}(\mathbf{r}) \tag{2.14}$$

where $\mathbf{R}$ represents the periodicity. The states are designated by the wave vector $\mathbf{k}$ and, in addition, with n specifying the kind of Bloch function or band. The eigenenergy $E_n(\mathbf{k})$ corresponding to $\psi_{n\mathbf{k}}(\mathbf{r})$ is a continuous function of **k**, namely, an electronic-energy band labeled by the quantum number n.

To apply our current understanding of solid-state physics to organic conductors, it is generally preferable to describe the electronic structure with this band description. Usually one uses the tight-binding band approximation in which only the overlap of LUMO and/or HOMO lying at the Fermi energy is used to find the system energy.

In one-dimensional molecular crystals with a periodic potential, the wave function of the π-electron is written as a linear combination of atomic orbitals by using Bloch's theorem

$$\psi_s(k, r) = \frac{1}{\sqrt{N}} \sum_{j=1}^{N} \sum_{\mu}^{AO} e^{ikja} C_{s\mu}(k) \chi_\mu(r-ja) \tag{2.15}$$

where N denotes the total number of cells, j the index of the cell, a the spacing between the cells, $\chi_\mu(r-ja)$ the μth Atomic Orbital (AO) for the jth cell, k the wavevector in the first Brillouin zone, and s the label of the Molecular Orbital (MO). With this wave function, by following (2.5-8), one can find the relation between k and the energy ϵ. This can be extended to three-dimensional crystals to find the relations between **k** and ϵ.

This calculation is complicated for a typical charge-transfer crystal such as $(TMTSF)_2 X$ since there are 52 atomic positions and 136 bonding orbitals, even if we neglect the counter anion X^-. Sometimes one can make simplifications such as replacing CH_3 by H, assuming that the intermolecular overlap is dominated by Se atoms. One also limits the calculation to HOMO and/or LUMO. Choosing this orbital as $\chi_\mu(r)$, one can perform the tight-binding band calculation in the way described above. The actual results will be described in Sects. 3.1 and 4.3.

2.3 Quasi One-Dimensional Conductors

2.3.1 Quasi One-Dimensional Electronic System

Quasi one-dimensional organic conductors are characterized by an anisotropy in the electrical conduction, strong electron-phonon coupling and strong Coulomb interaction.

When the electron conduction along a certain crystallographic axis is larger than that in another direction, we call it a **quasi one-dimensional conductor**. For organic conductors, this occurs when the electron transfer energy $t_\parallel$ for a particular direction is much stronger than that for the other direction $t_\perp$ ($t_\parallel \gg t_\perp$). This occurs in linear polymers such as polyacetylene and in charge-transfer salts composed of stacks of planar molecules such as TTF·TCNQ and $(TMTSF)_2 X$. For polyacetylene the electron-transfer energy between atoms is given by the resonance integral β_{qp} in (2.10). For columnar arrays of molecular stacks, the overlap of π-orbitals on adjacent molecules within a stack is usually stronger than that between stacks. The transfer energy between molecules is given in the extended Hückel approximation by

$$t = \frac{1}{2}K(E_i + E_j)S_{ij} \tag{2.16}$$

where E_i is the energy level of the ith atomic orbital, and K is a constant between 1 and 2 which is conventionally taken to have the value 1.75. S_{ij} is the overlap integral between the atomic orbitals i and j [2.5].

By applying the tight-binding band approximation to the quasi one-dimensional system, an energy band can be written as a cosine band

$$\epsilon(k_a) = 2t_a \cos(k_a a) . \tag{2.17}$$

Here we take the direction of highest conductivity as the a-direction, and a and k_a are the lattice spacing and the wave number of the a-direction, respectively. Figures 2.5a,b display the energy dispersion and the density of states $N(\epsilon)$ of the one-dimensional band, with a bandwidth of $4t_a$. The Fermi surface for this quasi one-dimensional system is given by a pair of flat planes, as shown by solid lines in the k_a-k_b plane (Fig. 2.6), since the

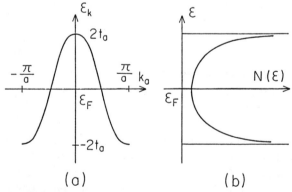

Fig. 2.5. (a) Energy-dispersion relation for a one-dimensional electronic system in the tight-binding band model, and (b) the corresponding density of states $N(\epsilon)$. The Fermi level is denoted by ϵ_F

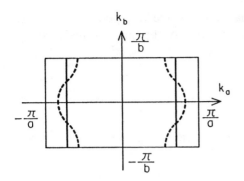

Fig. 2.6. Fermi surfaces for a one-dimensional (*solid lines*) and quasi one-dimensional metal (*dashed lines*) in the k_a-k_b plane

energy in this system is determined by k_a, and it is independent of both k_b and k_c.

When the interstack electron-transfer energy t_b in the b-direction is not negligible, the energy dispersion is modified to

$$\epsilon(\mathbf{k}) = 2t_a \cos(k_a a) + 2t_b \cos(k_b b) \tag{2.18}$$

where b and k_b are the lattice spacing and the wave number in the b-direction, respectively. When t_b is not negligible, the Fermi surface is warped, as shown by the dashed lines in Fig. 2.6. If t_b is much less than t_a, e.g., less than 10% of t_a, the Fermi surface cannot be closed within the Brillouin zone, whereas, if t_b reaches more than ca. 30% of t_a, it is closed and has a shape something like a cylinder parallel to the k_c-axis.

Let us now turn our attention to the molecular lattice underlying the electron system. The lattice is three-dimensional but it exhibits obvious anisotropy between the stacking direction and the perpendicular directions. As a result the molecular lattice has a variety of vibrational modes at very different energies. In particular, due to weak interstack couplings, the crystal is not very rigid. Thus, changes in the electronic system affect the lattice, and electron-phonon interactions may be strong and result in a Peierls instability, as will be described in the next subsection.

The screening of the electrons in one-dimensional organic conductors is weak compared to the ordinary three-dimensional metals because of the restrictions in the electron movement. The Coulomb repulsion has a tendency to become more appreciable, as will be described in Sect. 2.3.7.

2.3.2 The Peierls Transition

In quasi one-dimensional metals, the electronic system tends to become unstable against perturbations with the wave number $2k_F$, k_F being the Fermi wave number. Let $\delta\rho(x)$ be the deviation of the electron density $\rho(x)$ at the position x from its mean value by the influence of the external field $F(x)$,

$$\delta\rho(x) = - \int dx' \chi(x-x') F(x') \tag{2.19}$$

where $\chi(x)$ is the response function. The Fourier component of this relation for the wave number Q is written as

$$\delta\rho(Q) = - \chi(Q) F(Q) . \tag{2.20}$$

According to the perturbation treatment, the response function of an electronic system to external perturbations with the wave number Q is given by

$$\chi(Q) = \frac{2}{N} \sum_k \frac{f_{k+Q} - f_k}{\epsilon_k - \epsilon_{k+Q}}$$

with $\qquad\qquad\qquad\qquad\qquad\qquad\qquad\qquad\qquad\qquad\quad$ (2.21)

$$f_k = \frac{1}{\exp(\epsilon_k / k_B T) + 1}$$

where N is the number of electronic states, ϵ_k is the energy of the electrons with the wave number k, k_B is the Boltzmann constant, and T denotes the temperature. Taking the origin of the energy at the Fermi energy $\epsilon_F = 0$, energies near the Fermi level are approximated by straight lines, as shown in Fig. 2.7, and are given by

$$\epsilon_k = v_F \hbar(|k| - k_F) \tag{2.22}$$

where v_F is the Fermi velocity. With this relation, the response function for $Q = 2k_F$ can be written as

$$\chi(2k_F) = - \frac{a}{2\pi} \int_{k(-\epsilon_B)}^{k(+\epsilon_B)} \frac{dk}{\epsilon_k} [f(\epsilon_k) - f(-\epsilon_k)] \tag{2.23}$$

where $\epsilon_B (\gg k_B T)$ specifies the energy region where the electron distribution may be perturbed, and $k(\epsilon_B)$ is the wave number corresponding to ϵ_B.

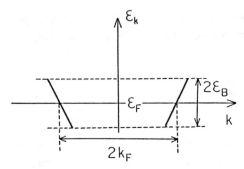

Then, if we get the effect of temperature explicitly by [2.6]

$$\chi(2k_F, T) = \frac{a}{\pi \hbar v_F} \int_0^{\epsilon_B} \left(\frac{d\epsilon}{\epsilon}\right) \tanh\left(\frac{\epsilon}{2k_B T}\right)$$

$$= \frac{a}{\pi \hbar v_F} \ln\left[\frac{1.13\epsilon_B}{k_B T}\right] . \tag{2.24}$$

If $T \to 0$, Eq.(2.24) diverges. That is, $\chi(2k_F)$ dominates in $\chi(Q)$, and the electron density $\rho(x)$ is modulated predominantly with the wave number $2k_F$.

In reality the electron system is more or less coupled to the underlying lattice system, and hence the lattice is also deformed when the electron density is perturbed through electron-phonon interactions. Let the electron-lattice interaction Hamiltonian be

$$\mathscr{H}_{e-p} = \frac{1}{\sqrt{N}} \sum_Q g(Q) u(Q) \rho_{-Q} \tag{2.25}$$

where g is the electron-phonon coupling constant, and u(Q) is the normal mode displacement. Then, the normal-mode frequency $\omega(Q)$ can be written in terms of the normal-mode frequency $\omega_0(Q)$ in the absence of an electron-phonon interaction:

$$\omega^2(Q) = \omega_0^2(Q) \left[1 - \left(2\frac{g^2(Q)}{\hbar\omega_0(Q)}\right) \chi(Q, T)\right] . \tag{2.26}$$

The lattice becomes unstable against perturbations with the wave number Q when $\omega^2(Q) < 0$, and the critical temperature T_P for this instability is given by

$$k_B T_P = 1.13 \epsilon_B \exp\left(\frac{-\pi \hbar^2 \omega_0 v_F}{2g^2 a}\right).$$ (2.27)

Below T_P the lattice is modulated with the wave number $2k_F$ and the phase transition to this lattice is called the **Peierls transition**. In this structure Bragg reflections occur for electrons with the modulation wave number $2k_F$ [2.7]. As a consequence an energy gap, called the **Peierls gap**, appears in the electron energy spectrum; in this situation the quasi one-dimensional metal turns into an insulator.

2.3.3 The Fröhlich Conduction

As mentioned in the previous subsection, at low temperature a quasi one-dimensional metal becomes unstable for external perturbations with the wave number $2k_F$. Due to electron-phonon interactions, the $2k_F$-modulated electronic structure accompanies lattice modulations of the same wave number. The related energy dispersion becomes insulator-like (Fig.2.8a): at $\pm k_F$ an energy gap of 2Δ appears, which is determined by the strength of the electron-phonon interaction.

However, it should be noted that the modulated structure is not necessarily static. In other words, the periodic electron modulation is not fixed to

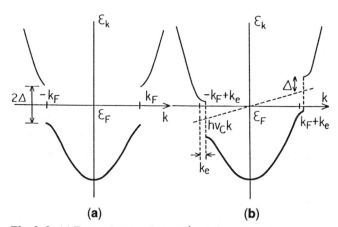

(a)　　　　　　　**(b)**

Fig. 2.8. (a) Energy band splitting 2Δ, at $\pm k_F$ due to lattice modulation. (b) Energy band for collective electron motion with velocity v_c and momentum k_e. After [2.9]

the lattice when $(k_F a/\pi)$ is not a rational number (a being the lattice constant) because the energy of the electronic system is independent of the relative position of the modulated structure with respect to the underlying lattice unless phase matching occurs [2.8]. Thus, if an external electric field E is applied to the modulated electronic system, the electrons are accelerated together with the accompanying $2k_F$ lattice modulation. If we let the energy of the electron system be $\epsilon(k)$ for $E = 0$, the resultant energy of a moving electron under the applied field becomes $\epsilon(k) + \hbar k v_c$, v_c being the speed of the electrons moving collectively (Fig.2.8b). As long as $\hbar k_F v_c < \Delta$, when there are no other obstacles (defects) the electrons continue to move and carry electrical current. When $\hbar k_F v_c$ exceeds Δ, individual electrons are excited to the upper band, and the collective motion is destroyed.

This was first pointed out by *Fröhlich*, who proposed that this might be a possible explanation for superconductivity [2.10]. The collective motion, however, is easily destroyed by lattice imperfections, which may act as pinning centers. Thus, the Fröhlich mode is very sensitive to crystalline defects or impurities.

So far we have considered a wave on a one-dimensional column. In reality similarly modulated waves exist on nearby columns in the close two- or three-dimensional neighborhood. They may interact with one another through, for example, Coulomb interactions, which may result in three-dimensional locking. To overcome pinning and locking, application of high electric fields or other perturbations is required.

2.3.4 Charge Density Waves

The $2k_F$-modulated electronic structure corresponds to an electron-density modulation, and hence is called a **Charge Density Wave** (CDW). Since this is usually accompanied by a lattice modulation through electron-phonon coupling, the CDW is regarded as a collective mode of coupled electron and lattice systems.

There are two kinds of CDW: One is the bond ordering wave in which the charge density is spatially modulated according to the distribution of bonding electrons [2.11]. A typical example is the polyacetylene, with alternating single and double bonds, as depicted in Fig.2.1a. The bond alternation splits the conduction band, resulting in an insulator, as mentioned in Sect. 2.2. The other CDW is the charge ordering wave which is found in one-dimensional charge-transfer salts, where charge density is determined from the distribution of conduction electrons. Hereafter, we discuss only the second type of CDWs.

The CDW is expressed by

$$\rho(x) = \rho_0 \cos(2k_F x + \phi) \tag{2.28}$$

where ρ_0 is the amplitude of the electron-density modulation, and ϕ is a phase factor. *Lee* et al. [2.12] investigated the dispersion relation for the CDW by using the Fröhlich Hamiltonian based on a continuum model. As a result two modes are derived, namely

$$\omega_+^2 = \lambda\omega_Q^2 + \frac{4}{3}\frac{m}{m^*}v_F^2 k^2 \tag{2.29}$$

$$\omega_-^2 = \frac{m}{m^*}v_F^2 k^2 \tag{2.30}$$

$$m^*/m = 1 + \frac{4\Delta^2}{\lambda\hbar^2\omega_Q^2} \tag{2.31}$$

$$\omega_Q = 2k_F v_s \tag{2.32}$$

where v_s is the sound velocity, m is the electron mass, m^* is the effective mass of the electrons, and λ is the coupling constant.

The respective dispersion relation for this is illustrated in Fig.2.9. The ω_+-mode corresponds to a spatial variation of the amplitude, whereas the ω_--mode stems from a modulation of the phase. Here, the ω_--mode, called a **phason**, corresponds to an extension of the acoustic mode (Fig.2.9) due to softening at $2k_F$ through the presence of **Peierls modulation**. From the resultant periodicity, the zone covering the range from k_F to $2k_F$ can be translated to that from $-k_F$ to 0, yielding a reduced zone from $-k_F$ to k_F. In a perfect crystal the ω_--mode can be excited by a very small energy (*gapless*), but in the presence of imperfections, the phason mode is pinned so that higher energies are required. Accordingly, the dispersion near $k = 0$ is corrected, as represented by broken lines in Fig.2.9. This corresponds to the pinning of the Fröhlich mode discussed in the previous subsection.

Thus, the dynamics of CDW at low excitation energies are strongly influenced by pinning. According to *Rice* [2.6], for the case of weak pinning, the dynamics of CDW can be expressed through the following equation of motion

$$M\ddot{\xi} + \gamma\dot{\xi} + \kappa\xi = e^*E , \tag{2.33}$$

where ξ is the displacement of the pinned CDW from an equilibrium position, $M = n_s m^*$ the effective mass of the CDW, γ the friction coefficient,

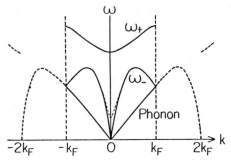

Fig. 2.9. Dispersion relation of CDW [2.8]

κ the restoring force, $e^* = n_s e$ the effective charge of CDW, and n_s the number of electrons contained in the CDW. Since the dipole moment formed by the pinned CDW is $p = e n_s \xi$, the frequency-dependent dielectric constant due to the motion of the CDW is given by

$$\epsilon_F(\omega) = 1 + \frac{\omega_p^2 (n_s/n)(m/m^*)}{\omega_F^2 - \omega^2 - i\omega\Gamma} \tag{2.34}$$

where $\omega_p^2 (= 4\pi n e^2/m)$ is the plasma frequency, and $\omega_F^2 = \kappa/n_s m^*$ is the pinning frequency. The corresponding conductivity can be expressed as

$$\sigma_F(\omega) = \frac{(n_s e^2 \omega^2)/(i\omega m^*)}{\omega_F^2 - \omega^2 - i\omega\Gamma} . \tag{2.35}$$

in the absence of a pinning restoring force, that is, $\omega_F \rightarrow 0$, at $\omega = 0$,

$$\sigma_F(0) = n_s e^2 \Gamma^{-1}/m^* . \tag{2.36}$$

Under the condition $\Gamma = 0$, $\sigma_F(0)$ tends to infinity, which means that the Fröhlich frictionless conduction (superconductivity) occurs.

The frequency dependence of the conductivity is depicted in Fig. 2.10, where $\sigma(0)$ shows a peak whose height is determined by $1/\Gamma$, and E_g is the energy gap due to the single-particle excitation of electrons. In the presence of pinning, the peak shifts to the pinning frequency ω_F.

2.3.5 Charge-Density-Wave Phenomena

A door to the quasi one-dimensional organic conductor was opened when the giant conductivity of TTF·TCNQ was discovered. This charge-transfer salt consists of arrays of columns of TTF and TCNQ. On average a charge

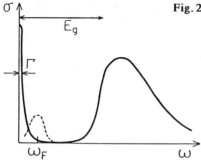

Fig. 2.10. Frequency dependence of CDW conductivity

of 0.59(−e) is transferred from each TTF molecule to each TCNQ molecule, hence the TTF column is hole-conducting whereas the TCNQ column is electron-conducting. Figure 2.11 exhibits the temperature dependence of the conductivity along the principal directions, where the conductivity σ_b along the b-direction which is the stacking axis, is more than 2 orders of magnitude higher than both σ_a and σ_{c*}. The conductivity shows that TTF· TCNQ is metallic down to 60 K and insulating below a phase transition point at 54 K. It has been concluded that in the metallic state CDWs are contributing to the electrical conduction as well as ordinary single electrons according to the following facts.

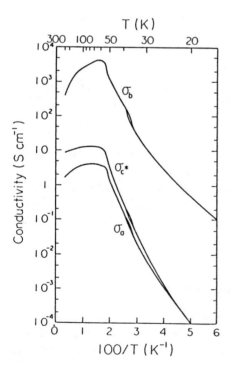

Fig. 2.11. Temperature dependence of the electrical conductivity of TTF· TCNQ for 3 principal crystalline axes. The hysterestis at 38 K is due to a structural transiton in the TTF column. After [2.13]

When we assume that single electrons are dominating the conductivity, they must move with a rather long mean-free path, compared to the lattice spacing, along the axis of highest conductivity, whereas in the perpendicular direction they make only short hops. The ratio of the conductivities $\sigma_\parallel$ and $\sigma_\perp$ parallel and perpendicular to the one-dimensional axis in the metallic region is [2.13]

$$\frac{\sigma_\perp}{\sigma_\parallel} = \frac{2a_\perp^2 t_\perp^2}{\hbar^2 v_F^2} \, , \tag{2.37}$$

while

$$\frac{\sigma_\perp}{\sigma_\parallel} = \frac{2m^* a_\perp^2 t_\perp}{\hbar^2 k_B T} \tag{2.38}$$

for the non-metallic region, where $a_\perp$ and $t_\perp$ are the spacing and electron-transfer energy between the columns, respectively. Equations (2.37, 38) reveal that in the metallic region the conductivity ratio should be independent of temperature, while in the non-metallic region, $\sigma_\perp/\sigma_\parallel$ should increase with decreasing temperature. Experimental results are far from this prediction; however, in the metallic region σ_b [which corresponds to $\sigma_\parallel$ in (2.37)] increases extraordinarily with decreasing temperature roughly as $T^{-2.3}$, whereas σ_{c^*} is proportional to $T^{-1.5}$. In the non-metallic region the experimentally derived $\sigma_\perp/\sigma_\parallel$ changes in the direction opposite to the prediction of (2.38). The difference is ascribed to the contribution of the CDW sliding motion. This is supported by the fact that the metallic behavior, in particular the high electrical conductivity near 60 K, is very sensitive to crystal imperfections caused by impurities or radiation damage.

Conclusive evidence for CDWs rather than single electrons to represent the vehicles of conductivity is the pressure dependence of σ_b, which exhibits a dramatic decrease near 20 kbar, as depicted in Fig. 2.12. The charge transfer between TTF and TCNQ molecules in TTF·TCNQ increases linearly from $\rho = 0.59$ at ambient pressure to 0.67 ($\approx 2/3$) under pressures of 15 kbar and up [2.15]. Since each TTF (or TCNQ) molecule can contain two holes (or electrons), the charge-transfer ratio indicates that the TTF (or TCNQ) band is emptied (or filled) by $\rho/2$. For a tight-binding band, the Fermi wave number k_F is given by

$$k_F = \frac{\pi}{b} \frac{\rho}{2} \, .$$

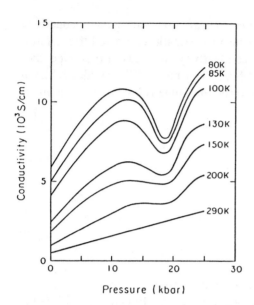

Fig. 2.12. Pressure dependence of the electrical conductivity of TTF·TCNQ at different temperatures. After [2.14]

Then, if ρ is equal to 2/3, $2k_F$ becomes commensurate with the reciprocal lattice of the wave number $2\pi/b$. This means that the CDW, if it exists, is locked by the lattice potential.

These facts (deviation of the temperature dependence from the single-particle prediction) indicate that a large fraction (possibly almost half) of the conduction is due to CDW transport. This is consistent with the pronounced decrease in the conductivity under a pressure of 18 kbar where the commensurate locking of CDW commences, as shown in Fig. 2.12.

The frequency dependence of the optical conductivity of TTF·TCNQ indicates that a pseudo gap with an energy of about 0.1 eV, as depicted in Fig. 2.10, exists even in the metallic region [2.16]. This gap has been interpreted to mean that CDW amplitude modulation already exists at room temperature but that the CDWs on each column are independent of one another and fluctuating by themselves. When the neighboring CDWs are in phase, the CDW gap becomes static due to three-dimensional locking. When the CDWs are so locked together, the system becomes insulating. The metal-insulator transition in TTF·TCNQ appearing at $T_p = 54$ K is attributed to this CDW phase locking.

Structural investigation of crystals through X-ray and neutron diffraction supports this interpretation exclusively. Above 54 K, diffuse scattering either of X-rays or neutrons is observed in a pattern corresponding to sheets in three-dimensional reciprocal lattice space, reflecting the lattice modulation in the b-direction of the wave number $0.295b^*$, b^* being a unit in reciprocal lattice space. This indicates that CDW modulations exist in columns but their phase is not correlated in the perpendicular directions. Below T_p, the sheet-like diffuse scattering pattern converges to spots, reflecting the three-dimensional (3-D) ordering of the CDWs with locked phases.

The critical temperature for CDW formation, T_p', is estimated to be 450 K from the observed optical gap E_g according to the mean-field relation $E_g = 3.5 k_B T_p'$. Although CDWs emerge at T_p' and their amplitude grows with decreasing temperature, their phases are not locked until T_p, at which the system becomes a real insulating state (**Peierls insulator**).

2.3.6 Open Fermi Surface

In quasi one-dimensional conductors, the Fermi surface consists of a pair of planar sheets (Fig.2.6). With increasing transverse interactions, the sheets are warped. If the degree of the warping is low, two unique features emerge: First, one sheet can be nested with the other by a shift of the wave vector **Q** given by $(\pm 2k_F, \pi/b, 0)$. The nesting is necessary to induce a CDW instability (Sect.2.3.2), but with increased warping the nesting is deteriorated, the CDW instability is weakened, and finally it is suppressed.

The second characteristic feature of the weakly warped Fermi surface is its openness. In ordinary metals, the Fermi surfaces are mostly closed within the Brillouin zone, or the surface is opened and connected to an adjacent zone in limited regions. For quasi one-dimensional metals a Fermi surface exists only in one direction; in other directions no surfaces which confine electrons in momentum space exist. Thus, the Fermi surface is thoroughly open in two directions, as indicated in Fig.2.13.

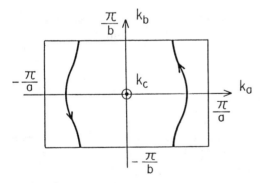

Fig. 2.13. Fermi surface of a quasi one-dimensional metal. The surface remains open to adjacent Brillouin zones in the k_b and k_c directions and closed in the k_a direction, being along the one-dimensional axis. *Thick lines* show the trajectory of the electron motion on the open Fermi surfaces in a magnetic field along the c-direction

We now describe features of the electron dynamics in a strong magnetic field. Let us express the band energy by (2.18) as

$$\epsilon(\mathbf{k}) = 2t_a \cos(ak_a) + 2t_b \cos(bk_b) , \quad t_a \gg t_b . \tag{2.39}$$

When a magnetic field H is applied along the c-axis perpendicular to the a-b plane, the equations of the electron motion can be written as

$$\frac{dk_a}{dt} = -\frac{eH}{\hbar c}v_b ,$$

$$\frac{dk_b}{dt} = \frac{eH}{\hbar c}v_a = \frac{eH}{\hbar c}\frac{2at_a}{\hbar}\sin(ak_a) , \tag{2.40}$$

$$\mathbf{v} = \frac{1}{\hbar}\frac{\partial\epsilon(\mathbf{k})}{\partial\mathbf{k}} , \tag{2.41}$$

where c is the velocity of light. In a quasi one-dimensional metal, the electrons traverse the Fermi surface in the b-direction into the adjacent Brillouin zone, as illustrated in Fig.2.13. The frequency of traversing successive Brillouin zones in the b-direction is given by

$$\omega_c = \frac{eH}{m_a^* c}k_F b \tag{2.42}$$

where m_a^* is the effective mass of the electrons in the k_a-direction at the Fermi surface.

Based on the relaxation approximation, under a strong magnetic field applied along the c-axis, the conductivity tensor $[\sigma_{ij}]$ within the a-b plane is

$$[\sigma_{ij}] \simeq ne\mu \begin{bmatrix} S & 1/\omega_c\tau \\ -1/\omega_c\tau & 1/(\omega_c\tau)^2 \end{bmatrix} \tag{2.43}$$

where μ is the electron mobility, τ is the carrier relaxation time, and S is a constant which is independent of H [2.17]. We omit here the c-component, since it is unaffected by the magnetic field. Then, the currents along the a- and b-directions, J_a and J_b, are related to the electric fields, E_a and E_b, by

$$J_a = \sigma_{aa}E_a + \sigma_{ab}E_b ,$$

$$J_b = \sigma_{ba}E_a + \sigma_{bb}E_b . \tag{2.44}$$

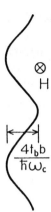

Fig. 2.14. Trajectory of an electron on an open Fermi surface in real space in a magnetic field. After [2.19]

Setting $J_b = 0$ and considering the case when the current is in the a-direction, the resistivity under the magnetic field is given by

$$\rho_{aa} = \frac{1/\sigma_{aa}}{1 - (\sigma_{ab}\,\sigma_{ba}/\sigma_{aa}\,\sigma_{bb})} \tag{2.45}$$

which is independent of H when we use (2.43). In other words, in quasi one-dimensional conductors, the conductivity along the one-dimensional axis saturates in a high magnetic field. On the other hand, one can check in a similar way that ρ_{bb} increases proportionally to H^2.

When the Fermi surface is closed, the electrons follow a cyclotron motion under a high magnetic field. When it is open, as in the above case, the electron oscillates strongly in the k_b-direction but hardly at all in the k_a-direction. In real space the electrons make a zig-zag-like motion with the amplitude $4t_b b/\hbar\omega_c$ [2.18], as shown in Fig.2.14. With an increase in the magnetic field, the width of the electron motion is narrowed, and when $\hbar\omega_c \gg 4t_b$ the electrons are almost confined to the one-dimensional columns. This means that at a certain point the electrons in a corrugated open Fermi surface can be squeezed to carry out a one-dimensional motion.

2.3.7 Coulomb Repulsion

So far we have treated the quasi one-dimensional metal by assuming that an electron-phonon interaction is dominating. However, in low-dimensional conductors, another phenomenon that needs to be considered is the limited screening of the Coulombic charge due to restraints on the electron movement.

The strength of these Coulomb interactions has been evaluated in TTF·TCNQ. When a magnetic field H is applied to a metal, the conduction electrons have average magnetic moments proportional to H, which modify the internal field to the nuclei through a contact correction. As a result, the NMR frequency shifts from the value measured in the absence of conduction electrons by an amount proportional to H (Knight shift). In the absence of the electron-electron correlation the NMR relaxation time T_1 due to the conduction electrons satisfies the following **Korringa relation**

$$T_1 K^2 = \frac{\mu_B{}^2}{\pi \gamma_N{}^2 k_B T} \tag{2.46}$$

where K is the Knight shift, μ_B is the Bohr magneton, and γ_N is the gyromagnetic ratio.

From measurements of the **Knight shift** and of T_1 in TTF·TCNQ, *Heeger* et al. [2.20] found that the value of $K^2 T_1 T$ is almost constant. On the basis of this fact and the relation also usually holding in ordinary metals where the screening is high, they claimed that the on-site Coulomb effect is not significant. Contrary to this result *Soda* et al. [2.21] found that the relation changes with frequency. From the frequency dependence of the NMR relaxation time, they estimated the value of the **on-site Coulomb repulsion** U between two electrons on the same molecule as

$$U/4t_b = 0.8 \div 1 \, .^1$$

where $4t_b$ is the bandwidth. Thus, *Soda* et al. claimed that one cannot neglect the Coulomb interaction. (Note that for TTF·TCNQ the one-dimensional axis lies in the b-direction).

The size of the Coulomb repulsion has also been evaluated by magnetic-susceptibility measurements. For non-interacting electrons in a tight-binding band the paramagnetic susceptibility χ_p at a low temperature is given by

$$\chi_p(0) = \frac{N_0 \mu_B{}^2}{\pi t_b \sin(\pi \rho / 2)} \quad \text{for} \quad U = 0 \tag{2.47}$$

where N_0 is Avogadro's number. In the simplest case, if we include the effects due to Coulomb correlations and take account only of the on-site Coulomb energy U, the susceptibility is increased to

---

[1] The symbol $\div$ is used throughout the text as a shorthand for "from - to" or "between".

$$\chi(T) = \frac{\chi_p(t_b, T)}{1 - U/[2\pi t_b \cdot \sin(bk_F)]} \quad \text{for} \quad U < 4t_b . \tag{2.48}$$

The effect of U was discussed, for more general cases, by *Bonner* and *Fisher* [2.22], and *Shiba* [2.23]. To explain the susceptibility in metallic regions *Torrance* et al. [2.24] estimated the effect of the on-site Coulomb repulsion as

$$U/4t_b \approx 1 ,$$

using a reasonable estimate for t_b based on measurement of the plasma frequency, the optical conductivity, the thermoelectric power, and so on.

Diffuse X-ray scattering showing a Peierls distortion in TTF·TCNQ revealed a modulation of the wave number $4k_F$, in addition to that at $2k_F$ [2.25]. It was demonstrated that the $4k_F$-modulation is not a higher harmonic of the $2k_F$-modulation through the fact that their temperature dependences are different. The occurrence of this $4k_F$ modulation has been interpreted through strong Coulomb repulsion [2.26]. The electrons in the TCNQ and/or TTF columns have a tendency to form a localized electron lattice, or a **Wigner crystal**, and the latter entails the $4k_F$ lattice distortion. The Coulomb repulsion has been observed to also affect optical absorption spectra from the infrared to the visible [2.27].

2.3.8 One-Dimensional Electron-Gas Approach

The electronic properties of the one-dimensional metal are determined by interactions of the electrons in the vicinity of the Fermi level. For this case the band energy is also given by (2.22). One of the most important interactions between electrons near the Fermi level is for a scattering process in which two electrons with the wave numbers $k_1 \simeq k_F$ and $k_2 \simeq -k_F$ collide transferring a wave number of $2k_F$. The final electron wave numbers are then $k_1' \simeq -k_F$ and $k_2' \simeq k_F$ (Fig. 2.15). The interaction coefficient for this backward-scattering process is denoted by g_1. The second limiting case is when two electrons exchange only a small wave number (right side of Fig. 2.15) so that after the collision each of them remains near the original k-branch of the Fermi surface. The interaction coefficient for this forward-scattering process is denoted by g_2. In the absence of any other interaction the ground state of this system is determined by the relative magnitudes of g_1 and g_2. This is done by comparing the strength of the divergence of the response functions for the various phases as in Fig. 2.16. Here SS denotes the Singlet Superconductivity in which the superconducting electrons make

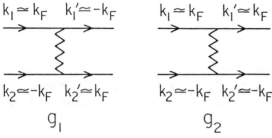

Fig. 2.15. Electron collision in a one-dimensional electron gas. The *left side* illustrates backward scattering and the *right side* forward scattering. After [2.28]

singlet-spin pairs. TS designates the Triplet Superconductivity in which electrons are paired with a spin quantum number equal to 1. This one-dimensional electron-gas approach is called **g-ology** and gives exact results for the phase diagram in the ground state of purely one-dimensional systems. For real organic conductors, an interaction with the lattice, transverse interactions, etc. must be taken into account. An application of the g-ology to actual organic conductors will be discussed in Sect. 4.2.

On the other hand, an exact solution has been given for a one-dimensional many-Fermion system by *Tomonaga* [2.29] and by *Luttinger* [2.30]. By describing the interacting one-dimensional electron gas in terms of the linear dispersion represented by (2.22), the Hamiltonian of the electron gas is expressed by

$$\mathcal{H} = \hbar v_F \sum_k k(a_k^+ a_k - b_k^+ b_k) + g \sum_{k,p_1,p_2} a_{p_1+k}^+ a_{p_1} b_{p_2-k}^+ b_{p_2} , \quad (2.49)$$

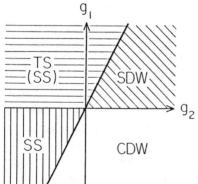

Fig. 2.16. Phase diagram of the ground state of a one-dimensional electron gas in the g_1-g_2 plane. After [2.28]

where a_k and a_k^+ are the anihilation and creation operators of the Fermions represented a k_1 in Fig.2.15, and b_k and b_k^+ are those of k_2. In a strongly interacting system collective motions in which many particles are involved, play important roles. It has been shown that $\mathcal{H}$, (2.49), is cast into a bilinear form in terms of the Boson operators representing the collective motion or density fluctuation as follows

$$\rho_q^a = \sum_k a_{k-q}^+ a_k$$

(2.50)

$$\rho_q^b = \sum_k b_{k-q}^+ b_q .$$

Then, the Hamiltonian is diagonalized with only sound-like excitations. The momentum distribution function no more has a jump at k_F but becomes smooth with a form represented by

$$\langle \rho_k \rangle \propto \left| k - k_F \right|^\alpha \, \text{sgn}(k - k_F) \quad (k \gtrless k_F)$$

(2.51)

where α is a positive exponent, although it is still singular at k_F. Because of the missing discontinuity, the system does not have a Fermi surface in the usual case.

This system describing the interacting one-dimensional electron gas is called a **Tomonaga-Luttinger model**. It applied not only to organic conductors but also to one-dimensinal quantum well and edge states of the fractional quantum Hall system [2.31]. With regard to a thorough overview on the electron correlation effects in molecular materials, the reader is referred to the monograph by *Fulde* [2.32].

2.4 Increase in Dimensionality

The quasi one-dimensional organic conductors have displayed previously unfamilar physical phenomena, such as the Peierls transition and the CDW phenomena. However, if one wants to design superconductors, one-dimensional conductors have a few serious shortcomings, such as the inability to maintain the long-range order thereby preventing phase transitions at finite temperatures. Furthermore, the quasi one-dimensionality has a strong tendency to drive Fermi-surface instabilities such as CDW and SDW which, if they freeze, result in an insulator.

With an increase of the dimensionality, e.g., if t_b is not negligible with respect to t_a in (2.18), the nesting of the Fermi surface is suppressed and a Peierls instability hardly occurs. Open Fermi surfaces exist in both TTF· TCNQ and $(TMTSF)_2 X$. On the other hand, a number of ET salts have closed Fermi surfaces. This is due to the fact that in these salts the intercolumn transfer of the electrons can occur at a rate similar to that in an intracolumn transfer, as will be discussed in Sects.4.3 and 5.1.3. This two-dimensionality stems from side-by-side interactions between planar molecules, which become comparable in strength to the face-to-face ones. As a result, two-dimensional conducting layers are formed and in many ET salts the metallic phase is maintained down to low temperatures.

When the Fermi surface is closed, in a strong magnetic field applied perpendicular to the two-dimensional plane the electrons follow a cyclotron motion, the eigenenergy of which is quantized in the Landau levels. Coupled to this quantization, the magnetization and the magnetoresistance also oscillate with the magnetic field. This is referred to as the **de Haas-van Alphen effect** and the **Shubnikov-de Haas effect**, respectively. Furthermore, an angle-dependent magnetoresistance oscillation can be found in two-dimensional metals. From the oscillation characteristic, one can deduce the cross section of the Fermi surface, the Fermi wave vector and other parameters on the dynamics of the electrons. These subjects will be discussed in Sect.5.1.4.

With regard to the electron correlation, features of the metal-nonmetal transition in a two-dimensional system have been developed quite extensively. It has been motivated by an understanding of new high-T_c materials where superconductivity takes place in the copper-oxide planes. The effect of the Coulomb repulsion has been studied based on the Hubbard model [2.33], in which the long-range Coulomb effect is neglected. The competition between the Coulomb repulsion energy U and the kinetic energy represented by the bandwidth W, which can be controlled by the intermolecular distance, determines the electronic conductivity. The electrons are allowed to move when the number of electrons is changed from the half-filling value. It is noteworthy that the spins in these systems can interact via an exchange interaction, and order antiferromagnetically.

The discovery of superconductivity in $K_3 C_{60}$ and its related compounds has extended the field of molecular superconductors to include the genuine three-dimensional system, as will be described in Chap.10.

3. TMTSF Salts: Quasi One-Dimensional Systems

The first organic material which was discovered to show superconductivity was $(TMTSF)_2PF_6$. This salt is one of the so-called **Bechgaard salts** with the general formula $(TMTSF)_2X$, where X^- is a monovalent anion, first synthesized by *Bechgaard* et al. [3.1]. These compounds are charge transfer salts; the 2:1 salt is formed by transferring one electron from two TMTSF molecules to one X. The crystal structures are isomorphous for various electron acceptors X. The structures remain the same even when the TMTSF molecule is replaced with a TMTTF molecule, which differs in that S atoms are substituted for the Se atoms of TMTSF (Fig.1.1). This has allowed the investigation of the role of the constituent molecules and the crystal structure on the superconducting behavior. For example, the effect of intermolecular distances has been studied by modifying the molecular spacings under varying pressure and also by changing X. Although their superconducting transition temperatures are rather low ($<3K$), they have been model systems for the study of superconductivity in organic charge transfer salts at early stages and have provided a basis for understanding the much more complex and varied ET salts.

In this chapter, the structure and electronic properties of Bechgaard salts are described with the emphasis on superconductivity. In the phase diagram $(TMTSF)_2X$ exhibits an SDW ordering in the vicinity of the superconducting phase, and thus their underlying interactions are considered to be related to the superconductivity. SDW and the related phenomena are also discussed at some length in this chapter. More detailed theoretical descriptions of SDW and its interplay with the superconductivity will be given in Chap.4.

3.1 Properties of $(TMTSF)_2X$ Salts

3.1.1 Molecules, and the Crystal Structures

The charge-transfer compound consisting of a TMTSF electron donor and a monovalent counter anion X^- such as PF_6^-, AsF_6^-, SbF_6^-, or NO_3^- exhibits metallic behavior down to ca. 20 K, below which it becomes insulating

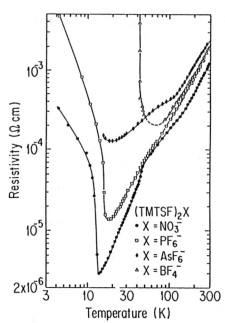

Fig. 3.1. Temperature dependence of the resistivity for various $(TMTSF)_2 X$. From [3.1]

(Fig. 3.1). By applying hydrostatic pressure *Jérome* et al. [3.2] found that the metal-insulator transition temperature (T_{MI}) for the PF_6 salt is reduced and a superconducting phase appears at a critical temperature (T_c) of 0.9 K for 12 kbar. $(TMTSF)_2 PF_6$ is the formula of the compound, **TMTSF** being the abbreviation of tetramethyltetraselenafulvalene, $(CH_3)_4 C_6 Se_4$ (the recommended IUPAC name is 4,4′,5,5′-tetramethyl-$\Delta^{2,2'}$-bi-1,3-diselenolyliden). Figure 3.2 depicts a space-filling model of TMTSF in which the diameters of the balls that represent atoms are proportional to the van der Waals radii. These planar molecules, which due to their thickness are better compared to bricks rather than thin plates, are stacked in a column but with a slight tendency toward dimerization. The highest electrical conduction is observed along the column axis. Figure 3.3 shows two views of the crystal structure of $(TMTSF)_2 PF_6$. The crystal has triclinic symmetry with the space group $P\bar{1}$ and the unit cell contains one chemical formula unit ($Z = 1$). The lattice parameters at room temperature are a = 7.297 Å, b = 7.711 Å, c = 13.522 Å, $\alpha = 83.39°$, $\beta = 86.27°$ and $\gamma = 71.01°$ [3.3].

The charge-transfer ratio ρ of TMTSF is fixed at 0.5, for $(TMTSF)_2 \cdot PF_6$. A pair of TMTSF molecules are of monovalence, i.e., $(TMTSF)_2^+$. The monovalent counter anion PF_6^- can be replaced by other anions with either octahedral symmetry, such as AsF_6^-, SbF_6^-, and TaF_6^-, or tetrahedral symmetry, ClO_4^- and ReO_4^-, keeping the crystal structure isomorphous. The atomic arrangements in the anions are displayed in Fig. 3.4. Furthermore, X is replaced by FSO_3 [3.5], $F_2 PO_2$ [3.6], NO_3, and $H_2 F_3$ [3.7]. It

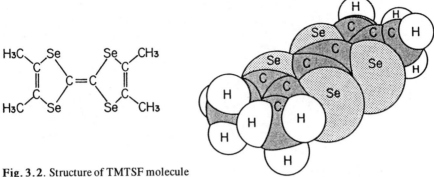

Fig. 3.2. Structure of TMTSF molecule

was found that TMTSF molecules produce salt with magnetic anions, such as $FeCl_4^-$, $MnCl_4^{2-}$, and $ZnCl_4^{2-}$ [3.8]. In these cases, however, very many different phases with different structures and stoichiometries have been recorded. We do not concern ourselves with these compounds further since they do not exhibit superconductivity.

3.1.2 The Electronic Structure

In $(TMTSF)_2 X$ the spacings between the molecules along the stacking direction are shorter than the sum of the van der Waals radii of the Se atoms, as given by *Pauling* (4.0Å) [3.9] or just beyond the Bondi value (3.8Å) [3.10]. In fact, the shortest intermolecular Se-Se distance is 3.87 Å for $(TMTSF)_2 PF_6$. Although the Se-Se spacings between the side-by-side and face-to-face adjacent molecules are comparable, as shown in Fig.3.3., the overlap of the π-orbitals is strongest within the stacks, i.e., along the a-direction. This good overlap leads to a transfer integral along the a-direction of the crystal much larger than along the b-direction. The weakest coupling is found in the c-direction, along which the anions and the methyl groups separate the main π-orbitals of the TMTSF molecules. As a result the electrons can move most easily along the column arrays, but appreciable mobility does appear in the b-direction.

Contrary to the TMTSF molecules, in the first approximation the counter anions X^- play no essential role in the electronic behavior. They stabilize the crystal structure by maintaining overall charge neutrality and separate the columns such that a low-dimensional structure results. According to the laws of thermodynamics, an ideal one-dimensional system cannot have a long-range order and undergo a phase transition. But the TMTSF salt does actually perform a phase transition like the superconducting and SDW phases. This is not surprising, however, because the columns are not at all

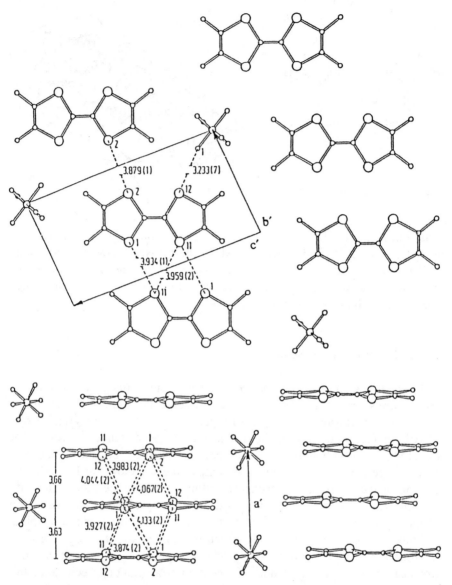

Fig. 3.3. Crystal structure of (TMTSF)$_2$PF$_6$. *Upper part*: View along the a-direction, b$'$ and c$'$ are the projections of b and c. *Lower part*: Side-view of stacks (tilted at 10°); a$'$ is the projection of a. Distances in Å. From [3.3]

independent of one another; they interact with each other appreciably through transverse interactions between adjacent TMTSF columns. Indeed, the phase-transition characteristics and their critical temperatures may apparently be governed by the size of X$^-$: the spacing between the columns affects the nature of the phase transition.

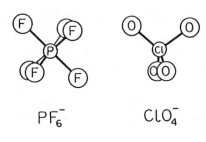

Fig. 3.4. Structure of PF_6^- and ClO_4^- anions

PF_6^- ClO_4^-

The electron band energy is expressed in the *tight-binding band approximation* as

$$\epsilon(\mathbf{k}) = 2t_a \cos(a_s k_a) + 2t_b \cos(b_s k_b) + 2t_c \cos(c_s k_c) , \qquad (3.1)$$

where $\epsilon(\mathbf{k})$ is the electron energy, a_s, b_s, and c_s are the intermolecular distances along the a-, b- and c-directions, t_i is the electron transfer energy, and k_i the electron wave vector along the i-direction. Here, for simplicity, we have set $\alpha = \beta = \gamma = 90°$, and both b_s and c_s correspond to the lattice parameters b and c, while $a_s = a/2$ since a relates to the thickness of the TMTSF dimer (Fig. 3.3).

The electron transfer energies t_a, t_b, and t_c along the a-, b-, and c-directions[1] are estimated to be 0.25, 0.025, and 0.0015 eV, respectively, from plasma-frequency measurements and the *extended-Hückel calculation* [3.11, 12]. Because the transfer energies are so different for the axes, as expected from the crystal structure, the electron energy band is regarded to be quasi one-dimensional in the first approximation: the electron energy $\epsilon(\mathbf{k})$ is mostly dominated by k_a, and the effects of k_b and k_c are minor. However, for example, for phase transitions the roles of k_b becomes essential. In fact, the $(TMTSF)_2 X$ system is sometimes regarded as showing quasi two-dimensional nature, as will be discussed in detail in Chap. 4.

The electron bands represented by (3.1) would be filled if each TMTSF molecule were neutral. In the 2:1 charge-transfer compound such as $(TMTSF)_2 X$, since only one electron is donated from two TMTSF molecules, the band is three-quarter-filled if we employ a reduced Brillouin zone of the size $(2\pi)^3 / a_s b_s c_s$. As a result, $(TMTSF)_2 X$ is metallic. It is found that the holes in the band contribute to the conductivity. This has been confirmed by the observation of positive thermoelectric power.

---

[1] To represent experimental results the principal axes are denoted either by a, b^*, c^*, or by a, b', c': the b^* axis is perpendicular to both the c- and a-axes, while the c^* axis is perpendicular to the a- and the b-axes, corresponding to the reciprocal axes. The quantities b' and the c' are the projections of the b- and c-axes onto the plane perpendicular to the a-axis.

In the $(TMTSF)_2X$ two TMTSF molecules are slid slightly toward each other, forming a sort of **dimerized structure**. Thus, in the electron band calculated with $a = 2a_s$ a dimerization gap opens up at $k_a = \pi/2a_s = \pi/a$. In this situation the Fermi level is situated at the half-filled level of the upper part of the split band.

The band structure of $(TMTSF)_2X$ was calculated by the tight-binding scheme [3.12] with the following simplifications: 1) the initial molecular orbitals belong to the isolated molecules, 2) only the Highest Occupied Molecular Orbital (HOMO) is included, 3) CH_3 groups are replaced by H, and 4) the Se-Se overlap is considered to be dominant. Then, the tight-binding dispersion relation can be written as

$$\epsilon(\mathbf{k}) = 2\left[t_{I3}\cos(\mathbf{k}\cdot\mathbf{b}) + t_{I4}\cos[\mathbf{k}\cdot(\mathbf{a}-\mathbf{b})]\right] \pm |T(\mathbf{k})| \, , \tag{3.2}$$

where

$$T(\mathbf{k}) = t_{S2} + t_{S1}\exp(-i\mathbf{k}\cdot\mathbf{a}) + t_{I2}\exp(-i\mathbf{k}\cdot\mathbf{b}) + t_{I1}\exp[-i\mathbf{k}\cdot(\mathbf{a}-\mathbf{b})] \, ,$$

a and **b** being the unit vectors along the a- and b-axes, and t_{Ii} and t_{Si} the electron transfer energies for the processes depicted in Fig. 3.5. A detailed discussion of this model is deferred to Sect. 4.3.3.

Numerical calculations show that the approximations

$$t_{S1} \simeq t_{S2} \, , \quad t_{I2} \simeq t_{I3} \gg |t_{I1}| \simeq |t_{I4}| \tag{3.3}$$

can be made, leading to the simplified cosine band approximating the k-dependence in the neighborhood of the Fermi energy, i.e.,

$$\epsilon(\mathbf{k}) \simeq 2[t_I\cos(\mathbf{k}\cdot\mathbf{b}) \pm t_S\cos(\tfrac{1}{2}\mathbf{k}\cdot\mathbf{a})] \, , \tag{3.4}$$

where

$$t_I \simeq t_{I3} + t_{I2}/2\sqrt{2} \, , \quad t_S = (t_{S1} + t_{S2})/2 \, .$$

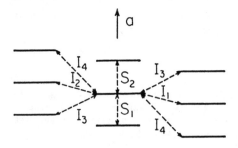

Fig. 3.5. Assignment of the transfer integrals. Solid lines represent TMTSF molecules, S_i the intracolumn transfer and I_i the intercolumn transfer. After [3.12]

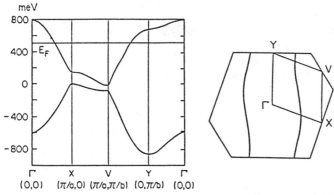

Fig. 3.6. Calculated band structure of (TMTSF)$_2$AsF$_6$ (*left*) and the Fermi surface (*right*). These are representative of all (TMTSF)$_2$X: the shape of the Fermi contour is almost indistinguishable for different X. From [3.12]

For (TMTSF)$_2$ClO$_4$ the values of t_S and t_I are calculated to be 366 and 21.6 meV, respectively, and 380 and 23.9 meV for (TMTSF)$_2$PF$_6$. The band structure and the Fermi surface in the first Brillouin zone of (TMTSF)$_2$AsF$_6$ are exhibited in Fig. 3.6.

The electronic band is based on the overlap of π-orbitals between TMTSF molecules. Therefore, it is interesting to know the spatial distribution of the dominant π-orbitals. This distribution has been evaluated by means of both ESR and high-resolution NMR measurements [3.13, 14]. In doing so the anisotropies in the g-value [3.13, 15] have been the subject of concern. The symmetry of the observed anisotropic shifts in the g-value in (TMTSF)$_2$X is not directly related to the crystalline axes but rather to the molecular axes of TMTSF, indicating that ESR represents the electronic structure of the constituent molecules. On the assumption that the g-shifts stem from spin-orbit coupling, the g-values of the cation have been calculated by using molecular orbitals simplified through the intermediate neglect of differential overlap-type approximation in the framework of the unrestricted Hartree-Fock method [3.13]. By fitting the calculated g-values to the experimental results for (TMTSF)$_2$ClO$_4$, the coefficients of the atomic orbitals which are related to their distribution have been deduced. Figure 3.7 exhibits the distribution of the spin density, derived from the coefficients of singly occupied molecular orbitals, for a TMTSF dimer.

3.1.3 Electrical Properties in the Metallic Region

The temperature dependence of the resistivity for various (TMTSF)$_2$X salts is plotted in Fig. 3.1. The room-temperature conductivity along the a-direction is confined to the range $400 \div 800$ S/cm. The carrier density in

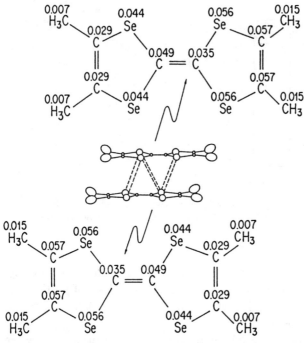

Fig. 3.7. Spin density distribution on the TMTSF dimerized cation. The distribution becomes symmetric as a dimer. Courtesy of N. Kinoshita, after [3.13]

$(TMTSF)_2PF_6$ is estimated to be $1.4 \cdot 10^{21}$ cm^{-3} since the unpaired electrons resulting from the charge transfer act as carriers. Then, together with the carrier relaxation time τ_{op} of $3 \cdot 10^{-15}$ s determined from infrared spectra [3.16] and the assumption that the effective mass of the carrier, m*, is that of the bare electron mass m, we can deduce that the a-axis conductivity $\sigma_a = ne^2\tau_{op}/m^*$ is ca. 10^3 S/cm, in agreement with the observed value.

With decrease in temperature, the conductivity increases monotonically up to an order of 10^5 S/cm at liquid helium temperatures. At low temperatures, e.g., 25 K, the carrier relaxation time τ_{DC} estimated by the DC conductivity reaches 10^{-13} s, whereas τ_{op} determined from infrared spectra based on the Drude model is 10^{-14} s, that is, lower by one order of magnitude than τ_{DC}. The infrared conductivity at 2 K and 200 cm^{-1} stays at ca. 2500 S/cm [3.17]. The difference between τ_{op} and τ_{DC} may indicate that the single-particle treatment is inappropriate for this system and that some correlation between the electrons is to be considered.

An anomalous behavior in the metallic conduction of Bechgaard salts is found in the magnetic-field dependence. Figure 3.8 shows the magnetic-field dependence of the change in resistivity ρ_{aa} along the a-axis (conducting axis) and ρ_{bb} along the b-axis for $(TMTSF)_2ClO_4$, where the magnetic

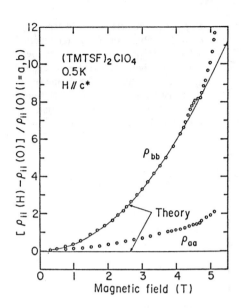

Fig. 3.8. Magnetic-field dependence of ρ_{aa} and ρ_{bb} of $(TMTSF)_2ClO_4$ at 0.5 K. The circles denote observed values and the *solid lines* denote calculated. The magnetic field was applied in the c^* direction. From [3.18]

field is applied along the c^*-direction [3.18]. The transverse resistance ρ_{bb} changes almost proportionally to the square of the magnetic field in agreement with the discussion in Sect. 2.3.6. According to that model, however, ρ_{aa} should be almost independent of the magnetic field in contrast to the observation. This discrepancy demonstrates that the metallic conductivity of $(TMTSF)_2ClO_4$ cannot be described satisfactorily in terms of normal electrons in the relaxation approximation.

By the way, the anistropy in the DC conductivity is remarkable. The conductivity ratio σ_a/σ_b reaches 25 for $(TMTSF)_2ClO_4$ and 200 for $(TMTSF)_2PF_6$ at room temperature [3.19-21]. By using the values $a_s = (a/2) = 3.6$ Å and $b = 7.7$ Å and the relation

$$\sigma_a/\sigma_b \simeq (a_s t_a/bt_b)^2 \; , \tag{3.5}$$

the ratio of the transfer energy t_a/t_b is found to be of the order of 10. Here a_s represents the average intermolecular spacing in the dimerized structure along the a-direction. Due to separation by the counter anions in the c-direction, the conductivity ratio σ_b/σ_c reaches $3 \cdot 10^4$ for $(TMTSF)_2PF_6$ and $9 \cdot 10^2$ for $(TMTSF)_2ClO_4$ [3.17]. Thus, the t_b/t_c ratio is also estimated to be larger than 10, and we may set $t_a:t_{b*}:t_{c*} \simeq 10:1:0.1$ in agreement with the band calculation [3.12]. A detailed discussion will be given in Sect. 4.3.

The *optical polarized reflectance* also shows this unique anisotropy, as depicted in Fig. 3.9. When the electrical vector **E** is parallel to the conducting axis, the spectra at 300 K and 25 K are fitted with the Drude dielectric functions by using the values $t_a = 0.25$ eV and 0.28 eV, repsectively.

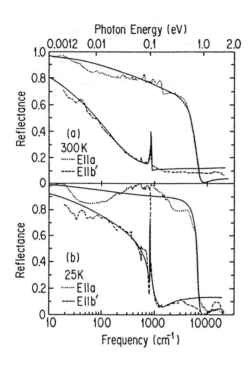

Fig. 3.9. Polarized reflectance of $(TMTSF)_2 PF_6$ for $\mathbf{E} \,||\, \mathbf{a}$ (conducting axis) and $\mathbf{E} \,||\, \mathbf{b}'$ (perpendicular to the a-axis in the sheets of TMTSF molecules) at two temperatures (**a**) 300 K and (**b**) 25 K. Least-squares fits to the reflectance with a Drude dielectric function are shown by the *solid lines*. From [3.16]

When $\mathbf{E}$ is perpendicular to the most-conducting axis but still in the TMTSF plane (b′-direction), one gets a good fit with $t_b = 24$ meV for spectra measured at 25 K, although the band is smeared out by thermal fluctuations at 300 K (Sect. 4.3.2). This implies that the system is almost one-dimensional at room temperature in the sense that the band description is valid only along the stacking axis. It crosses over to a quasi two-dimensional metallic behavior with decreasing temperature.

The *thermopowers* S_a and S_b along the a- and the b-directions exhibit an anomaly around 100 K [3.22]. The longitudinal thermopower S_a changes almost linearly with temperature down to 12 K with a slight change in slope near 100 K. On the other hand, the transverse thermopower S_b is only slightly temperature-dependent above 100 K but it is approximately linear below. This thermopower anisotropy is explained by considering that the conduction along the a-axis is metallic whereas the conduction along the b-axis is dominated by phonon-induced intercolumn electron hopping above 100 K.

3.1.4 Geometric Magnetoresonance Oscillation

In the angle dependence of the magnetoresistance a series of dips at magic angles between field and crystal orientations are found for $(TMTSF)_2 ClO_4$

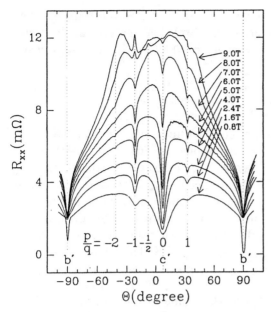

Fig. 3.10. Angular sweep of magnetoresistance of $(TMTSF)_2 PF_6$ under 10 kbar of pressure at 0.5 K. At the lowest field angular sweep of 0.8 T, partial superconductivity is observed near the b-axis. From [3.23]

and $(TMTSF)_2 PF_6$. For magnetic fields rotated in the $b'c^*$ plane, the angle dependence of ρ_{aa} in the metallic region of $(TMTSF)_2 PF_6$ are found for different fields, as shown in Fig. 3.10. It is noteworthy that the angle θ of the dips is independent of the field intensity and given by

$$\tan\theta = \frac{p}{q} \frac{b\sin\gamma}{c\sin\beta\sin\alpha} - \cot\alpha \qquad (3.6)$$

where b, c, α, β, γ are the lattice parameters, and p, q are integers. The relationship determined by the lattice parameters without the band parameters suggests that the dips are related to the **geometric commensurate motion** of the electrons on the Fermi surface of $(TMTSF)_2 PF_6$ that are induced by the application of a magnetic field tilted to the magic angle [3.23, 24]. A similar effect has been observed for $(TMTSF)_2 ClO_4$ in the second angle derivative of the resistance with more fine structures [3.25] and also in magnetic-torque measurements [3.26].

An oscillatory and sharply anisotropic behavior with respect to the field direction was predicted by *Lebed* [3.27] in connection with the decrease of the threshold for instability of the metallic phase under the influence of electron-electron interactions. Despite an extensive search no such an instability has been established experimentally, but the series of the above-mentioned dips has been found. Later, *Lebed* explained the dips at the magic angles on the basis of a three-dimensional band model with the two transfer integrals t_b and t_c as

$$\epsilon = \hbar v_F (|k_x| - k_F) - 2t_b \cos(k_y b) - 2t_c \cos(k_z c) \tag{3.7}$$

and an electron-electron scattering into the direction of the magnetic field, leading crossover between one-dimensionality and two-dimentionality [3.28]. On the other hand, *Maki* ascribed the dip to an approximate damping constant of the fluctuations [3.29]. In later work, he proposed a possible formation of new quantized states at a magic angle that exhibit doubly quantized Hall conductivity [3.30]. In order to explain the dip, *Chaikin* suggested that the regions in which the electron scattering rate is anomalously high, may be located on the Fermi surface [3.31]. The microscopic origin has been investigated and understood in terms of an electron-electron scattering in quasi one-dimensional conductors, in a way analogous to the reason for the existence of the van Hove singularities in the density of states. However, it is not strong as far as the effect of the magnetic field is concerned semiclassically [3.32]. In the meantime an explanation has been proposed by taking into account strong correlation in a quasi one-dimensional system, describable by a *Luttinger liquid*, where the incoherent single particle hopping between spin-charge-separated states is considered. It was claimed that a field-dependent renormalization of the coherent part of the interchain hopping is related to the anomalous dip [3.33].

Another type of striking angular dependence has been found in $(TMTSF)_2 ClO_4$ from resistance measurements along the least conducting direction (c-axis), when the magnetic field is tilted from the a-axis to the c-axis (Fig.3.11). The figure depicts traces of the c-axis resistance as the field is rotated in the ac plane through the a-axis for different magnetic fields. In the simplest picture, no magnetorestance is expected when the field is parallel to the current direction, and maximum magnetoresistance when the current and the field are perpendicular. At high fields a maximum is found when the field direction is close to the a-axis with a series of peak structures, the angular positions of which are independent of the field. For fields below 2 T, minima due to superconductivity appear near $\theta = 0$. This geometric resonance has been interpreted in terms of semiclassical orbital averaging of the c-axis velocity v_c given by $2t_c c \sin(k_z c)$. The Fermi surface of $(TMTSF)_2 ClO_4$ consists of two sheets extended in the $k_y - k_z$ plane (Fig. 3.12) whose k_x values are close to $\pm k_F$. There are weak modulations in k_x determined by the transfer integrals t_b and t_c, respectively, as experssed by (3.7). The orbits of the fields near the a-axis in the ac plane are plotted in Fig.3.12. As the field is tilted away from the a-axis, the orbits show large

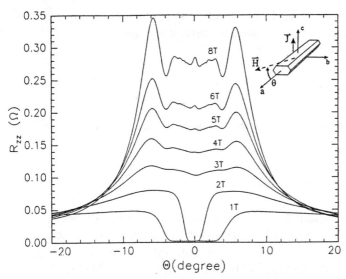

Fig. 3.11. Angular dependence of the c-axis resistance R_{zz} for rotation in the ac plane at different fields for $(TMTSF)_2 ClO_4$ at 0.5 K. The 2 T and 1 T traces exhibit superconductivity at smaller angles. The *inset* presents a drawing of a crystal with the axis, field and current directions as shown. From [3.34]

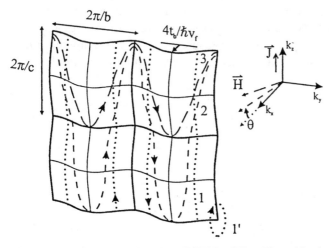

Fig. 3.12. The Fermi surface for $(TMTSF)_2 ClO_4$. The orbits 1 and 1′ (*dots*) indicate the approximate paths for electrons when the field is parallel to the a-axis for closed (1′) and open (1) orbits. The orbits 3 (*long dashes*) and 2 (*short dashes*) represent the paths when the angle of the magnetic field is near, respectively, the primary maxima and the first secondary maxima of the data in Fig. 3.11. From [3.34]

57

amplitude oscillations in k_z. In the absence of scattering, the average of v_z becomes zero and hence ρ_{zz} becomes maxima at θ_n where

$$\tan\theta_n \simeq \frac{2t_b c}{n2\pi \hbar v_F} .$$ (3.8)

By fitting the relation to the experimental data, $t_b = 0.012$ eV has been deduced [3.34].

3.1.5 Spin-Density-Wave Ordering

In a quasi one-dimensional electron system, the metallic state becomes unstable and changes to become insulating, with associated lattice modulations having the wave number $2k_F$ (k_F is the Fermi wave number), as mentioned in Sect. 2.3. This is the so-called **Peierls transition**. For $(TMTSF)_2 PF_6$ and $(TMTSF)_2 AsF_6$, a metal-insulator transition is observed at 12 K at ambient pressure (Fig. 3.1). This transition, however, is not ascribed to the Peierls transition accompanying the Charge Density Wave (CDW), since the relevant lattice modulations have not been observed. Furthermore, the magnetic susceptibility does not decrease, except that for the b-direction. For CDW insulators, the spin susceptibility decreases in any direction, since the carrier density decreases exponentially with the opening of the energy gap.

The magnetic-susceptibility χ_i measurements reveal a unique anisotropy (Fig. 3.13). Below the metal-insulator transition temperature T_{MI}, χ_b'

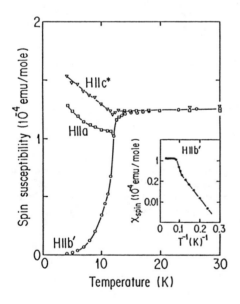

Fig. 3.13. $(TMTSF)_2 AsF_6$ spin susceptibility, as deduced from static measurements with field parallel to a, b', and c*, respectively. The magnetic field H = 3 kOe used for the measurement is less than the critical spin-flop field. After [3.35]

decreases quickly with decreasing temperature, while both χ_a and χ_{c*} show a slight increase. In addition, with increase in the magnetic field H, $\chi_b{}'$ increases dramatically and becomes comparable with χ_a and χ_{c*} above 4.5 kOe. These features demonstrate conclusively that below T_{MI} the spins are ordered antiferromagnetically, i.e., alternating along the b'-direction. The threshold magnetic field for the recovery of $\chi_b{}'$ corresponds to the spin-flop field where the polarization direction of antiferromagnetically ordered spins is flopped into the a-direction to reduce the energy.

The presence of the **antiferromagnetic ordering** is confirmed by both the ESR and the NMR spectra of either ^{77}Se or ^{1}H for $(TMTSF)_2PF_6$. Below T_{MI} the ESR signals observed in the metallic phase disappear due to a shift in the resonance frequency [3.36], while the NMR signal of ^{77}Se disappears due to an extreme inhomogeneous broadening and the proton line shape is broadened [3.37, 38]. These features are assigned to the appearance of inhomogeneous local magnetic fields due to the spin ordering. In the ESR experiment the shifted resonance fields reveal remarkable angular dependence. This is interpreted in terms of the anisotropic antiferromagnetic resonance, due to the dipole-dipole interactions among the spins. The amplitude of the SDW was estimated to be $10 \div 20\%$ of the total spins, and the antiferromagnetic interaction constant J was estimated to be 604 K for $(TMTSF)_2AsF_6$ [3.39].

The good **nesting property** of the Fermi surface expected for the quasi one-dimensional band structure of $(TMTSF)_2X$ is consistent with the appearance of SDW. The metal-insulator transitions for $(TMTSF)_2PF_6$ and $(TMTSF)_2AsF_6$ are thus attributed to the SDW ordering.

Proton NMR experiments have gone further to determine the wave vector **Q** and the amplitude σ of the undulation of the spin density [3.40, 41]. According to the relaxation experiments, protons in the methyl groups of TMTSF molecules stop their rotation at temperatures below 30 K. The complicated line shapes of the NMR spectra at lower temperatures include contributions from all the protons of the methyl groups and current carriers. This line shape was well fitted by taking account of the contact field and the dipolar field coming from the spin density in each TMTSF molecule. The spin density reflects the spin distribution in the HOMO of each molecule (as in Fig. 3.7). Since Q_a is given by $2k_F$ and the Q_c-dependence is practically negligible, the fitting parameters are Q_b and σ. *Delrieu* et al. [3.40] obtained $Q_b = 0.20b^*$ and $\sigma = 0.085\mu_B$ (μ_B is the Bohr magneton), while *Takahashi* et al. [3.41] obtained $Q_b = 0.24b^*$ together with $Q_c \simeq -0.06c^*$ and $\sigma = 0.08\mu_B$. The SDW wave vector derived is in good agreement with the value obtained from the nesting vector of the calculated band. The amplitude σ is much smaller than unity and assures the delocalized nature of the SDW.

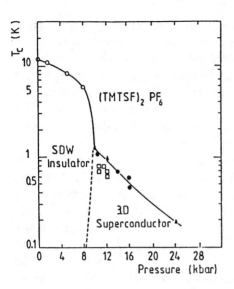

Fig. 3.14. Phase diagram of (TMTSF)$_2$ · PF$_6$. From [3.42]

The application of hydrostatic pressure suppresses the metal-insulator transition in (TMTSF)$_2$PF$_6$. T_{MI} decreases with pressure and ultimately tends to zero. Then a transition to a superconducting phase appears at $T_c =$ 0.9 K under 12 kbar [3.2]. For (TMTSF)$_2$PF$_6$ and (TMTSF)$_2$AsF$_6$ the pressure dependences of T_{MI} and T_c are shown in Figs. 3.14 and 15, respectively. The lowest pressure above which superconductivity appears is called the **critical pressure** P_c. For the (TMTSF)$_2$X with anions of octahedral

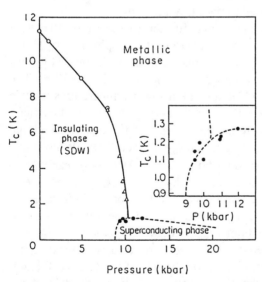

Fig. 3.15. Phase diagram of (TMTSF)$_2$AsF$_6$. The metal-insulator transition temperature is defined either by the maximum of $-d\ln R/dT$ (P $\leq$ 8 kbar *open circles*) or by the criterion $R(T_{MI})/R_{min} = 2$ (P > 8 kbar *open triangles*). From [3.43]

Table 3.1. Characteristic temperatures and pressures of phase transitions in $(TMTSF)_2 X$

		Metal-insulator	Super-conducting	Super-conducting	Anion ordering
X (Anion)	Anion symmetry	T_{MI} [a] [K]	T_{cm} [K]	P_c [kbar]	T_{AO} [a] [K]
PF_6	Octahedral	12	1.1	6.5	-
AsF_6	Octahedral	12	1.1	12	-
SbF_6	Octahedral	17	0.4	11	-
TaF_6	Octahedral	11	1.4	12	-
ClO_4 [b]	Tetrahedral	-	1.2 ± 0.2	0	24
ClO_4 [c]	Tetrahedral	6.5	-	-	-
ReO_4	Tetrahedral	180	1.3	9.5	180
FSO_3	Tetrahedral-like	86	3	5	86
F_2PO_2	Tetrahedral-like	137	-	-	
NO_3	Triangular	12	-	-	41
H_2F_3	V-shape	63	-	-	63

[a] At ambient pressure.
[b] Slowly cooled; anions are ordered.
[c] Rapidly cooled; anions are disordered.

symmetry, similar phase diagrams are obtained. In Table 3.1 T_{MI}, T_{cm} and P_c values are listed for various X (T_{cm} is the maximum value of T_c). A detailed theoretical description of an SDW is given in Sect. 4.4.

3.1.6 Anion Ordering

So far we have discussed salts with anions of octahedral symmetry. In contrast to these compounds, when we consider the anions of non-centrosymmetric tetrahedral symmetry, a complication comes in. The anion of tetrahedral symmetry can take two orientations with respect to the crystalline system. The anions are situated at the inversion center in a cage made by surrounding TMTSF molecules at room temperature. The averaged symmetry belongs to the triclinic $P\bar{1}$ space group. On cooling, anion-ordering transitions occur at T_{AO} (*anion-ordering temperature*).

For $(TMTSF)_2 ClO_4$, diffuse X-ray scattering studies revealed that the anion-ordering occurs at 24 K to form a **superlattice** with the wave number $Q = (0, \frac{1}{2}, 0)$ [3.44]. However, when $(TMTSF)_2 ClO_4$ is rapidly cooled from a higher temperature to 15 K the orientations of anions are frozen at

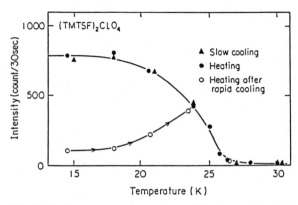

Fig. 3.16. Temperature dependence of the intensities of the X-ray diffraction peak corresponding to the $Q = (0, \frac{1}{2}, 0)$ superstructure in $(TMTSF)_2 ClO_4$. The rate of rapid cooling was 60 K/min from 30 K to 15 K, and the slow cooling was 0.1 K/min. The *closed marks* represent the observation during slow cooling and heating. The *open mark* illustrates the observation during heating after rapid cooling. From [3.45]

random directions, as indicated by open circles in Fig. 3.16. In the slowly cooled (0.1 K/min) case, at 24 K, the conductivity increases with the onset of anion ordering (Fig. 3.17). This is due to the reduction of the scattering by the randomly distributed anion potential. Upon further cooling of the crystal, superconductivity appears at $T_c = 1.2 \pm 0.2$ K. This is the first discovered ambient-pressure organic superconductor [3.46]. By rapid cooling (>50 K/min), the disordered state existing above T_{AO} is quenched near 24 K and the temperature dependence of the resistivity becomes more gradual than the case for the slow cooling, as shown by the Q state curve down to 6

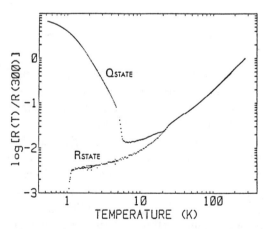

Fig. 3.17. Temperature dependence of the resistivity of $(TMTSF)_2 ClO_4$. The *lower curve* is obtained by slow cooling (denoted by R state) while the *higher one* is by rapid cooling (denoted by the Q state). Courtesy of H. Bando. After [3.47]

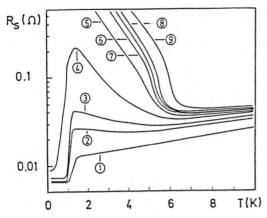

Fig. 3.18. Temperature dependence of the resistivity of $(TMTSF)_2 ClO_4$. The curves are numbered in order of increasing cooling rates. From [3.48]

K in Fig.3.17. An insulating phase appears below 6 K. A mixture of these behaviors is observed when the *cooling conditions* are varied and the transitions between phases become less sharp (Fig.3.18). The intermediate state with only a partial degree of anion ordering was produced by quenching (rapid cooling) of the sample from various temperatures T_Q. When T_Q is well below 24 K the anions are ordered, while for $T_Q \geq 24$ K the degree of the disorder increases with T_Q.

On the basis of results of NMR experiments as well as of ESR and static magnetic susceptibility studies, the insulating state below 6 K for $(TMTSF)_2 ClO_4$ produced by rapid cooling is assigned to an SDW phase [3.49, 50]. The activation energy of the resistivity R for the insulating phase near T_{MI} is derived from

$$2\Delta(T) = k_B T \cdot \ln(R/R_{300K}) .$$ (3.9)

With $\Delta(0) = 11$ K [3.48] we find $2\Delta(0)/k_B T_{MI} = 3.64$ at $T_{MI} = 6.05$ K. This is in good agreement with the mean-field value of 3.53.

For $(TMTSF)_2 ReO_4$ an anion ordering transition occurs at $T_{AO} \simeq 180$ K with the superstructure of $(\frac{1}{2}, \frac{1}{2}, \frac{1}{2})$ and leads to an insulating state [3.51]. It was also found that the anion ordering parallels the modulation of TMTSF stacking. In the insulating state the spin susceptibility decreases exponentially below T_{MI}, irrespective of the crystal orientation, indicating that this is not to be assigned to the SDW state. This is because the periodic potential is made by ordered anions, which forms a gap, and the only current carriers that exist are thermally excited. The insulating state is suppressed under pressure above 12 kbar, however, and superconductivity is observed below 1.3 K [3.52]. It has been found that above 12 kbar the ordered phase with $Q = (\frac{1}{2}, \frac{1}{2}, \frac{1}{2})$ changes to $(0, \frac{1}{2}, \frac{1}{2})$. At intermediate pressures $(9 \div 12 \text{kbar})$ a complex behavior is observed with different low-temperature states being produced depending on the thermal treatment, just as in glassy phases [3.53]. This is attributed to two competing ordered phases characterized by different superstructures.

A similar anion ordering transition has been found in $(\text{TMTSF})_2 \cdot \text{FSO}_3$ [3.4]. At ambient pressure this salt undergoes a metal-insulator transition at 86 K through an **antiferroelectric dipole ordering** of the anions with $Q = (\frac{1}{2}, \frac{1}{2}, \frac{1}{2})$, since FSO_3^- is not tetrahedral and has a permanent dipole [3.51]. In this case, too, an applied pressure suppresses the insulating state and a gradual transition to superconductivity is observed near 3 K: In the pressure region between 5 and 6 kbar, a glassy behavior is observed as in $(\text{TMTSF})_2 \text{ReO}_4$. In the glassy state, the resistivity increases with cooling, but the thermopower remains ungapped [3.4].

In $(\text{TMTSF})_2 \text{H}_2 \text{F}_3$, a structural transition occurs at 63 K. An X-ray scattering study has revealed two kinds of superstructures reflecting a periodic modulation of the TMTSF stack with one-dimensional precursor scattering and an order-disorder transition of the $\text{H}_2 \text{F}_3^-$ anions [3.7]. The metal-insulator transition and the spin susceptibility drop occur at T_{AO}, but the low-temperature thermopower exhibits metallic character, although an anomalous behavior is observed near T_{AO}.

In the case of $(\text{TMTSF})_2 \text{NO}_3$, the conductivity increases with the anion ordering for $Q = (\frac{1}{2}, 0, 0)$ occuring at 41 K [3.51], presumably due to the reduction in random potentials caused by disordered anions. At 12 K, this salt becomes an insulator [3.7].

In summary, there are two types of phase transitions for $(\text{TMTSF})_2 X$: one stems from the electronic nature of the TMTSF stacks such as the SDW ordering, and the other is due to the ordering of anions lacking the inversion center. The characteristic parameters for these transitions are listed in Table 3.1.

3.1.7 TMTTF Salts

The variety of organic molecular salts can be expanded by simply substituting chemical groups and atoms with similar species. In $(\text{TMTSF})_2 X$ salts this has been done by varying the counter anion X^-. The TMTSF molecule can itself be modified by replacing the Se atoms with S atoms to produce TMTTF (tetramethyltetrathiafulvalene) (Fig. 1.1). This molecule, developed previously to TMTSF [3.54], can also be combined with the same counter anions X^- to form $(\text{TMTTF})_2 X$ which is isomorphous to $(\text{TMTSF})_2 X$ [3.55].

Table 3.2. Characteristic parameters of (TMTTF)$_2$X at ambient pressure (SP: spin-Peierls state, SDW: spin-density-wave state)

X (Anion)	Anion symmetry	Temperature of maximum conductivity T_{max} [K]	Anion ordering transition T_{AO} [K]	Modulation wave number Q	Ground state at low T
PF$_6$	Octahedral	230	-	-	SP($<$15K)
AsF$_6$	Octahedral	230	-	-	SP($<$11K)
SbF$_6$	Octahedral	180	-	-	SDW($<$9K)
ClO$_4$	Tetrahedral	250	75	($\frac{1}{2}$,0,0)	
ReO$_4$	Tetrahedral	230	160	($\frac{1}{2}$,$\frac{1}{2}$,$\frac{1}{2}$)	
BF$_4$	Tetrahedral	210	40		SP($<$40K)
NO$_3$	Triangular	210	50	($\frac{1}{2}$,0,0)	
SCN	Linear	240	160	(0,$\frac{1}{2}$,$\frac{1}{2}$)	SDW($<$7K)
Br	Spherical	100	-	-	SDW($<$17K)

In contrast to the (TMTSF)$_2$X salts which are metallic at room temperature, the (TMTTF)$_2$X salts are semiconducting with a shallow maximum in the conductivity near $T_{max} \simeq 200$ K (Table 3.2). The resistance minimum is assumed not to be due to the metal-insulator transition but rather to a competition between the mobility which decreases as $T^{-\alpha}$ (α depends on the scattering mechanism) and the number of charge carriers excited across the energy gap, which increases with T. *Mortensen* et al. [3.56] claimed that, even above room temperature, a gap opens at the Fermi energy. This is based on thermopower measurements showing a 1/T dependence without any structure around T_{max}. Another interpretation is a gradual change from the metallic state to a *Mott-Hubbard insulator* phase. The difference in the electronic structure of (TMTSF)$_2$X and (TMTTF)$_2$X is attributed more to the difference in the degree of dimerization and one-dimensionality.

TMTTF salts with non-centrosymmetric anions undergo *order-disorder transitions* just like (TMTSF)$_2$X salts, the temperatures of which (T_{AO}) are listed in Table 3.2, but substantial changes in the electrical properties do not emerge since the salts are already in non-metallic states. Magnetic measurements in the nonmetallic region of (TMTTF)$_2$X have revealed phase transitions at low temperatures. These transitions cannot be ordinary Peierls transitions since the system is not in a metallic state. They are of magnetic origin. There are two possibilities: one is a transition to the antiferromagnetic ordering as for (TMTSF)$_2$X and the other is a transition to a spin-Peierls state. In a one-dimensional spin system with antiferromagnetic interaction, the electrons tend to make singlet pairs by alternating the spacing between them, since the exchange interaction between spins changes with their separation. As a result, at low temperature, in the quasi one-dimensional conductor, the spin-Peierls state accompanied by a lattice modulation may become more stable than the antiferromagnetically ordered state [3.57, 58].

Through magnetic measurements it has been revealed that $(TMTTF)_2 \cdot PF_6$ and $(TMTTF)_2 AsF_6$ undergo a **spin-Peierls transition** to a state with singlet-paired electrons [3.59]. On the other hand, the low-temperature ground state of $(TMTTF)_2 SbF_6$ has been identified to convert to an SDW phase [3.60]. For $(TMTTF)_2 PF_6$, the proton NMR relaxation data have revealed that at higher pressures (>13kbar) the SDW state emerges although the salt is nonmagnetic and in a spin-Peierls state at ambient pressure [3.61]. This demonstrates that SDW states become more stable than spin-Peierls states with increase in the dimensionality, which, as for $(TMTSF)_2 \cdot X$, is caused by the applied pressure. The characteristic properties related to the transition of $(TMTTF)_2 X$ salts are summarized in Table 3.2. Note that for smaller X such as Br and SCN, the low-temperature ground state is the SDW rather than the spin-Peierls state, being consistent with the stabilization of the SDW ordering by pressure for these salts.

Because the $(TMTTF)_2 X$ salts are structurally similar to $(TMTSF)_2 X$ salts, it is natural to expect that they will also be superconductors. Actually, the superconductivity transition at 0.8 K was found in $(TMTTF)_2 Br$ under 26 kbar [3.62]. The electrical anisotropy and the magnetic ordering indicate that $(TMTTF)_2 X$ is more one-dimensional than $(TMTSF)_2 X$. This is sup-

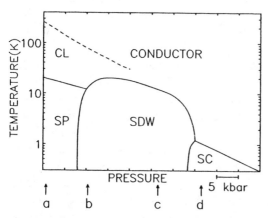

Fig. 3.19. Generalized phase diagram for the $(TMTTF)_2 X$ and $(TMTSF)_2$ series. The notation CL, SP, SDW, and SC refers to charge-localized, spin-Peierls, spind density wave, and superconducting states, respectively. The lower-case letters designate compounds and indicate their location at atmospheric pressure in the generalized diagram. [(a) $(TMTTF)_2 PF_6$, (b) $(TMTTF)_2 Br$, (c) $(TMTSF)_2 PF_6$, (d) $(TMTSF)_2 ClO_4$]. From [3.62]

ported by the fact that $(TMTTF)_2 PF_6$ undergoes a spin-Peierls transition at ambient pressure but changes to an SDW phase at higher pressures. A generalized phase diagram of spin-Peierls, SDW and superconductivity has been proposed by combining $(TMTSF)_2 X$ and $(TMTTF)_2 X$ salts vs. intermolecular spacing or pressure (Fig. 3.19).

3.1.8 Resistance Jumps and Mechanical Kinks

Crystals of $(TMTSF)_2 X$ are grown in the shape of a thin needle with a hexagonal cross section. This shape is appropriate for electrical-conductivity measurements via the four-terminal method. During the measurement by cooling, however, one often faces an embarrassing problem, that is, sudden step-wise increases in the resistivity (Fig. 3.20). Both the size and the position of the "jumps" change from sample to sample and measurement to measurement, although they do seem to occur only above 60 K. Curve A in Fig. 3.20 shows the behavior of a cooled virgin sample of $(TMTSF)_2 ClO_4$. This was followed by measurements taken during the heating process (curve B). The temperature dependence is quite different and there are no signs of resistance jumps. Above ca. 220 K the resistance gradually drops to nearly the value of the virgin sample, almost as if the sample were "cured" by heating. Since the estimated Debye temperature, found by a specific heat measurement [3.65], is 213 K, the curing above that temperature might be related to some lattice property. However, it is clear that the resistance jumps are not due to mechanical fracture but rather to some reproducible threshold-limited mechanical processes. The trace denoted by C represents the temperature dependence of the resistivity during a second cooling. Although the trace is not the same as for the first cooling, an essential similarity is found. This is also true for the first and second heating process (curve

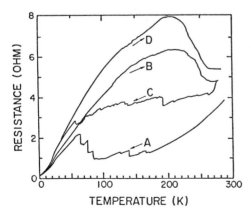

Fig. 3.20. Four different measurements of the temperature dependence of the resistance on the same $(TMTSF)_2 \cdot ClO_4$ sample. Note the resistance "jumps". From [3.64]

D). The size of the resistance jumps depends on sample and cooling conditions. The data in Fig.3.20 are for a typical case, and we should mention that samples that can be cooled to the superconducting transition without resistance jumps are rare.

It has been found that the temperatures of superconducting transitions are not strongly affected by resistance jumps [3.64]. The resistivity becomes almost zero well below T_c even if the resistance jumps are large and frequent. One effect of the resistance jumps is the broadening of the transition. One way to evaluate the effect of a resistance jump is to compare the ratio of the resistance at room temperature to the resistance just above T_c. When there is no jump the resistance ratio in $(TMTSF)_2 ClO_4$ reaches ca. $2 \cdot 10^3$. With the resistance jumps, the value is reduced to as low as 3. For a resistance ratio of 10, a transition to zero resistance can be found but the transition region becomes very broad; the onset is at 1.2 K and a resistance near zero is reached around 0.3 K [3.64].

The origin of the resistance jump has not yet been uncovered, although it appears quite generally for $(TMTSF)_2 X$ and $(TMTTF)_2 X$ crystals. It must have some appreciable effect on the transport properties at low temperatures. The heat-cycle measurements mentioned above are reminiscent of the effects observed due to mechanical stress that are caused by localized defects or electrode contacts.

One possible clue is the mechanical kink effect unique to these salts. When one carefully pushes a point of a thin needle of $(TMTSF)_2 ClO_4$ in the transverse direction, the crystal makes a pair of kinks, which can be removed without fracture and even moved along the needle axis by carefully applying appropriate pressures [3.66,67]. Through X-ray diffraction analysis by means of a precession photograph, it has been seen that the kinks are due to a mechanical twinning with the boundary on the (210) plane, which is close to the plane of the TMTSF molecular plane. Figure 3.21 depicts a pair of such kinks.

The annealing characteristics of the resistance jumps, the reversible mechanical distortions and the twinning may be related phenomena. However, so far no direct connections have been found, and the intrinsic properties of the resistance jumps have kept themselves open to investigation.

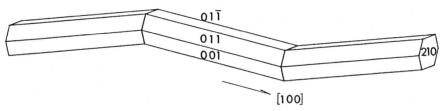

Fig. 3.21. Kink in a $(TMTSF)_2 ClO_4$ by mechanical twinning. From [3.66]

3.2 Superconducting Properties

3.2.1 The Superconducting Transition

At ambient pressure below 12 K the organic metal $(TMTSF)_2 PF_6$ becomes an SDW insulator. When hydrostatic pressure is applied, the SDW phase is suppressed and the superconducting state is stabilized at low temperatures above the critical pressure P_c. The pressure-phase diagrams of $(TMTSF)_2$ $\cdot PF_6$ and $(TMTSF)_2 AsF_6$ are exhibited in Figs.3.14 and 15, respectively. In $(TMTSF)_2 AsF_6$ the superconducting state "intrudes" into the SDW state near P_c. This intruding part is called a **re-entrant state**. The transition characteristics are explained in terms of a free-energy consideration, and the transition must be first order, as will be discussed in Chap.4. The role of the pressure in stabilizing the superconductivity is attributed to the decrease in the intercolumn distance and the accompanying increase in the transverse transfer energies resulting in a suppression of Fermi-surface nesting.

In $(TMTSF)_2 ClO_4$ the superconductivity can be stabilized at ambient pressure by carefully cooling the samples across the anion ordering temperature T_{AO} of 24 K at a slow rate, e.g., less than 0.2 K/min. Under these conditions the anions are well ordered below T_{AO} and superconductivity appears at $T_c = 1.2 \pm 0.2$ K. When the sample is cooled rapidly, e.g., with a rate of 50 K/min, the anion disorder is frozen and, instead of superconductivity, an insulating state is formed at $T_{MI} \simeq 6.1$ K (Fig.3.17). With an intermediate cooling rate one gets a mixture of anion-ordered and the disordered phases (Fig.3.18). Thus, the superconducting ground state is in competition with an insulating phase.

The **cooling-rate dependence** of the transition characteristics of $(TMTSF)_2 ClO_4$ was discovered by *Takahashi* et al. [3.49] during NMR measurements on [77]Se and [1]H. Under fast cooling, the ^{77}Se-NMR intensity decreases and the resonance line is broadened. This is attributed to inhomogeneous broadening of the resonance due to the onset of a strong local field of hyperfine origin. In concert with this, the spin lattice relaxation rate $1/T_1$ displayed a sharp peak at the same temperature where the intensity starts to decrease. These experimental results together with the data of an antiferromagnetic resonance study [3.39,68] indicate that for $(TMTSF)_2$ $\cdot ClO_4$, the SDW in the insulating phase is stabilized by fast cooling. In the following discussion we describe only the data obtained under slow-cooling conditions.

The **specific-heat** measurements have shown a large anomalous effect in the contribution from the electronic energy at 1.22 K (Fig.3.22), which

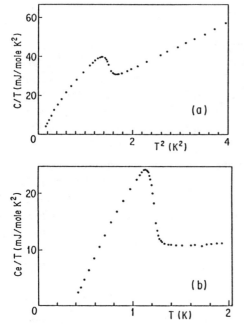

Fig. 3.22. Temperature dependence of the $(TMTSF)_2 ClO_4$ molar specific heat in zero magnetic field for lattice and electronic energy contributions (**a**) and for electronic energy contribution only (**b**). From [3.65]

can well be interpreted as being due to the onset of a three-dimensional superconducting order [3.65]. Above T_c, the specific heat C obeys the relation

$$C/T = \gamma + \beta T^2 \tag{3.10}$$

with $\gamma = 10.5$ mJ/mole K^2 and $\beta = 11.4$ mJ/mole K^4. The increment of C at T_c, ΔC, is equal to 21.4 mJ/mole K. With these data one gets $\Delta C/\gamma T_c = 1.67$, which is in fair agreement with the BCS value of 1.43. From the free-energy difference between the normal and superconducting states one gets the condensation energy which equals $H_c^2/8\pi$, where H_c is the **thermodynamic critical field** and is given by, in the present case, $H_c = 44 \pm 2$ Oe.

The superconducting phase in $(TMTSF)_2 X$ is characterized by its close contact with the insulating phase in the phase diagram and emerges at a pressure P above P_c. However, increased pressure does not necessarily enhance the superconductivity: the increase of P above P_c lowers T_c at a rather rapid rate, e.g., $dT_c/dP = -(8 \pm 1) \cdot 10^{-2}$ K/kbar [3.69].

A way to interpret the suppression of T_c by pressure is to assume the **BCS relation,**

$$T_c = 1.13 \hbar \omega_D e^{-1/\lambda} \,,$$

$$\lambda = \frac{N_F \langle I^2 \rangle}{M \langle \omega^2 \rangle} \,, \tag{3.11}$$

where ω_D is the Debye frequency, N_F is the density of states at the Fermi surface, $\langle I^2 \rangle$ is the Fermi surface average of the electron-phonon coupling, and M is the lattice mass. The phonon frequency $\langle \omega^2 \rangle$ is remarkably increased by an applied pressure in molecular crystals with rather weak intermolecular coupling [3.69, 70]. Thus, the large negative pressure dependence of T_c may be understood in terms of the decrease in λ, due to a pressure-induced lattice hardening.

The phase diagram for the quenched sample of $(TMTSF)_2 ClO_4$, with respect to the proximinity of SDW and the superconductivity phases, is the same as the salts with centrosymmetric counter anions. In this compounds, however, the direction of the ordered spins in the SDW phase is tilted from the crystallographic principal axes [3.68]. In the ReO_4 salt, metal-nonmetal transitions are observed in connection with the anion ordering (Table 3.1). The transition is suppressed by pressure, and the ReO_4 salt exhibits superconductivity with $T_c \simeq 1.3$ K.

The noncentrosymmetric FSO_3 anion deserves special attention because it contains a small electric dipole moment. In both the ReO_4 and FSO_3 salts, the insulating phase involves anion ordering instead of SDW ordering which is found in other superconducting $(TMTSF)_2 X$ salts. This view is due to the magnetization which decreases exponentially with temperature for any direction of the magnetic field. At ambient pressure, the FSO_3 salt exhibits anion ordering below 86 K, but the application of pressure reduces T_{AO}. By means of DC resistivity measurements, indications of a superconducting state are detected at temperatures as high as 3 K at ca. 5 kbar, while T_{AO} is around 40 K, as shown in Fig. 3.23. The resistivity is semiconductor-like below 90 K, but the thermoelectric power suggests that a gap does not open at the Fermi surface and that the main cause of the semiconducting behavior is the mobility of the conduction electrons. The

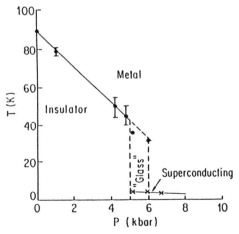

Fig. 3.23. Phase diagram showing T_c for $(TMTSF)_2 FSO_3$. From [3.5]

resistance increase may be due to the scattering of the electrons from disordered charges and polarized species in a frozen dipole glass formed by FSO_3 anions. The superconducting transition spreads out over a large temperature range, but the onset temperature and the critical field are the highest in the TMTSF family. Static spin-susceptibility measurements have shown that the *diamagnetic shielding signal* amounts to 2% of that of a perfect superconducting state at 0.3 K around 6 kbar [3.71]. The *Meissner signals* range typically from 50% to 70% of the shielding signals. This suggests that although the fraction of the sample that is in a superconducting state is small, the superconductivity in this region is well developed to expel magnetic fluxe lines.

Many of the characteristics of superconductivity such as the specific heat jump and the energy gap deduced from tunneling spectroscopy (Sect. 3.2.3) are consistent with simple BCS theory. However, the **nuclear relaxation rate** probing the dynamical aspects of the conduction electrons through the interaction of their magnetic moments with those of the nuclei has some anomalous features. The proton spin-lattice relaxation rate $(1/T_1)$ in the superconducting state of $(TMTSF)_2 ClO_4$ has been measured by the field-cycling method which allows measurements at zero magnetic field. *Takigawa* et al. [3.72] found that the $1/T_1$ of protons in the methyl groups decreases rapidly just below T_c (Fig. 3.24). This contrasts remarkably with typical BCS superconductors where $1/T_1$ increases below T_c reaching a maximum at $T \simeq 0.9 T_c$. The enhancement of $1/T_1$ below T_c is associated with the divergence of the density of states at the edge of the energy gap, as well as the coherence factor characteristic to superconductivity [3.73].

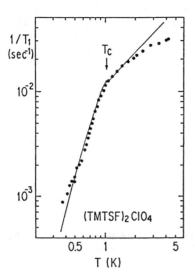

Fig. 3.24. Temperature dependence of $1/T_1$ of protons at zero field in methyl groups of $(TMTSF)_2 ClO_4$. The *solid curve* shows the calculation for the d-wave or p-wave superconducting states normalized to the experimental data at $T_c = 1.06$ K. From [3.72]

These features are reminiscent of the behavior of heavy electrons in $CeCu_2Si_2$, UBe_{13} etc., in which they have been proposed to be responsible for the vanishing of the anisotropic order parameter along certain lines on the Fermi surface. By extending the heavy electron model, *Hasegawa* and *Fukuyama* [3.74] revealed that a similar temperature dependence may hold in $(TMTSF)_2ClO_4$ and that the NMR relaxation rate in the superconducting phase can be explained in this context. They employed the following model: the on-site Coulomb repulsion is large, while there exist attractive interactions between neighbouring sites both along the a and the b axes as well as on the same site. The origin of the attractive force could be antiferromagnetic fluctuations. With such spin fluctuations, the possibility of an **anisotropic singlet state** with lines of zero energy gap on the Fermi surface for the superconducting order parameter, as well as of a **triplet state**, arises. In Fig.3.24, the calculated temperature dependence of $1/T_1$, with the wave-number-dependent order parameter is indicated for either $\Delta_s(k) = \Delta_s \times \cos(bk_y)$ (for singlet) or $\Delta_t(k) = \Delta_t \sin(bk_y)$ (for triplet), where it has been assumed that t_b is negligible compared to t_a. The calculation reproduces the experimental results fairly well, except for the small enhancement in the calculated $1/T_1$ value near T_c: at and below T_c the calculated values lie slightly above the experimental data (Sect.8.2.2).

For $(TMTSF)_2ClO_4$, the effects of **spin fluctuations** due to the quasi one-dimensionality have been the subjects of interest. *Bourbonnais* et al. [3.75] have interpreted the enhancement of ^1H and ^{77}Se NMR relaxations in terms of the one-dimensional $2k_F$ spin correlations for repulsive short-range interchain electron-electron interactions. In the high-temperature part ($T > 25K$), the Korringa behavior ($T_1T \simeq$ const.) holds, but it deviates in the low-temperature region (Fig.3.25). The enhancement of the SDW fluc-

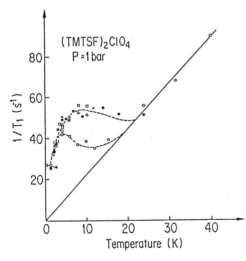

Fig.3.25. Temperature dependence of ^{77}Se spin-lattice relaxation rate $1/T_1$ for $(TMTSF)_2ClO_4$ at various fields: different marks respresent the data at different fields. The line is a guide for the eyes. From [3.75]

tuation has been claimed to be an indication of the importance of the one-dimensional electron gas scheme.

3.2.2 Anisotropic Behavior

Low-dimensionality is the unique feature of the $(TMTSF)_2 X$ family. This is manifested in the superconducting properties such as pronounced anisotropy. The orientational dependence of these compounds has been most thoroughly investigated in $(TMTSF)_2 ClO_4$ in the superconducting state at ambient pressure.

$(TMTSF)_2 ClO_4$ exhibits several highly anisotropic properties. The zero resistance in the superconducting state appears in all three directions. This was shown by the observation of zero resistivity for currents along the c-axis [3.20] which is the weakest coupling direction, as well as along the other axes. The critical current for the c-direction, however, is 0.1 ± 0.05 A/cm^2 at 0.5 K which is one order of magnitude smaller than for the a-axis, reflecting the weak coupling.

Superconductivity appears to be a volume effect according to measurements of the Meissner effect. The measurement was first carried out for $(TMTSF)_2 PF_6$ under pressure, and the results are displayed in Fig. 3.26. The sample was first cooled below T_c in zero magnetic field. Then, a 0.115 Oe field was applied perpendicular to the conducting a-axis, and the diamagnetic susceptibility χ_{dia} was measured. The highest value obtained is almost $-1/2\pi$ (note that the demagnetization factor for the needle-like

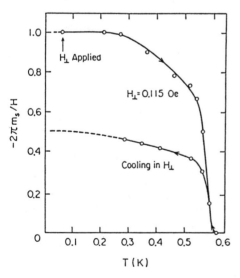

Fig. 3.26. Diamagnetic susceptibilities $\chi_{dia} = -2\pi m_s /H$ of $(TMTSF)_2 PF_6$ under 17.7 kbar as a function of temperature, which are normalized to $-1/2\pi$ appropriate for the diamagnetization factor of a cylinder in the transverse magnetic field $H_\perp$ (rather than $-1/4\pi$ for the longitudinal magnetic filed $H_\parallel$). The diamagnetic magnetization is denoted by m_s. *Upper curve*: Cooling in $H = 0$ followed by an application of 0.115 Oe during heating. *Lower curve*: Cooling below T_c in an applied field of 0.115 Oe. From [3.76]

sample is taken into account) with an uncertainty of 15 % below 3 Oe for the field perpendicular to the a-axis. The *Meissner effect* was measured by cooling the sample under the magnetic field, and the value of the Meissner signal reached about 50 % of the above-mentioned diamagnetic shielding signal. By increasing the field further, the diamagnetic signal is rapidly reduced but the superconductivity remains as in conventional type-II super-conductor state, where the magnetic flux penetrates into the superconductor below the critical field H_{c2}. At 0.2 K the **lower critical magnetic field** H_{c1}, above which the magnetic flux starts to penetrate, is estimated to be $\approx$ 5 Oe, while the **upper critical magnetic field,** H_{c2}, is 500 to 1000 Oe. With the magnetic field parallel to the a-axis, the Meissner effect remained below 8 %, while H_{c1} was ≈ 0.5 Oe and H_{c2} reached 7.2 ± 0.2 kOe.

The anisotropies of the Meissner effect and the diamagnetic shielding effect have been measured in more detail for $(TMTSF)_2 ClO_4$ which exhibits superconductivity without applying pressure [3.77, 78]. The shielding is almost complete in the three principal field directions at a low field at 50 mK. The Meissner signals, however, exhibit different behaviors according to the field direction. When the field is along the c^*-axis, the Meissner signal decreases slowly from 80 % in the lowest field to 50 % at 0.8 Oe. Along the b^*-axis, it drops from 55 % at 0.01 Oe to 24 % at 0.1 Oe, and decreases slowly below 15 % in 0.8 Oe. In the a-direction, the Meissner signal is around 1 % from 0.01 Oe to 0.8 Oe. From the field dependence of the shielding magnetization, *Mailly* et al. [3.77] determined H_{c1} for the three directions as $H_{c1a} = 0.2$ Oe, $H_{c1b} = 1.0$ Oe, $H_{c1c} = 10$ Oe at T = 50 mK.

The field dependence of the magnetization is hysteretic reflecting the pinning of the magnetic flux. Combining the H_{c1} value and the H_c determined through the specific-heat measurement by *Garoche* et al. [3.65], the **Ginzburg-Landau (G-L) parameters** κ_3, the ratio of the London penetration length λ and the coherence length ξ can be found by

$$H_{c1i} = \frac{H_c \ln \kappa_{3i}}{\sqrt{2} \, \kappa_{3i}} \quad \text{if} \quad \kappa_{3i} > 10 \tag{3.12}$$

where the subscript i denotes the field direction [3.79]. Table 3.3 lists the values obtained for the three crystallographic axes. κ_{3a} is anomalously high and indicates a very short coherence length in the b- and c-directions. One can further compare the results for H_{c2} deduced from experiments by using the Gor'kov relation [3.79]

$$H_{c2i} = \sqrt{2} \, \kappa_{1i} H_c \tag{3.13}$$

Table 3.3. Critical fields and Ginzburg-Landau parameters for the three crystallographic axes in $(TMTSF)_2ClO_4$. After [3.77, 80]

Axis	a	b	c
H_{c1} (50 mK) [Oe]	0.2	1.0	10
κ_3	1020	157	9
H_{c2} (0 K) (cal.) [kOe]	63.5	9.77	0.56
H_{c2} (0 K) (obs.) [kOe]	28	21	1.6

where κ_1 is the G-L parameter associated with H_{c2}. The estimated H_{c2} values calculated by assuming $\kappa_3 \simeq \kappa_1$ (usually the difference is less than 25%) [3.77]) are listed in Table 3.3.

The temperature dependence of H_{c2} has been measured by *Murata* et al. [3.80] and the results are displayed in Fig. 3.27. The estimated H_{c2}'s at 0 K are given in Table 3.3. The differences between calculated and observed H_{c2}'s are in part attributed to uncertainties in the measurement, i.e., both H_{c1} and H_{c2} transitions are broad, and samples used are strongly affected by cooling. In addition to this, the experimental definition of H_{c1} (onset of the flux penetration) and H_{c2} (midpoint of the transition) could differ somewhat from the true thermodynamical value.

For H_{c2a} the difference between the calculated and observed values is enlarged due to the decrease in the slope of H_{c2} vs. temperature relation with decreasing temperature (Fig. 3.27). The value listed in Table 3.3 is that obtained by extrapolating the data towards 0 K. Note that a saturation tendency is seen for the a-axis, which gives the highest value. This may be due to the **Pauli limit** in H_{c2} which is a critical value for breaking of singlet superconductivity due to the Zeeman energy of the spins, and for a free-electron gas model it is given by [3.81]

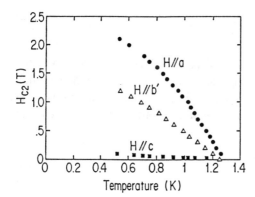

Fig. 3.27. Temperature dependence of H_{c2} for $(TMTSF)_2ClO_4$ for three principal crystal directions [3.80]

$$H_{Pauli} = \Delta_0 \frac{1}{\sqrt{2}\,\mu_B} \tag{3.14}$$

where μ_β is the Bohr magneton, and Δ_0 is the gap parameter at 0 K. Assuming that the BCS relation $(2\Delta_0 = 3.53 k_B T_c)$ holds for $(TMTSF)_2 ClO_4$, we get $H_{Pauli} = 23$ kOe when $T_c = 1.25$ K.

The anisotropic G-L **coherence length** $\xi(0)$ is found by utilizing the relation [3.82]

$$\left. \frac{dH_{c2i}}{dT} \right|_{T=T_c} = \frac{\phi_0}{2\pi} \frac{1}{\xi_j(0)\xi_k(0)T_c}$$

$$\xi_i(T) = \xi_i(0)\sqrt{T_c/(T_c - T)} \tag{3.15}$$

where $\phi_0 = hc/2e = 2.07 \cdot 10^{-17}$ Gauss·cm^2 $(2.07 \cdot 10^{-15}$ Wb) is the flux quantum. The derived values of $\xi(0)$ for two samples with different T_c are listed in Table 3.4. Note that $\xi_c(0)$ is short and comparable with the lattice parameter c = 13.5 Å, demonstrating that the superconductivity is obviously two-dimensional, in a sense that the coherence is almost limited within planar sheets. At the same time the anisotropy within the plane reflects that of the Fermi velocity and thus the transfer energy.

From the sample-thickness dependence of the diamagnetic susceptibility at low fields, *Schwenk* et al. [3.83] deduced the **London penetration depth** to be as large as 40 μm. As one may infer from Table 3.3, the Meissner effect which is the expulsion of the magnetic flux is very anisotropic, indicating that the flux cannot be expelled, and remains pinned in the specific orientation. This is interpreted by noting that the effective cross section of a vortex along b is on the order of $\approx \xi_a(0)\xi_c(0)$ which is much smaller than the cross section $\approx \xi_a(0)\xi_b(0)$ for a vortex along c, based on the ξ values in Table 3.4.

Table 3.4. Critical temperature and coherence lengths of the superconductivity for two samples. After [3.80]

	Sample #1	Sample #2
T_c [K]	1.25	1.04
$\xi_a(0)$ [Å]	706	837
$\xi_b(0)$ [Å]	335	385
$\xi_c(0)$ [Å]	20.3	22.7

3.2.3 Tunneling and Infrared Spectroscopies

The superconductivity is characterized by the energy gap and the related density of states. These are evaluated from electron tunneling spectra across interface boundaries and infrared reflectance spectra. In the first case, the tunneling current I(V) is given by

$$I(V) = P(V) \int d\epsilon \, N_1(\epsilon - eV) N_2(\epsilon) [f(\epsilon - eV) - f(\epsilon)] \tag{3.16}$$

where P(V) is the bias-dependent tunneling probability, $N_1(\epsilon)$ and $N_2(\epsilon)$ the state densities in the adjacent superconductors, and $f(\epsilon)$ the Fermi distribution function. If N_1 is known, the bias-voltage V dependence of I(V) provides the unknown density of states N_2.

For $(TMTSF)_2 X$ salts, the first observation of the tunneling characteristics was performed by using a Schottky barrier formed by evaporating GaSb onto $(TMTSF)_2 PF_6$ [3.84]. To stabilize the superconductivity at 1 K, a pressure of 11 kbar was applied. The pairing energy was estimated from tunneling spectra to be as high as 3.6 meV. This was regarded as evidence for the presence of a large pseudo-gap and a justification of the one-dimensional superconducting fluctuation at high temperatures [3.85]. Similar experimental data were reported for $(TMTSF)_2 ClO_4$ [3.86].

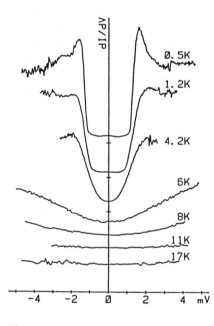

Fig. 3.28. Dependence of dI/dV on V of a $(TMTSF)_2 ClO_4$-amorphous Si-Pb junction at various temperatures [3.47]

These experimental results, however, could not be reproduced by later experiments with a $(TMTSF)_2 ClO_4$-amorphous Si-Pb junction [3.47]. The amorphous Si film was used as an insulating layer to prevent $(TMTSF)_2 \cdot ClO_4$, which contains a very strong acid, from oxidizing the Pb electrode. The actual composition of the silicon film that *Bando* et al. adopted was chemically-inactive H-doped amorphous Si with a low potential tunneling barrier of ≈ 0.1 eV: an about 20 nm film was deposited on needle-like crystals by a glow discharge in SiH_4 gas.

The experimental results obtained under conditions of slow cooling are depicted in Fig. 3.28. Below $T_c = 1.2$ K, both electrodes are superconducting since T_c of Pb is 7.2 K. The bias dependence of the differential conductance dI/dV at 0.5 K gives the sum $\Delta + \Delta_{Pb} \simeq 1.60$ meV, Δ and Δ_{Pb} being the **gap parameters** of $(TMTSF)_2 ClO_4$ and Pb, respectively. By subtracting $\Delta_{Pb} = 1.38$ meV, one gets $2\Delta \simeq 0.44$ meV which is much less than the previously reported value [3.85] and in fair agreement with the BCS relation. The gap difference signal expected to appear at $V = \Delta_{Pb} - \Delta$ is smeared out. This can be explained, for example, if one assumes that Δ is broadened by a 10 % spread in T_c due, presumably, to internal strains in the sample. Thus, as far as the tunneling spectra under consideration are concerned, the superconductivity is not inconsistent with BCS theory.

Intriguing data in relation to the above-mentioned large pseudo energy gap had been obtained by far-infrared magneto-absorption spectra [3.87] and *thermal-conductivity* measurements for $(TMTSF)_2 ClO_4$ [3.88]. In the former case, the application of a field of 2 kOe at 1.2 K leads to an increase of the absorption below 3.8 meV suggesting the presence of an energy gap. This value appears in good agreement with that expected for the proposed model of short-range one-dimensional superconducting pairing, which was estimated from a Schottky-type tunneling experiment [3.85].

Following up this *infrared-spectra* experiment, *Eldridge* et al. [3.89] measured the far-infrared absorption spectra for a series of $(TMTSF)_2 X$, where $X = PF_6$, AsF_6, SbF_6, ClO_4, BF_4 and ReO_4. When X is octahedral, the infrared spectra are analyzed in terms of three lattice modes and one internal mode. When X is tetrahedral, a sharply activated spectral structure below the anion-ordering temperature is observed in the spectra. The additional feature is interpreted as the result of zone folding, due to the superlattice formed in the anion-ordered state. In particular, a feature around 30 cm^{-1} common to all three tetrahedral X compounds has been thought to be a transverse-acoustic zone-boundary phonon which couples very strongly to the electrons. It has been concluded that this is the absorption which had been found to be sensitive to the magnetic field and to suggested fluctuating superconductivity.

As mentioned, yet another piece of evidence for fluctuating high-T_c superconductivity exists in the intriguing experimental results on thermal conductivity. For $(TMTSF)_2 ClO_4$, a striking decrease in the thermal conductivity was observed to occur below 50 K, which was recovered by applying a magnetic field [3.88]. These results can be explained if one assumes that the decrease is due to a transition to superconductivity with short-range order. However, these experimental results could not be reproduced by later experiments which employed a careful technique where the heat flow due to radiation or residual gases could be determined. The results showed, contrary to the previous results [3.88], that the thermal conductivity monotonically increased with decreasing temperature from 100 to 2 K [3.90]. Furthermore, there was a slight reduction in the thermal conductivity at low temperatures in a magnetic field of 8 T. These results can be interpreted in terms of phonon-mediated transport, which may interact with the electrons as in conventional metals.

3.2.4 Effect of Disorder

In slowly cooled $(TMTSF)_2 ClO_4$, the resistivity continues to decrease down to ca. 2 K, just above T_c, without exhibiting resistance saturation. Thus, the ratio of the room-temperature resistance to that just above T_c reaches as high as $2 \cdot 10^3$ [3.64]. In ordinary metals the resistivity decrease saturates at some temperature below 20 K due to scattering by defects and impurities.

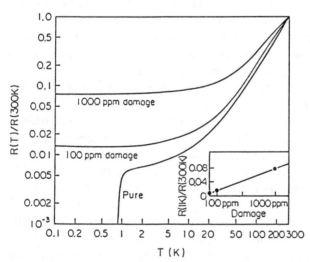

Fig. 3.29. Temperature dependence of the resistivity of $(TMTSF)_2 PF_6$ under pressure for several values of damage. The pressure was set at ≈ 14 kbar at room temperature and was measured to be about 11 kbar at 4.2 K. The inset shows the damage depencence of the 1-K resistance on a linear plot. No superconducitvity was observed in the damage sample down to 20 mK. From [3.91]

In quasi one-dimensional conductors, it is thought that the electrons are strongly scattered or localized by disorder, contrary to the experimental results mentioned above. This may indicate that the one-dimensional conducting TMTSF columns may free themselves from defects by ejecting impurities to anion sites, if we do not take account of any unusual electron transport.

In order to check the effect of the disorder in $(TMTSF)_2 X$ intentionally, experiments have been conducted on species where random ordering has been introduced by either **radiation damage** or **alloying**. The effect of radiation damage with 2.5 MeV protons on the low-temperature transport has been investigated for $(TMTSF)_2 PF_6$, which undergoes a superconducting transition under pressure [3.91]. The results are presented in Fig.3.29. The damage fraction was estimated from the proton flux by assuming that the all absorbed energy is used to damage molecules. Figure 3.29 shows that the superconducting transition observed in the pure, nonirradiated, sample is completely removed upon introduction of a small amount of damage, and the resistance exhibits a residual value which increases with increasing dose. This indicates that the radiation-induced defects also increase the electron scattering in $(TMTSF)_2 X$. The defects caused by proton irradiation might be on the TMTSF molecules and the damaged molecule may become magnetic by losing electrons or atoms. The magnetic defects break the electron pair and thereby destroy the superconductivity.

It is known that in ordinary superconductors singlet superconductivity is rather insensitive to nonmagnetic defects. However, when weak nonmagnetic disorder was introduced into $(TMTSF)_2 ClO_4$ by alloying it with $(TMTSF)_2 ReO_4$ or $(TMTTF)_2 ClO_4$, surprisingly the superconductivity is sharply suppressed with increasing alloying ratio and no superconductivity was observed for $(TMTSF)_2 (ClO_4)_{1-x} (ReO_4)_x$ for $x > 0.03$ [3.92]. The superconducting state is also suppressed by disorder in the conduction chain in $[(TMTSF)_{1-x} (TMTTF)_x]_2 ClO_4$ with $x > 0.02$ [3.93]. It should be recalled that by rapid cooling of $(TMTSF)_2 ClO_4$, the superconductivity is suppressed, partly due to weak random potentials caused by the random orientation of the ClO_4 anions. These facts exhibit conclusively that the superconductivity in $(TMTSF)_2 X$ is very sensitive to **nonmagnetic defects**, in contrast to ordinary metal superconductors. This sensitivity may be explained if the superconductivity is associated with *triplet pairs*, but there is limited support for them. Instead, it can be interpreted in terms of *electron localization* in low-dimensional systems according to *Hasegawa* and *Fukuyama* [3.94]. This will be described in more detail in Sect.5.2.7 for the case of the ET salts.

4. Spin-Density-Wave Phase and Reentrant Superconducting State

The SuperConducting (SC) and Spin Density Wave (SDW) phases lie next to each other in the pressure-temperature phase diagram of the $(TMTSF)_2 X$ compounds. Except for the ambient pressure $(TMTSF)_2 ClO_4$ superconductor, the SC phase appears only after the SDW phase is suppressed by pressure. This chapter demonstrates that a rather simple framework based on a tight-binding band and a mean-field approximation is able to give a consistent theoretical description of SDW and to clarify why the SDW phase vanishes under pressure with the reentrant SC phase appearing.

First we describe the properties of the SDW phase in $(TMTSF)_2 X$ salts. In the next section, we examine the results derived from the one-dimensional (1-D) electron-gas theory, or g-ology (Sect.4.2). The electronic band of $(TMTSF)_2 X$, however, has two-dimensional (2-D) features, in addition to the one-dimensional ones. We will see band-structural studies of these salts in the tight-binding scheme and find that the transverse transfer energy t_b is about one tenth of the longitudinal transfer energy t_a. Since t_c in the third direction is smaller by more than a factor of 10, we take a 2-D tight-binding band with t_a and t_b as our basic electronic framework. Assuming that the on-site Coulomb energy at each molecule drives the SDW, we derive a mean-field theory of SDW for the Bechgaard salts. A small amount of t_c, which is neglected, decreases the thermal fluctuations enough to justify the mean-field treatment. On the basis of this framework we can successfully obtain the fundamental properties of the SDW state. The Fermi-surface nesting of the $(TMTSF)_2 X$ salts is just on the brink of losing the SDW. We find a pressure-sensitive band parameter, an increase of which under pressure etc. breaks the Fermi-surface nesting and results in disappearance of the SDW ordering. Then, the superconducting state with T_c on the order of 1 K appears without changing the strengths of the various coupling constants. We show that these results are obtained by the mean-field treatment on the basis of a model of an anisotropic two-dimensional tight-binding band. The theoretical stability condition of the SDW is verified by a comparison with structural and band-calculation studies, etc. The reentrant superconducting phase does not coexist with the SDW phase, and both orders compete with each other. This is shown theoretically when the total energies of both phases are compared, and also by the unstable nature of the coexistent phase. The framework for the Bechgaard salts under consideration is also supported by the success in applying it to the magnetic-Field-Induced SDW (FI-SDW), to be treated in Chap. 9.

4.1 Experimental Investigations of the Spin Density Wave

The SDW state is a kind of antiferromagnetic state with the electronic spin density forming a static wave. The density varies periodically as a function of the position with no net magnetization in the entire volume. Specifically, an SDW occurs with spatial spin-density modulation due to delocalized (or itinerant) electrons rather than localized ones [4.1,2]. In the normal state the density $\rho_\uparrow(\mathbf{r})$ of electron spins polarized upward with respect to any quantization axis is completely canceled by the density $\rho_\downarrow(\mathbf{r})$ of downward polarized spins. In the SDW state, however, the difference $\sigma(\mathbf{r})$ between $\rho_\uparrow(\mathbf{r})$ and $\rho_\downarrow(\mathbf{r})$ is finite and undulates in space as a function of the position vector $\mathbf{r}$ in the SDW state.

The Bechgaard salts $(TMTSF)_2 X$ with X: PF_6, AsF_6, SbF_6, NO_3, BF_4, etc. become insulating at temperatures around 12 K, after showing a decrease of three orders of magnitude in resistivity with the temperature decreasing from room temperature (Fig. 3.1). This insulating state was first suspected to be due to a CDW observed at low temperatures in most previously found synthetic metals. However, an X-ray study [4.3] disclosed that in the insulating state there is no lattice distortion which otherwise always accompanies the CDW state. NMR experiments with Se confirmed the magnetic nature of the state by checking that the resonance line suddenly disappears at the transition temperature due to the broadening of the resonance line by local magnetic fields [4.4]. Proton NMR experiments also confirmed this broadening effect [4.5].

The insulating state was found to be antiferromagnetic by the observation of the characteristic anisotropy of the magnetic susceptibility, as depicted in Fig. 3.13 for $(TMTSF)_2 AsF_6$ [4.6]. Although the slight increase in χ_a and χ_{c*} with decreasing temperature is anomalous (they should remain constant), the figure clearly reveals an antiferromagnetic spin configuration polarized in the b'-direction, i.e., with the easy axis being along the b'-direction. When the field applied along this direction is intensified, the spin direction flops to the a-direction at a spin-flop field H_{sf}, e.g., $H_{sf} = 0.45$ Tesla in $(TMTSF)_2 AsF_6$. This leads to the conclusion that the intermediate axis is the a-axis, with the c^* axis being the hard one.

Experiments on the antiferromagnetic resonance of the electron spin decisively proved the presence of an antiferromagnetic state [4.7,8]. In an antiferromagnetic state, there are two branches of dynamical spin waves.

The behavior of the spins is analyzed by a model of antiferromagnetic, localized spins S_i in a linear chain. The effective Hamiltonian for such a system is given by

$$\mathcal{H}_{eff} = 2J \sum_i S_i \cdot S_{i+1} - g\mu_B H \cdot \sum_i S_i$$

$$+ E \sum_i (S_i^z S_{i+1}^z - S_i^y S_{i+1}^y) - D \sum_i S_i^x S_{i+1}^x , \qquad (4.1)$$

where J is the antiferromagnetic interaction between the spins along the chain, D and E represent the anisotropy energies, and J, D, E > 0. The easy, intermediate, and hard axes are z, y and x, respectively. In terms of this model, the above-mentioned branches are expressed as

$$\hbar^2 \Omega_+^2 = \pi^2 (D + E)J \quad \text{and} \quad \hbar^2 \Omega_-^2 = 2\pi^2 EJ . \qquad (4.2)$$

The localized model Hamiltonian should be just taken as a convenient representation. As a function of the field direction and intensity the anisotropy parameters were determined to be [4.7]

$$J = 604 \, k_B , \quad D = 4.5 \cdot 10^{-4} \, k_B , \quad \text{and} \quad E = 0.3 \cdot 10^{-4} \, k_B , \qquad (4.3)$$

where k_B is the Boltzmann constant.

Experiments using proton NMR resonance went further to determine the wave vector Q and the amplitude σ of the undulation of the spin density of $(TMTSF)_2 PF_6$, as described in Sect. 3.1.5 [4.9-12]. The SDW vector Q is in good agreement with the value obtained for the nesting vector of the calculated band, which reveals a good nesting property of the Fermi surface, as will be shown in Sect. 4.3. On the basis of these results, the antiferromagnetic state in the Bechgaard salts was confirmed as being an SDW state.

4.2 An Approach from g-ology

For an interpretation of the phase diagram of the Bechgaard salts, the first theory to be applied was the g-ology [4.13, 14], i.e., the theory of one-dimensional electron-gas phase diagrams which take into account the electron-electron interactions in the vicinity of the Fermi surface, as introduced in Sect.2.3.8. The basis for this assumption was the one-dimensionality of the electronic properties in the salts. We also begin our analysis with this theory since it provides a basic knowledge for quasi one-dimensional metals.

First, the Beckgaard salts were interpreted as being situated on the boundary of the SDW and TS (Triplet Superconductivity) phase regions in the phase diagram of Fig.2.16, since only here do the magnetic and super-conducting phases share a common boundary. It was believed that pressure would move the phase point from the SDW side to the superconducting side. This interpretation has the following shortcomings: If it were correct, then for the superconducting phase T_c must increase as pressure increases. On the contrary, T_c has a tendency to decrease as the pressure increases. Furthermore, as will be described later, the superconducting state is now considered to be a Singlet Superconducting (SS) state rather than TS which neighbors the SDW phase in the diagram. (For a detailed explanation of SS and TS, see Sect.8.1.4).

Horovitz et al. [4.15, 16] developed a more elaborate theory of phase diagrams for quasi one-dimensional metals, by taking into account other features of the (TMTSF)$_2$ X salts. The model band energy is

$$\epsilon(\mathbf{p}) = v_F(|k_x| - k_F) - t_\perp[\cos(bk_y) + \cos(ck_z)],\qquad(4.4)$$

where v_F is the Fermi velocity, b and c are the lattice constants in the trans-verse directions, and $t_\perp$ is the energy due to the interchain coupling. In this section we use the atomic unit convention: $\hbar = k_B = 1$. The interesting range of values of $t_\perp$ is

$$4T_P \leq t_\perp \leq 3(T_P T_F)^{1/2},\qquad(4.5)$$

where T_F is the Fermi temperature and T_P the mean-field Peierls or SDW transition temperature for $t_\perp = 0$. The lower limit is determined by the ap-plicability of the mean-field theory, and the upper limit is chosen such that a sufficient Fermi-surface nesting with the wave vector $\mathbf{Q}_0 = (2k_F, \pi/b, \pi/c)$ is ensured.

In this region, the involved set of diagrams reduce to diagrams of the Hartree-Fock scheme for possible types of ordering. When one takes into account only non-retarded interactions in the g_1-g_2 plane, the four phases, SDW, CDW, SS and TS, are separated by the lines

$$g_1 = 2g_2 + |g_3| \quad \text{and} \quad g_1 = 0,\qquad(4.6)$$

where g_1 and g_2 are the interaction coupling constants for the backward and forward scatterings, respectively, and g_3 is the coupling constant for the

Umklapp scattering. For example, in the last process two electrons both with the wave numbers around k_F collide thereby changing their wave numbers to $-k_F$, losing $4k_F$ which is equal to the reciprocal lattice in the $(TMTSF)_2X$ salts. When $g_3 = 0$, the above result is the same as that of the g-ology diagram for the one-dimensional case ($t_\perp = 0$) in Fig.2.16. For $g_3 \neq 0$ the oblique phase boundary $g_1 = 2g_2$ is shifted upwards, i.e., to $g_1 = 2g_2 + |g_3|$.

Emery et al. [4.17, 18] argued that in such a phase diagram g_3 should decrease under pressure, changing the SDW state into the TS state in the region just above $g_1 = 2g_2$. This interpretation is basically the same as the one above, thus sharing the same shortcomings. They asserted that the critical pressure P_c for the superconducting state correlates with the degree of dimerization of the TMTSF stack and that the latter is proportional to g_3. They argued that $(TMTSF)_2ClO_4$ is superconducting at ambient pressure because the anions are ordered with the wave vector $(0,\frac{1}{2},0)$ [4.18]; here this vector means $0 \cdot \mathbf{a}^* + \frac{1}{2}\mathbf{b}^* + 0 \cdot \mathbf{c}^*$. In this situation the potential exerted on the TMTSF molecules by the anions at each molecular site is equal so that the dimerization is weak, resulting in a small g_3. Later, however, an example contradicting their argument was found in $(TMTSF)_2ReO_4$ [4.19]. The superconducting state of $(TMTSF)_2ReO_4$, when placed under pressures above ≈ 10 kbar, was found to have an anion ordering with $\mathbf{Q} = (0,\frac{1}{2},\frac{1}{2})$, while the insulating phase has $Q = (\frac{1}{2},\frac{1}{2},\frac{1}{2})$. The $(0,\frac{1}{2},\frac{1}{2})$ phase has a stronger dimerization and, consequently, a larger g_3 than the $(\frac{1}{2},\frac{1}{2},\frac{1}{2})$ state.

Horovitz et al. extended the calculation to include the retarded interactions, i.e., inter-electron interactions mediated by phonons, so that the interaction is not instantaneous [4.15]. In other words, it is non-zero only for a restricted range of exchanged energy. They neglect g_3, since the $(TMTSF)_2X$ salts have approximately a glide-plane symmetry which makes each TMTSF molecule in the dimerized stack have an identical environment around it. They used the electron-phonon coupling constants g_{1p} and g_{2p} for the exchanged wave number $q \simeq 2k_F$ and 0, respectively, and the non-retarded, or direct, electron-electron forward (g_{2e}) and backward (g_{1e}) interactions. Dimensionless coupling constants can be defined by

$$\bar{g}_{ie} = g_{ie}N(0)/2 , \quad \lambda_i = g_{ip}^2 N(0)/\omega_0 \quad (i = 1, 2) , \tag{4.7}$$

where $N(0) = 2/\pi v_F$ is the total density of states, and ω_0 is the phonon frequency which is assumed to be constant.

The gap equations for the different types of order are obtained in the mean-field approximation as

$$\Delta_n = \frac{1}{2}\bar{g}^N \int_{-E_c}^{+E_c} d\epsilon\, T \sum_m \frac{\Delta_m}{\omega_m^2 + \epsilon^2 + \Delta_m^2}$$

$$+ \frac{1}{2}\bar{g}^R \int_{-E_c}^{+E_c} d\epsilon\, T \sum_m \frac{\Delta_m}{\omega_m^2 + \epsilon^2 + \Delta_m^2} \frac{\omega_0^2}{(\omega_n - \omega_m)^2 + \omega_0^2} \quad ,(4.8)$$

where $\omega_n = \pi T(2n+1)$ is the Matsubara frequency for an electron with integer n, E_c is the electron cutoff energy, ϵ is the band energy, and Δ_m is the order parameter for the frequency ω_m; the retarded ($\bar{g}^R$) and non-retarded ($\bar{g}^N$) coupling constants due to the two sources are listed in Table 4.1. The first term on the right-hand side is the non-retarded term and the second the retarded one. There is also a contribution from an Umklapp process specific to the (TMTSF)$_2$X compounds. The acceptor anions X$^-$ produce a periodic potential with the wave number $4k_F = \pi/a_s$, a_s being the spacing between the TMTSF molecules. Expanding this potential into powers of the displacement of the anionic position due to phonons, one finds that this potential adds to the Hamiltonian the term

$$\mathscr{H}_U = W_0(4k_F)^2 \sum_q \frac{1}{M\omega_q}(b_q + b_{-q}^\dagger)\left[b_{q+4k_F} + b_{-q-4k_F}^\dagger\right] + \text{H.c.} , \quad (4.9)$$

where b_q ($b_q^\dagger$) is the phonon annihilation (creation) operator, and W_0 is the amplitude of the $4k_F$ periodic potential. Through a process, such as pictured in Fig. 4.1, this interaction adds to (4.8) the term

Table 4.1. Composite non-retarded ($\bar{g}^N$) and retarded ($\bar{g}^R$) coupling constants [4.15]

	$\bar{g}^N$	$\bar{g}^R$
SS	$-\bar{g}_{1e} - \bar{g}_{2e}$	$\lambda_1 + \lambda_2$
TS	$\bar{g}_{1e} - \bar{g}_{2e}$	$-\lambda_1 + \lambda_2$
CDW	$-2\bar{g}_{1e} + \bar{g}_{2e} + 2\lambda_1 + 2\lambda_3$	$-\lambda_2 - \lambda_3$
SDW	$\bar{g}_{2e}$	$-\lambda_2 + \lambda_3$

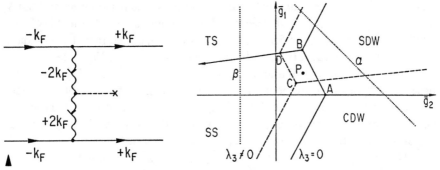

Fig. 4.1. Retarded Umklapp process via two phonons in the presence of a crystal potential of period $4k_F$ [4.15]. *Straight* and *wavy lines* represent electrons and phonons, respectively

Fig. 4.2. Phase diagram in the $(\bar{g}_1, \bar{g}_2)$-plane $(\bar{g}_1 = \bar{g}_{1e} - \lambda_1, \bar{g}_2 = \bar{g}_{2e} - \lambda_2)$ [4.15]. *Full lines* are the coexistence lines in the case of $\lambda_3 = 0$. *Dashed lines* are for $\lambda_3 \neq 0$. *Dotted lines* labeled α and β are the boundary lines of the region where both SS and SDW have finite transition temperatures

$$\pm \lambda_3 \int_{-E_c}^{+E_c} T d\epsilon \sum_m \frac{\Delta_m}{\omega_m^2 + \epsilon^2 + \Delta_m^2} \left[\frac{\omega_0^2}{(\omega_n - \omega_m)^2 + \omega_0^2} \right]^2 ; \qquad (4.10)$$

the $+$ and $-$ signs attached to λ_3 are for SDW and CDW, respectively, and

$$\lambda_3 = \frac{|\lambda_1| 8 |W_0| k_F^2}{M \omega_0^2} . \qquad (4.11)$$

This interaction also contributes the coupling constant $2\lambda_3$ to the non-retarded coupling $\bar{g}^N$ (Table 4.1). Since $\lambda_1 \propto 1/\omega_0^2$, we find that $\lambda_3 \propto 1/\omega_0^4$ so that under pressure λ_3 rapidly decreases with an increase of ω_0 due to hardening. Thus, λ_3 is taken as the most pressure-sensitive parameter in this theory.

Equations (4.8, 10) are treated analytically and numerically to obtain the transition temperatures for the different types of order. The phase diagram showing the order with the highest transition temperature is presented in Fig. 4.2. The solid lines are the phase boundaries in the case of $\lambda_3 = 0$. The dotted lines labeled α and β are the boundary lines of the region in which both SS and SDW occur. For SS T_c is zero on the line α and increases as one moves to the left, while for SDW the transition temperature T_{SDW} is zero on the line β and increases as one moves to the right. This phase diagram has a different feature which does not show up in Fig. 2.16 for non-retarded interactions. The degeneracy at the point where four phases coexist is lifted, and one finds a boundary line between SS and SDW labeled AB. When $\lambda_3 \neq 0$, the transition temperature for the SDW and CDW are increased while SS and TS are not changed, so that the boundary lines are shifted to the left, as indicated by dashed lines in Fig.4.2.

Horovitz et al. [4.15] assumed that the change from the SDW to the SS ordering implies a movement of the CD line from left to right. As mentioned above, λ_3 rapidly decreases under pressure. Therefore, if the phase point at ambient pressure is located, for example, at the point P in Fig.4.2, the SDW phase turns into the SS phase under an applied pressure.

This scheme reveals that SDW and SS can have a mutual boundary if one takes into account both Coulomb and electron-phonon interactions. However, for the Umklapp process to work in the formation of SDW and CDW, the wave vector $\mathbf{Q}$ of the ordered phase must be commensurate with the reciprocal lattice, $\mathbf{a}^*$, $\mathbf{b}^*$, and $\mathbf{c}^*$, for example, $2\mathbf{Q} = h\mathbf{a}^* + k\mathbf{b}^* + \ell\mathbf{c}^*$ must be satisfied (h, k, ℓ being integers). This turns out not to be the case, since the model band energy in (4.4) is actually modified, e.g., by replacing $\cos(bk_y)$ by $\cos(bk_y \mp \phi)$ with ϕ/π being an irrational value characteristic of each system, as will be demonstrated in Sect.4.3. We shall see in Sect.4.4.3 that the wave vector $\mathbf{Q}$ for SDW is shifted as a function of pressure, and in Sects.9.2.2 and 9.3.3 as a function of the magnetic field. The $(TMTSF)_2 X$ compounds turn out to have, roughly speaking, a value of $t_\perp$ close to the upper bound in (4.5), as discussed in Sects.4.3 and 4. This makes the present scheme of g-ology inapplicable to the $(TMTSF)_2 X$ salts.

4.3 Band Structure of $(TMTSF)_2 X$ Salts

4.3.1 Two-Dimensional Band Model

Since it had been believed that a good metal of the charge-transfer complex must have a uniform stack of molecules, the persistence of the metallic state in the $(TMTSF)_2 X$ salts down to about 12 K, despite the observed dimerization, was surprising. The dimerized system has a period of $2a_s$ in the stacking a-direction, so that the Brillouin-zone boundary with a gap lies at the wave number equal to $\pi/2a_s$. Assuming that the band is one-dimensional, the Fermi wave number k_F is found to be $\pi/4a_s$, because for every two TMTSF molecules one electron is removed from the TMTSF stack to X. Holes occupy half of the Brillouin zone. Therefore, the dimerization actually does not induce a gap at the Fermi level. However, since the system has a half-filled band, the on-site Coulomb energy on the order of the band width is usually considered to turn the system into a Mott-Hubbard insulator. This puzzle is solved by recognizing that the band structure is actually two-dimensional.[1]

Jérome et al. [4.20] who discovered superconductivity in (TMTSF)$_2$ ·PF$_6$ asserted that the band structure is one-dimensional with the stacking axis being the direction of highest conductivity and the conduction along the other directions being very poor. According to the crystal structure (Fig. 3.3), however, transfer energies between side-by-side adjacent molecules located on the nearest neighbor columns in the b-direction are quite appreciable [4.21]. The closest distance betweeen Se atoms of adjacent molecules is 3.88 Å, which is close to twice the Se van der Waals radius of 1.90 Å, and is as short as the distance between face-to-face adjacent molecules in a column. However, the transfer energies between molecules neighboring in the c-direction must be very small because of the methyl groups at both ends of the molecules and because of the closed shell anions in the cages which are made of surrounding TMTSF molecules. Both of these constituents make up an insulating layer between the conducting layers of closely packed TMTSF molecules

From thermopower experiments the band width in the a-direction was found to be about 1 eV [4.22]. The band width due to the transfer energies in the b-direction was undetermined for a long time. Considering it unknown, one employed the following two-dimensional band as the simplest model:

$$\epsilon_{\mathbf{k}} = 2t_a \cos(a_s k_a) + 2t_b \cos(b k_b) , \qquad (4.12)$$

where t_a is the average transfer energy between the nearest neighbor TMTSF molecules along the stack and is about 0.25 eV; t_b is the effective transfer energy in the b-direction, whose value is unknown but assumed to be much smaller than t_a.

4.3.2 Effective Transfer Energies for the Two Directions

Contrary to the viewpoint in the previous subsection that t_b should be much smaller than t_a, the vanishing of SDW in (TMTSF)$_2$PF$_6$ under pressure could also be ascribed to a deterioration in the nesting property of the quasi one-dimensional Fermi surface due to an increase in t_b. A mean-field theory of SDW for the Bechgaard salts based on this band model suggests that $t_a/t_b \simeq 10$ [4.23]. This means that $t_\perp$ in (4.4) is near the upper bound in (4.5).

---

[1] When the band dispersion width in the b-direction is larger than the dimerization gap, as it is the case for Bechgaards salts, the argument in favor of the Mott-Hubbard insulator becomes invalid.

A direct means of finding t_a and t_b comes from optical data on the plasma frequencies [4.24]. Typical structures of the plasma frequency are observed in the optical conductivities at low temperatures for both the a- and b-directions, although at room temperature the structure is discernible only for the a-direction.

The expressions for the plasma frequencies of an anisotropic system are qualitatively different for the case with a spherical Fermi surface and that with an open Fermi surface. In the latter, the expressions are

$$\omega_{pl,a}^2 = 16Ne^2 a_s^2 t_a \sin(a_s k_F) , \qquad (4.13)$$

$$\omega_{pl,b}^2 = \frac{8Ne^2 b^2 t_b^2}{t_a \sin(a_s k_F)} , \qquad (4.14)$$

where N is the density of the electronic sites or the TMTSF molecules; the a- and b-directions are assumed to be orthogonal, and the Fermi surface is approximately given by

$$k_a \simeq \pm\left[k_F + \frac{t_b \cos(bk_b)}{a_s t_a \sin(a_s k_F)}\right] . \qquad (4.15)$$

The Fermi surface is displayed in Fig.4.3. The above results are obtained by expanding the RPA expression of the dielectric constant in powers of t_b/t_a in the small wave-number limit [4.25], or by directly calculating the formula of the plasma frequency [4.26, 27]. The ratio of (4.14 to 13) gives the ratio of t_b/t_a through

$$\left(\frac{\omega_{pl,b}}{\omega_{pl,a}}\right)^2 = \frac{1}{2}\left[\frac{b}{a_s \sin(a_s k_F)}\right]^2 \left(\frac{t_b}{t_a}\right)^2 . \qquad (4.16)$$

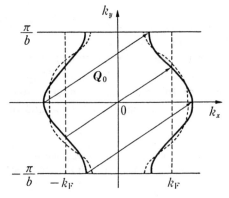

Fig.4.3. *Solid curves* represent the Fermi surface given by (4.15), where k_x and k_y replace k_a and k_b, respectively, and Q_0 is the nesting vector. *Dashed curves* show the deviation of the Fermi surface from this simple form as a result of a higher harmonic component proportional to $\cos 2bk_y$

Table 4.2. The transfer energies t_a and t_b of $(TMTSF)_2 X$ salts from anisotropic plasma frequency data. The expression for $\omega_{pl,a}^2$ in (4.13) is multiplied by $\{1 - [t_b^2/4t_a^2 \sin^4(a_s k_F)] \cdot [1 - 2b^2 \sin^2(a_s k_F) \cos^2(\gamma)/a_s^2]\}$, where γ is the angle between the a- and b-axes, in order to take account of the obliqueness of the lattice [4.29]. The results are in reasonable agreement with [4.27]

	$\omega_{pl,a}$ [1/cm]	$\omega_{pl,b}$ [1/cm][c]	t_a [meV][d]	t_b [meV]
ClO_4	10170[a]	2020	258	25.4[d]
PF_6	10185[b]	1830	264	23.4[d]
AsF_6	10470[c]	1670	283	22.3[d]
SbF_6		1510		18[c]

[a] [4.28]
[b] As suggested in [4.27], extrapolated to low temperatures from 300 K value and increased by 5%.
[c] [4.27]
[d] [4.29]

Substituting into this equation the observed values for $(TMTSF)_2 PF_6$, i.e., $\omega_{pl,a} = 11400$ cm^{-1} and $\omega_{pl,b} = 2360$ cm^{-1} together with the crystallographic data, one gets $t_a/t_b \simeq 10$. If one employs the plasma-frequency formula in terms of the effective mass by assuming a spherical Fermi surface, one finds $t_b/t_a \propto (\omega_{pl,b}/\omega_{pl,a})^2$ and $t_a/t_b \simeq 100$; this value of the ratio contradicts the assumption of a spherical Fermi surface. Results of the analysis on $(TMTSF)_2 X$ salts are listed in Table 4.2.

The ratio of t_a/t_b, thus derived, is consistent with the experimental anisotropy ratio of $\sigma_a/\sigma_b \propto (t_a/t_b)^2$ [4.30, 31]. However, the experimentally determined ratio of the Ginzburg-Landau coherence lengths [4.32, 33] in three directions is not fully consistent with the theoretical expressions given in [4.34-36] and the above value of t_a/t_b. This is considered to be due to inadequacies in the present theories of superconductivity in organics.

4.3.3 Multiple-Transverse-Transfer Model

Most results of band calculations [4.37-40] for the $(TMTSF)_2 X$ salts yield a 2-D tight-binding band in which the one-dimensional band is warped by transverse transfers of the electron. This scheme relies on transfer energies calculated between neighboring molecules for the Highest Occupied Molecular Orbitals (HOMO). Because it gives much insight into the electronic structure, we introduce this type of band calculation.

According to *Grant* [4.37], the transfer energies are labeled, as shown in Fig. 3.5; and the tight-binding band energy $\epsilon_{\mathbf{k}}$ is given by

$$\epsilon_{\mathbf{k}}^{(\pm)} = 2t_{I3}\cos y + 2t_{I4}\cos(2x-y)$$

$$\pm \left| t_{S1}e^{ix} + t_{S2}e^{-ix} + t_{I1}e^{i(x-y)} + t_{I2}e^{-i(x-y)} \right| , \tag{4.17}$$

where $x = \mathbf{a}\cdot\mathbf{k}/2$ and $y = \mathbf{b}\cdot\mathbf{k}$, $\mathbf{a}$ and $\mathbf{b}$ being the lattice vectors with $a = 2a_s$. The transfer energies t_j are defined by

$$t_j = \int \psi^*(\mathbf{r}-\mathbf{R}_j)\mathcal{H}\,\psi(\mathbf{r})\,d\mathbf{r} , \tag{4.18}$$

where $\psi(\mathbf{r})$ is the HOMO of TMTSF, $\mathbf{R}_j$ is the position vector of the molecule labeled by j with respect to some origin, and $\mathcal{H}$ is the Hamiltonian of the system. The intrastack transfer energies t_{S1} and t_{S2} are on the order of 0.3 eV. The interstack transfer energies t_{Ii}, $i = 1$ to 4, are smaller by one order of magnitude. Due to dimerization along the stack, t_{I1} and t_{I2} are different, which brings in two eigenvalues for each value of the wave vector $\mathbf{k}$ in (4.17). In the neighborhood of the Fermi surface, being important in the following calculation, x is close to $\pm\pi/4$. Therefore, the leading term in the expression inside the absolute-value symbol in (4.17) is $(t_{S1}+t_{S2})\cos x$; $i(t_{S1}-t_{S2})\sin x$ and terms with t_{I1} and t_{I2} are smaller by one order. By expanding the upper branch of (4.17) in powers of $t_{S1}-t_{S2}$ and t_{Ii} divided by

$$t_a = (t_{S1} + t_{S2})/2 , \tag{4.19}$$

and putting the y-dependent terms into one (4.17), one obtains

$$\epsilon_{\mathbf{k}} \simeq 2t_a\cos x + 2t_b\cos(y-\phi) , \tag{4.20}$$

with

$$t_b = \sqrt{[t_{I3} + t_{I4}\cos(2x) + t_T\cos(x)]^2 + [t_{I4}\sin(2x) + t_T\sin(x)]^2} , \tag{4.21}$$

$$\phi = \tan^{-1}\frac{t_{I4}\sin(2x) + t_T\sin(x)}{t_{I3} + t_{I4}\cos(2x) + t_T\cos(x)} , \tag{4.22}$$

and

$$t_T = (t_{I1} + t_{I2})/2 . \tag{4.23}$$

Since $x = \mathbf{a} \cdot \mathbf{k}/2 \simeq \pm\pi/4$ in the vicinity of the Fermi surface, and since this region is physically the most important one, as a reasonable approximation one can set $x = \pm\pi/4$ in the expressions for t_b and ϕ. This leads to

$$\epsilon_{\mathbf{k}} \simeq 2t_a \cos(x) + 2t_b \cos(y \mp \phi), \quad x \gtrless 0 \tag{4.24}$$

where $\mp$ is for $x \gtrless 0$, respectively, and t_b and ϕ should be taken as the values fixed at $x = \pi/4$. The angle ϕ plays an important role in the incommensurability of the Fermi-surface nesting. Except in this problem, it is not important.

Thus, the simple model (4.12) has a broad applicability to $(TMTSF)_2 X$ compounds. Calculated results for the band parameters are sensitive to the value of the ζ-parameter of the Slater orbitals, especially for the d-orbitals of Se. The values of the band widths obtained in the a- and b-directions seem to have been the first things to be checked in band calculations.

It is worth making a few remarks on the signs of the transfer energies and the dependence of t_b on them. The HOMO is a π-orbital consisting of atomic p- and d-orbitals, as shown in Fig.4.4, in which the projections of molecules along their longest axes are shown by horizontal lines. The molecules B, C and D are arbitrarily drawn in reference to A. The transfer energy t_{AB} between A and B is positive since the overlap integral S_{AB} is negative, as seen from Fig.4.4, and t_{AB} is given by

$$t_{AB} = K E_\gamma S_{AB} \tag{4.25}$$

with $K = 1.75$ in the extended Hückel approximation [4.41]; the eigenenergy E_γ for the HOMO is negative. Since S_{AC} is negative, t_{AC} is positive. However, since $S_{AD} > 0$, t_{AD} is negative.

This is the reason why all the transfer energies labeled by the scheme in Fig.3.5 are positive except for t_{I1} and t_{I2}, which are negative. Their absolute magnitudes are usually a few times smaller than t_{I3} despite the closeness of the related molecules [4.40]. As can be inferred from (4.21), the negative value of $t_T = (t_{I1} + t_{I2})/2$ cancels out a part of t_{I3} in the expression for t_b. If the relative positions of the related molecules were slightly shifted

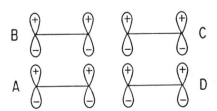

Fig.4.4. Dependence of the overlap integral S on the molecular positions B, C and D relative to position A. The *lines* represent the projections of molecules along their longest axes. The *propeller-shape objects* symbolize the atomic p- and d-orbitals

so that the sign of t_{I1} and t_{I2} became positive, t_b would be much larger. Such a situation occurs in β–$(ET)_2X$ salts which possibly have a sufficiently large value of t_b to realize a closed Fermi surface in the conducting plane. This suggests the possibility that more 2-D systems than the $(TMTSF)_2X$ salts can be found by small modifications of these salts.

4.4 Mean-Field Theory of the Spin Density Wave

It had been known [4.42] that a CDW in quasi 1-D systems is suppressed when the transverse transfer energies are increased. Then it was thought of that the disappearance of the SDW in $(TMTSF)_2X$ under pressure is caused by the increased t_b. In order to verify this idea and to get the related value of t_b, anisotropic 2-D systems with the band given by (4.12) have been studied [4.23]. Since it had long been of interest to suppress the metal-insulator transition in synthetic metals, the mechanism of the disappearance of the SDW was an important problem. Incidentally, the reason for the absence of a CDW in the $(TMTSF)_2X$ compounds will be discussed in Sect.8.3.

For a three-quarter-filled band it is more convenient to treat the problem by considering holes instead of electrons as particles. Then, we treat the equivalent problem, i.e., that of the quarter-filled state of a band:

$$\epsilon_{\mathbf{k}} = -2t_a\cos(a_s k_x) - 2t_b\cos(bk_y) - \mu ,\tag{4.26}$$

with $t_a \gg t_b > 0$, and μ is the chemical potential. The Fermi surface is open in the k_y-direction and is given by

$$k_x \approx \pm(1/a_s)\cos^{-1}[\cos(a_s k_F) - (t_b/t_a)\cos(bk_y)]$$

$$\approx \pm[k_F + 2(t_b/\hbar v_F)\cos(bk_y) - (\epsilon_0/\hbar v_F)\cos(2bk_y)] ,\tag{4.27}$$

where k_F is the Fermi wave number in the limit of $t_b = 0$; $v_F = 2a_s t_a$ $\times\sin(x_F)/\hbar$ with $x_F = a_s k_F$ and

$$\epsilon_0 = \frac{t_b^2\cos(x_F)}{2t_a\sin^2(x_F)} .\tag{4.28}$$

If one neglects the second-harmonic term in (4.27), the Fermi surface is completely sinusoidal like the continuous curves in Fig.4.3. Then, the part of the Fermi surface with $k_x \approx -k_F$ completely nests with that for $k_x \approx k_F$ when the former is moved by

$$\mathbf{Q}_0 = (2k_F, \pi/b) . \tag{4.29}$$

The part of the Fermi surface which extends outside the Brillouin zone can be shifted by $(0, \pm 2\pi/b)$ to form its equivalent part within the Brillouin zone. The wave vector $\mathbf{Q}_0$ is called **optimal** and this nesting property and the Coulomb interaction cause the *SDW ordering*. When the $\cos(2bk_y)$ term in (4.27) is taken into account, the deformed Fermi surface indicated by dashed curves in Fig.4.3 does not nest completely any more. The term comes from the nonlinearity of ϵ_k with respect to k_x, i.e.,

$$\partial^2 [-2t_a \cos(a_s k_x)]/\partial k_x^2 \neq 0 \quad \text{at} \quad k_x \simeq \pm k_F .$$

The increase in the coefficient of the $\cos(2bk_y)$ term with increase in t_b is considered to break the SDW.

4.4.1 Mean-Field Treatment

The model Hamiltonian for the SDW can be expressed as

$$\mathcal{H} = \sum_{k,\sigma} \epsilon_k c_{k\sigma}^\dagger c_{k\sigma} + I \sum_{k,k',q} c_{k\uparrow}^\dagger c_{k+q\uparrow} c_{k'\downarrow}^\dagger c_{k'-q\downarrow} , \tag{4.30}$$

where $c_{k\sigma}^\dagger (c_{k\sigma})$ is the creation (annihilation) operator of an electron having the wave number k and spin σ. The first term represents for the one-electron energy. The second term denotes the on-site Coulomb interaction; $I = U/N$ with U being the on-site Coulomb energy and N the number of the TMTSF molecules. This is the *Hubbard model for an anisotropic 2-D band*. We consider an SDW ordering with the wave vector $\mathbf{Q} = \mathbf{Q}_0$ and spin polarization in the xy-plane. Then, we find the static average as

$$M = I \sum_k \langle c_{k-Q\downarrow}^\dagger c_{k\uparrow} \rangle = I \sum_k \langle c_{k-Q\uparrow}^\dagger c_{k\downarrow} \rangle , \tag{4.31}$$

where, for simplicity, M is assumed to be real, although it restricts the freedom of phase of the SDW. This average appears to the electrons as a periodic potential with the amplitude M, called the *mean-field potential*. We bilinearize the Hamiltonian by taking it out from the interaction term in (4.30) in the standard mean-field procedure, i.e.,

$$AB = [(A-\langle A\rangle) + \langle A\rangle][(B-\langle B\rangle) + \langle B\rangle]$$

$$\simeq \langle A\rangle B + \langle B\rangle A - \langle A\rangle\langle B\rangle , \tag{4.32}$$

where A and B are operators, and the term $(A-\langle A\rangle)(B-\langle B\rangle)$ is neglected. The average $\langle \dots \rangle$ is to be calculated with $\mathcal{H}_{av}$. Then, we get the following average Hamiltonian:

$$\mathcal{H}_{av} = \sum_{\mathbf{k},\sigma} \epsilon_{\mathbf{k}} c^{\dagger}_{\mathbf{k}\sigma} c_{\mathbf{k}\sigma} - M \sum_{\mathbf{k}} (c^{\dagger}_{\mathbf{k}\uparrow} c_{\mathbf{k}-\mathbf{Q}\downarrow} + c^{\dagger}_{\mathbf{k}-\mathbf{Q}\downarrow} c_{\mathbf{k}\uparrow} + c^{\dagger}_{\mathbf{k}\downarrow} c_{\mathbf{k}-\mathbf{Q}\uparrow}$$

$$+ c^{\dagger}_{\mathbf{k}-\mathbf{Q}\uparrow} c_{\mathbf{k}\downarrow}) + 2M^2/I . \tag{4.33}$$

The change in sign is due to the change of the order of the operators. The amplitude M of the SDW potential is obtained from $\mathcal{H}_{av}$ in the following way:

$$M = I\cdot\mathrm{Tr}\{\exp(-\beta\mathcal{H}_{av})(c^{\dagger}_{\mathbf{k}\uparrow} c_{\mathbf{k}-\mathbf{Q}\downarrow} + c^{\dagger}_{\mathbf{k}-\mathbf{Q}\downarrow} c_{\mathbf{k}\uparrow} + c^{\dagger}_{\mathbf{k}\downarrow} c_{\mathbf{k}-\mathbf{Q}\uparrow}$$

$$+ c^{\dagger}_{\mathbf{k}-\mathbf{Q}\uparrow} c_{\mathbf{k}\downarrow})/4\}/\mathrm{Tr}\{\exp(-\beta H_{av})\}$$

$$= \frac{I}{4\beta} \frac{\partial}{\partial M}\left[\ln\mathrm{Tr}\{\exp(-\beta\mathcal{H}_{av})\} + \frac{2\beta M^2}{I}\right] , \tag{4.34}$$

where Tr stands for the summation of all the diagonal elements, and $\beta = 1/k_B T$ with T being the temperature.

With the appearance of the periodic potential having the wave vector Q, the one-particle eigenvalues of $\mathcal{H}_{av}$ are reorganized into the bands $E_{i\sigma}(\mathbf{k})$ in the Brillouin zone defined by $|k_x| \leq Q_x/2 = k_F$ and $|k_y| \leq \pi/b$. Then, one gets

$$M = \frac{I}{4M} \sum_i \sum_{\mathbf{k},\sigma} \frac{\partial}{\partial M}\ln\left[1 + \exp[-\beta E_{i\sigma}(\mathbf{k})]\right] , \tag{4.35}$$

where the **k**-sum is over the Brillouin zone. Calculating $E_{i\sigma}(\mathbf{k})$ up to terms on the order of M^2 and taking account of the three lowest bands, one obtains the following self-consistency equation:

$$1 = \frac{I}{2} \sum_{|k_x - k_F| \leq 2k_F, \, |k_y| \leq \pi/b} \frac{f(E_-(\mathbf{k})) - f(E_+(\mathbf{k}))}{\sqrt{(\epsilon_\mathbf{k} - \epsilon_{\mathbf{k-Q}})^2/4 + M^2}} \, , \qquad (4.36)$$

where $E_\pm(\mathbf{k})$ is the approximate one-particle energy in the SDW state given by

$$E_\pm(\mathbf{k}) = \frac{1}{2}(\epsilon_\mathbf{k} + \epsilon_{\mathbf{k-Q}}) \pm \sqrt{\tfrac{1}{4}(\epsilon_\mathbf{k} - \epsilon_{\mathbf{k-Q}})^2 + M^2} \, . \qquad (4.37)$$

For simplicity, the ordering expressed by (4.31) is chosen to give the spin polarization aligned in the x-direction. We can also choose the y-, z-, or any other, direction of the spin polarization by putting an appropriate phase unto M and get the same gap equation as in (4.36).

When $\epsilon_{\mathbf{k-Q}}$ is equal to $-\epsilon_\mathbf{k}$ and consequently the **k** point on the normal-state Fermi surface, defined by $\epsilon_\mathbf{k} = 0$, satisfies $\epsilon_\mathbf{k} = \epsilon_{\mathbf{k-Q}} = 0$, the Fermi-surface nesting is perfect, i.e., the piece of the Fermi surface in the nega-tive k_x branch perfectly coincides with that in the positive k_x branch when the former is shifted by **Q**. In such a case $E_\pm(\mathbf{K})$, (4.37), is constant and equal to $\pm M$ for the **k** point on the normal-state Fermi surface. In the pre-sent case one finds

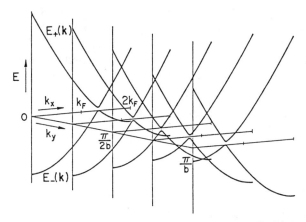

Fig. 4.5. k_x-dependence of the reorganized bands $E_\pm(\mathbf{k})$ for five fixed values of k_y. Note the relative shifts of the narrow gaps around $E = 0$. From [4.61]

$$E_+(k) \approx \epsilon_0 \cos(2bk_y)$$

$$\pm \sqrt{[v_F \hbar(k_x-k_F) - 2t_b \cos(bk_y)]^2 + M^2} - \mu' , \qquad (4.38)$$

with $\mu' = \mu + 2t_a \cos(x_F)$. The **k**-dependence of $E_+(k)$ is depicted in Fig. 4.5. $E_+(k)$ has a weak **k** dependence for the **k** point on the normal-state Fermi surface.

Let us see here from the viewpoint of energy why quasi-1-D conductors have an SDW ordering? When we calculate the total energy E_{SDW} in reference to the normal-state energy E_n for H_{av} in (4.33) utilizing (4.38), we get

$$E_{SDW} - E_n = e \sum_k \left\{ - \sqrt{[v_F \hbar(k_x-k_F) - 2t_b \cos(bk_y)]^2 + M^2} \right.$$

$$\left. + [v_F \hbar(k_x-k_F) - 2t_b \cos(bk_y)] \right\} + \frac{2M^2}{I}$$

$$\simeq - \frac{Na}{\pi v_F \hbar} M^2 \ln\left(\frac{2\sqrt{e}\, v_F \hbar k_F}{M}\right) + \frac{2M^2}{I} . \qquad (4.39)$$

The region of summation is $0 \leq k_x < k_F + (2t_b/v_F \hbar)\cos(bk_y)$, $-\pi/b < k_y < \pi/b$. The ϵ_0 term in (4.38) is neglected since it vanishes in the summation. The second equality holds to the order of M^2 and neglects the order of $t_b/v_F \hbar k_F$. The first term represents the energy gain due to the lowering of the single-electron energy due to the formation of the SDW gap M mainly in the neighborhood of the Fermi level. The second term denotes the increase of the average interaction energy. If we choose an appropriately small M, the first term always wins over the second term and renders the sum of the two terms negative. This is the energetic reason why SDW is formed in the 1-D case and also the quasi-2-D case with good nesting. When we take account of a finite value of ϵ_0, how about the SDW formation? This is our problem under consideration.

For a fixed value of k_y the extrema are given approximately when k_x satisfies $v_F \hbar(k_x-k_F) - 2t_b \cos(bk_y) = 0$, i.e.,

$$[E_+(k)]_{ex} \approx \epsilon_0 \cos(2bk_y) \pm M - \mu' , \qquad (4.40)$$

with ϵ_0 defined by (4.28). When the gap parameter M_0 at absolute-zero temperature is larger than ϵ_0, the upper and lower bands of $E_+(k)$ have no semimetallic band overlap, as shown in Fig.4.6. Then, at absolute-zero temperature, the gap equation of (4.36) is reduced to

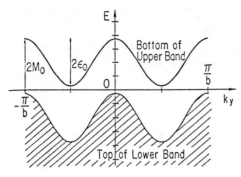

Fig. 4.6. Top of the lower and the bottom of the upper band as functions of k_y. The *hatched* area is occupied by electrons; M_0 is the value of M at absolute-zero temperature and ϵ_0 is defined in (4.28)

$$1 = N(0)I \cdot \ln(2D/M_0) \quad \text{or} \quad M_0 = 2D \cdot \exp[-1/N(0)I] , \tag{4.41}$$

where

$$N(0) = \frac{N}{4\pi t_a \sin(x_F)} \tag{4.42}$$

is the state density per one k_x-branch per spin, and

$$D = \frac{4t_a \sin^2(x_F)}{\cos(x_F)}$$

with $x_F = a_s k_F$. This is the same result as for the gap in a 1-D system.

When $M_0 = 2D \exp[-1/N(0)I]$ is smaller than ϵ_0, the band structure is semimetallic. Then, electron pockets in the upper band and hole pockets in the lower band make large negative contributions to the right-hand side of (4.36), cancelling $\ln(1/M_0)$ in (4.41), and leading to the gap equation:

$$1 = N(0)I \cdot \ln(2D/\epsilon_0) . \tag{4.43}$$

Since $\epsilon_0 > M_0$ as assumed, the right-hand side is smaller than $N(0)I \times \ln(2D/M_0)$ and cannot be equal to 1; the equality is never satisfied. Therefore, the stability condition for the SDW with the wave vector $\mathbf{Q} = \mathbf{Q}_0$ is given by [4.23]

$$\epsilon_0 < M_0 = 2De^{-1/N(0)I} . \tag{4.44}$$

If one assumes the BCS relation $2M_0 = 3.53 k_B T_{SDW}$ with $T_{SDW} = 11.5$ K, one finds $M_0 = 20.3 k_B$. With $t_a = 0.25$ eV, the upper bound for t_b given by (4.44), $t_{b,cr}$, is 24.9 meV. We use 1 eV $= 11605.7 k_B$. This is close to the observed value of t_b listed in Table 4.2. However, as will be shown in Sect.4.4.2, this condition of the SDW stability is drastically modified by another term in a more realistic model.

With the above set of parameters, the values of M_0 requires that $U \simeq 0.344$ eV.[2] The spin density $2\langle S_x(\mathbf{k})\rangle$ is given in units of μ_B by

$$2\langle S_x(\mathbf{k})\rangle = 4(M_0/U)\cos(\mathbf{Q}\cdot\mathbf{R}) , \qquad (4.45)$$

the amplitude of which is nearly equal to $0.020\,\mu_B$ with the above parameter values.

The transition temperature T_{SDW} is plotted as a function of t_b in Fig. 4.7. It vanishes when t_b becomes equal to $t_{b,cr}$, where ϵ_0 becomes equal to M_0, i.e., when the inequality (4.44) is violated. The temperature dependence of the gap parameter M is depicted in Fig.4.8. At absolute zero the gap parameter $M_0 = 2D\exp[-1/N(0)I]$ is independent of t_b and thus also of T_{SDW}. When the temperature is close to T_{SDW} so that M is much smaller than M_0, the lower and upper bands in Fig.4.6 overlap and, as a result, a

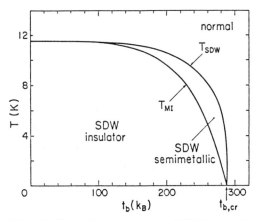

Fig. 4.7. Transition temperature of SDW phases T_{SDW} as a function of transverse transfer energy t_b. The *curve* labeled T_{MI} divides the SDW phases into a semimetallic band overlap region and a semiconductor region having no such overlap. The values $t_a = 0.25$ eV and $U = 0.3441$ eV were chosen so that $T_{SDW} = 11.5$ K. From [4.23]

---

[2] This is significantly smaller than the value of $U = 1.25$ eV obtained by optical experiments for TTF in TTF salts [4.43]; for TMTTF and TSeF salts slightly smaller values for U were suggested. However, since the spin susceptibility of $(TMTSF)_2 AsF_6$ is only enhanced by $10 \div 30\%$ from the non-interacting Pauli susceptibility [4.6], suggesting the same order for the ratio of U to the band width, the above-mentioned smaller value of U of the order of 0.4 eV is considered more appropriate for the SDW properties. The difference is regarded to arise mainly from the smaller scale of energy involed in the SDW ordering etc., since for the processes with the smaller energy exchange the screening effect of surrounding electrons must be more efficient.

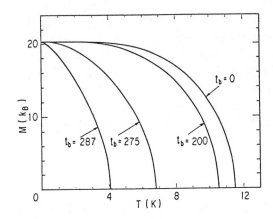

Fig. 4.8. Amplitude M of the SDW periodic potential as a function of temperature T for several values of t_b. Other parameter values are the same as in Fig. 4.7. From [4.23]

semimetallic band structure appears. This is possible only at finite temperatures. In Fig. 4.7 the curve labeled T_{MI} represents the boundary between the semimetallic and the semiconducting regions.

If we calculate the difference between the total energy in the SDW state E_{SDW} and that in the normal state E_n at absolute zero, the stability condition in (4.44) is more understandable. After a tedious but straightforward calculation, one obtains [4.25]

$$E_{SDW} - E_n = N(0)(\epsilon_0^2 - M_0^2) . \tag{4.46}$$

The energy E_{SDW} is given by $\langle \mathcal{H}_{av} \rangle$, and E_n is obtained by considering the non-interacting part in (4.30). One has to be careful to keep the total electron number constant by adjusting the chemical potential μ and to take account of it in (4.46). The difference is sketched in Fig. 4.9. The requirement for (4.46) to be negative leads to the condition (4.44).

The above stability condition can be more easily obtained by transforming the band energy in (4.26) to

$$\widetilde{\epsilon}_{\mathbf{k}} = v_F \hbar(|k_x| - k_F) - 2t_b \cos(bk_y) + \epsilon_0 \cos(2bk_y) - \widetilde{\mu} . \tag{4.47}$$

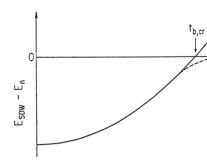

Fig. 4.9. Difference between the total energies in the SDW and the normal states $E_{SDW} - E_n$ as a function of t_b. *Solid curve* is for SDW wave vector **Q** fixed at $\mathbf{Q}_0$, *dashed curve* illustrates the lowered energy of a transient SDW state

103

Here the first term reproduces the first derivative with respect to k_x at the Fermi energy, but higher-order derivatives are neglected. To compensate for it, the third term is added so that $\tilde{\epsilon}\mathbf{k}$ reproduces the $k_x - k_y$ curves of the Fermi surface. This transformation of the band energy for the SDW problem can be performed in a quasi-classical approximation [4.44].

The wave vector $\mathbf{Q}_0$ is called *optimal* since it gives the lowest total energy in good nesting systems. When one applies the above argument to the more realistic band in (4.24), one gets

$$\mathbf{Q}_0 = (2k_F, (\pi + 2\phi)/b) . \tag{4.48}$$

Since the angle ϕ is not, in general, commensurate with π, $\mathbf{Q}_0$ is incommensurate with the reciprocal lattice [4.23], although $2k_F = \pi/2a_s$ is commensurate with the x-component of the latter.

In the present subsection we have treated the linear SDW. The spiral SDW polarization was found to have an energy higher than the linear one because the SDW gap is formed only on one side of the Fermi surface for each spin [4.23].

4.4.2 The Case of Multiple Transverse Transfers

In the preceding subsection, the $\cos(2bk_y)$ component of the Fermi surface was harmful to the SDW stability. The approximate form of $\epsilon_{\mathbf{k}}$ in (4.24) contains only a $\cos(bk_y \mp \phi)$ term but there must be higher harmonic terms as corrections. These are the terms up to the order of t_b^2/t_a. Here, we derive an appropriate band energy giving the $\cos(2bk_y)$ dependence explicitly in the neighborhood of the Fermi energy. When one takes out $2t_a \cos(x)$, with $t_a = (t_{S1} + t_{S2})/2$, from the $+$ branch $\epsilon_{\mathbf{k}}^{(+)}$ in (4.17), by expanding in powers of t_{Ii} or $t_{S1} - t_{S2}$ over t_a, the remaining terms are at most in the order of t_b. Since one is most interested in preserving the original form of the Fermi surface and keeps the terms up to the order of t_b^2/t_a, the factor $x = (\mathbf{a} \cdot \mathbf{k})/2$, which appears in combination with y in (4.17), can be replaced by its value at the Fermi surface, i.e.,

$$x \approx \pm \left[x_F + \frac{t_b \cos(y \mp \phi)}{t_a \sin(x_F)} \right] . \tag{4.49}$$

Then, one gets the following model band energy [4.45] which reproduces the form of the Fermi surface with precision up to the order of $(t_b/t_a)^2$ and gives an appropriate $\mathbf{k}$-dependence near the Fermi energy:

$$\epsilon_k \simeq 2t_a \cos x + 2t_b \cos\eta + 2\tau_{cos} \cos 2\eta \pm 2\tau_{sin} \sin 2\eta , \qquad (4.50)$$

where

$$\eta = y \mp \phi \quad (\text{for } x \gtrless 0),$$

$$\tau_{cos} = -\frac{t_T (t_{13} + t_{14})\sin(x_F) + 2t_{13}t_{14}}{2t_a \sin(x_F)} - \frac{(\Delta t_T)^2}{4t_a \cos(x_F)}\sin 2\phi , \qquad (4.51)$$

$$\tau_{sin} = \frac{t_T^2 + 2t_{14}^2 + t_T [t_{13}\cos(x_F) + 3t_{14}\sin(x_F)]}{2t_a \sin(x_F)}$$
$$- \frac{(\Delta t_T)^2}{4t_a \cos(x_F)}\cos 2\phi \qquad (4.52)$$

with $\Delta t_T = (t_{11} - t_{12})/2$; ϕ is given by (4.22) with $x = \pi/4$. Recalling the last paragraph of the preceding subsection, this can be further simplified to

$$\epsilon_k = -v_F \hbar(|k_x| - k_F) + 2t_b \cos\eta + 2t_b' \cos 2\eta$$
$$\pm 2\tau_{sin} \sin 2\eta , \quad (k_x \gtrless 0) \qquad (4.53)$$

where

$$t_b' = \tau_{cos} - \epsilon_0/2 , \qquad (4.54)$$

and τ_{cos} and τ_{sin} are on the order of $t_b{}^2/t_a$, thus on the same order as ϵ_0 defined by (4.28). $v_F = 2a_s t_a \sin(x_F)/\hbar$.

Now let us proceed with the argument for SDW. The optimum nesting vector $\mathbf{Q}_0$ is again given by (4.48). The $2t_b' \cos 2\eta$ term deteriorates the Fermi-surface nesting but the $\pm 2\tau_{sin} \sin 2\eta$ term does not so much. Here we treat holes as particles instead of electrons, as performed in Sect.4.4.1. We assume that $-\epsilon_k$ is the new band energy. Then the one-particle energies in the SDW state with $\mathbf{Q} = \mathbf{Q}_0$ are given by

$$E_\pm(\mathbf{k}) = -2t_b' \cos 2\eta$$

$$\pm \sqrt{[v_F \hbar(k_x - k_F) - 2t_b \cos\eta - 2\tau_{sin} \sin 2\eta]^2 + M^2} + \text{const} . \quad (4.55)$$

The extrema optimized with respect to k_x are expressed as functions of k_y by

$$[E_\pm(\mathbf{k})]_{ex} = -2t'_b\cos2\eta \pm M + \text{const} . \tag{4.56}$$

Therefore, by analogy with the simplest case, the stability condition of the SDW with $\mathbf{Q} = \mathbf{Q}_0$ is reduced from the requirement for a non-semimetallic band overlap and given in the leading term by

$$2|t'_b| < M_0 , \tag{4.57}$$

or

$$|2\tau_{\cos} - \epsilon_0| < M_0 . \tag{4.58}$$

This replaces (4.44). Since $2\tau_{\cos}$ is also on the order of $t_b{}^2/t_a$, this term drastically modifies the stability condition. Therefore, the deformation of the Fermi surface due to the multiple transverse transfer energies is severer than due to ϵ_0. Still (4.58) implies that when t_b becomes as large as $t_a/10$, the SDW becomes unstable.

4.4.3 Comparison with Experiments

According to crystallographic experiments at low temperatures and under pressure, carried out by *Gallois* and coworkers [4.46], and the calculation of band parameters by *Ducasse* et al. [4.40], one can check if the vanishing of SDW in the (TMTSF)$_2$X salts is due to a violation of the stability condition given by (4.58).

Table 4.3 lists the values of intermolecular transfer energies t_j calculated by *Ducasse* et al. [4.40] and by *Grant* [4.37], and the values of ϵ_0, $\tau_{\cos}$, and $2\tau_{\cos}-\epsilon_0$ derived from (4.28, 50). The accuracy of the obtained values for the latter quantities is difficult to evaluate, as can be seen when one compares the results of the two calculations at 300 K. However, as seen for X: PF$_6$, there is a clear tendency in $\tau_{\cos}$ to rapidly increase as the temperature goes down and as the pressure is increased, while ϵ_0 is not so clearly affected. Consequently, $2t'_b = 2\tau_{\cos}-\epsilon_0$ also grows markedly first with a decrease of temperature and then again significantly increases under pressure.

The tendency of $\tau_{\cos}$ to increase under pressure stems from a structural reason. According to the results on (TMTSF)$_2$PF$_6$ [4.46c], the short Se-Se contacts between the pairs of molecules corresponding to t_{I1} and t_{I2} (Fig. 3.5) are shortened but the angle which they make with the molecular plane of TMTSF is kept constant. This results in an increase in the absolute magnitudes of t_{I1} and t_{I2} or $t_T = (t_{I1}+t_{I2})/2$ but their negative sign is preserved.

Table 4.3. Band parameters calculated for the refined one-electron energy model in (4.24) for $(TMTSF)_2X$ using data of [4.37,40], as indicated. Transfer energy symbols are defined in Fig.3.5 and in the text. Energies in meV

X	Temp. [K]	Press. [bar]	Ref.	t_{s1}	t_{s2}	t_{I1}	t_{I2}	t_{I3}	t_{I4}	t_a	t_b	ϕ [deg]	ϵ_0	τ_{cos}	τ_{sin}	$2\tau_{cos}-\epsilon_0 = 2t'_b$
PF_6	300	1	4.40	252	209	-6.4	-13.3	40.2	2.7	231	33.5	-7.3	3.44	0.26	-0.71	-2.92
PF_6	4	1	4.40	280	254	-17.8	-47.9	46.9	5.6	267	29.5	-36.7	2.31	2.13	-0.98	1.95
PF_6	1.7	7000	4.40	290	272	-20.6	-53.5	49.8	6.2	281	30.9	-40.3	2.41	2.47	-0.92	2.53
ClO_4	300	1	4.40	258	221	-11.6	-28.3	41.2	3.6	240	29.1	-21.2	2.49	1.06	-0.99	-0.37
ClO_4	7	1	4.40	287	266	-34.0	-64.1	46.2	7.5	277	29.5	-67.0	2.23	3.20	0.55	4.17
PF_6	300	1	4.37	395	334	-9.5	-36.2	41.5	9.9	365	26.1	-13.9	1.32	0.10	-0.99	-1.12
ClO_4	300	1	4.37	393	339	-15.3	-54.4	45.1	11.5	366	24.3	-32.7	1.14	1.03	-1.09	0.92
ReO_4	300	1	4.37	390	338	-15.0	-54.5	43.2	11.8	364	22.6	-34.4	0.76	1.00	-1.00	1.01

Their relative changes are the largest among all the t_j's. Therefore, one gets from (4.51, 28)

$$\frac{\partial}{\partial P}(2\tau_{\cos} - \epsilon_0) \approx -\frac{2(t_{I3} + t_{I4}) + \sqrt{2}t_T}{t_a}\left[\frac{\partial t_T}{\partial P}\right] > 0 \ , \tag{4.59}$$

where $2(t_{I3} + t_{I4}) + \sqrt{2}t_T > 0$ and $\partial t_T / \partial P < 0$ have been employed. Thus, the increase of t_b' under pressure can theoretically be verified.

The value of M_0 is estimated at 1.75 meV from the BCS relation $2M_0 = 3.53k_B T_{SDW}$ with $T_{SDW} = 11.5$ K. Infrared data for $(TMTSF)_2 PF_6$ suggest $2M_0 = 45$ cm^{-1} or $M_0 = 2.8$ meV, which looks more reliable than the above value of 1.75 meV $(1\text{cm}^{-1} = 1.434k_B)$ [4.47]. Although *Timusk* et al. [4.48] criticized that the absorption peak at 45 cm^{-1} may be due to some lattice vibration, they agreed with the opinion that $2M_0$ is in the range of $40 \div 60$ cm^{-1} for $(TMTSF)_2 PF_6$ and $(TMTSF)_2 AsF_6$ from the decrease of optical conductivity at temperatures below the SDW transition temperature. For the latter salt, *Eldridge* et al. [4.49] and *Degiorgi* et al. [4.50] assigned $2M_0 = 70$ cm^{-1}. Thus, it seems safer to say that the infrared data gives $2M_0 = 40 \div 70$ cm^{-1} for the Bechgaard salts. The optical features around $\hbar\omega = 2M_0$ are not so clear-cut as in the case of the CDW. This is, presumably, because the contribution of the single-particle excitations across the SDW gap is almost canceled by that of the collective excitations. With both contributions, the features have a peak weight proportional to $(1 - m_{band}/m^*)$, m^* being the mass of the collective mode and is close to the band mass m_{band} in the case of SDW [4.51].

Susceptibility data [4.6] of $(TMTSF)_2 AsF_6$ give $E_A = 2.7$ meV, where E_A is the activation energy in the SDW state and should be smaller than M_0, as will be explained in this section. Transport data for $(TMTSF)_2 PF_6$ give $E_A = 2.1$ meV [4.52]. Since $(TMTSF)_2 X$ has only a weak coupling in the c-direction, thermal fluctuations are thought to suppress T_{SDW} although slightly [4.23]. Moreover, since t_b' is close to its upper bound of stability, the BCS relation may not hold. Thus, the optical value of $M_0 = 2.8$ meV becomes even more appropriate. Even if we assume that both quantities are not so accurately obtained, it is reasonable to say that $2t_b'$, which is smaller than M_0 at ambient pressure, surpasses M_0 under pressure, violating the criterion given by (4.57 or 58).

In the case of a slowly cooled sample of $(TMTSF)_2 ClO_4$ the disappearance of SDW is assumed to be due to a violation of the same stability condition since its value of $2t_b'$ at 7 K in Table 4.3 clearly exceeds the value of M_0. M_0 of $(TMTSF)_2 ClO_4$ must be smaller than that of $(TMTSF)_2 PF_6$, since T_{SDW} of the former never goes beyond 6 K even under applied magnetic fields, which should help the SDW formation, as will be discussed in

Chap.9. The anion ordering with $\mathbf{Q} = (0, \frac{1}{2}, 0)$ seems to be also playing a role.

There are also experimental data suggesting that the value of $\partial \tau_{\cos} / \partial P$ is in fair agreement with the calculated value. The activation energy E_A in the SDW state of $(TMTSF)_2 PF_6$ at ambient pressure has been reported to be 2.1 meV [4.52]. This activation barrier must vanish under a pressure of 7 kbar where SDW vanishes. From the value of the indirect gap displayed in Fig.4.6, we obtain

$$E_A = M_0 - |2\tau_{\cos} - \epsilon_0| . \tag{4.60}$$

By assigning the decrease of E_A to the increase of $\tau_{\cos}$, one gets $\partial \tau_{\cos} / \partial P \simeq 0.15$ meV/kbar. In the SDW state of the quenched state of $(TMTSF)_2 ClO_4$, one finds $E_A = 0.5$ meV at ambient pressure. Since the SDW vanishes under a pressure of 2 to 3 kbar, one gets $\partial \tau_{\cos} / \partial P \simeq 0.08$ to 0.13 meV/ kbar. These values are not so far from the experimental value of 0.05 meV/kbar obtained for the PF_6 salt from Table 4.3. The value of $t_b' = \tau_{\cos} - \epsilon_0/2$ was obtained by another experiment which will be described in Sect.4.6.3.

Analyses of proton NMR line shapes in terms of a dipolar field coming from the SDW clearly support the present nesting model of SDW. *Delrieu* et al. [4.12] obtained the b-component Q_b of the SDW wave vector $\mathbf{Q}$ for $(TMTSF)_2 PF_6$, which is found to be $(0.20 \pm 0.05) \cdot 2\pi/b$. *Takahashi* et al. [4.12] gave 0.24 ± 0.03 in units of $2\pi/b$. Equation (4.48) yields the theoretical value 0.30 together with $\phi = -36.7°$ listed in Table 4.3 in good agreement with these derivations.

According to *Roger* et al. [4.53], the anisotropy energy of the SDW spin polarization is obtained from the dipole and spin-orbit interactions. It is very sensitive to the Q_b component of the SDW wave vector. Theoretically, to obtain the easy axis along the b'-axis, the intermediate axis along the a-axis, and the hard axis along the c^*-axis, they showed that the value of $Q_b = 0.20 \cdot (2\pi/b)$ is necessary, in agreement with the observation.

A remarkable feature of the present model is that while $\mathbf{Q}$ becomes $\mathbf{Q}_0$, M_0 is independent of T_{SDW} which varies as a function of pressure. This has been observed in $(TMTSF)_2 PF_6$ [4.54].

The above-mentioned two groups also obtained the spin amplitude $0.085 \, \mu_B$ and $0.080 \, \mu_B$, respectively. Equation (4.45) yields $0.03 \mu_B$ for the theoretical value of the spin amplitude with $M_0 = 2.8$ meV, although the latter value has an uncertainty of the order of 50%. Part of this discrepancy comes from the neglected attractive interaction which is mediated by phonons and suppresses the SDW gap parameter in the vicinity of the Fermi energy.

If we assume that our model system in (4.30) has, in addition to the Coulomb repulsion, a Bardeen-Cooper-Schrieffer-type of attractive interaction [4.55] which works between electrons only when their band energies are within a certain cutoff energy E_c from the Fermi energy, then the SDW gap parameter M_k becomes dependent on the wave vector $\mathbf{k}$ in the following way [4.25]:

$$M_k = M_0 \eta(\mathbf{k})\eta(\mathbf{k}-\mathbf{Q}) + M_\infty [1 - \eta(\mathbf{k})\eta(\mathbf{k}-\mathbf{Q})] ; \qquad (4.61)$$

here M_0 and M_∞ are constants; we define $\eta(\mathbf{k}) = 1$ when $|\epsilon_\mathbf{k}| < E_c$ and $\eta(\mathbf{k}) = 0$ otherwise. M_0 determines the SDW gap in the vicinity of the Fermi energy obtained from the optical gap and the activation energy, while the spin amplitude is given by $4M_\infty /U$. The value of M_∞ is larger than M_0 by the factor

$$\frac{M_\infty}{M_0} = \frac{I}{I - V_{BCS}[1 - N(0)I \cdot \ln(D/E_c)]} , \qquad (4.62)$$

where V_{BCS} is the coefficient of the attractive interaction defined in (8.5), and D is an average half band width. This factor can be on the order of 1.4 and eliminates a considerable part of the discrepancy between the observed and calculated values of the SDW spin amplitude.[3]

The approximate band expression in (4.24) is considered indispensable to understand the metal-insulator transition accompanying the anion-ordering in some Bechgaard salts as well. For example, $(TMTSF)_2 ReO_4$ has a first-order anion-ordering transition at 180 K with the wave vector ($\frac{1}{2}a^*, \frac{1}{2}b^*, \frac{1}{2}c^*$) [4.57]. The most prominent lattice deformation due to the transition is the jump-wise increase of the lattice parameter γ by 0.5° [4.58]. In the approximate band scheme of Sect. 4.3.3, this change causes the decrease of the absolute magnitude of t_T, which is negative, and, therefore, the decrease of the absolute magnitude of ϕ according to (4.22) from the room temperature value around −34.4° in Table 4.3. This γ-dependence of ϕ in $(TMTSF)_2 X$ has been demonstrated in [Ref.4.58, Fig.1]. The decrease of

---

[3] If we introduce the Electron-Molecular-Vibration (EMV) interaction, which will be treated in Sect. 8.3, the above factor becomes about 2.0 and eliminates a substantial part of the discrepancy in view of the uncertainty of the infrared data. The remaining discrepancy of a factor of 1.5 may be partly due to a possible energy dependence of U. It becomes as large as 1 eV for the processes involving exchange of large energy on the order of 1 eV [4.43], although it may be around 0.5 eV for processes related only with electrons in the vicinity of the Fermi level. Another factor is due to results of thermal fluctuations; one effect is due to the quasi 1-D band structure, while another effect is that which accounts for the itinerant electron magentism with a small spin amplitude [4.56].

$|\phi|$ makes the nesting vector $(2k_F, (\pi + 2\phi)/b)$ in (4.48) in the 2-D conducting plane close to the observed anion-ordering wave vector $(\frac{1}{2}a^*, \frac{1}{2}b^*)$. This makes the deformed system an insulator. In this system the one-electron energy is lowered by the band gap formed by the periodic potential of the ordered anions. Therefore, not only the structural energy gain but also the electronic one are important in this anion-ordering transition.

4.5 Transient SDW State and Reentrant Superconducting Phase

4.5.1 The Transient SDW State

In the band model expressed by (4.50), when the stability condition $|2\tau_{\cos} - \epsilon_0| < M_0$ is violated with increasing $\tau_{\cos}$, the SDW with the optimum wave vector Q_0 given by (4.48) becomes unstable. SDW still survives in a transient form, however, which has a wave vector Q slightly shifted from Q_0 and a smaller gap parameter M. This transient SDW gains energy by partly well nesting parts of the Fermi surface with Q. It has a semimetallic band structure even at absolute zero. This state occurs in an energy range on the order of $6k_B$ in $\tau_{\cos}$ in the temperature vs. $\tau_{\cos}$ phase diagram. In terms of this state one can well understand the puzzling properties of the SDW in the vicinity of the phase boundary and the *reentrant* superconducting phase in the T-P phase diagram. It also facilitates the understanding of the magnetic-field-induced SDW, as will be discussed in Chap. 9. A transient type of SDW was also reported in [4.60].

This state was first investigated by a variational method that applied the band model of (4.26) with increasing t_b [4.61] and later by using the band energy of (4.50) with increasing $\tau_{\cos}$ [4.62]. Here the features of the transient SDW are described mainly on the basis of results from the latter model. We introduce three variational parameters. Two parameters s and ϕ_0 define the SDW wave vector Q shifted from the optimal $Q_0 = (2k_F, Q_b^{(0)})$ in the following way:

$$Q = (2k_F(1+s), Q_b^{(0)} - 2\phi_0/b), \tag{4.63}$$

with $k_F = \pi/4a_s$. We treat holes as particles instead of electrons. The third parameter is the gap parameter M which is no longer equal to M_0 even at absolute zero. According to a variational formulation [4.23], the ground-state energy E_{SDW} of the SDW state is given by

$$E_{SDW} = \sum_i \sum_\sigma \sum_k [E_{i\sigma}(k) + \mu]f(E_{i\sigma}(k)) + 2M^2/I \,, \qquad (4.64)$$

where the k-summation is carried out in the reduced Brillouin zone in the SDW state defined by $|k_x| \le Q_x/2$ and $|k_y| \le \pi/b$; $E_{i\sigma}(k)$ is for the different reorganized band energies. These are given by the eigenvalues of the following matrix:

$$\hat{H}(k) = \begin{bmatrix} \epsilon_{k+Q\bar\sigma} & -M & 0 \\ -M & \epsilon_{k\sigma} & -M \\ 0 & -M & \epsilon_{k-Q\bar\sigma} \end{bmatrix} . \qquad (4.65)$$

The function f(x) is unity if x < 0 and zero otherwise; $\bar\sigma$ denotes the inverse of σ. For a hole the band energy ϵ_k is defined by

$$\epsilon_k = -2t_a\cos(a_s k_x) - 2t_b\cos(bk_y) - 2\tau_{cos}\cos(2bk_y)$$

$$\mp 2\tau_{sin}\sin(2bk_y) - \mu \,, \qquad (4.66)$$

where μ is the chemical potential, adjusted so that the total number of carriers is conserved. Here, we assumed $\phi = 0$, and $\mp$ refers to $k_x \gtrless 0$.

As τ_{cos} approaches the critical value $\tau_{cos}^{(cr)} \equiv (\epsilon_0 + M_0)/2$ at which (4.58) is first violated, Q and M, which minimize the variational total energy in (4.64), start to deviate from Q_0 and M_0, respectively, at a value slightly smaller than $\tau_{cos}^{(cr)}$, as in Fig.4.10. In the case illustrated in the figure, $\tau_{cos}^{(cr)} =$

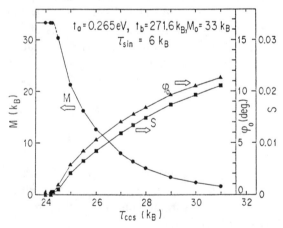

Fig. 4.10. Gap parameter M, ϕ_0, and s defining $Q_y = (\pi - 2\phi_0)/b$ and $Q_x = 2k_F(1+s)$, respectively, as functions of τ_{cos} in the transient SDW (SDW2) state. From [4.62]

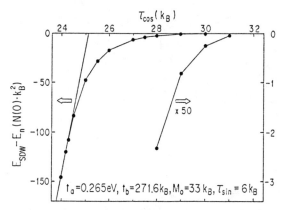

Fig. 4.11. Difference between the total energy E_{SDW} in the transient SDW (SDW2) state and E_n in the normal state as a function of τ_{cos}. The corresponding parameter values are shown in Fig. 4.10. At $\tau_{cos} \geq 28\,k_B$, the curve is replotted on a 50 times magnified ordinate scale. The *straight line* in the left-hand side is for the case when $\mathbf{Q} = \mathbf{Q_0}$. From [4.62]

$25.0k_B$. The difference between the total energy E_{SDW} in the SDW state and E_n in the normal state is plotted in Fig. 4.11. The energy E_{SDW} of this variational state is lower than $E_{SDW}^{(0)}$ for $\mathbf{Q_0}$ and M_0 expressed by the nearly straight line. The latter is given by

$$E_{SDW}^{(0)} - E_n = N(0)[(2\tau_{cos} - \epsilon_0)^2 - M_0^2] \qquad (4.67)$$

in the leading-term approximation. This can easily be obtained by employing a model in which the k_x dependence is transformed into a linear expression as in (4.53). This also clearly provides the condition of (4.57) for the stability of SDW with $\mathbf{Q_0}$. The transient SDW persists even when $\tau_{cos} > \tau_{cos}^{(cr)}$ in an interval of $\Delta\tau_{cos} \approx 6k_B$.

We can get an idea about the mechanism of the energy gain in the transient SDW state by examining the extremum curves $[E_{\pm}(\mathbf{k})]_{ex}$ vs. k_y, that is, the lower bound of the upper band and the upper bound of the lower band, as in (4.56). In the leading term approximation, they are given by

$$[E_{\pm}(\mathbf{k})]_{ex} = -2t_b' \cos(2y) - 2t_b \sin(\phi_0)\sin(y) \pm M + \text{const.}, \qquad (4.68)$$

where $t_b' = \tau_{cos} - \epsilon_0/2$ and $y = bk_y + \phi_0$. The extremum curves for $\mathbf{Q} = \mathbf{Q_0}$ or $\phi_0 = 0$ when $t_b' > 0$, as in the present case, are plotted in Fig. 4.12a. The minima of the curves are at $y = bk_y + \phi_0 = 0$, while in the case of $t_b' < 0$ the maxima are at $y = 0$ as in Figs. 4.12d and 4.6. When t_b' approaches M_0, $\mathbf{Q}$ and M start to shift. When ϕ_0 is finite, say $4°$, the second term, which is odd with respect to y, makes the extremum curves asymmetric with

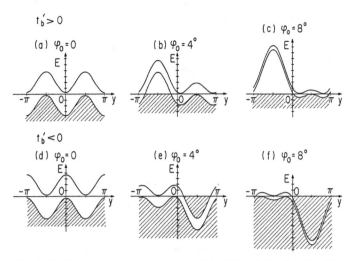

Fig. 4.12. Extremum curves showing $[E_+(\mathbf{k})]_{ex}$ defined in (4.68) as functions of $y = bk_y + \phi_0$ for various t'_b and ϕ_0 (see text). t'_b is defined in (4.54), ϕ_0 gives $Q_y = (\pi - 2\phi_0)/b$. *Hatching* indicates areas occupied by electrons

respect to the ordinate line. We can confirm with optimized values of $\mathbf{Q}$ and M that the Fermi level lies between the flat parts of the two extremum curves (Fig. 4.12b). The energy gain is obtained from this region arising from a partial Fermi-surface nesting. As $\tau_{\cos}$ or t'_b increases further, ϕ_0 also increases. The extremum curves become like those in Fig. 4.12c; the region of the partial nesting is slightly decreased with decrease in M. With further increase of t'_b, this transient SDW finally disappears. It is not clear if M vanishes continuously or discontinuously with increase in t'_b.

The unoccupied part below the lower extremum curve in Fig. 4.12b,c makes a pocket for carriers of the type opposite to that in the normal state. Since the area $Q_x 2\pi/b$ of the lower band is the sum of the occupied and unoccupied k-space, this means that $Q_x = 2k_F(1+s)$ is slightly enlarged, i.e., s takes a positive finite value, as is indicated in Fig. 4.10. With the SDW wave number Q_x, the volume of the Brillouin zone in the k_x–k_y plane is $(2\pi)^2 Q_x/bc$. Since it is full when $Q_x = 2k_F$, the density ρ of the new vacancies is given by

$$\rho = \frac{2k_F s}{\pi bc} = \frac{Ns}{2}. \tag{4.69}$$

In the case when $t'_b < 0$, the extremum curves with the optimal wave vector $\mathbf{Q} = \mathbf{Q}_0$ and with a shifted wave vector $\mathbf{Q} \neq \mathbf{Q}_0$ are shown in Fig. 4.12d-f. In this case, the transient SDW state has a carrier pocket with carriers of the same type as in the normal state.

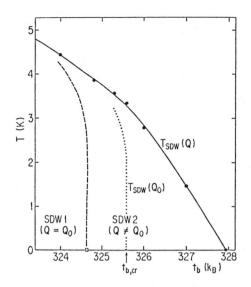

Fig.4.13. T_{SDW}-t_b phase diagram for the transient SDW (SDW2) with wave vector **Q** shifted from the optimal **Q**$_0$. The SDW2 region extends both beyond the upper bound of $t_{b,cr}$ for the SDW with **Q** = **Q**$_0$ (SDW1) (*dotted curve*) and also into the region slightly below $t_{b,cr}$ down to the *dashed curve*

A phase diagram in the T-t_b plane for the band model in (4.26) is displayed in Fig.4.13. The transition temperature T_{SDW} of the transient SDW was obtained by minimizing the free energy. A similar study was carried out by *Hasegawa* et al. [4.63] for the band model of (4.53). The transient type of SDW described above appears in any boundary region of both band models (4.26 and 52).

4.5.2 Experimental Results

The change in the carrier type accompanying the SDW transition is observed in the Hall resistance measurements of $(TMTSF)_2PF_6$ in the vicinity of the critical pressure P_c. The Hall resistance, given by the ratio of Hall voltage to the current, changes sign at P = 6.1 kbar between 4.0 and 2.1 K, as shown in Fig.4.14. From a series of ESR experiments the state at T = 4.0 K is found to be metallic while at T = 2.1 K it is in the SDW state (Fig.4.15). According to Sect.4.4.3, $(TMTSF)_2PF_6$ loses the SDW state under pressure mainly due to the increase in τ_{cos} so that $t'_b = \tau_{cos} - \epsilon_0/2 >$ M_0. Therefore, the observed different signs of the Hall resistance between T = 2.1 K and 4.0 K can be understood as being due to a change of the type of current carrier caused by the transition from the normal state to the transient SDW state; the extremum curves of the latter state are displayed in Fig.4.12b.

The upturn in the curve for T = 4.0 K at H > 9 T in Fig.4.14 is interpreted as being a transition into an SDW phase induced by magnetic fields, which is treated in Chap.9. This state also has a band structure essentially

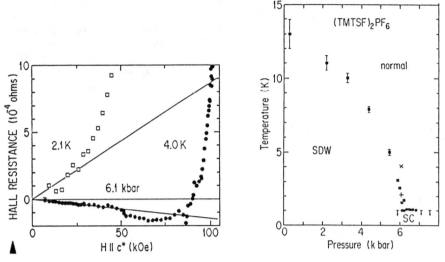

Fig. 4.14. Hall resistance as a function of applied magnetic field for $(TMTSF)_2 PF_6$ near the metal-SDW phase boundary at 6.1 kbar. At 4 K the sample is metallic, but at 2.1 K it is an SDW semimetal. The high-field data at 4 K show the onset of the magnetic-field-induced SDW transition. The absolute sign of R_H, as written in the text, is controvertal. From [4.64]

Fig. 4.15. Temperature-pressure phase diagram obtained in an ESR experiment. *Squares* mark the metal-SDW transition, *dots* the superconducting transition. The points indicated by *arrows* represent the pressures at which no superconductivity was observed down to the temperature shown. *Cross* and *plus* mark the (T, P) values for the two curves in Fig. 4.14. From [4.65]

the same as illustrated in Fig. 4.12b,c for the transient SDW state. Therefore, the change of the sign of the Hall voltage for H > 9 T can be understood in the same way. The transition occurs even at higher temperatures with an increased magnetic field. This is understood on the basis of the stabilization of SDW in the field, and was qualitively treated in [4.66] (Sect. 4.6.3).

A kinetic theory of the Hall constant R_H in the normal state for quasi 1-D conductors of this type gives [4.67]

$$R_H = - \frac{a_s k_F}{nec \cdot \tan(a_s k_F)} \,, \qquad (4.70)$$

where e is the absolute value of electronic charge, and n is the density of holes. In real systems the value of $a_s k_F$ should be set to $3\pi/4$ so that $R_H > 0$ in the normal state in agreement with the results of some researchers [4.68, 69]. However, some others reported a negative sign [4.64, 70]. This seems to be due to subtle technical problems in the experiment. Everyone agrees with the observed change of sign in the Hall voltage accompanying the appearance of the magnetic-field-induced SDW. Here, we have focused on the change of the sign of R_H.

4.5.3 Reentrant Superconducting Phase

As is clearly seen in Fig. 3.13, the T-P phase diagram for $(TMTSF)_2 AsF_6$ [4.71], the superconducting phase intrudes into the SDW phase in the pressure region of ≈ 1 kbar above the critical pressure P_c. Within this pressure range, as the temperature decreases at constant pressure the system traverses metallic, insulating, and superconducting, i.e., again metallic regions, judging from the resistivity behavior. We call this part of the superconducting phase *reentrant* [4.72]. The reentrant phase was observed in $(TMTSF)_2 AsF_6$ [4.71] and $(TMTSF)_2 PF_6$ [4.73] soon after *Jérome* et al. [4. 20] found superconductivity in $(TMTSF)_2 PF_6$ under pressure.

This reentrant phase has attracted the interest of theoreticians who noted the possibility of coexistence of magnetism and superconductivity. At present, however, this phase is thought to be purely superconducting without a coexisting SDW. The superconducting phase intrudes into the SDW phase because its free energy is lower than that of the SDW phase. Therefore, the phase boundary between the SDW and the superconducting phases was predicted to be of first order [4.25]. This was verified by ESR experiments [4.65]. In the reentrant pressure region, with decreasing temperature and a fixed pressure the ESR signal first showed a resonance typical of the normal metallic state, then vanished suddenly because of line broadening due to the SDW. Finally it reverts, although over a very narrow temperature interval, into the signal typical of a superconducting state. The data shown in Fig. 4.15 were obtained by this experiment [4.65].

The rather large width in pressure of the reentrant phase, about 1 kbar, can be understood in terms of the transient SDW as in the following. Thus, the occurrence of such a phase provides further support for the transient nature of SDW in the boundary region of the phase diagram.

In a superconducting state with a superconducting gap parameter at absolute zero of Δ_0, the total energy E_{SC} in the superconducting state is lower than the energy in the normal state E_n by [4.25]

$$E_{SC} - E_n = -N(0)\Delta_0^2 \qquad (4.71)$$

with the state density $N(0)$ defined by (4.42). For the state with $T_c \simeq 1$ K, this difference is about $3N(0)k_B^2$. If the system depicted in Fig. 4.11 has a superconducting state with $T_c \simeq 1$ K, by an additional attractive electron-

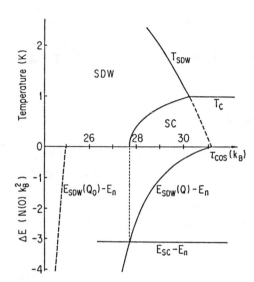

Fig. 4.16. *Lower part: solid curves* represent the energy differences $E_{SDW} - E_n$ and $E_{SC} - E_n$ as functions of τ_{cos} at $T = 0$ K. *Dashed curve* shows $E_{SDW} - E_n$ for **Q** fixed at $\mathbf{Q}_0$. *Upper part*: phase diagram in the temperature-τ_{cos} plane illustrating the superconducting (SC) phase intruding into the SDW phase since the former has a lower free energy than the latter. It is assumed that T_{SDW} rises sharply with decreasing τ_{cos}, based on the result shown in Fig. 4.13 for the T-t_b plane

electron interaction, E_{SC} is lower than E_{SDW} in the region $28k_B \lesssim \tau_{cos} \lesssim 31k_B$, as illustrated in Fig. 4.16. In this region, the transient SDW is replaced by the superconducting phase, as exhibited in the upper part of Fig. 4.16. Since $\partial \tau_{cos}/\partial P \approx 0.15$ meV/kbar, as stated in Sect. 4.4.3, this corresponds to about 2 kbar, in fair agreement with the width of the reentrant region in Fig. 3.12b.

If we consider the possibility of an SDW with only the optimal wave vector $\mathbf{Q}_0$, the total energy $E_{SDW}^{(0)}$ in the SDW state is given by (4.67). Then the width of the intrusion of the superconducting state into the SDW state is expressed by

$$\Delta\tau_{cos} = \frac{\Delta_0^2}{4(2\tau_{cos} - \epsilon_0)} \approx 10^{-2} \, k_B \,, \tag{4.72}$$

which is too small for the phase to be reentrant.

Incidentially, the additional mechanism, for example, mediated by phonons can drive superconductivity even if it is weaker than the on-site Coulomb interaction, as will be discussed at the end of Sect. 8.1.3, once the SDW vanishes. Even if the superconducting mechanism would be non-phononic, the above argument holds similarly.

4.5.4 Competition Between SDW and Superconducting Phases

Machida et al. [4.74] performed a precursory analysis on the possibility of coexisting superconducting and SDW orders in the reentrant region, and suggested such a possibility. Their model includes both the Coulomb and BCS-type attractive interactions. Both orders compete with each other in taking the Fermi surface for forming their own ordering. They showed that when one order, e.g. SDW, appears at a higher temperature than the other, i.e. superconducting order, and extends the band gap over the entire Fermi surface, the other is completely suppressed. If a part of the Fermi surface remains in the SDW phase, superconductivity appears with a finite value of T_c, which is given with a reduced dimensionless coupling constant λ due to the reduced state density. They assumed, however, that the BCS-type attractive interaction remains the same after the band reorganization due to the preceding ordering, i.e., the SDW transition.

When one takes into account that the one-particle state is drastically modified in the SDW phase, however, one finds this possibility questionable in the case of the Bechgaard salts. Starting from a model having both on-site Coulomb and BCS-type attractive interactions, one finds that in the presence of the SDW order the modification of the one-particle state leads to an equation for the superconducting gap for singlet superconductivity with the attractive coupling coefficient reduced by the factor [4.25, 75]

$$\frac{w_{\mathbf{k}}^2 - M^2}{w_{\mathbf{k}}^2 + M^2}, \tag{4.73}$$

with

$$w_{\mathbf{k}} = (\epsilon_{(\mathbf{k} \pm \mathbf{Q})/2} - \epsilon_{\mathbf{k}}) + \sqrt{\tfrac{1}{4}(\epsilon_{\mathbf{k} \pm \mathbf{Q}} - \epsilon_{\mathbf{k}})^2 + M^2}.$$

Since this factor becomes very small in the vicinity of the Fermi energy, together with the effect of the reduced state density in the SDW state, it drastically reduces T_c. If one takes into account the Coulomb interaction, T_c would vanish or at least become very small.

In the case of triplet superconductivity, neither the reducing factor in (4.73) nor the Coulomb term appears. Therefore, one obtains a finite value of T_c. However, this coexisting state, with a self-consistent solution for Δ and M, is found to be unstable because it has a larger free energy than the pure SDW state [4.25]. Furthermore, the above value of T_c turns out to be much lower than that for the first-order phase transition boundary between the SDW and superconducting phases [4.25].

Although the above calculations were performed with the SDW wave vector fixed at Q_0, the results definitely suggest that both order parameters compete for the Fermi surface and cannot coexist when there is only one type of anisotropic electron band, as is the case for the Bechgaard salts.

4.6 Developments on SDW

4.6.1 A New Phase Boundary in the SDW Phase of (TMTSF)$_2$X

Takahashi et al. [4.76] found a peak at 3.5 K and a sudden disappearance for temperatures below 3.5 K of the NMR relaxation rate in (TMTSF)$_2$PF$_6$ in the SDW phase. This seems as if there would occur another phase transition accompanied by the vanishing of low-energy modes that helps the spin relaxation. This new *phase boundary* slowly decreases under pressure in such a way as it were connected smoothly to the SC phase boundary for pressures above the critical pressure P_c [4.10, 76].

A similar phenomenon was found in (TMTSF)$_2$AsF$_6$ and the rapidly cooled in (TMTSF)$_2$ClO$_4$ in which SDW appears. The ratio of T_{SDW} to the peaking temperature is nearly equal to 3.5 in both materials [4.77]. A slight kink of the specific heat at the transition was observed at the transition [4.78].

One possible origin of this phenomenon is a transition from the incommensurate SDW at higher temperatures to a commensurate SDW at lower temperatures. However, the SDW wave vector Q has been reported to be incommensurate in the low-temperature phase from the NMR spectrum. Another possibility is a change of the subsidiary SDW order parameters coupled to the main SDW component. Such a phase change will be described in Sect. 9.5.3 as a multi-order-parameter phase change.

Takahashi et al. also reported that the temperature dependence of the NMR relaxation rate becomes slower below 2 K, indicating another transition [4.76]. The glass nature of SDW below 2 K was observed in the measurements of the dielectric relaxation [4.79].

4.6.2 Sliding of SDW

Sliding of the Charge-Density Wave (CDW) has long been known in quasi-1-D conductors such as NbSe$_3$. SDW is very close to CDW since the spin density $\rho_\sigma(\mathbf{r})$ for spin σ is modulated in space; a difference is that in SDW the net spin density $\rho_\uparrow(\mathbf{r})-\rho_\downarrow(\mathbf{r})$ oscillates in space with the net charge density $\rho_\uparrow(\mathbf{r})+\rho_\downarrow(\mathbf{r})$ unmodulated while in CDW the net charge density oscillates with no net spin density. Therefore, sliding of $\rho_\sigma(\mathbf{r})$ as a whole is conceivable in the SDW state as in the CDW state [4.80]. Sliding of $\rho_\sigma(\mathbf{r})$ is accompanied by the whole spin σ charge. Therefore, sliding SDW carries a current and it is accelerated by the electric field. The impurity potential was shown to pin SDW by slightly shifting the relative phase of $\rho_\uparrow(\mathbf{r})$ and $\rho_\downarrow(\mathbf{r})$ so that the net charge $\rho_\uparrow(\mathbf{r}) + \rho_\downarrow(\mathbf{r})$ gets an energy gain in the neighboorhood of the impurity potential [4.81, 82].

Sliding of SDW was found in a few Bechgaard salts. First it has been reported in $(TMTSF)_2 ClO_4$ in the field-induced SDW state [4.83]. In the absence of a magnetic field, *Tomic* et al. first found it in $(TMTSF)_2 NO_3$ [4.84]. Then, it was observed in $(TMTSF)_2 ClO_4$ which had been rapidly cooled [4.85], $(TMTSF)_2 PF_6$ [4.86, 87], and $(TMTSF)_2 AsF_6$ [4.88]. With an increase of the electrical field applied to these salts in the SDW state, the pinning of SDW ceases to work, allowing the sliding of SDW. This gives rise to a nonlinear component of the conductivity as a function of the field. Narrow-band noise which is a typical effect of sliding CDW was observed when CDW slides [4.89]. Narrowing of a proton resonance line due to SDW sliding was noticed as well [4.90]. The reader should refer to [4.91] for detailed properties of sliding SDW.

4.6.3 Further Test of the Mean-Field Results

From the measurements of the proton NMR line width in $(TMTSF)_2 PF_6$, *Takahashi* et al. gave a slow decrease of M with increasing temperature in a qualitative agreement with the mean-field result presented in Sect.4.4.1 [4.12, 92]. They observed that the SDW amplitude at absolute-zero temperature stays constant under pressure although T_{SDW} is largely suppressed [4.92]. This is a feature in good agreement with the mean-field result presented in Fig.4.8.

The mean-field result discussed in Sect.4.4.1 suggusts that the band structure in the SDW state at finite temperatures is semimetallic since the lower and upper bands overlap when M has not grown larger than ϵ_0 [4.23]. This is true also in the case of a realistic multiple-transverse-transfer model when M stays smaller than $|2t_b'| = |2\tau_{cos} - \epsilon_0|$ at finite temperatures. The presence of such a semimetallic state was verified by transport studies on $(TMTSF)_2 AsF_6$ and $(TMTSF)_2 PF_6$ under pressure [4.93].

The theories for the magnetic-field-induced SDW in Chap.9 suggest that even if the Fermi-surface nesting is incomplete so that T_{SDW} is lower than $T_{SDW}^{(max)}$ of the complete nesting, the applied magnetic field recovers the T_{SDW} to $T_{SDW}^{(max)}$. Actually T_{SDW} of $(TMTSF)_2 PF_6$ increases up to ≈ 13 K under high magnetic fields. *Montambaux* et al. obtained the result that an increase of T_{SDW} is proportional to H^2 with the proportionality coefficient related with the denesting parameter $t'_b = \tau_{cos} - \epsilon_0/2$ in (4.54) [4.66].

Biskap et al. made measurements of J_{SDW} of $(TMTSF)_2PF_6$ under pressure and verified the imperfect nesting theory [4.94]. They further confirmed the H^2-dependent increase of T_{SDW} in an applied magnetic field. *Danner* et al. determined this coefficient experimentally for $(TMTSF)_2PF_6$ [4.95]. The obtained value of $t'_b = 4.5$ K for ambient pressure is in qualitative agreement with the value obtained from band-calculation data in Table 4.3, although smaller by a factor of ≈ 2.

By closely analysing the proton NMR lineshape in $(TMTSF)_2PF_6$ near $T_{SDW} = 12.1$ K, *Clark* et al. showed that the metallic and SDW phases coexist with a discontinuous jump in the order parameter at the transition [4.96]. This is evidence of the first-order nature of the transition although the deviation from the second-order nature is small. This suggests that there is some interaction energy other than that taken in our mean-field scheme such as the commensurability energy, coupling to another order, etc. This may be related with the observation that slightly defected samples of $(TMTSF)_2PF_6$ have higher T_{SDW} than the clearest samples [4.97].

Recently, *Pouget* et al. observed very weak $2k_F$ and $4k_F$ satellite X-ray reflections in the SDW state of $(TMTSF)_2PF_6$ [4.98]. *Kobayashi* et al. advanced a mean-field theory that shows such a coexistence of $2k_F$ SDW and CDW to occur since the TMTSF column is dimerised and coupled to the longitudinal phonons [4.99].

4.6.4 SDW in $(TMTTF)_2Br$, $(DMET)_2Au(CN)_2$, $(ET)_2MHg(SCN)_4$

The Orsay group found SDW in $(TMTTF)_2Br$ [4.100]. They reported that it becomes superconducting at a pressure higher than 26 kbar [4.101]. Therefore, this material lies in between $(TMTSF)_2X$ and $(TMTTF)_2PF_6$ in the physical properties; the latter is a spin-Peierls insulator at low temperatures. The SDW wave vector was found to be $\mathbf{Q} = (\frac{1}{2}, \frac{1}{4}, 0)$, commensurate but very close to that of $(TMTSF)_2PF_6$ [4.102]. Concerning the other $(TMTTF)_2 \cdot X$ salts, see Sect.3.1.7 and Table 3.2.

The ET molecule was also found to be able to produce the SDW state in β-$(ET)_2ICl_2$ according to the ESR study of the Argonne group [4.103].

The asymmetrical molecule DMET, a derivative of TTF shown in Fig.1.1, yields a variety of organic conductors $(DMET)_2X$ having properties similar to those of 1-D $(TMTTF)_2X$ to κ-$(ET)_2X$, as will be described in Chap.6. $(DMET)_2Au(CN)_2$ was concluded to have the SDW transition at 25 K from the spin susceptibility, proton NMR linewidth and relaxation measurements, etc. [4.104]. The salts with X = PF_6 and BF_4 were found to have very similar temperature dependences of the spin susceptibility, starting to fall around 25 K, although the temperature dependences of the resitivity are different. Another group of salts $(DMPD)_2 AsF_6$ and $(DMPD)_2$ $\cdot PF_6$ were reported to have a similar behavior of the spin susceptibility and the divergence on the NMR linewidth around 20 K, evidencing an antiferromagnetic nature of the transition [4.105]. Here DMPD is a derivative of DMET whose ethylene group is replaced by propylene. Another asymmetric molecule MDT-TTF, depicted in Fig.1.1, was verified to yield an SDW salt $(MDT-TTF)_2 Au(CN)_2$ with $T_{SDW} \approx 20$ K [4.106].

α-$(ET)_2 MHg(SCN)_4$ with M = K, Rb, Tl was reported to have a kind of the density wave transition at about 8 K [4.107]. Spin susceptibility [4.108] and μSR [4.109] measurements had suggested it to be an SDW transition. In this kind of ET salts there are a pair of flat planes and a cylinder as its Fermi surface. The flat planes are considered to vanish at the 8 K transition. These materials exhibit complicated oscillatory transport and magnetization properties, which will be described in Sect. 9.6.3.

Another material reported to have SDW is β''-$(ET)_2 AuBr_2$ [4.110]. This also has planar and cylindrical pieces of the Fermi surfcae and shows interesting oscillations of resistivity.

5. ET Salts: Quasi Two-Dimensional Systems

A second family of organic materials that exhibit superconductivity was found in ET salts, ET representing bisethylenedithio-tetrathiafulvalene (alternatively abbreviated as BEDT-TTF). Its chemical structure is displayed in Fig.1.1. These compounds have characteristics quite different from the first generation of superconductors, $(TMTSF)_2 X$. The ET molecules form salts with various monovalent anions, but their compositions are remarkably varied, as represented by the shorthand $(ET)_m X_n$. However, most of the ET superconductors are of the 2:1 composition ratio like $(ET)_2 X$, although their crystal structures are of a large variety. Among the ET salts, there are several quasi one-dimensional conductors, but most of them are already non-metallic at room temperature. The salts exhibiting metallic behavior down to low temperatures are mostly quasi two-dimensional.

In this chapter we first present the various characteristics of ET salts, such as the crystal and electronic structures, electrical and optical properties, etc. in the normal state. Then, in Sects.5.2 and 3 the superconducting properties of those salts with monovalent linear and polymeric anions, respectively, that extend the T_c limit of organic superconductors to higher temperatures will be discussed. Other types of ET and ET-derivative superconductors are presented in Sects.5.4 and 5.5 with the emphasis on their unique features. With regard to characteristics of superconductivity, some of the ET salts exhibit typical two-dimensional behavior.

5.1 ET Charge-Transfer Salts

5.1.1 Molecular Stacking

For a quasi one-dimensional system such as $(TMTSF)_2 X$, the dimensionality can partly be raised by introducing side-by-side interactions between the one-dimensional columns of TMTSF molecules. This can be accomplished by applying pressure to bring the columns into closer contact with one another. One can also increase the number of adjacent atoms which supply π-electrons in the plane of the molecule so that the side-by-side overlap integrals between adjacent molecules are increased. On the basis of this idea,

Saito et al. developed charge-transfer salts based on bisethylenedithio-tetra-thiafulvalene $[(CH_2)_2]_2 C_6 S_8$ (in short either ET or BEDT-TTF) as an electron donor molecule to form two-dimensional conductors [5.1]. The molecule is formed by adding rings containing S atoms to the outer ends of the TTF molecule. The structure of ET is depicted in Fig.1.1a.

The neutral ET is nonplanar, twisted at the central double C=C bond. When it forms the charge-transfer complex, the molecule fairly untwists and the packing density is increased. The central fragment of two C and four S atoms flattens to a plane but the outer rings remain nonplanar. The end $-(CH_2)_2-$ groups can take on one of two degenerate out-of-plane conformations (Fig.5.1) for β-$(ET)_2 I_3$, as will be described in Sect.5.2.4. Thus, the overlap integrals within the stacks (or columns) are partly reduced while the intercolumn molecular interaction becomes appreciable. Some of the intercolumn S...S distances become shorter than the intracolumn molecular distances, leading to an increased dimensionality.

Saito et al. [5.1] synthesized the first metallic ET-compound $(ET)_2 \cdot ClO_4 (TCE)_{0.5}$, TCE denoting 1,1,2-trichloroethane $(C_2 H_3 Cl_3)$ used as a solvent in crystal growth. The crystal and the stacking structures of ET molecules are depicted in Fig.5.2. In this crystal ET molecules look stacked along the [102] direction. However, larger side-by-side intercolumn interactions due to strong S...S overlaps are induced along the [−102] direction. In fact, the conductivity ratio $\sigma_{\parallel}/\sigma_{\perp}$ becomes 0.4 at 3 K, $\sigma_{\parallel}$ and $\sigma_{\perp}$ being the conductivities parallel and perpendicular to the stacking axis, respectively. As a result the compound retains metallic conductivity down to 16 K, below which the resistivity slightly increases. The absence of the apparent metal-insulator transition is attributed to the increase of the intercolumn interaction, i.e., an increase in the dimensionality.

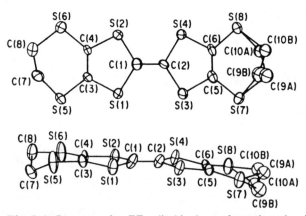

Fig. 5.1. Geometry of an ET radical in the conformation taken in a charge-transfer complex (hydrogen atoms are not shown). The C(9)-C(10) ethylene group can have two different positions corresponding to C(9A)-C(10A) and C(9B)-C(10B). From [5.2]

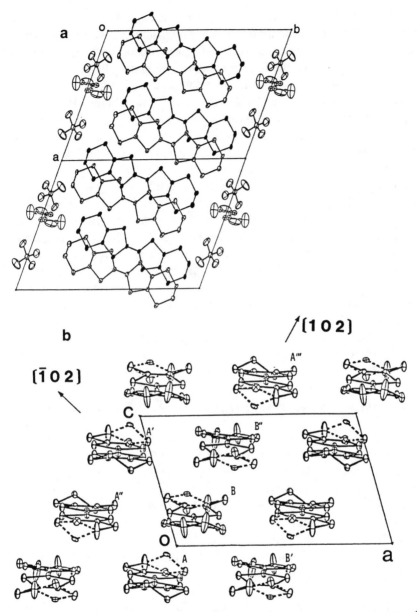

Fig. 5.2a, b. Structure of $(ET)_2 ClO_4 (TCE)_{0.5}$. Crystal structure viewed along the c^* axis. Crystallographically independent two ET molecules are distinguished by open and filled marks (**a**). Arrangement of ET molecules in the (010) plane (**b**). Courtesy of H. Kobayashi. From [5.3]

The nonplanar structure of ET together with the large thermal vibration of -CH$_2$- acts against building a good π overlap along the stacking axis. The S...S network increases the intercolumn interaction and these two competing factors determine the dimensionality of the crystal. Taking a specific example, β-$(ET)_2 PF_6$ exhibits quasi one-dimensionality with a conductivity ratio between the highest and second-highest conductive axes as high as 50. However, in this case the largest conductivity is not along the stacking axis but rather in the direction of the S...S intercolumn network [5.4].

The ET compounds have a wide variety of compositions. The valence of ET, and therefore the charge-transfer ratio, can be altered rather easily. For example, in $(ET)_m (ReO_4)_n$ the molecular ratio of ET to ReO_4 can be either 2:1 or 3:2. Furthermore, even for a particular molecular composition such as $(ET)_3 (ReO_4)_2$, three types of crystal morphologies (α-, β-, and γ-types) are known to exist [5.5]. The 2:1 salt $(ET)_2 ReO_4$ is described at some length in Sect.5.4.1. It exhibits superconductivity under a pressure of 4 kbar with $T_c = 2$ K. In this salt, planes of ET molecular stacks are separated by planes of ReO_4 anions as in $(TMTSF)_2 ReO_4$. There are, however, important differences between $(ET)_2 ReO_4$ and $(TMTSF)_2 ReO_4$, in addition to the nonplanarity of ET molecules. In the former the anions are ordered at room temperature, but not in the latter. In the primitive unit cell conventionally chosen for $(ET)_2 X$, the anion ordering can be described by the wave vector $Q = (0, \frac{1}{2}, \frac{1}{2})$. This is the same anion arrangement which drives the insulating transition in $(TMTTF)_2 SCN$ below 160 K (Table 3.2), although $(ET)_2 ReO_4$ is metallic.

5.1.2 Crystal Structures of Salts with Linear or Polymeric Anions

From the viewpoint of superconducting systems, the $(ET)_2 X$ salts with a linear or polymeric anion X^- such as I_3^-, or $Cu(NCS)_2^-$, are the most interesting ones. The crystals can be grown with a variety of morphologies, as mentioned above. The crystal data for four crystal structures of $(ET)_2 I_3$ are listed in Table 5.1. Among them, α-$(ET)_2 I_3$ exhibits two-dimensional metallic behavior from room temperature down to 135 K, where it undergoes a metal-insulator transition [5.6]. The arrangement of ET molecules in a α-$(ET)_2 I_3$ crystal is depicted in Fig.5.3. The other three crystals in Table 5.1 exhibit superconductivity at ambient pressure.

The β-$(ET)_2 I_3$ salt is the second ambient pressure-superconductor that was found with $T_c \simeq 1.5$K [5.10]. The linear stack alignment of the ET molecules in β-$(ET)_2 I_3$ is shown in Fig.5.4. The ET cation radicals are packed face-to-face, as it is typical for quasi one-dimensional organic crystals. The average ET planes are approximately parallel to the $(0, 2, 2)$ plane with a dihedral angle of $7°$ for adjacent molecules. It should be noted here that all the intracolumn S...S distances are longer than both the shortest intercolumnar separation (3.568Å) and that the sum of the van der Waals radii of 3.70 Å: the average interplanar ET-ET distances in the columns are

Table 5.1. Crystal data for different morphologies of $(ET)_2 I_3$

	α-type	β-type	θ-type	κ-type
a [Å]	9.211	6.615	10.076	16.387
b [Å]	10.850	9.097	33.85	8.466
c [Å]	17.488	15.291	4.964	12.832
$\alpha°$	96.95	94.35	90	90
$\beta°$	97.97	95.55	90	108.56
$\gamma°$	90.75	109.75	90	90
V [Å^3]	1690.3	855.9	1693	1687.6
Space group	P$\bar{1}$	P$\bar{1}$	Pnma	P2$_1$/c
Z	2	1	2	2
Reference	5.6	5.7	5.8	5.9

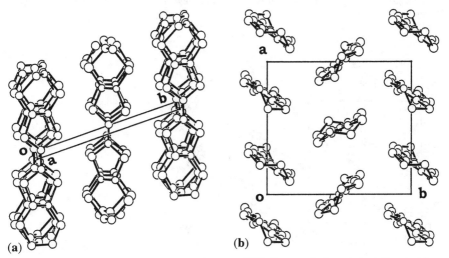

Fig. 5.3. Arrangement of ET molecules in an α-$(ET)_2 I_3$ crystal viewed along the stacking axis (**a**), and along the axis of the ET molecule (**b**). Courtesy of H. Yamochi

equal to 3.75 Å for I-II molecules and 3.86 Å for I-III (Fig. 5.4a). Thus, the stacks of cation radicals form sheets parallel to the ab-plane with the centrosymmetric linear anions I_3^- separating them by forming an insulating layer.

The θ-$(ET)_2 I_3$ salt also consists of linear stacks (Fig. 5.5). The data given in Table 5.1 are averaged ones. The crystal essentially belongs to the monoclinic system [5.8].

In contrast, the κ-$(ET)_2 I_3$ salt reveals a quite different packing from the traditional one-dimensional system. Figure 5.6 shows one of the ET sheets, projected in the direction perpendicular to the long axis of ET. Pairs of ET molecules contact face-to-face, but adjacent pairs are almost orthog-

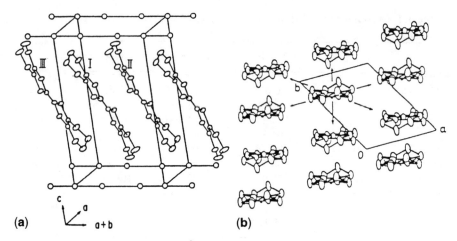

(a)

(b)

Fig. 5.4a,b. Crystal structure of β-(ET)$_2$I$_3$. Arrangement of ET molecules within a column (**a**). Stacking arrangement of ET molecules viewed from the molecular long axis (**b**). From [5.13]

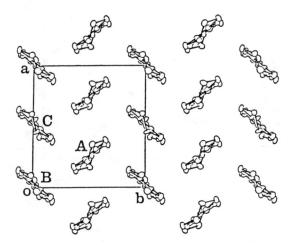

Fig. 5.5. Arrangement of ET molecules of θ-(ET)$_2$I$_3$. Courtesy of H. Kobayashi. From [5.8]

onally aligned. Like in the other configurations, the crystal consists of sheets of ET molecules and sheets of I$_3$ anions alternating along the a-axis.

The ET molecule also yields salts with I$_3$ in other composition ratios: γ-(ET)$_3$(I$_3$)$_{2.5}$ with the space group Pbnm and ϵ-(ET)$_2$I$_3$(I$_8$)$_{0.5}$ with the space group P2$_1$/c [5.11]. The γ-salt is only characterized by an averaged structure. In these crystals part of the I$_3$ anions form a separating layer while others are dispersed within the ET layer. We will not go into the details of these compounds because the nature of superconductivity has not yet

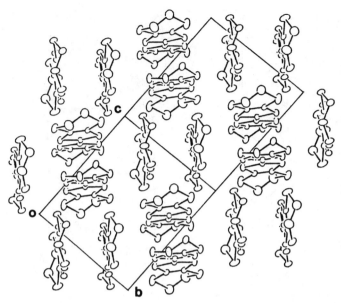

Fig. 5.6. Arrangement of ET molecules of κ-$(ET)_2 I_3$. Courtesy of H. Kobayashi. From [5.9]

been understood, except for the observation that the γ-type salt undergoes a superconducting transition at 2.5 K.

The crystal structures of κ-$(ET)_2 Cu(NCS)_2$ and κ-$(ET)_2 Cu[N\cdot(CN)_2]Cl$, with T_c reaching 10 K, are illustrated in Figs. 5.7 and 8, respectively. The molecular packing pattern is analogous to that in κ-$(ET)_2 I_3$: Two ET molecules are paired with their central tetrathioethylene planes almost parallel, and adjacent pairs are almost perpendicular to one another in the bc-plane; this results in a donor sheet. For κ-$(ET)_2 Cu(NCS)_2$ the insulating layer is formed by arrays of V-shaped $Cu(NCS)_2$ anions. The anion arrays exhibit the characteristic features: The anion molecules are weakly bonded to one another through the Cu...S interaction, so as to form a planar polymer-like structure, represented as $[-(SCN)Cu(NCS)-]_n$, which lacks a center of symmetry. The structure belongs to the $P2_1$ space group. The lattice parameters of the κ-type salts with polymeric anions are listed in Table 5.2.

The crystal structure and the arrangement of the ET molecules of α-$(ET)_2 RbHg(SCN)_4$ are displayed in Fig. 5.9 and the lattice parameters of α-$(ET)_2 MHg(XCN)_4$ (M: NH_4, K, Rb, Tl; X: S, Se) are compiled in Table 5.3.

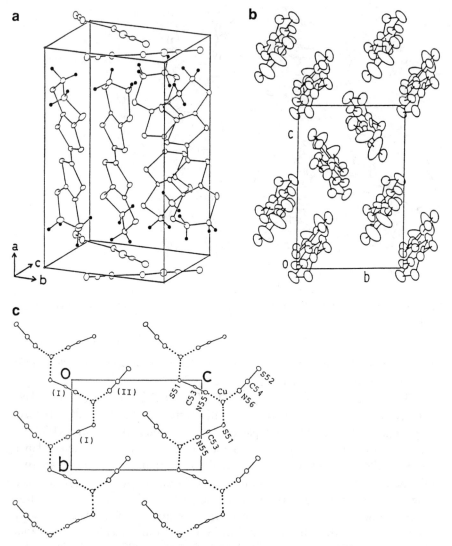

Fig. 5.7. Crystal structure of κ-(ET)$_2$ Cu(NCS)$_2$ (**a**). Arrangement of ET molecules (**b**) and Cu(NCS)$_2$ molecules (**c**). Courtesy of H. Urayama. From [5.14]

5.1.3 Band-Structure Calculation

As mentioned in connection with TMTSF salts, the intermolecular π-π interactions in the stacking direction dominate the electronic structure which induces a quasi one-dimensional band. On the other hand, for ET compounds the steric conformation of the end ethylene groups prevents close face-to-face stacking of the molecules. Thus, the strength of the side-by-

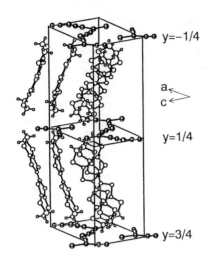

Fig. 5.8. Crystal structure of κ-$(ET)_2 Cu \cdot [N(CN)_2]Cl$. Care must be taken that the frame does not correspond to a unit cell. Ordinate axes represent parts of crystalline b-axis. The anion layers are parallel to ac planes which are located at $y = -1/4$, $1/4$ and $3/4$. Courtesy of H. Yamochi

side interaction becomes comparable to that of the face-to-face interaction. Incidentally, we should mention an important point with regard to the transverse transfer energy. If multiple transverse transfer energies have the same sign, the effective total transverse energy becomes large, but their signs change with respect to the direction of close contacts between neighboring molecules. In β-$(ET)_2 X$, important transverse transfer energies take the same sign, making the transverse bandwidth comparable to the longitudinal one (Sect. 4.3).

In order to understand the role of the molecular arrangement in the electronic structure, *Mori* et al. [5.20], carried out quantum-chemical calcu-

Table 5.2. Crystal data for κ-type superconductors with polymeric anions at room temperature

	Structure	a [Å]	b [Å]	c [Å]	β [Å]	v	z	Ref.
κ-$(ET)_2 Cu(NCS)_2$	Monocline P2$_1$	16.248	8.440	13.124	110.30	1688.0	2	5.14
κ-$(ET)_2 Cu(CN)[N(CN)_2]$	Monocline P2$_1$	16.00	8.631	12.90	110.97	1663	2	5.15
κ-$(ET)_2 Cu[N(CN)_2]Br$	Orthorhombic Pnma	12.949	30.016	8.539	–	3317	4	5.16
κ-$(ET)_2 Cu[N(CN)_2]Cl$	Orthorhombic Pnma	12.977	29.977	8.480	–	3299	4	5.16
κ-$(ET)_2 Cu[N(CN)_2]I$	Orthorhombic Pnma	12.928	30.356	8.683	–	3408	4	5.16

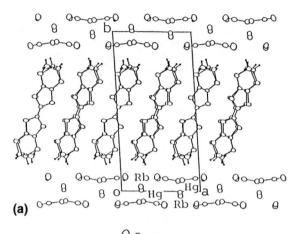

(a)

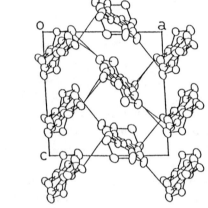

(b)

Fig. 5.9. Crystal structure of α-(ET)$_2$ RbHg(SCN)$_4$ (**a**), and stacking arrangement of ET molecules viewed from the molecular long axis (**b**). From [5.18]

Table 5.3. Crystal data for α-(ET)$_2$ MHg(XCN)$_4$ (M: NH$_4$, K, Rb. Tl; X: S, Se) at room temperature. Each crystal has a triclinic structure of the P$\bar{1}$ type

	a [Å]	b [Å]	c [Å]	α [deg]	β [deg]	γ [deg]	v [Å]	z	Ref.
α-(ET)$_2$ NH$_4$ Hg(SCN)$_4$	10.091	20.595	9.963	103.65	90.53	93.30	2008.1	2	5.18
α-(ET)$_2$ KHg(SCN)$_4$	10.082	20.565	9.933	103.70	90.91	93.06	1997.0	2	5.18
α-(ET)$_2$ RbHg(SCN)$_4$	10.050	20.566	9.965	103.57	90.57	93.24	1998.5	2	5.19
α-(ET)$_2$ TlHg(SCN)$_4$	10.051	20.549	9.934	103.63	90.48	93.27	1990	2	5.17
α-(ET)$_2$ TlHg(SeCN)$_4$	10.105	20.793	10.043	103.51	90.53	93.27	2047.9	2	5.17

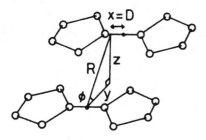

Fig. 5.10. Alignment of a molecular dimer as a function of ϕ: $\phi = 0°$, side-by-side arrangement; $\phi = 90°$, face-to-face arrangment. From [5.20]

lations. They computed the intermolecular overlap between the adjacent molecules including the limiting cases of face-to-face and side-by-side contacts for pairs of TTF as well as ET molecules. In these calculations the relative rotation of molecules within each pair is neglected. The respective arrangement is depicted in Fig. 5.10, where R stands for the shortest distance, connecting the long axes, of two molecules, D is the molecular slipping distance, and ϕ is the angle between the molecular plane and the direction of the adjacent molecules.

The overlap integral S of the Highest Occupied Molecular Orbitals (HOMO) has been calculated as a function of D and ϕ for R = 3.80 Å. The main factor affecting D is the core repulsion between the two molecules. The ϕ dependence of the overlap integrals is depicted in Fig. 5.11 for both TTF and ET pairs at D = 1.60 Å The largest overlap at $\phi = 90°$ corresponds to the normal face-to-face stacking. At $\phi = 0°$ the interaction is

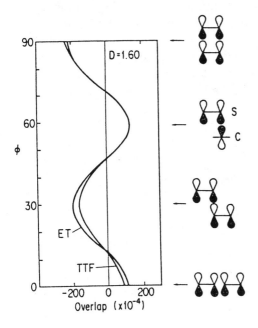

Fig. 5.11. Overlap integrals of HOMO between two molecules of TTF and ET as a function of ϕ. In the right-hand side, the HOMOs viewed along the long axes of the molecules are shown. From [5.20]

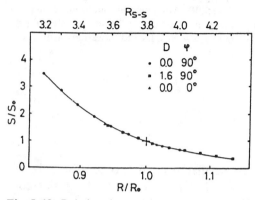

Fig. 5.12. Relative change of the overlap integral S as a function of the intermolecular sulfur-to-sulfur distance R, compared with the overlap S_0 at $R_0 = 3.80\,\text{Å}$, at three different values of D and ϕ. From [5.20]

due to a π-like overlap of the p_z (and d) orbitals. The peak around $\phi = 30°$ corresponds to a σ-like interaction which mainly comes from the components of the p_z orbital projected parallel to the bonding direction. The peak at $\phi = 60°$ mainly derives from the overlap between the central carbon and sulfur. Figure 5.12 reveals that the $R = 3.80\,\text{Å}$ estimate is reasonable to test the behavior of the overlap integral since its R dependence is similar for various ϕ.

From the overlap integral S one can deduce the transfer energy t according to (2.16), where the proportionality constant is on the order of the energy of HOMO, ca. 10 eV. Then by using the standard tight-binding approximation, the structure of the conduction band and the shape of the Fermi surface are obtained.

The calculated band structure of β-(ET)$_2$I$_3$ is shown in Fig. 5.13. When two electrons are removed from a unit cell consisting of ET's, the conduction band formed by HOMO of ET is half-filled, and the electronic state is

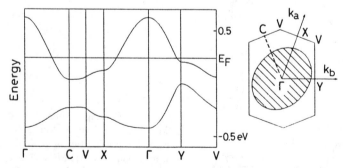

Fig. 5.13. The energy-band structure and Fermi surface of β-(ET)$_2$I$_3$. The shaded region indicates the hole-like part. From [5.21]

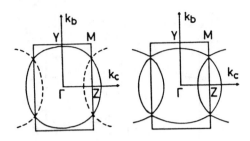

Fig. 5.14. Fermi surfaces of κ-$(ET)_2 I_3$ (*left*) and κ-$(ET)_2 Cu(NCS)_2$ (*right*). Courtesy of H. Mori

expected to be metallic. In addition, reflecting the rather isotropic interactions within the ab-plane, the Fermi surface is closed within the first Brillouin zone. This is consistent with the two-dimensional nature of this material.

By computing the band structure of β-$(ET)_2 PF_6$ in a similar way, *Mori* et al. [5.22] have demonstrated that in this case the Fermi surface is not closed, and the material shows quasi one-dimensional behavior in the side-by-side packing direction, as described in the previous subsection. In fact, this salt undergoes a Peierls transition at 297 K.

Tight-binding band calculations using HOMOs were also performed for κ-$(ET)_2 I_3$ and κ-$(ET)_2 Cu(NCS)_2$ by *Kobayashi* et al. [5.23] and by *Mori* et al. [5.24], respectively. The Fermi surfaces are depicted in Fig. 5.14. For κ-$(ET)_2 I_3$ the Fermi surface is closed but intersects the zone boundary in the k_c-direction. For κ-$(ET)_2 Cu(NCS)_2$, the surface is similar to κ-$(ET)_2 I_3$ except that it opens a gap on the zone boundary due to the lack of a center of symmetry. Accordingly, an open Fermi surface and a closed hole pocket

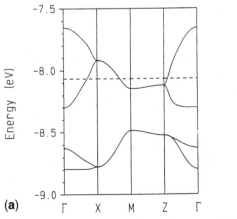
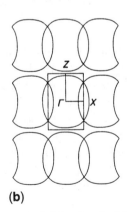

(a)   **(b)**

Fig. 5.15. Dispersion relations of the two highest occupied bands calculated for a single donor-molecule layer of κ-$(ET)_2 Cu[N(CN)_2]Cl$, where $\Gamma = (0, 0)$, $X = (a^*/2, 0)$, $z = (0, c^*/2)$, $M = (a^*/2, c^*/2)$ (**a**), and Fermi surface associated with the half-filled band (**b**). From [5.16]

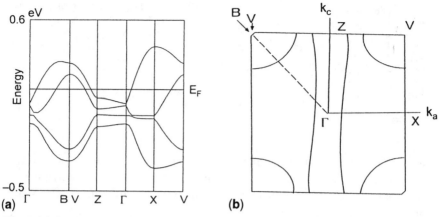

Fig. 5.16. Band structure (**a**), and Fermi surface (**b**) of α-(ET)$_2$ NH$_4$ Hg(SCN)$_4$. Courtesy of H. Yamochi

are formed. The energy-band structures and the Fermi-level trajectories of κ-(ET)$_2$ Cu[N(CN)$_2$]Cl and α-(ET)$_2$ NH$_4$ Hg(SCN)$_4$ are illustrated in Figs. 5.15 and 16, respectively.

The tight-binding band calculation for (ET)$_2$ ReO$_4$ reveals that the band structure contains both one-dimensional and two-dimensional Fermi surfaces [5.25]. The two-dimensional Fermi surface is nearly isotropic while the one-dimensional Fermi surface is open along the interstack direction. Thus, the anisotropic behavior observed in the conducting plane of (ET)$_2$ ReO$_4$ [5.5] is probably dominated by the one-dimensional band.

5.1.4 Measurement of the Fermi Surface

In order to obtain experimental information on the two-dimensional Fermi surface, magneto-quantum oscillation measurements have been carried out. When a strong magnetic field is applied in the direction penetrating the two-dimensional conducting plane at low temperatures, the elctron energies are quantized into Landau levels. The periodic oscillations in the magnetization (**de Haas-van Alphen effect**) and the magetoresistance (**Shubnikov-de Haas effect**) arise from the periodic oscillation of the density of states at the Fermi level E_F as a function of the inverse of the magnetic field, $1/H$. From the oscillation period $\Delta(1/H)$, the extremal area of the cross section of the surface S_F is derived from

$$S_F = \frac{2\pi e}{\hbar c} \frac{1}{\Delta(1/H)} , \qquad (5.1)$$

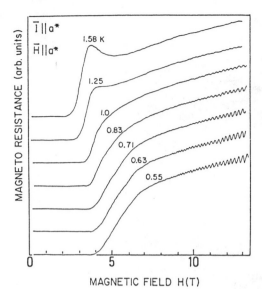

$\overline{I} \parallel a^*$

$\overline{H} \parallel a^*$

1.58 K

1.25

1.0

0.83

0.71

0.63

0.55

MAGNETO RESISTANCE (arb. units)

MAGNETIC FIELD H(T)

Fig. 5.17. Shubnikov-de Haas oscillation for κ-$(ET)_2 Cu(NCS)_2$, where the longitudinal magnetic field is applied perpendicular to the conducting plane ($\mathbf{H} \parallel \mathbf{a}$). The sample is superconducting at low fields. The oscillation is visible in the normal state below 1K. From [5.24]

where $-e$ and c are the electron charge and the light velocity, respectively. The Shubnikov-de Haas oscillation was found in the organic conductor κ-$(ET)_2 Cu(NCS)_2$ (Fig. 5.17) by *Oshima* et al. in 1988. Following this, similar oscillations have been noticed for various ET salts. Figure 5.18 depicts an example of the de Haas-van Alphen effect, together with the Shubnikov-de Haas signal in κ-$(ET)_2 I_3$. It has been confirmed that the cyclotron effective mass is proportional to $1/\cos\theta$, θ being the angle between the direction of the magnetic field and the normal to the conducting plane, reflecting the two-dimensionality of the elctronic state. The measured $\Delta(1/H)$ and the deduced Fermi-level cross-section area S_F for various ET salts are listed in Table 5.4, as the ratio S_F/S_B, S_B being the cross section of the basal plane of the first Brillouin zone. The different values of S_F/S_B correspond to different closed orbits. In most cases the observed S_F agrees fairly well with the calculation based on the extended Hückel method. Concerning a detailed description about the Fermi-surface study, readers are referred to the monograph by *Wosnitza* [5.40].

From the temperature and the magnetic-field dependence of the oscillation amplitude D, one can derive the cyclotron effective mass m_c and the Dingle temperature T_D, which represents the degree of scattering by sample imperfections, through the relation

$$D \propto \frac{T}{\sqrt{H}} \exp\left(\frac{-\lambda m_c T_D}{m_e H}\right) \sinh\left(\frac{\lambda m_c T}{m_e H}\right), \qquad (5.2)$$

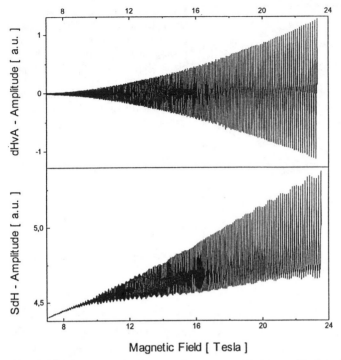

Fig. 5.18. Shubnikov-de Haas and de Haas-van Alphen oscillations for κ-(ET)$_2$I$_3$. From [5.26]

where $\lambda = 2\pi^2 m_e ck_B /e\hbar$, m_e and c denoting the electron mass and the velocity of light, respectively.

In addition to the magneto-quantum oscillation for which the orbital quantization is essential, a new type of oscillation involving the direction of the applied magnetic field with respect to the conducting plane has been found in a strong magnetic field at low temperatures, as depicted in Fig. 5.19 for β-(ET)$_2$IBr$_2$ and in Fig. 5.20 for θ-(ET)$_2$I$_3$. The magnetoresistance oscillates regularly as the constant magnetic field is inclined from the normal to the conducting plane. The angle under which the magnetoresistance is minimum, is well described by the relation $\tan\theta = sN$ (s = 0.39, N = 0, 1, 2,...), θ being the angle between the c* axis and the magnetic field tilted toward the a-axis in the case of Fig. 5.20.

These **Angle-dependent MagnetoResistance Oscillations** (AMRO) are explained in terms of semiclassical electron orbits in the reciprocal lattice in a magnetic field [5.43] (Appendix A). When the Fermi surface of the cylindrically closed form has a weakly modulated cross section along its axis, the distribution of the orbital areas at the Fermi energy has a finite width. When the magnetic field is inclined, this width vanishes at certain

Table 5.4. Shubnikov-de Haas or de Haas-van Alphen parameters. S_F and S_B are the cross-sectional areas of the Fermi surface and the basal plane of the first Brilluin zone, respectively

	$\Delta(1/H)$ $[10^{-5}/\text{kOe}]$	S_F/S_B [%]	m_c/m_e	T_D [K]	Ref
$\beta_H\text{-}(ET)_2I_3$	2.65	51	4.65	0.53	5.27
$\beta\text{-}(ET)_2IBr_2$	2.60	51	4.2	1.1	5.28
$\kappa\text{-}(ET)_2I_3$	17.5	15	1.9		5.26
	2.57	102	3.9		
$\kappa\text{-}(ET)_2Cu(NCS)_2$	16.7	15.7	3.5	0.5	5.29
	2.55	105	6.5		
$\kappa\text{-}(ET)_2Cu[N(CN)_2]Br$	64.2	4.4	0.95	3.5	5.30
under 9 kbar	2.63	106	6.4		
$\kappa\text{-}(ET)_2Cu[N(CN)_2]Cl$	16.9	15.5			5.31
under 7.7 kbar	2.51	102			
$\kappa\text{-}(ET)_2Cu_2(CN)_3$	83.3	3.1			5.32
under 8 kbar	2.64	96	4.5		
$\kappa\text{-}(ET)_2Ag(CN)_2H_2O$	14	17	2.7	0.3	5.33
$\theta\text{-}(ET)_2I_3$	12.8	19	1.8		5.34
	2.36	102	3.5		
$\alpha\text{-}(ET)_2NH_4Hg(SCN)_4$	18	13	2.1	1.4	5.35
$\alpha\text{-}(ET)_2KHg(SCN)_4$ *	15	16	1.4	4.0	5.35
	17.6	13	2.53	0.6	5.26
$\alpha\text{-}(ET)_2RbHg(SCN)_4$ *	14.7	16.5	1.5		5.37
$\alpha\text{-}(ET)_2TlHg(SCN)_4$	15.1	16	1.5		5.38
	2.34	103	4.0		
$\alpha\text{-}(ET)_2TlHg(SeCN)_4$	15.3	16	1.8	6	5.39

* For fields higher than 200 kOe

angles, leading to a complete discretization of the eigenenergies into Landau levels and to an increase of the magnetoresistance at low temperatures: This is due to the vanishing of the group velocity along the field direction. These angles are in agreement with the observed peak angles of the magnetoresitance. AMRO can explicitly be obtained in resistivity tensor calculations for a quasi two-dimensional metal [5.44]. Taking into account an interlayer hopping integral, the conductivity tensors are calculated as functions of the angle of rotation of the constant magnetic field direction (Appendix A). Based on this type of analysis, the transverse cross section of the Fermi surface of $\beta\text{-}(ET)_2IBr_2$ is obtained, as it appears in Fig. 5.21.

The transverse cross section of the Fermi surface of $\kappa\text{-}(ET)_2Cu\cdot[N(CN)_2]Cl$ under a pressure of 7.7 Kbar, at which superconductivity is

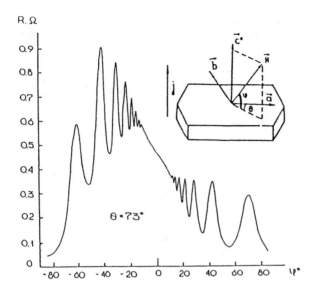

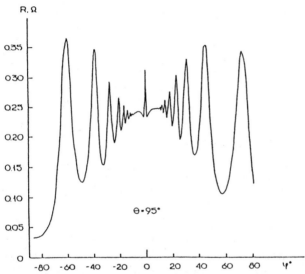

Fig. 5.19. Angule-dependent magnetoresistance oscillation (AMRO) in β-(ET)$_2$IBr$_2$: The angles φ and θ are shown in the inset [5.41]

suppressed, has been obtained by AMRO measurements (Fig.5.22). Similar AMRO studies were carried out for α-(ET)$_2$MHg(SCN)$_4$, (M: NH$_4$, K, Rb, Tl) and α-(ET)$_2$TlHg(SeCN)$_4$ [5.35, 44-47]. For α-(ET)$_2$NH$_4$Hg(SCN)$_4$ the AMRO data are consistent with the cross section of the cylidrical Fermi surface evaluated from the Shubnikov-de Haas effect. In the remaining salts, for example α-(ET)$_2$KHg(SCN)$_4$, the patterns of AMRO are modified dra-

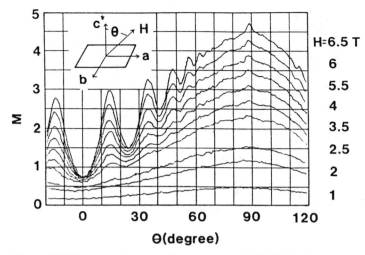

Fig. 5.20. Magnetoresistance against the magnetic field direction rotated in the ac* plane, as shown in the inset for θ-$(ET)_2 I_3$ at 4.2 K, where M represents $\Delta R/R$ (ΔR is the increment of the resistance R due to the magnetic field). From [5.42]

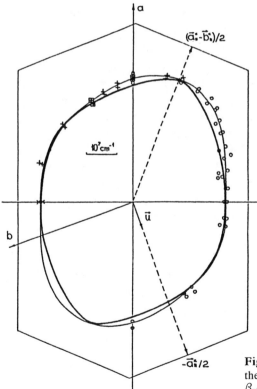

Fig. 5.21. Transverse cross-section of the Fermi surface (*thick line*) in β-$(ET)_2 IBr_2$ deduced from the AMRO. From [5.41]

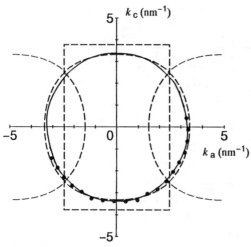

Fig. 5.22. Transverse cross-section of the Fermi surface (*solid line*) in κ-(ET)$_2 \cdot$ Cu[N(CN)$_2$]Cl under 7.7 kbar. The *dashed line* represents the Brillouin zone and the calculated Fermi surface contour. Note that the trajectory corresponding to the β orbit (Table 5.5) is seen. From [5.31]

matically with respect to a density-wave formation, presumably the antiferromagnetic ordering, as explained in Sect. 5.1.5.

Using a far-infrared laser to provide radiation with a large number of discrete energies in the range $1 \div 40$ meV and a magnetic field of 20 T, **cyclotron-resonance experiments** have been carried out by monitoring either transmission or reflectance signals for several kinds of ET salts. The results are listed in Table 5.5. It is surprising to see a large difference between the effective mass obtained from Shubnikov-de Haas oscillations and the cyclotron-resonance mass. The Shubnikov-de Haas effect yields an effective mass between $2.0m_e$ and $3.5m_e$ (m_e: electron mass), while the cyclotron resonance leads to low carrier masses between $0.4m_e$ and $1.2m_e$. *Singleton* et al. claimed that the difference between the transport and cyclotron masses is understood in terms of the **mass-enhancement effect** due

Table 5.5. Effective masses of carriers determined by cyclotron resonance, m_{CR}, and by the Shubnikov-de Haas effect, m_F. m_e deneotes the electron mass. From [5.48]

	m_{CR}/m_e	m_F/m_e
κ-(ET)$_2$Cu(NCS)$_2$	1.18	3.5
α-(ET)$_2$NH$_4$Hg(SCN)$_4$	1.17	2.5
α-(ET)$_2$KHg(SCN)$_4$	0.40	2.0
	0.94	2.4

to electron-electron interactions [5.48, 49]. The effective mass m_{CR} measured by cyclotron resonace in a three-dimensional translationally invariant electronic system is independent of an electron-electron interaction, yielding a band mass renormalized only by electron-phonon coupling, while the mass measured by the Shubnikov-de Haas effect is additionally influenced by electron-electron interactions. For α-$(ET)_2 KHg(SCN)_4$ two effective masses are reported; this is probably due to the magnetic-field-induced phase transition in this material. The lower mass is for the low-temperature phase with antiferromagnetic ordering. The higher effective mass is for the intermediate phase.

The band structure was also examined by measurement of the **thermo-electric power**. From room temperature to about 200 K, the thermopower of β-$(ET)_2 I_3$ in the a direction is positive and roughly proportional to the temperature, reflecting a metallic state with hole carriers [5.50]. In the b-direction it gradually becomes negative between 200 and 2 K. The anisotropic temperature dependence of the thermopower has been explained using the Boltzmann-equation method [5.51] by taking into account band calculations and by parametrizing the size of the transverse transfer energy in a reasonable range.

The electronic properties studied with regard to magneto-quantum oscillations and other transport studies show that the electronic phases are of a likely degenerate electron gas. However, the **photoemission** studies carried out on κ-$(ET)_2 Cu[N(CN)_2]Br$, κ-$(ET)_2 Cu(NCS)_2$ and β-$(ET)_2 I_3$ have reve-

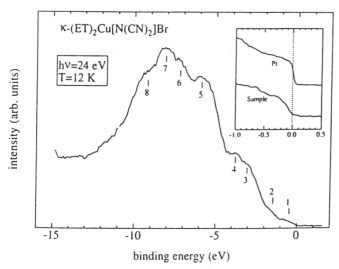

Fig. 5.23. Energy Distribution Curve (EDC) measured on κ-$(ET)_2 Cu[N(CN)_2]Br$ using $h\nu$ = 24 eV. The inset shows an EDC measured on κ-$(ET)_2 Cu[N(CN)_2]Br$ (*lower curve*) and an EDC measured on a platinum foil (*upper curve*) for the energy region near E_F. From [5.52]

aled that a metallic edge typical for metals is not found in any of the three compounds [5.52]. An energy-distribution curve measured for κ- $(ET)_2$ Cu·$[N(CN)_2]$Br at $h\nu$ = 24 eV is seen in Fig.5.23. In a metallic system such as Pt, the Fermi level lies at the midpoint of the rising edge, as shown in the inset. For κ-$(ET)_2$Cu$[N(CN)_2]$Br, the photoemission intensity is near zero at exactly E_F. A metallic edge is also lacking in κ-$(ET)_2$Cu$(NCS)_2$ and β-$(ET)_2$I$_3$. Electron-electron correlations and the excitation of phonons are considered to be the likely causes.

The momentum distribution of metals can be measured by **positron annihilation** experiments. Using a two-dimensional angular correlation of positron annihilation radiation, the measurement of the Fermi surface of κ-$(ET)_2$Cu$[N(CN)_2]$Br has been carried out [5.53]. However, the experimental results are not consistent with the prediction based on the extended Hückel molecular-orbital calculation. The discrepancy may be due to the neglect of positron wave-function effects in the interpretation of the experimental results.

5.1.5 Electrical Properties

ET salts have a wide variety of electrical properties ranging from insulation to superconductivity. In addition, the dimensionality changes from one to two, depending on the relative strength of the face-to-face and side-by-side interactions between adjacent ET molecules. In the charge-transfer process, electrons are transferred from ET molecules to form the counter anions. Thus, depending on the composition ratio, for example $(ET)_2$X, $(ET)_3$X$_2$, or (ET)X, X$^-$ denoting the counter molecule to form the monovatent anion, the valence of ET can vary from 0.5 to 1. For $(ET)_2$X, because each neutral ET molecule can accommodate two electrons in its HOMO, the $(ET)_2$ cation HOMO is 3/4-filled. When the ET molecules in the stack are dimerized and the conduction band is split into two parts, the highest occupied band is half-filled and the system can be metallic according to the band theory. However, if the electron-electron interaction is very strong due to the strength of the on-site Coulomb energy in comparison with the transfer energy, the electrons are localized on each site, and the system exhibits the so-called **Mott-Hubbard insulating state**. The effect of the on-site Coulomb energy and the dimer structure in $(ET)_2$X were suggested to give rise to the transition based on the Hartree-Fock approximation [5.54].

The first discovered metallic ET cation radical salt $(ET)_2$ClO$_4$(TCE)$_{0.5}$ exhibits a positive-temperature derivative of the resistivity down to 1.4 K without undergoing a distinct nonmetallic transition. The calculated band structure of this material has a two-dimensional closed Fermi surface

[5.20], which is consistent with the absence of a nonmetallic phase induced by Fermi-surface nesting.

In contrast to the $(TMTSF)_2 X$ salts, a series of $(ET)_2 X$ salts with varying X (PF_6, AsF_6, SbF_6) do not give a set of isomorphous systems. $(ET)_2 \cdot PF_6$ crystallizes into more than two forms, a triclinic α-type in the P1 space group and an orthorhombic β-type in the Pnna space group [5.55]. In both these crystal structures, however, the side-by-side interactions of the ET molecules are stronger than the face-to-face interactions. In fact, the conductivity ratio near 300 K is $\sigma_c : \sigma_a : \sigma_b \simeq 10^4 : 200 : 1$ for β-$(ET)_2 PF_6$, where the a-axis is the stacking direction, and the c-axis the side-by-side axis. The electrical conductivity decreases rapidly with temperature after reaching a maximum at 297 K, and the activation energy at low temperatures is about 0.2 eV. The thermoelectric power shows that the carriers are hole-like; it becomes negative below the metal-insulator transition. The α-type crystals are semiconducting below room temperature with an activation energy of about 0.1 eV.

Both $(ET)_2 AsF_6$ and $(ET)_2 SbF_6$ salts crystallize to a monoclinic system in the C_2/c space group [5.56]. In these cases the side-by-side intermolecular interaction is stronger than the face-to-face one. The electrical conductivity decreases upon cooling although an anomalous peak at $T_c = 264$ and 273 K for the $(ET)_2 AsF_6$ and $(ET)_2 SbF_6$ salts, respectively, is clearly seen in a plot of $d(\ln\sigma)/d(1/T)$ $(= -E_a/k_B)$ vs. T. The low-temperature extrapolated value of the gap is $E_a \approx 0.1$ eV. In accordance with the phase transition detected from measurements of the electrical conductivity, the spin paramagnetism χ_p evaluated by an integration of the ESR signal exhibits a strong decrease: χ_p decreases monotonically to 100 K where it is less than 1% of the room temperature value. Over the $100 \div 200$ K range, an estimate for the activation energy $(E_a \approx 0.1 eV)$ is derived by fitting χ_p, namely,

$$\chi_p = \frac{C}{T} \exp\left(\frac{-E_a}{k_B T}\right). \tag{5.3}$$

This value is in agreement with the estimated energy gap for the electrical conductivity cited above. Thus, although in the calculation the band is half-filled, the system is nonmetallic.

Most of the 2:1 ET cation radical salts with linear or polymeric anions, such as triiodide and $Cu(NCS)_2^-$, are of a two-dimensional structure, where a conducting layer formed by stacked ET molecules is alternately separated by an insulating layer of anions. The stacking patterns of the ET molecule are typically classified by α, β, κ, and θ types (Sect. 5.1.2). The conductivity is consistent with the two-dimensional nature, with the intralayer (in-

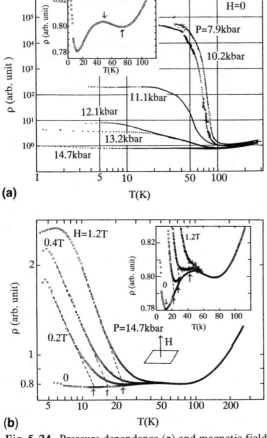

Fig. 5.24. Pressure dependence (**a**) and magnetic field dependence under 17.4 kbar (**b**) of the resistivity for α-(ET)$_2$I$_3$. From [5.57]

plane) conductivity being larger than the interlayer (out-of-plane) conductivity. The 2:1 ET salts produce various kinds of metals and ambient-pressure superconductors, but they include semiconducting and magnetically ordered phases as well.

As representatives of the α-type salts, we consider here α-(ET)$_2$I$_3$ and α-(ET)$_2$MHg(SCN)$_4$ (M: NH$_4$, K, Rb). The **α-(ET)$_2$I$_3$** salt undergoes a metal-insulator transition at $T_{MI} = 135$ K. With the pressure P, T_{MI} of α-(ET)$_2$I$_3$ decreases, as seen in Fig. 5.24a. Under a pressure of 14.7 kbar, the insulating phase is suppressed, but no indication of an onset of superconductivity is detected down to 0.1 K. It is considered that the electronic state for P > 12 kbar is semimetallic [5.57]. It is also noteworthy that the metal-insulator transition suppressed by pressure revives dramatically with the aid of a magnetic field (Fig. 5.24b). This phenomenon is understood as the

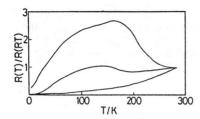

Fig. 5.25. Temperature dependences of the resistivity for three different α-(ET)$_2$NH$_4$Hg·(SCN)$_4$ samples. From [5.59]

opening of the energy gap by the magnetic field, which is due to an almost nested quasi one-dimensional Fermi surface [5.58].

The resistivity of α-(ET)$_2$NH$_4$Hg(SCN)$_4$ gives various temperature dependences (Fig.5.25), where the origin of the difference in the temperature dependence or of the resistance peak is not well understood, but presumably ascribed to crystal imperfections. Below 100 K, the resistivity decreases monotonically on cooling and becomes zero below 1.8 K [5.59, 60]. Similar temperature dependences are found for α-(ET)$_2$KHg(SCN)$_4$ and α-(ET)$_2$RbHg(SCN)$_4$. However, these salts do not undergo superconductivity transitions down to 0.3 K. Their resistivities exhibit a step-like change near 8 K (Fig.5.26) which is ascribed to the formation of a kind of density wave, probably SDW [5.62]. The system is of a highly two-dimensional nature due to the triple insulating anion layer with the Hg^{2+} layer sandwiched by two sheets containing the SCN groups (Fig.5.9). For α-(ET)$_2$TlHg·(SeCN)$_4$, the resistance anisotropy at room temperature is given by $\sigma_a/\sigma_b/\sigma_c = 1/0.17/(2.8 \div 50) \cdot 10^{-6}$ with $\sigma_a \simeq 140$ S/cm [5.63]. The salts of this family have attracted much attention due to the low-temperature electronic phase with density-wave ordering, and dramatic and intriguing magneto-oscillatory phenomena.

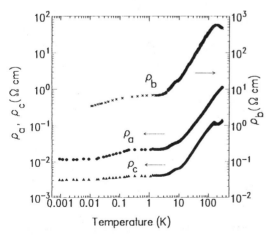

Fig. 5.26. Temperature dependence of the resistivity for α-(ET)$_2$KHg(SCN)$_4$ for three principal crystalline directions. From [5.61]

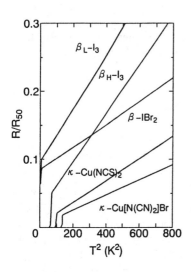

Fig. 5.27. The low-temperature dependence of normalized resistivity R/R_{50} (R_{50} is the resistivity at 50K) in the β_L and β_H phases $(ET)_2 I_3$ (Sects. 5.2.3 to 5), $\beta\text{-}(ET)_2 IBr_2$, $\kappa\text{-}(ET)_2 Cu \cdot (NCS)_2$ and $\kappa\text{-}(ET)_2 Cu[N(CN)_2]Br$

In the case of β-type salts such as $\boldsymbol{\beta\text{-}(ET)_2 I_3}$ and $\boldsymbol{\beta\text{-}(ET)_2 IBr_2}$, the resistivity decreases monotonically by cooling and undergoes a superconductivity transition. Figure 5.27 illustrates the T^2-dependence of the resistivity in the normal state of the ET salts, suggesting a dominant role of the **electron-electron scattering** below 20 K [5.64]. For a given compound the slope does not depend on the residual resistivity.

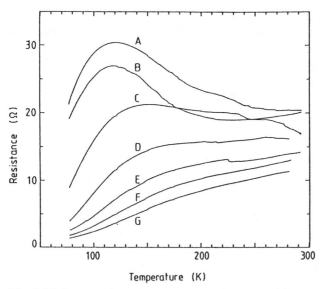

Fig. 5.28. Pressure dependence of the resistivity of $\kappa\text{-}(ET)_2 Cu(NCS)_2$ under various pressure. (A: 1 bar, heating; B: 1 bar, cooling; C: 0.5 kbar, heating; D: 1 kbar, heating; E: 1.5 kbar, cooling; F: 2 kbar, cooling; G: 2.5 kbar, heating). From [5.65]

For κ-$(ET)_2 Cu(NCS)_2$, as a representative example of the κ-type salts, the resistivity displays a pronounced broad maximum near 90 K at ambient pressure. Figure 5.28 plots the temperature dependence of the resistivity under various pressures. It is worth noting that the resistivity peak being very sensitive to pressure is suppressed by a pressure of less than 2 kbar, while it is enlarged dramatically by uniaxial elongation applied along the conducting plane [5.66]. In the low-temperature region (< 40 K), the resistivity decreases, exhibiting a T^2-dependence down to the superconductivity transition, as with the β-type salt.

In order to clarify the origin of the **resistance maximum**, intensive investigations have been carried out. Measurements of the lattice parameters have revealed an anomaly near 100 K, indicating a conformational change in the ordering of the terminal ethylene, $-C_2 H_4-$, of an ET molecule, in other words, a structural change in the hydrogen bonding between the terminal ethylene and the counter anion molecule, $-SCN-Cu-SCN-$ [5.67]. A structural anomaly has also been detected by the Extended X-ray Absorption Fine Structure (EXAFS), which reflects the local structure around the Cu atom [5.68]. Taking into account the interlayer spacing observed in the X-ray study, not only a conformational change but also an emergence of a valence mixed state of Cu^{1+} and Cu^{2+} in the polymer are postulated in the high-temperature phase [5.69]. It is interesting that the magnetic susceptibility is almost constant from room temperature to 100 K, but by further cooling it starts to decrease, accompanied by $\approx 20\%$ reduction in the spin susceptibility. On the other hand, the proton spin-lattice relaxation signals a phase transition near 50 K, accompanied by a change in the density of electron states at the origin of the Fermi level, although no appreciable change is found near 100 K [5.70]. The resistance maximum has also been studied with respect to the electron correlation. *Toyota* and *Sasaki* [5.71] postulated that the intersite Coulomb interaction splits the bands. By decreasing the temperature or increasing the pressure, the transfer integral, which is sensitive to the intermolecular spacing, increases. An increase in bandwidth results in a decrease in the splitting energy. In this case, the nonmetallic behavior is ascribed to the Mott-Hubbard-like insulating state. It has also been found that the logarithmic derivative of the resistivity exhibits a peak at 45 K, accompanied by a change in the anisotropy; similar anomalies have been noted in the spin susceptility and the thermopower [5.72].

The family of κ-$(ET)_2 Cu[N(CN)_2]X$ (X: Cl, Br, I) is isostructural. The κ-$(ET)Cu[N(CN)_2]Br$ salt has the in-plane conductivity $\sigma_\parallel (300 K) \simeq 48$ S/cm and is an ambient-pressure superconductor with $T_c = 11.2$ K. Its resistivity exhibits a temperature dependence similar to that of κ-$(ET)_2 Cu \cdot (NCS)_2$; it possesses a maximum around 90 K, which tends to disappear under pressure. The spin susceptibility obtained from ESR experiments is almost constant between 300 and 50 K but shows a drop below 50 K, indicating a strong decrease in the density of states at the Fermi level. This suggests the formation of a narrow gap or pseudo-gap [5.73]. Based on the NMR of ^{13}C isotope, which is substituted into the central double-bonded carbon site, it is claimed that **antiferromagnetic fluctuation** due to Coulomb repulsion is enhanced by nesting at a low temperature, leading to a pseudo-gap in the electronic state. The large variation of the Knight shift is interpreted as the presence of electronic correlations. The ratio of the on-site Coulomb energy to the bandwidth is estimated to be of the order of unity [5.74]. A large enhancement of the relaxation rate and a drop of the Knight shift, which are ascribed to antiferromagnetic fluctuations [5.75], are observed near 50 K, but they tend to disappear under pressure. A similar behavior has been observed in κ-(ET)$_2$Cu(NCS)$_2$.

κ-**(ET)$_2$Cu[N(CN)$_2$]Cl**, with a room-temperature conductivity of ≈ 2 S/cm, shows a semiconductor-like temperature dependence at ambient pressure, but exhibits a metallic one under pressure. Figure 5.29 displays the temperature dependence of the resistivity as a function of pressure for κ-(ET)$_2$Cu[N(CN)$_2$]Cl. It becomes superconducting with T$_c$ = 12.8 K under an applied pressure of 0.3 kbar [5.77] but becomes resistive below $6 \div 7$ K. The pressure-temperature phase diagram derived from the temperature vs.

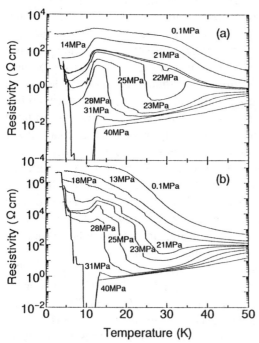

Fig. 5.29. Temperature dependences of in-plane (**a**) and interplane resistivity (**b**) of κ-(ET)$_2$Cu[N(CN)$_2$]Cl under pressure. From [5.76]

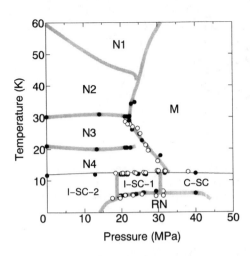

Fig. 5.30. Pressure-temperature phase diagram of κ-(ET)$_2$Cu[N(CN)$_2$]Cl (M: metallic phase, N1, N2, N3, and N4: nonmetallic phases, RN: reentrant nonmetallic phase, C-SC: complete superconducting phase, I-SC-1, I-SC-2: incomplete superconducting phase). From [5.76]

resistivity curves is displayed in Fig. 5.30. Through the observation of the proton nuclear spin-lattice relaxation, a transition to an **antiferromagnetic ordering state** is observed in the $26 \div 27$ K range at ambient pressure [5.78]. It was claimed that the lineshape splits into a finite number of lines, indicating the emergence of a discrete distribution of the local field due to antiferromagetism and its commensurability to the lattice. The evaluated moment per dimer becomes $0.4 \div 1$ μ_B. A **weak ferromagnetic hysteresis** with a saturation moment of $8 \cdot 10^{-4}$ μ_B/dimer has been observed below 22 K [5.79]. This phenomenon is ascribed to canting of the antiferromagnetic moments. However, there is a speculation that the insulating behavior stems from lattice behavior due to the conformational disorder of the ethylene groups of the ET molecule, leading to a localization of the electron states. The lone spins at the localized site may couple antiferromagnetically but due to incomplete compensation they result in a ferromagnetic-like state [5.80]. On the other hand, the **magnetic viscosity phenomenon**, which is a long-term relaxation of resistance on change of an applied magnetic field and maximized near 30 K, was ascribed to the antiferromagnetic phase with domain structure [5.81].

The κ-(ET)$_2$Cu[N(CN)$_2$]I salt has $\sigma_{||}(300\text{K}) \simeq 11$ S/cm and it exhibits weak metallic resistivity from 300 to about 150 K and weak semiconducting behavior below 150 K [5.77]. ESR experiments display a decrease in the spin susceptibility, indicating the formation of a gap below 50 K. Superconductivity is not observed even under an applied hydrostatic pressure of 5 kbar.

The κ-(ET)$_2$Cu(CN)[N(CN)$_2$] salt can be classified into the κ-(ET)$_2$Cu[N(CN)$_2$]X family according to its chemical formula, but it belongs to the same crystal type as κ-(ET)$_2$Cu(NCS)$_2$, compiled in Table 5.2. Its resistivity decreases monotonically from room temperature to the super-

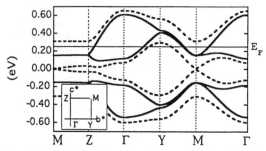

Fig. 5.31. The calculated band structure of κ-(ET)$_2$Cu(NCS)$_2$ in the paramagnetic metallic state derived with U = 0.75 eV (*solid lines*) and in the antiferromagnetic insulating phase derived with U = 0.78 eV (*broken lines*). E_F shows the Fermi energy in the paramagnetic metallic state. From [5.54]

conducting transition at 11.2 K; it does not exhibit a resistance maximum in the intermediate temperature region [5.15].

We comment here on the insulating state found in some (ET)$_2$X salts, such as β'-(ET)$_2$ICl$_2$, β'-(ET)$_2$BrICl and α'-(ET)$_2$AuBr$_2$. They emerge already at room temperature, although it is believed that they possess a half-filled band structure [5.82]. In these salts, the susceptibility remains at approximately $5 \cdot 10^{-4}$ emu/mole. A phase transition to some magnetically ordered state is observed below 200 K. Such an insulating state is likely due to electron correlation resulting in a Mott-Hubbard insulating state.

The origin and interplay of various electronic states appearing with respect to the arrangement of ET molecules as well as intermolecular distances have been explained in terms of the effect of the **on-site Coulomb interaction** by taking account of the transfer integrals between ET molecules [5.54]. The extended Hückel calculations predict that κ-(ET)$_2$X-type salts are metals, for which the bands crossing the Fermi level are partly filled, as represented in Fig. 5.31 with solid lines. Solving a Hubbard-type Hamiltonian with in the Hartree-Fock self-consistent technique, *Kino* and *Fukuyama* [5.54] have shown that when the on-site Coulomb energy U exceeds a critical value the band structure drastically changes to exhibit an antiferromagnetic insulating phase, as depicted in Fig. 5.31 with broken lines. These researchers asserted that the transfer integrals between ET molecules are determined by both the arrangement of the ET molecules and the intermolecular spacing, and influence the effective Coulomb energy of the dimer and hence the electronic state. The relationship of electronic states among α-(ET)$_2$I$_3$, α-(ET)$_2$MHg(SCN)$_4$ and κ-(ET)$_2$X is explained on this basis (Fig. 5.32). In the molecular arrangment without dimerization like the κ-type arrangement, each ET molecule has an average of 1/2 holes, and the energy band can be three-quarter-filled. When dimerization becomes appreciable, a dimer can be regarded as a unit, and the system can be consi-

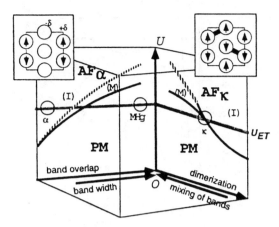

Fig. 5.32. A schematic view of phases of α-(ET)$_2$I$_3$, α-(ET)$_2$MHg(SCN)$_4$ and κ-(ET)$_2$X type salts derived by the results of Hartree-Fock calculations. AF$_\alpha$ and AF$_\kappa$ are the antiferromagnetic ordred phase for α- and κ-type compounds. (I) and (M) denote insulating and metallic phases, respectively. The insets show the alignmentof the spin moments (*arrows*) and charge transfer (δ) in the ordered phase. U$_{ET}$ is the onsite Coulomb energy of the ET molecule. From [5.83]

dered half-filled. In this case, the insulating state is induced by on-site Coulomb repulsion. It has been suggested that the strong antiferromagnetic fluctuation due to Coulomb repulsion, observed through the nuclear relaxation in κ-(ET)$_2$Cu[N(CN)$_2$]Br [5.74], is enhanced further by nesting of a quasi one-dimensional electron Fermi surface coexisting with a quasi two-dimensional hole pocket [5.84].

5.1.6 Optical Properties

Infrared reflectance or conductivity spectra provide important information on band parameters for metals, such as the effective mass, the bandwidth, the collision time of carriers, and so on. By using polarized light one can also elucidate anisotropic features with respect to the orientation of crystals. In fact, the reflectance spectra show a distinct difference whether the polarization is parallel or perpendicular to the conducting direction. The reflectance spectra of the ET salts were analyzed in terms of the Drude model which assumes that the plasma frequency ω_p and the relaxation rate τ are anisotropic [5.85, 86]. *Kuroda* et al. [5.87] claimed that in ET salts a contribution from the transition between the split bands, which are formed as a result of the strong HOMO interaction within the ET dimer, is superimposed on the broad component of the **intraband spectra**.

These optical properties have been elucidated through a systematic study of **reflectance spectra** for various (ET)$_2$X salts. For β-(ET)$_2$I$_3$, the

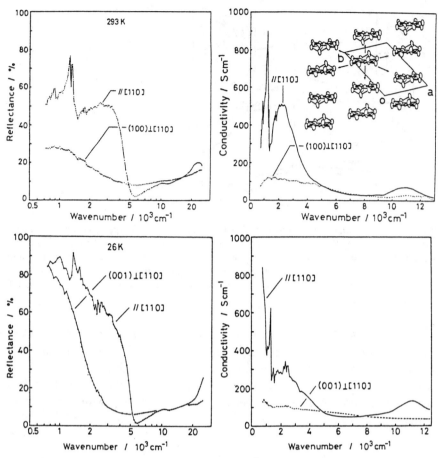

Fig. 5.33. Reflectance and conductivity spectra of β-(ET)$_2$ I$_3$ at 293 K and at 26 K. From [5.89]

conductivity spectra converted from reflectance spectra exhibit a peak at 2000 cm^{-1} accompanied by a broad intraband Drude-like feature (Fig. 5.33). The conductivity peak is found for light polarized along the [-110] direction but not for that perpendicular to it. Thus, it must be related to the molecular stacking structure, in particular, to the dimerization. The conductivity peak is attributed to the **Electron-Molecular Vibration** (EMV) **coupling**. It causes the peaks at the frequencies of the totally symmetric intramolecular vibrations through a mechanism proposed by *Rice* and *Lipari* [5.88].

To separate intraband and interband transitions, the reflectance spectra have been analyzed by the following Drude-Lorentz model

$$\epsilon(\omega) = \epsilon_c - \frac{\omega_p{}^2}{\omega(\omega + i\gamma)} - \sum_j \frac{\Omega_{pj}{}^2}{(\omega - \omega_j)^2 - i\Gamma_j\omega} , \qquad (5.4)$$

where ϵ_c is the frequency-independent dielectric constant, ω_p and γ are the plasma frequency and the relaxation rate of the charge carriers; ω_j, Γ_j, and Ω_{pj} are the parameters of the Lorentz oscillators for simulating the **inter-band transitions** and the vibrational structures [5.89].

As shown in Fig.5.33, the intraband conductivity increases with decreasing temperature. This is due to the suppression of the carrier scattering rate and is observed for any salt exhibiting metallic behavior down to low temperature. Furthermore, a comparison of the spectra for two polarization directions, i.e., parallel and perpendicular to the [110] direction (the stacking direction) shows a difference in the effective masses of the carriers through the **plasma frequencies**. The derived plasma frequencies ω_p and effective masses m* of metallic ET salts for two principal directions are listed in Table 5.6.

A systematic investigation of the possibility of an interband transition in the dimerized molecules has been carried out by comparing β-(ET)$_2$I$_3$, κ-(ET)$_2$I$_3$, and θ-(ET)$_2$I$_3$ [5.87]. Here, β-(ET)$_2$I$_3$ has dimerized linear stacks where the interband transitions can occur in only one direction (the [110] direction), whereas κ-(ET)$_2$I$_3$ has a checkered pattern of dimerized molecules along the [011] and [01-1] directions. The θ-(ET)$_2$I$_3$ salt has no dimerized stack units. The differences in these molecular arrangements should lead to different behavior for the interband transitions if they exist. The experimental results are as follows: A transition is observed in β-(ET)$_2$I$_3$ for a polarization along the [110] direction and also in κ-(ET)$_2$I$_3$ for both the [011] and [01-1] directions, but it is very weak in θ-(ET)$_2$I$_3$. This confirms the hypothesis of interband transitions in dimerized molecules in ET salts and must be taken into account in an analysis of reflectance spectra.

Table 5.6. Plasma frequencies and effective masses of metallic ET salts. From [5.89]

	ω_{p1} [eV]	ω_{p2} [eV]	m_1^*/m	m_2^*/m
β-(ET)$_2$I$_3$	0.89($\parallel$[110])	0.48($\perp$[110])	2.0	7.0
θ-(ET)$_2$I$_3$	1.05($\parallel$a)	0.73($\parallel$c)	1.5	3.0
κ-(ET)$_2$I$_3$	0.84($\parallel$c)	0.74($\parallel$b)	2.2	3.0
κ-(ET)$_2$Cu(NCS)$_2$	0.64($\parallel$c)	0.55($\parallel$b)	4.1	5.5

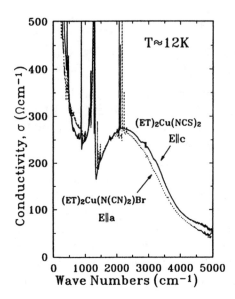

Fig. 5.34. The optical conductivity of κ-(ET)$_2$Cu[N(CN)$_2$]Br for **E** $||$ **a** compared with that for κ-(ET)$_2$Cu(NCS)$_2$ for **E** $||$ **c**, at $\approx$12 K. From [5.91]

Measurements of the polarized reflectivity of κ-(ET)$_2$Cu(NCS)$_2$, κ-(ET)$_2$Cu[N(CN)$_2$]Br and κ-(ET)$_2$Cu[N(CN)$_2$]Cl from 200 to 5500 cm^{-1} at temperatures between 10 and 295 K were presented by *Eldridge* et al. [5.90-92]. The optical conductivities, obtained after a Kramers-Kronig analysis, are very similar for Cu(NCS)$_2$ and Cu[N(CN)$_2$]Br salts, with the exception of a narrower electronic band in the latter (Fig. 5.34). The far-infrared conductivity rises dramatically at temperatures below 100 K, in agreement with the temperature dependence of the DC conductivity. The far-infrared peak (less than 1000 cm^{-1}) is due to intraband transitions of free carriers, while the mid-infrared peak (1000 ÷ 4000 cm^{-1}) is due to interband transitions, superimposed on the free-carrier tail. The vabrational frequencies exhibit some differences due to deviations in the symmetry class of the salts: the Cu(NCS)$_2$ salt has a monoclinic space group P2$_1$ with two dimer units per unit cell (Z = 2), while the Cu[N(CN)$_2$]Br salt is orthorhombic, Pnma, with Z = 4. The major points that emerge from the comparison are that, for **E** $||$ **a** in the Cu[N(CN)$_2$]Br salt, the a$_g$ modes of ET, which are normally inactive but are activated by electron coupling, all increase in frequency relative to the corresponding variations in the Cu·(NCS)$_2$ salt. Conversely, the frequencies of the normally infrared-active b$_u$ modes all decrease. The a$_g$ increase is due to decreased electronic coupling, since it is known that this coupling acts to reduce the vibrational frequencies. The b$_u$ decrease in frequency implies that the lattice contribution to the intermolecular force constants has decreased, indicating a softer lattice. This is due to the increased size of the unit cell.

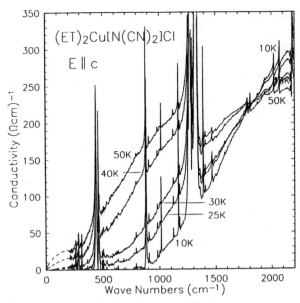

Fig. 5.35. The optical conductivity of κ-$(ET)_2 Cu[N(CN)_2]Cl$ at various temperatures below the metal-semiconductor transition for $\mathbf{E} \parallel \mathbf{c}$ below $2200\ cm^{-1}$. From [5.92]

For κ-$(ET)_2 Cu[N(CN)_2]Cl$, the optical conductivity in the low-energy region decreases below 50 K, giving a large temperature-dependent semiconductor energy gap with a value of $\approx 900\ cm^{-1}$ at 10 K (Fig. 5.35). The optical-conductivity spectra obtained from α-$(ET)_2 NH_4 Hg(SCN)_4$ [5.93] differ from those reported for α-$(ET)_2 I_3$ [5.94], mainly because of the much higher conductivity of the former. Via Drude fits, the parameter values of $\omega_p = 7000\ cm^{-1}$, $\gamma = 700\ cm^{-1}$, $\epsilon_c = 3.5$ and the effective mass of $1.8 m_e$ have been deduced for the former.

The parameters for electron-electron vibrational coupling have been determined for $(ET)_4 Hg_{2.89} Br_8$ from an analysis of the spectra in the frequency range of $650 \div 40,000\ cm^{-1}$ based on a phase-phonon model [5.95].

Using the **infrared** and **Raman spectra** of several **isotope analogs**, an assignment of the normal modes of the ET electron-donor molecules was performed [5.96]. The analogs used are: $^{13}C(2)$-ET (two central carbon atoms are substituted), $^{13}C(6)$-ET (six central carbon atoms are substituted), $^{34}S(8)$-ET (all sulphur atoms are substituted), and d(8)-ET (all hydrogen atoms are substituted). Lists of frequencies and assignments for the infrared and Raman vibrational features of ET and isotope substituted ET's were compiled in [5.96]. Assignments of over half the mode features agree with those given by *Kozlov* et al. [5.97], but nearly half of them are assigned differently.

5.2 Superconductivity of β-$(ET)_2X$

5.2.1 Overview of ET Superconductors

The sulfur-based ET cation radical has been used as a building block for a large number of organic superconductors, most of which are superconducting under ambient pressure (Table 1.1). The crystal morphology of ET salts is manifold, but most of the crystals that exhibit superconductivity have quasi two-dimensional features. For example, superconducting $(ET)_2I_3$ has three morphologies, β, κ and θ, but all of them are two-dimensional. Although this diversity of the crystal structures offers a rich variety of problems to be investigated, at the same time, it dramatically increases their complexity. In such situations the experience gained through the study of comparatively simple $(TMTSF)_2X$ superconductors has been a useful guide to the nature of ET superconductors.

The first ET superconductor to be observed was $(ET)_2ReO_4$. This salt is rather exceptional among the ET superconductors, however, in that its phase diagram resembles that of TMTSF superconductors. $(ET)_2I_3$ salts are quite unique and hence are the most extensively investigated in the early stages of the ET-superconductor study. They form several ambient-pressure superconductors with T_c as high as 8 K under certain conditions. The salts with polymeric anions such as κ-$(ET)_2X$ (X: $Cu(NCS)_2$, $Cu[N(CN)_2]Br$, $Cu[N(CN)_2]Cl$, $Cu(CN)[N(CN)_2]$) have brought about superconductivity with T_c reaching and exceeding 10 K.

Two-dimensionality characterizes the superconductivity of ET salts, which manifests by itself in the presence of a magnetic field. That is, the resistivity transition becomes too broad to determine the transition temperature due to a thermal fluctuation phenomenon under a strong magnetic field. This effect becomes more obvious with higher T_c. Theoretical devices to analyze the data have been offered through the development of theories to describe high-T_c two-dimensional cuprate superconductors. As a result, the high-T_c κ-type ET superconductors can be understood reasonably well as two-dimensional superconductors, in other words, superconducting-sheet systems coupled by the Josephson effect. The high anisotropy found in the magnetic response is explained in terms of the vortex dynamics in intrinsic two-dimensional superconductors. In addition, the localization phenomena of superconductivity and the Kosterlitz-Thouless transition characteristics specific to two-dimensional systems are found in κ-$(ET)_2Cu[N(CN)_2]Cl$ and α-$(ET)_2MHg(SCN)_4$ (M: K, Rb, Tl).

It is not so strange that the superconductivity is very sensitive to the applied pressure if we take into account the softness of organic materials. However, real systems are much more sensitive than expected. For example, the superconductivity of β-$(ET)_2I_3$ is drastically raised even under pressures as low as about 1 kbar. In the case of κ-$(ET)_2Cu[N(CN)_2]Cl$, super-conductivity phase diagrams are affected dramatically by a pressure of less than 0.5 kbar. However, these two cases are exceptional. In general, T_c by itself decreases with increased pressure at a rate higher than that for TMTSF superconductors.

Superconductivity is also sensitive to nonmagnetic defects. In ordinary metals, superconductivity is sensitive to magnetic defects but not to non-magnetic ones. To interpret this difference, the possibility of triplet super-conductivity has been proposed, although there is not yet a strong support-ing result. On the other hand, some ET salts have been found to exhibit an-tiferromagnetism ascribable to electron correlation in a reduced two-dimen-sional system, and the anisotropy of the superconductivity wave function is considered to be d-like.

In the following section, the superconductivity of ET salts is described with the emphasis on the behavior of β-$(ET)_2I_3$ and κ-$(ET)_2X$ (X: Cu·$(NCS)_2$, $Cu[N(CN)_2]Br$, and $Cu[N(CN)_2]Cl$) in Sects 5.2 and 3, respective-ly. Other salts such as $(ET)_2ReO_4$, $(ET)_4(Hg_2Cl_6)HgCl_2$, $(ET)_2M(CF_3)_4$ (solvent), α-$(ET)_2MHg(SCN)_4$, $(ET)_3Cl_2 2H_2O$ and $(ET)_4M(CN)_4H_2O$ are discussed in Sect. 5.4. The superconducting salts with BETS or BO molec-ules, i.e., ET molecules that are modified by substitution of some of the sulfur atoms with selenium or oxygen, are described in Sect. 5.5.

5.2.2 Superconductivity in β-$(ET)_2I_3$ Under Ambient Pressure

Crystals of β-$(ET)_2I_3$ can be grown in shapes of either plates or needles by electrochemical methods [5.98]. At room temperature, the conductivity measured along the long axis of the needles, and in an arbitrary direction in the plane of the plates, has the same order of magnitude ($\approx 30\,S/cm$).

Figure 5.36 exhibits the temperature dependence of the resistance of two plates and two needles of β-$(ET)_2I_3$ below 30 K under ambient pressure [5.10]. The transition to zero resistance begins at $1.6 \div 1.7$ K. If the critical temperature T_c is taken as the midpoint of the transition, one obtains $T_c = 1.4 \div 1.5$ K. At lower temperatures, by applying a magnetic field of 20 kOe along the b-axis and measuring the current along the a-axis, the resistivity resumes the value which would be obtained by a smooth extrapolation from those values above 4 K.

Measurements of the AC susceptibility clearly indicated the onset of a diamagnetic shielding current below 1.05 K [5.99, 100]. In the DC magneti-zation experiment, one can distinguish between two kinds of diamagnetic effects (Fig. 5.37). When a sample is cooled well below T_c in zero magnetic fields and then a DC field is applied, supercurrents on the sample surface are induced to screen the magnetic field from the inside of the sample.

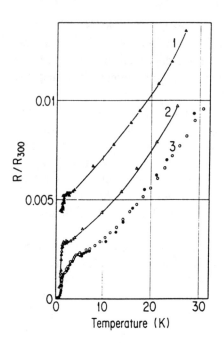

Fig. 5.36. Temperature dependence of the resistance of β-(ET)$_2$I$_3$ at ambient pressure. *Triangles*: plate-like crystals; *circles*: needle-shaped crystals. Filled and open marks distinguish the samples used. From [5.10]

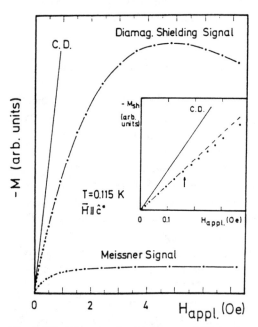

Fig. 5.37. Diamagnetic shielding and Meissner magnetization of β-(ET)$_2$I$_3$ vs. field (applied perpendicular to the ab plane) at T = 0.115 K. The *straight line* (C.D.) is what should be obtained in the case of complete diamagnetism, i.e., the shielding behavior of a perfect superconductor. The *insert* shows the low-field region of the diamagnetic shielding magnetization, the *arrow* indicates H$_{c1}$. From [5.99]

Upon warming up, the decay of the currents screening the superconductivity can be monitored by the magnetization change, which is the **diamagnetic shielding signal**. On the other hand, when the sample is cooled in a magnetic field, the magnetic flux is expelled from the sample yielding the **Meissner signal**. The largest diamagnetic shielding and Meissner signals are observed when the field is oriented normal to the highly conducting ab-plane ($H \parallel c^*$). From the magnetization curve, one can deduce the field at which the flux starts to penetrate the sample, being the **lower critical field** H_{c1}. This field is largest in the orientation $H \parallel c^*$: H_{c1c^*} is equal to 0.36 Oe (which corresponds to 0.036 mT when $\mu = 1$), $H_{c1a} = 0.05$ Oe and $H_{c1b'} = 0.09$ Oe [5.99]. These magnetization measurements conclusively prove that the superconductivity in β-(ET)$_2$I$_3$ is a bulk property.

The temperature dependences of the **upper critical field** H_{c2} along the a, b' and c* directions are shown in Fig. 5.38. The measurement was carried out for a sample with a ratio of the room-temperature (300 K) resistance to the low-temperature (4.2 K) resistance $R_{300K}/R_{4.2K}$ as high as 10^3 [5.101]. The data in Fig. 5.38 reveal that the critical fields are $H_{c2a} = 9.7$ kOe, $H_{c2b'} = 9.3$ kOe and $H_{c2c^*} = 0.46$ kOe at 0.5 K, also reflecting the two-dimensional nature of β-(ET)$_2$I$_3$.

The Ginzburg-Landau (GL) **coherence lengths** $\xi_i(T)$ at $T = 0$ K are found with (3.15) to be

$$\xi_a(0) = 633\text{Å}, \quad \xi_{b'}(0) = 608\text{Å}, \quad \xi_{c^*}(0) = 29\text{Å}. \tag{5.5}$$

Note that the coherence length along the c*-direction is short and only twice the lattice constant in that direction. It should be noted here that, although the bulk nature of superconductivity in β-(ET)$_2$I$_3$ is not in ques-

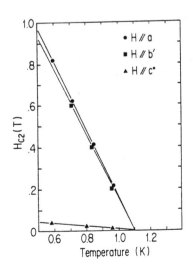

Fig. 5.38. Temperature dependence of the upper critical field H_{c2} along the a, b' and c* directions. From [5.101]

163

tion from an observation of the Meissner effect, it is dominated by interactions within layers with weak interlayer coupling.

The **paramagnetic limit** due to spin-pair breaking for the critical field of a superconductor is related to its T_c by

$$H_p[kOe] = 18.4 T_c[K] \tag{5.6}$$

in the limit of small spin-orbit scattering, based on a free-electron model [5.102]. For the sample used ($T_c = 1.1 K$), this value is 20.2 kOe at 0 K, and should be compared with $H_{c2a}(0) = 17.8$ kOe, which is estimated from a linear extrapolation to $T = 0$ K. The value of $H_{c2}(0)$ is in the region of the paramagetic limit.

On the basis of a tight-binding band model, where the Fermi surface which is closed in the ij-plane but open along the k-direction (i.e., a cylinder-like structure), the coherence-length anisotropy due to the orbital effect is related to the band-structure anisotropy by

$$\xi_i(0) : \xi_j(0) : \xi_k(0) = a_i\sqrt{t_i E_F} : a_j\sqrt{t_j E_F} : a_k t_k \quad \text{for} \quad t_i \simeq t_j \gg t_k \tag{5.7}$$

where a_i, t_i and E_F are the lattice spacing, the transfer integral, and the Fermi energy, respectively. From (5.7) combined with (5.5), the anistropy of the transfer integral is estimated as

$$t_a : t_{b'} : t_{c*} \simeq 45 : 25 : 1$$

using $a_{b'} = 8.55 Å$, $a_{c*} = 15.09 Å$ and $E_F = 1.15 t_a$, where the value of E_F corresponds to a Fermi energy filling half of the first Brillouin zone of a two-dimensaional band [5.101].

To conclude this subsection we would like to comment on the specific-heat measurements of β-(ET)$_2$I$_3$. Although the bulk nature of superconductivity was clearly demonstrated by the observation of the Meissner effect, as mentioned above, a measurement of the specific heat from 0.7 to 18 K in zero and applied magnetic fields showed no bulk-superconductivity features at T_c, which could be determined by an AC magnetization measurement on the same sample. To rule out sample dependences, the Meissner effect was also determined on the same crystals as used for the specific-heat measurement [5.103]. The low-temperature specific heat was represented by

$$C/T = \gamma + \beta T^2 , \tag{5.8}$$

where $\gamma = 24 \pm 3$ mJ/(mol·K^2) and $\beta = 19$ mJ/(mol·K^4).

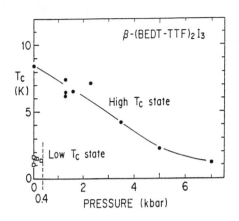

Fig. 5.39. Pressure dependence of T_c for β-$(ET)_2 I_3$

5.2.3 Superconductivity of β-$(ET)_2 I_3$: Pressure Effect

A drastic change in the superconductivity is observed upon application of hydrostatic pressure by the clamp-cell method [5.104, 105]. Figure 5.39 depicts the T_c-vs.-pressure diagram for β-$(ET)_2 I_3$. T_c which decreases with P at low pressure (<0.5 kbar) [5.106] increases discontinuously to $7 \div 8$ K near 1 kbar and then decreases again monotonically at high pressure. This abrupt increase in T_c near 1 kbar is quite anomalous unless accompanied by some crystal-phase change. It should also be noted that the pressure dependence of T_c above 1 kbar is unusually large, $dT_c/dP \simeq -1$ K/kbar, which is an order of magnitude greater than even that of $(TMTSF)_2 PF_6$, with $dT_c/dP = -0.09$ K/kbar [5.107]. Henceforth, we distinguish the superconductivity observed above 1 kbar, with T_c reaching 8 K, from that below 0.5 kbar, with $T_c \leq 1.1 \div 1.6$ K, and call the former a **high-T_c state** and the latter a **low-T_c state**.

The filled circle at zero pressure in the pressure phase diagram (Fig. 5.39) represents the high-T_c superconducting state occurring under ambient pressure. It was first observed as a large resistance drop after removal of the pressure at room temperature. Figure 5.40 depicts such a resistance drop near 8 and $2 \div 3$ K in a pressure-released sample. For comparison, the superconducting transition observed under a pressure of 1.3 kbar is indicated by a dashed line. It appears that the pressure-induced superconducting transition at 7.4 K exists, at least partially, after removal of the pressure. The remaining resistivity is ascribed to the structure undergoing the superconducting transition at $2 \div 3$ K. That this drop is due to a transition to a superconductivity state was confirmed via measurements on samples in magnetic fields: the temperature of the first resistance drop is shifted to 3 K when a magnetic flux density of 1.6 T is applied in the c^*-direction, while the second resistance drop was almost removed by a mag-

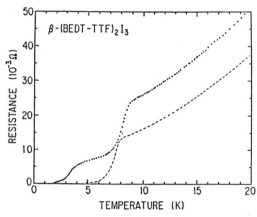

Fig. 5.40. Temperature dependence of resistance near T_c after pressure release. The *broken line* represents a superconducting transition under 1.3 kbar, drawn on an arbitrary vertical scale. From [5.108]

netic field with 0.6 T. This reveals that structures of the two superconducting states are independent of one another, that is, the sample is an inhomogeneous mixture of the two.

So what discriminates the high-T_c state from the low-T_c one under ambient pressure? A clue to this is the observation of the structural phase transition associated with an incommensurate lattice modulation below 200 K.

5.2.4 Incommensurate Lattice Modulation in β-(ET)$_2$I$_3$

In single-crystal X-ray diffraction and time-of-flight neutron diffraction experiments, superlattice peaks with indices of (hkl) $\pm$**q** with **q** = (0.08, 0.27, 0.205) were found at low temperatures [5.109]. The modulation period is incommensurate with the fundamental lattice. The ET molecules have two conformations of the terminal ethylene groups (Fig. 5.1). The linear anion I_3^- is encapsulated via hydrogen atoms of the ethylene groups.

The amplitude of the incommensurate lattice modulation increases with decreasing temperature (Fig. 5.41). The precise modulation configuration has been determined at 120 K. For each fundamental Bragg reflection, there are two first-order satellite reflections which are symmetrically displaced from the main peak by the vector $\pm$**q** = (0.076(2), 0.272(4), 0.206(3)). At low temperature, these satellite reflections have narrow linewidths suggesting a long-range order in the modulated structure.

A structural analysis has shown that the interactions of the triiodide ions with the surrounding ethylene groups are responsible for the modulation. However, the displacement vectors for the triiodide and the ET molecules are different in direction and magnitude [5.109]. The major compo-

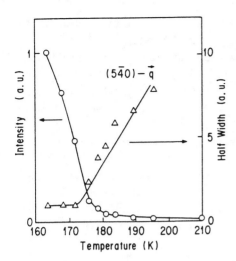

Fig. 5.41. Temperature dependence of the incommensurate superstructure represented by the X-ray scattering intensity and line width in β-(ET)$_2$I$_3$. From [5.110]

nent of the I$_3$-anion displacement is along the [100] direction with an amplitude of ≈ 0.27 Å. For the ET molecule, the major component is nearly perpendicular to the column axis [110] and is in the molecular plane. The displacement amplitude (≈ 0.11 Å) for the ET molecule is considerably smaller than that for the I$_3$ ion. If one includes the modulation in the S...S contacts in a two-dimensional cationic network, the fluctuations at the contacts are up to ± 0.2 Å away from the average.

The source of the modulation is supposed to be anion-cation interactions. ET molecules with disordered ethylene groups in the conformations (Fig. 5.1) which yield unfavorable H...I contacts are placed into sites in the modulated structure, which minimize the H...I interactions [5.109].

It has also been supposed [5.111] that the rate of thermal shrinks in the ET stacks is larger than that for I$_3$ anions. Since the anions have enough space to adopt any orientation, it is the arrangement of the cations that is adjusted or ordered when the free energy of the crystal is minimized. At low temperatures, the size of I$_3$ should be critical in the formation of the β-type stacking structure, which seems to relate that the superstructure appears in salts with the largest trihalide linear anions.

This lattice modulation plays an interesting role in the superconductivity. It has been found that above a critical pressure of 0.5 kbar the incommensurate lattice modulation slightly shifts the wave vector and disapears [5.112]. This fact, together with the pressure-phase diagram of T_c, indicates almost conclusively that the difference between the low-T_c state and the high-T_c one is directly related to whether or not the lattice is modulated incommensurately. In other words, the lattice modulation suppresses the high-T_c superconductivity, although the precise reason is still open to questions. Hereafter, we designate the β crystal with lattice modulation as β_L **crystal** and that without modulation as β_H or β^* **crystal**.

The absence of a modulated structure at low temperatures could mean either that the ethylene groups are completely ordered in the high-T_c state or that they are randomly distributed like at room temperature. The structural studies under pressure [5.113] revealed a new completely ordered phase with the ET molecules in the A configuration (Fig.5.1).

The emergence of incommensurate lattice modulation is also observed as an anomaly in the plot of dR/dT versus T, R being the resistivity [5.114, 115]. When this characteristic is observed, one can monitor the emergence of the lattice modulation without neutron or X-ray diffraction, which can damage samples by irradiation. Furthermore, measuring dR/dT allows monitoring the emergence of a superstructure in every conductivity measurement.

The difference in T_c of the β_L and β_H phases has been tried to be explained in terms of the suppression of superconductivity by a random scattering potential which arises in the β_L phase, because of the incommensurate structure-induced lattice distortion. The resistivity of the β_L and β_H phases are given by a combination of the temperature-independent parts, $\rho_L(0)$ and $\rho_H(0)$, and T^2-dependent parts. The T^2-dependent terms are almost the same for the β_L and β_H phases, but the temperature-independent term of the β_L phase is 10 to 20 times larger than that of the β_H phase [5.116]. Alternatively, possible occurrence of electronic structure modification below $20 \div 23$ K is claimed from anomalies found in thermoelectric-power [5.54], magnetoresistance [5.117], the Hall-effect [5.118], and crystal structure [5.119] measurements. Further, an anomalous temperature dependence of the Hall coefficient has been observed below 40 K in the β_L phase and is interpreted taking into account electron-electron umklapp scattering and the influence of the Fermi surface near the Brillouin-zone boundary [5.120].

In order to examine the role of molecular vibrations on the superconductivity, the effect of deuteration of the ET molecules has been studied [5.121]. Crystals of β_L-(ET)$_2$I$_3$ and its deuterated derivative β_L-(d-ET)$_2$I$_3$ were grown under equal crystallization conditions. The values of $H_{c2}(0)$ and $\xi(0)$ of these crystals are listed in Table 5.7. T_c of β_L-(d-ET)$_2$I$_3$ rose to 1.43 K from that of β_L-(ET)$_2$I$_3$ at 1.15 K. This is the opposite of what one would expect from a simple BCS model, namely, a decrease of T_c. On the other hand, the high-T_c state at 1.5 kbar seems to show that β_H-(d-ET)$_2$I$_3$ has a slightly lower T_c than β_H-(ET)$_2$I$_3$, although the value of this "normal" **isotope shift** contains considerable uncertainty.

Table 5.7. $H_{c2}(0)$ and $\xi(0)$ for β_L-$(ET)_2 I_3$ and β_L-$(d\text{-}ET)_2 I_3$. From [5.127]

Material	Direction of H	$H_{c2}(0)$ [kOe]	$\xi(0)$ [Å]
β_L-$(ET)_2 I_3$	$\parallel$ stack	20.9	587
	$\perp$ stack	24.8	696
	$\parallel c^*$	0.81	22.6
β_L-$(d\text{-}ET)_2 I_3$	$\parallel$ stack	25.8	506
	$\perp$ stack	25.5	500
	$\parallel c^*$	1.3	25.5

5.2.5 High-T_c State of β-$(ET)_2 I_3$

Under a pressure of about 1 kbar, the high-T_c state of β_H-$(ET)_2 I_3$ emerges, while the incommensurate lattice modulation disappears. It is therefore clear that the high-T_c state is suppressed by the lattice modulation, and any means to suppress the lattice modulation may generate the high-T_c state, even under ambient pressure. This has indeed been accomplished by the following method [5.122]: first, a sample is cooled to well below 175 K, e.g., 34 K, under a pressure of 1.5 kbar. Then, the pressure is released. The sample is thereafter further cooled, and a complete superconducting transition is observed near 8 K. This high-T_c state at ambient pressure can be generated reproducibly, even if the sample is temporarily heated, as long as the maximum temperature does not exceed 125 K.

The pressure versus temperature phase diagrams for the metastable β_H and β_L states are displayed in Fig.5.42. Figure 5.42a has been derived by the resistivity changes either as a function of T at constant P or as a function of P at constant T. The β_L state occurs at 180 K under ambient pressure without hysteresis. With growing pressure, the transition disappears above ≈ 400 bar. The β_L phase which exists under low pressure becomes the β_H phase on passing across the curve labeled 1 in Fig.5.42a. The β_H phase produced by cooling under pressure is stable as long as it is kept below the curve labeled 2. The positions of these curves, however, change slightly from sample to sample. The equilibrium line between the phases must be somewhere between the curves 1 and 2. Its presumed position is indicated by the dashed line. Point K on the phase diagram is the critical point of the second-order phase transition. On passing to the β_H phase from the ambient conditions along a path from the right of point K we should intersect the ethylene group order-disorder phase boundary without forming a

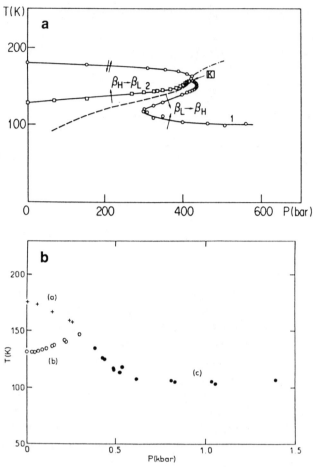

Fig. 5.42. Pressure-temperature phase diagram of β_L and β_H conformations of β-(ET)$_2$I$_3$ obtained from (**a**) resistivity measurements and (**b**) differential thermal analysis. From [5.123, 124]

superstructure. The presumed position of this boundary is marked by the dash-dot line in Fig. 5.42a.

The phase diagram deduced from a differential thermal analysis is exhibited in Fig. 5.42b. In this study it was asserted that the β_H-phase is produced at a low temperature by clockwise P-T cycling about the critical point ($P_c \approx 345$ bar and $T_c \approx 150$ K). It was also shown that anticlockwise P-T cycling with $P < P_c$ allows stabilization of the β_L phase at low temperature. The limits of stability of both the β_H and β_L phases in the P-T diagram are summarized in Fig. 5.42b with data denoted by b and c. The properties of the metastability of the phases are qualitatively interpreted in terms of a Landau expansion of the free energy with appropriate P and T dependences in the second- and fourth-order terms [5.124]. The difference between the P-T phase diagrams shown in Figs. 5.42a,b that appear just to the right of the critical point, however, is open to further study.

During an X-ray diffraction study on the superstructure in relation to the P-T phase diagram, it was found [5.125] that the wave vector of the superstructure changes when the sample at ambient pressure is kept at $100 \div 120$ K for a long time, e.g., 24 hours. The sample annealed for 65 hours becomes inhomogeneous, consisting of superconductors with T_c of ≈ 8 and ≈ 2 K. The associated change in the wave vector was considered to be due to the repositioning of the ethylene groups in the ET molecules.

The pressure-temperature phase diagram reveals that the minimum pressure needed to suppress the incommensurate lattice modulation is 0.35 kbar rather than the originally observed ≈ 1 kbar. *Schirber* et al. [5.126] have asserted that the high-T_c state in β-$(ET)_2 I_3$ cannot be accessed by only hydrostatic pressure but requires, in addition, a substantial shear component. When a pure hydrostatic pressure was applied to a sample through careful isobaric freezing of He^4, a high-T_c state was not observed. On the other hand, fast freezing of liquid He^4 or freezing cyclopentane produces a shear component in the pressure applied to the crystal, and under these conditions the high-T_c state was produced. When electrical contacts are glued onto the crystal with Au paste, *Schirber* et al. insisted that shear stress is caused by the contacts which induces the high-T_c state. So far no acceptable explanation has been found for the role of the shear stress in either the superconductivity or the lattice modulation.

Figure 5.43 illustrates the temperature dependence of the upper critical field H_{c2} of β-$(ET)_2 I_3$ with a T_c of 6.0 K at 1.3 kbar for the orientation of the magnetic field parallel to the ab-plane [5.127]. In the figure it is shown that the upper critical field is evaluated by the mid-point and the onset of

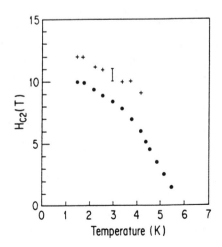

Fig. 5.43. Temperature dependence of H_{c2} for the magnetic field in the directions parallel to the ab-plane for β-$(ET)_2 I_3$ at 1.3 kbar. ($\bullet$: determined from the midpoint of the resistive transitions; +: determined from the onset). From [5.127]

Table 5.8. Measured and derived superconducting properties of β-(ET)$_2$X.

Material	T_0 [K]	H_{c1} (T) [Oe]				H_{c2} (0) [kOe][a]			$\xi(0)$ [Å]			Ref.
		a	b	c	(T [K])^2	a	b	c	a	b	c	
β_L-(ET)$_2$I$_3$	1.5	0.05	0.09	0.36	(1.15)							5.99
	1.1					17.8	17.0	0.8	633	608	29	5.101
β_H-(ET)$_2$I$_3$												
at 1.6 kbar	7.2					250^c		27	127		10	5.128
3.5 kbar	3.06					41		2.6	355^c		22.7	5.128
5.0 kbar	2.18					21.8		1.4	488^c		31	5.128
β-(ET)$_2$IBr$_2$	2.25					33.6	36.0	15	463	444	18.5	5.129
	2.42	3.9		16	(5)							5.130
β-(ET)$_2$AuI$_2$	4.2	4.0		20.5	(12)	66.3		5.1	249^c		19.2	5.130

The superscripts mean:
- [a] Extrapolated to 0 K
- [b] Measurement temperature of H_{c1}
- [c] No direction specified in the ab-plane

the resistive transition. As is described in Appendix B, due to the two-dimensionality of the material, the **thermal fluctuation** dominates in materials with high-T_c particularly under a magnetic field. As a result the upper critical field, as a parameter denoting the onset in the long-range order, cannot be defined in principle. This is demonstrated clearly by an extreme broadening in the resistive transition. For cases with coherence lengths much larger than the interlayer spacings, for example for β-(ET)$_2$I$_3$ salts, however, the derived results are considered not to be inaccurate, since the system can be treated by a three-dimensional approximation. Assuming that H_{c2} is isotropic in the ab-plane, the GL coherence lengths at 0 K are estimated from the slope of H_{c2} versus T near T_c to be $\xi_a(0)$ [$=\xi_b{}'(0)$] $= 125 \div 130$ Å and $\xi_{c*}(0) = 10$ Å. The derived values of H_{c2} and ξ for β_H-(ET)$_2$I$_3$ under various pressures are listed in Table 5.8 together with the values for β_L-(ET)$_2$I$_3$, β-(ET)$_2$IBr$_2$ and β-(ET)$_2$AuI$_2$. *Laukhin* et al. asserted that H_{c2} parallel to the conducting ab plane exceeds the **paramagnetic limit** provided that a free-electron model can be applied [5.131]. This implies that the superconductivity may be triplet or, if it is in a singlet state, the Fulde-Ferrell-Larkin-Ovchinikov state is formed, where the superconductivity becomes inhomogeneous with spatial oscillations of the gap function [5.132]. The applicability of the free-electron model needs further evaluation.

To see the role of the density of states in the pressure dependence of T_c, the magnetic susceptibility has been measured under pressure of up to 10 kbar [5.133]. It was found that the pressure derivative $d\ln\chi/dP$ is moderate in magnitude (-3.3% per kbar) and is independent of temperature below

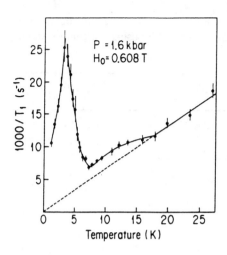

Fig. 5.44. Temperature dependence of proton NMR $1/T_1$ for β-$(ET)_2 I_3$. From [5.134]

300 K. This shows that the change in the density of states is not a dominant factor in determining the pressure dependence of T_c.

In order to clarify the nature of superconductivity, the spin-lattice relaxation rate $(1/T_1)$ in proton NMR has been measured as a function of temperature. When $1/T_1$ forms a peak just below T_c, representing the coherence effect (**Hebel-Slichter peak**), the superconductivity is judged to be BCS-like. For β_H-$(ET)_2 I_3$, the presence of the coherence peak is not found but an anomalous enhancement of the relaxation rate near the superconducting transistion is noted [5.134]. The result at 6.08 kOe under 1.6 kbar is indicated in Fig. 5.44: the continuous decrease in $1/T_1$ with temperature breaks down at 7.5 K where a rapid increase occurs down to 3.5 K. Since powdered samples were used in this measurement, the broadness of the temperature domain between 3.5 and 7.5 K might be explained through an anisotropy in H_{c2}, i.e., each crystal becomes a superconductor at a slightly different temperature depending on its orientation with respect to the applied magnetic field. When the applied field is reduced, the increase occurs in a narrower region ($7.5 \div 5$ K) but reaches a much higher value at 5 K, and then decreases to the same level as the high-field data at 3.5 K. A similar temperature dependence of $1/T_1$ is found in κ-$(ET)_2 Cu(NCS)_2$ and in κ-$(ET)_2 Cu[N(CN)_2]Br$ (Sects. 5.3.1 and 2).

5.2.6 Anion Substitution

The appreciable pressure dependence of T_c in β-$(ET)_2 I_3$ leads to the idea that substituting the triiodide anion with other, smaller, polyhalide anions, such as $I_2 Br^-$, IBr_2^-, etc. and similar anions having linear structure such as AuI_2^-, would generate crystals having the same pressure effect with respect

Table 5.9. Lattice paramter change ($\Delta K/K$) for β-(ET)$_2$X as compared with β-(ET)$_2$I$_3$, and effective lattice pressure (ΔP) estimated by using the compressibility (dK/KdP) for β-(ET)$_2$I$_3$. From [5.137]

X	I$_3$	I$_2$Br		IBr$_2$		AuI$_2$	
	dK/KdP [%/kbar]	$\Delta K/K$ [%]	ΔP [kbar]	$\Delta K/K$ [%]	ΔP [kbar]	$\Delta K/K$ [%]	ΔP [kbar]
a [100]	−0.45	−0.045	0.10	−0.33	0.74	−0.18	0.40
b [010]	−0.33	−0.84	2.55	−1.37	4.19	−0.93	2.86
c [001]	−0.27	−0.62	2.32	−1.26	4.75	0.77	−2.89
V		−1.03	−1.59	1.54	−3.18	3.09	−1.25	1.21
a+b [110]	−0.48	−1.00	2.11	−1.95	4.11	−1.74	3.66
a−b [1−10]	−0.32	−0.33	1.04	−0.53	1.68	−0.13	0.41

to the intermolecular spacing. This we call the **lattice pressure effect**. Indeed it was found that β-(ET)$_2$IBr$_2$ and β-(ET)$_2$AuI$_2$ undergo a superconducting transitions at T_c = 2.8 K and T_c = 4÷5 K, respectively [5.135, 136]. However, β-(ET)$_2$I$_2$Br does not exhibit superconductivity although a transition at $T_c \approx 6$ K is expected, based on the lattice pressure effect as described in the following.

Table 5.9 compiles the relative change in lattice parameters of β-(ET)$_2$I$_2$Br, β-(ET)$_2$IBr$_2$, β-(ET)$_2$AuI$_2$ compared to those of β-(ET)$_2$I$_3$ and the corresponding effective lattice pressure along the crystal axes estimated by means of the compressibility [5.138] of β-(ET)$_2$I$_3$ in each direction. The values of the effective lattice pressure estimated in this way show a large anisotropy. Namely, a large lattice parameter change is observed along the b and c axes while only a small change is seen in the a direction. The small change in the a-axis is reasonable because the lattice constant in the a-direction is mainly governed by the side-by-side networks of ET molecules. On the other hand, the lattice parameters in the b- and c-directions are closely related to the anion sizes, that is, along the b-axis they are related to the anion length and along the c-axis to the anion width or thickness. Thus, by applying the lattice-pressure model to the change in the lattice parameter along the b-axis, one finds that β-(ET)$_2$I$_2$Br, β-(ET)$_2$IBr$_2$, and β-(ET)$_2$AuI$_2$ are similar to β_H-(ET)$_2$I$_3$ when it is placed under a pressure of ca. 2, 3 and 4 kbar, respectively. This is also consistent with the pressure dependence of β-(ET)$_2$IBr$_2$ [5.129] and β-(ET)$_2$AuI$_2$ [5.130] (Fig.5.45). Among these, β-(ET)$_2$I$_2$Br is thought to be a very attractive salt, but in reality it does not show superconductivity due to anion disorder [5.139, 140], as described in Sect.5.2.7.

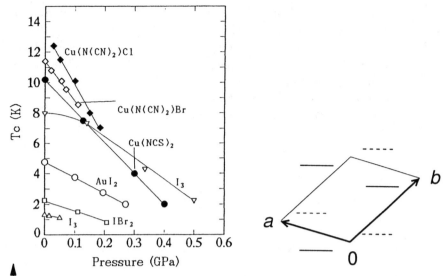

Fig. 5.45. Pressure dependence of T_c for $(ET)_2 X$ salts. X represents a counter anion.

Fig. 5.46. Two-dimensional array of ET molecules viewed along the molecular long axis. Two independent ET molecules are distinguished by *solid* and *dashed* lines

Table 5.9 lists the values obtained with the reduced lattice parameter $|a+b|$ which represents the spacing along the stacking direction (Fig. 5.46), **a** and **b** being the unit vectors in the a- and the b-directions, respectively. The lattice spacing for $a-b$ also reveals rather good agreement with the pressure dependence of T_c for β-$(ET)_2 I_3$, therefore indicating that the face-to-face spacing between ET molecules can be used to parameterize T_c.

In combination with the measured H_{c1} and H_{c2}, superconducting parameters such as the thermodynamic critical field H_c and the GL parameter κ are estimated on the basis of the effective mass model by assuming, for simplicity, that the ab-plane is isotropic [5.141, 142]. κ is an index denoting the easiness of the magnetic-field penetration in the superconductor. When $\kappa > 2^{-1/2}$, the superconductor is of the type II. For β-$(ET)_2 AuI_2$ at 1.2 K, from the relations

$$\kappa = \frac{H_{c2}}{\sqrt{2} H_c}, \tag{5.9}$$

or

$$\kappa = \frac{H_c}{\sqrt{2} H_{c1}} (\ln \kappa + 0.497) \tag{5.10}$$

175

the following values are deduced: $\kappa_{c*} = 16.6\pm2$, $H_c = 145\pm 10$ Oe for $H\|c^*$ and $\kappa_a = 177\pm15$, $H_c = 176\pm15$ Oe for $H\|a$. For β-$(ET)_2 IBr_2$ at 0.5 K, $\kappa_{c*} = 10.5\pm1.5$, $H_c = 87\pm10$ Oe for $H\|c^*$ and $\kappa_{\|(110)} = 143\pm 20$, $H_c = 144\pm 20$ Oe for $H\|(110)$, where H_c is somewhat dependent on the direction of the applied field H.

The **London penetration depths** parallel and perpendicular to the ab plane, $\lambda_\|$ and $\lambda_\perp$, for β-$(ET)_2 AuI_2$ are estimated from

$$\kappa_\perp = \lambda_\|/\xi_\| , \tag{5.11}$$

$$\kappa_\| = \frac{\sqrt{\lambda_\|\lambda_\perp}}{\sqrt{\xi_\|\xi_\perp}} , \tag{5.12}$$

where $\lambda_\perp = 4.1\cdot10^4$ Å and $\lambda_\| = 5.1\cdot10^3$ Å has been obtained at 1.2 K [5.141].

The **specific heat** of β_L-$(ET)_2 I_3$ measured down to 0.7 K shows a linear dependence, but no indication of the expected anomaly at T_c. In contrast, in β-$(ET)_2 AuI_2$ with $T_c = 5$ K a distinct peak in the specific heat is found although it is somewhat broadened [5.143]. This study was followed up by a more detailed investigation [5.144] where the difference in specific heats, when the sample is kept at a magnetic field of 30 kOe and zero field, was examined. By assuming that the broadness in T_c is due to sample inhomogeneity and by utilizing the BCS-like relation for the specific-heat part due to the electronic motion, $\gamma = 0.024$ J/(K^2 ·mol) could be determined.

The **superconducting energy gap** for β-$(ET)_2 AuI_2$, measured by the tunneling method, gives $2\Delta/k_B T_c \approx 4$ and is consistent with the BCS model [5.145]. However, it should be noted that in tunneling measurements using point contacts, significantly larger values (almost 4 times) than those predicted by the weak-coupling BCS theory were obtained in a given direction [5.146]. At present no explanation has been found for the anisotropic behavior.

5.2.7 Effect of Disorder

A comparison of the β-$(ET)_2 X$ with $X = (I_3)_{1-x}(IBr_2)_x$ $(0\leq x\leq 1)$ is expected to show the consistency of the lattice-pressure model. For these salts the lattice pressure changes continuously with x. However, the experimental results for mixed anions clearly reveal the effect of disorder due to the random distribution of the different anions. The randomness rather than the lattice-pressure effect [5.115] dominates the salt properties for $x \geq 0.1$.

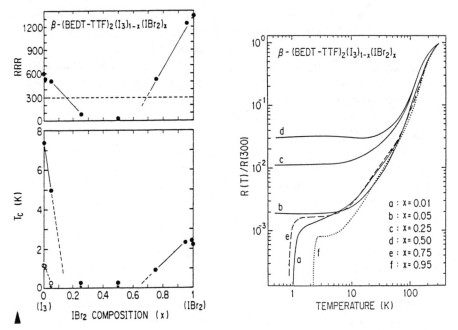

Fig. 5.47. Composition dependence of T_c and RRR (ratio of residual resistance near T_c to resistance at room temperature) for the binary mixed anion system β- $(ET)_2 (I_3)_{1-x}(IBr_2)_x$. (○: low-$T_c$ state. *Arrow* indicates a nonsuperconducting statedown to 0.5 K). From [5.115]

Fig. 5.48. Temperature dependence of the normalized resistivity of β-$(ET)_2$X. From [5.115]

Since iodine is the largest halogen atom, a substitution of iodine with smaller halogen atoms, such as bromine, results in lattice contraction. By mixing together two out of the three anions I_3^-, $I_2 Br^-$ and IBr_2^-, effects due to lattice contraction can systematically be studied when the crystal structure remains of the β-$(ET)_2 I_3$-type. The growth of β-$(ET)_2 X_{1-x} Y_x$, where X, Y are I_3, $I_2 Br$, or IBr_2, and $0 \leq x \leq 1$ has been carried out by electrochemical methods [5.147]. X-ray analysis for these mixed anion ET salts has confirmed that the lattice parameters change monotonically with x.

First, we examine the results of β-$(ET)_2 (I_3)_{1-x}(IBr_2)_x$, where salts with the two limiting compositions β-$(ET)_2 I_3$ and β-$(ET)_2 IBr_2$ are superconducting (Fig. 5.47). At ambient pressure, β-$(ET)_2 I_3$ exhibits an incommensurate superstructure below 175 K which can be seen as an anomaly in the dR/dT-vs.-T dependence. The superstructure is detected for x = 0.05, but not for x = 0.25; the phase boundary must be between x = 0.05 and 0.25. For x = 0.05, superconductivity is not observed down to 0.5 K. Repeating the measurements under pressure revealed that the high-T_c state is left, with T_c = 5 K at 1.5 kbar. On the other hand, β-$(ET)_2 I_3$ alloyed with 25% β-$(ET)_2 IBr_2$ shows neither a superconducting transition nor the 175 K transition under ambient pressure, suggesting that although the superstructure is suppressed, superconductivity is also destroyed by **alloying**. In fact, again repeating the measurements under an applied pressure confirmed the absence of a superconducting phase. This indicates that although the incommensurate lattice modulation is eliminated, the suppression of T_c by alloying is so fast that no superconductivity is observed, at least down to 0.5 K. If one varies x by decreasing it from 1.0, T_c also decreases and superconductivity is not observed down to 0.5 K for x = 0.5.

If the idea of a lattice-pressure effect is valid for mixed-anion systems, T_c should change monotonically from x = 0 to 1.0. In this case it is natural to take the value of β_H-(ET)$_2$I$_3$ for x = 0 that is denoted by the filled circles in the figure, since it is on the same line with the anion-substituted salts in a sense that they do not exhibit the incommensurate lattice modulation. However, the experimental results are far from this expectation. A clue to this discrepancy may be the composition dependence of the residual resistance at low temperatures. The temperature dependence of the resistivity normalized to the room-temperature value is depicted in Fig. 5.48. Since all the room-temperature resistivities are in the same range within the experimental uncertainty $\rho = (5.0 \pm 2.5) \cdot 10^{-2} \, \Omega \cdot cm$, the ratio of the resistances at low temperatures almost directly reflects the relative ratio of the absolute resistivity. Accordingly, the samples with a high alloy ratio give higher residual resistance and do not exhibit superconductivity. The x dependence of the ratio of the Residual Resistance at T_c to the Room-temperature value (RRR in Fig. 5.47), clearly shows a correlation between T_c and RRR, indicating that the random potentials introduced by alloying suppress superconductivity.

The absence of superconductivity for β-(ET)$_2$I$_2$Br mentioned in Sect. 5.2.6 can be understood on the same basis. The I-I-Br anions have no centrosymmetry and are randomly oriented in this salt, thereby generating random potentials which can cause electron localization.

It has been argued that T_c of the singlet superconductor should not be affected by **nonmagnetic defects** [5.148]. Based on the sensitivity to the disorder, the possibility of spin-triplet superconductivity has been discussed. However, the sensitivity to disorder has also been explained by *Hasegawa* and *Fukuyama* [5.149] in terms of an enhancement of the Coulomb repulsion due to randomness. They discussed the phenomenon of **electron localization** on superconductivity in low-dimensional systems in terms of a dirty singlet superconductor model assuming **s-wave** superconductivity. By using this model it can be shown that in the weakly localized regime, the enhancement of the Coulomb repulsion due to randomness results in the reduction of T_c. By representing the degree of localization through the scattering rate of the conduction holes, $1/\tau$, the residual-resistivity dependence of T_c was calculated. It was found that T_c is reduced to zero for β-(ET)$_2$X when the residual resistance becomes higher than $1.6 \cdot 10^{-4}$ $\Omega \cdot$cm. According to Fig. 5.47 the superconductivity is suppressed when RRR becomes less than 300. If the room-temperature conductivity is taken to be ca. 20 S/cm, the residual resistivity needed to suppress the superconductivity is $1.7 \cdot 10^{-4}$ $\Omega \cdot$cm, in excellent agreement with the calculated value. For the binary $(ET)_2 \cdot (IBr_2)_{1-x}(I_2Br)_x$ and $(ET)_2(I_2Br)_{1-x}(I_3)_x$ systems, a similar correlation between RRR and T_c has been found. This implies that low-dimensional organic superconductors are very sensitive to crystal imperfections and that T_c is remarkably reduced due to the electron localization by random potentials.

Radiation damage defects also create random potentials in crystals. Experiments with doped and with irradiated crystals provide complementary results because the chemical method produces a large number of weak random potentials, while irradiation makes large defects in concentrations as low as desired [5.150]. Low doses of irradiation have been seen to suppress superconductivity in $(TMTSF)_2PF_6$, as described in Sect. 3.2.4. The question is, how sensitive are ET salts to irradiation? Another problem of interest is whether the incommensurate lattice modulation is prevented by radiation damage. Such questions have been addressed in a study of the phenomenon of proton irradiation at levels corresponding to the estimated defect concentrations between 10 and 4800 ppm [5.151]. The effects on the electrical conductivity are summarized in Fig. 5.49. The superconducting transition of sample #1 with 10 ppm damage under ambient pressure is almost identical to the non-irradiated samples, with $T_c = 1.1$ K. In sample

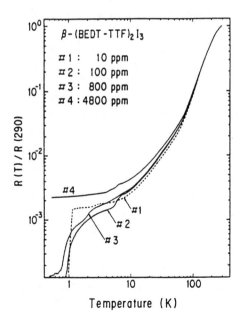

Fig. 5.49. Temperature dependence of resistances normalized to the value at 290 K for irradiated β-$(ET)_2I_3$. From [5.151]

#3 with 800 ppm damage, irradiation results in an extreme broadening of the superconducting transition on the higher-temperature side but T_c is reduced to 0.8 K. In sample #4 with 4800 ppm damage, no indication of a low-T_c transition appears at least down to 0.5 K under ambient pressure. On the other hand, a stepwise increase is observed around 7 K, which is possibly related to the high-T_c superconducting transition. This was confirmed by repeating the measurements with an application of 1.5 kbar pressure under which the superconducting transition was indeed observed. As for the incommensurate lattice modulation, the resistivity measurement revealed that the lattice modulation is not suppressed even at the 4800 ppm damage level. The increase in the residual resistance at low temperature is also moderate (Fig.5.49). Thus, in ET salts the effect of irradiation is very small compared to $(TMTSF)_2 PF_6$, where a 100 ppm damage level remarkably modifies both the low-temperature electron transport and T_c [5.152]. It is supposed that the ET molecule is more stable against irradiation than TMTSF because its weakest chemical bonds are stronger than in TMTSF. In TMTSF the damage is thought to occur at $-CH_3$ groups, whereas in ET they are replaced with $-CH_2-$.

5.2.8 Tempered ET Salts

ET salts are unique in that they are polymorphous. There is a wide variety of grown crystal types. Furthermore, some crystal types are drastically changed by either heat treatment or pressure: when crystals are kept under various heating conditions, not only the products vary in composition but also in structure.

When the ET salt $(ET)_2 I_3 (I_8)_{0.5}$, called ϵ-**type**, which belongs to the $P2_1/c$ space group [5.11], is kept at 70°C for 8.5 hours, the X-ray diffraction pattern of the ϵ-type crystal vanishes and a pattern corresponding to the β-$(ET)_2 I_3$ crystal appears [5.153]. The resistivity measurement of this tempered crystal revealed that it undergoes a superconducting transition near $7 \div 8$ K. This ϵ to β conversion was monitored by observing the X-ray diffraction after each $1.5 \div 2$ hour heating period. The product has a twinned structure with a twofold axis coinciding with a monoclinic axis of the original ϵ-phase. The twin junction lies in the (100) plane of the original ϵ-phase. The small number of reflections and their diffuse nature show that the conversion from ϵ- to β-types produces low-quality **mosaic crystals**.

Baram et al. [5.154] also found a surprising structural transformation of the α-$(ET)_2 I_3$ crystals into the β-phase by tempering at $70° \div 100°$C for about 10 to 20 hours. The stacking of the ET molecules is quite different in these crystal types; the unit cell of the α-$(ET)_2 I_3$ at room temperature is about twice as large as the unit cell of the β-$(ET)_2 I_3$ (1717Å versus 849Å)

(Sect. 5.1.2). The derived crystal is referred to as α_t-$(ET)_2I_3$ to distinguish it from the ordinary electrochemically prepared crystals. This tempered crystal exhibits superconductivity at 8 K under ambient pressure, just like β_H-$(ET)_2I_3$. The AC susceptibility measurement shows that this is a bulk superconducting phase, and the critical field H_{c2} is also similar to that observed for β_H-$(ET)_2I_3$ [5.155].

The emergence of 8-K superconductivity in the two tempered crystals indicates that the β_H-$(ET)_2I_3$-like structure is maintained down to low temperatures without adopting incommensurate lattice modulation. This is either due to the presence of twinning or domain boundaries which are inevitably formed during the conversion, or due to some ordering process of the ethylene groups in the ET molecules. In these cases the boundaries between the domains introduce macroscopic imperfections but presumably no microscopic ones which may directly relate to the localization of the electrons.

Stabilization of the β_H-like phase in the tempered crystal suggests that tempering electrochemically grown β-$(ET)_2I_3$ might also produce a stable β_H-$(ET)_2I_3$. A trial tempering period of twelve days at 85°C was found to be only partly successful: although the resistivity curve at low temperatures of the tempered β_t-crystal is very similar to the α_t-crystals, the AC susceptibility measurements indicated a superconducting transition below 4.5 K and that, at 1.3 K, at least 50% of the material is superconducting [5.155]. This transition temperature of 4.5 K is below the value for α_t-crystals but clearly above that for β_L-$(ET)_2I_3$.

Finally, the structural phase transition between the $\beta \leftrightarrows \alpha$ phases for $(ET)_2IBr_2$ [5.156] should be mentioned. Applying pressure at room temperature changes β-$(ET)_2IBr_2$ into the α-type crystal at 18 kbar. This was found by resistivity measurements and confirmed by the X-ray diffraction method. The converted crystal phase is stable and has been kept more than two years. To convert the crystal from the α-phase to the β-phase it must be heated to a temperature of $153° \div 155°C$ under ambient pressure. The resulting crystal undergoes a superconducting transistion at T_c identical to that grown electrochemically. The $\alpha \rightarrow \beta$ phase transition in the $(ET)_2IBr_2$ occurs with an absorption of heat, but the latent heat is an order of magnitude lower than that for the $\alpha \rightarrow \beta$ transition in $(ET)_2I_3$. Furthermore, the phase transition $\alpha \rightarrow \beta$ in $(ET)_2IBr_2$ occurs abruptly and apparently is of a martensitic nature. The inverse $\beta \rightarrow \alpha$ transition under pressure is of the same nature.

5.3 Superconductivity of κ-(ET)$_2$X

5.3.1 κ-(ET)$_2$Cu(NCS)$_2$

The β-(ET)$_2$X and (TMTSF)$_2$X organic superconductors are dissimilar in dimensionality and crystal morphology, but they have similar intermolecular configurations in the face-to-face packing. This linear stacking configuration had been regarded as one of the necessary prerequisites, not only for organic superconductivity, but also for organic metallic conductivity. This, however, was disproved by the κ-type salts such as κ-(ET)$_2$I$_3$ salt. The structure of κ-(ET)$_2$I$_3$ is the checkered arrangement of ET molecules (Fig.5.6), as described in Sect.5.1.2. Molecular orbital calculations have shown that this configuration gives a quasi-two-dimensional band structure and a potentially closed Fermi surface in the conducting plane.

The pressure phase diagram of β_H-(ET)$_2$I$_3$ indicates that T_c tends to increase with decreasing P, and if the pressure were applied negatively, T_c should continue to increase. In reality, it is difficult to apply negative pressure, but the concept of the lattice-pressure effect can be applied. In this case, the intermolecular distances are expanded by using counter anions which force the separation between adjacent ET's. For example, anions of elongated structure such as (SCN)-M-(NCS)$^-$ [5.98], M being a transition-metal atom, have been used. *Urayama* et al. first developed the ET salt with Cu(NCS)$_2^-$ as the counter anion, and, in fact, it produced $T_c = 10.4$ K in the resistive measurement [5.157]. The stacking structure of κ-(ET)$_2$Cu$\cdot$(NCS)$_2$ crystal is quite different from that of β_H-(ET)$_2$I$_3$. The former utilizes checkered packing as in κ-(ET)$_2$I$_3$: two ET molecules form a dimerized pair and the pairs are linked almost perpendicularly to one another to construct two-dimensional conducting sheets. Every conducting layer is sandwiched between two insulating Cu(NCS)$_2^-$ layers (Sect.5.1.2). The bc-plane corresponds to the two-dimensional conducting plane. The calculated Fermi surface is composed of a two-dimensionally closed cylinder (Fig.5.14), and modulated in one-dimensional-like planes [5.24]. The room temperature conductivities are $\sigma_b = 17\pm3$ S/cm, $\sigma_c = 32\pm8$ S/cm and $\sigma_a = 0.022 \pm 0.003$ S/cm [5.158].

Resistive Transition to Superconductivity

Figure 5.50 presents the temperature dependence of the resistivity along the b-axis. At ambient pressure the resistivity reveals a pronounced maximum around 90 K. After passing through the peak, the resistivity decreases rapidly and shows a sharp resistance drop at the onset of superconductivity (11.2K). The transition temperature determined from the midpoint of the resistivity drop is 10.4 K. With increased pressure, the resistivity peak is

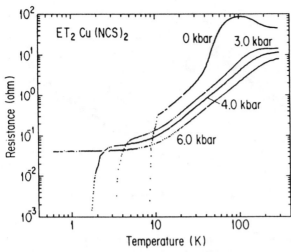

Fig. 5.50. Temperature dependence of the resistivity along the b-axis for κ-$(ET)_2 Cu(NCS)_2$ at several pressures. From [5.158]

suppressed, and T_c is reduced. At 3 kbar the high-temperature peak disappears while the residual resistance decreases at low temperature. When the pressure reaches 6 kbar, the superconductivity disappears, indicating that the overall pressure dependence is as high as $dT_c/dP = -1.3$ K/kbar, which is even higher than that for β_H-$(ET)_2 I_3$ (Sect. 5.2.3). (If we see an initial slope near ambient pressure, $dT_c/dP \simeq -3$ K/kbar [5.159]).

The width of the superconductivity transition, between the onset and the offset of a resistive change, is rather large ≈ 1 K, even at zero magnetic field. When a magnetic field is applied perpendicularly to the conductive plane, the transition region is remarkably broadened and shifts to lower temperatures. As shown in Fig. 5.51a for the in-plane resistivity, the resistive transition extends over more than 2 K at 0.4 T, where the mean-field transition point is difficult to be determined. In early stages, either onset or mid-point (or other appropriate points) of the transition had been adopted to derive the mean-field H_{c2}, for convenience. Its temperature dependence was analyzed to find the GL coherence length. However, due to the two-dimensionality, which is demonstrated, for example, by a torque magnetometry leading to the effective mass anisotropy of $4 \cdot 10^4$ [5.161] and the rather high T_c, the effect of **thermal fluctuation** cannot be neglected, in particular, under the influence of a magnetic field [5.162, 163]. In this situation, H_{c2}, the onset point of the long-range ordering, becomes difficult to be defined. The dominance of the thermal fluctuation in two-dimensional superconductors has been recognized through a study of high-T_c cuprate superconductors [5.164].

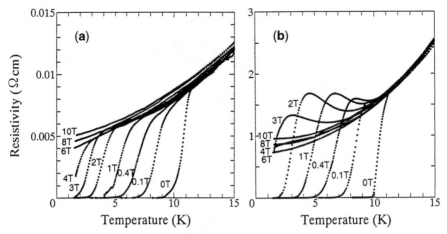

Fig. 5.51. Temperature dependence of the in-plane (**a**) and interplane (**b**) resistivities of κ-(ET)$_2$ Cu(NCS)$_2$ under the magnetic field applied perpendicular to the conducting plane. From [5.160]

In κ-(ET)$_2$ Cu(NCS)$_2$, the resistivity along the normal to the two-dimensional plane exhibits an upturn just above the superconducting state (Fig. 5.51b). A similar resistive-transition behavior is sometimes observed even in the in-plane resistivity when it is considered to contain stacking faults such as dislocations within the conductive plane [5.165]. The local maximum of the resistivity near the superconductivity transition has been found even in other κ-type ET superconductors, and its origin has been argued with respect to magnetic scattering [5.166] and localization [5.167] for normal electrons. However, since the local maximum is suppressed by a high magnetic field enough to destroy superconductivity, it is evident that the phenomenon is related to superconductivity. It can be ascribed to the contribution from **Josephson-coupled** superconductivity layers, existing in the direction normal to the plane in the stacking-fault region [5.160].

Magnetic Transition to Superconductivity

The magnetization measurements bring to light that superconductivity is a bulk phenomenon. The temperature dependence of the lower critical field H_{c1} is displayed in Fig. 5.52. The temperature dependence of the DC magnetization in the reversible region, where the magnetization is not affected by vortex pinning, is plotted in Fig. 5.53 as a function of the magnetic field applied perpendicular to the conductive plane. The data can be analyzed on the basis of the fluctuation-renormalization theory (Appendix B). The resultant superconductivity parameters, the transition temperature in the absence of a magnetic field, T_{c0}, GL coherence lengths parallel to plane at 0 K, $\xi_\parallel(0)$, and perpendicular to the plane, $\xi_\perp(0)$, are compiled in Table

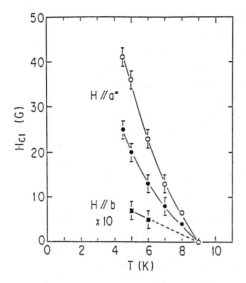

Fig. 5.52. Lower critical field measured with the applied field directed along the a and b axes as a function of temperature for κ-(ET)$_2$ Cu(NCS)$_2$. *Closed marks* correspond to the fields where the initial magnetization starts to deviate from linearity and *open circles* to the filed of maximum magnetization. From [5.168]

5.10, together with those of the isotope-substituted salts. The values evaluated based upon the mean-field treatment [5.170] are also listed. It is notable that T_{c0} is determined somewhat lower by a two-dimensional treatment, which agrees with the result obtained via specific heat measurements [5.171, 172]. The GL coherence lengths become shorter than those obtained by applying the mean-field model.

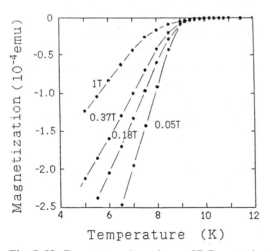

Fig. 5.53. Temperature dependence of DC magnetization for κ-(ET)$_2$ Cu(NCS)$_2$ under the magnetic field applied perpendicular to the conducting plane. Lines are guide for eyes. From [5.162]

Table 5.10. Listing of GL coherence length $\xi(0)$ values for κ-(ET)$_2$Cu(NCS)$_2$ and κ-(ET)$_2$Cu[N(CN)$_2$]Br

Material	T_{c0}	$\xi_\perp(0)$ [nm]	$\xi_\parallel(0)$ [nm]	Ref.
κ-(ET)$_2$Cu(NCS)$_2$	8.7 ± 0.2	0.31 ± 0.05	2.9 ± 0.5	5.162
	9.4	0.3	6.5	5.172
	10.4	0.96	18.2	5.166
	10.5	0.77	14.3[a]	5.158
κ-(d$_8$-ET)$_2$Cu(NCS)$_2$	9.0 ± 0.2	0.32 ± 0.05	2.9 ± 0.5	5.162
κ-(^{13}C-ET)$_2$Cu(NCS)$_2$	8.6 ± 0.2	0.31 ± 0.05	2.9 ± 0.5	5.162
κ-(ET)$_2$Cu[(N(CN)$_2$]Br	10.9 ± 0.2	0.58 ± 0.1	2.3 ± 0.4	5.162
	10.8 ± 0.05	0.4	3.7	5.163
κ-(d$_8$-ET)$_2$Cu[N(CN)$_2$]Br	10.6 ± 0.2	0.57 ± 0.1	2.3 ± 0.4	5.162
κ-(^{13}C-ET)$_2$Cu[N(CN)$_2$]Br[b]	10.9 ± 0.2	0.58 ± 0.1	2.3 ± 0.4	5.162
κ-(^{13}C-ET)$_2$Cu[N(CN)$_2$]Br[c]	12.2		6.0	5.169

[a] This value is given by $\sqrt{\xi_b(0)\xi_c(0)}$, where $\xi_b(0) = 17.4$ nm and $\xi_c(0) = 11.8$ nm

[b] Enriched ^{13}C at CH$_2$ sites

[c] Enriched ^{13}C at central C sites

Specific Heat

Specific heat measurements using a collection of single crystals revealed a distinct anomaly, when compared to normal-state data obtained by suppressing superconductivity with a magnetic field [5.171]. The ratio of the specific-heat jump ΔC at T_c was estimated to be larger than 50 mJ/ (K$^2 \cdot$mol). By evaluating γ from the normal-state specific heat to be 25 ± 3 mJ/(K$^2 \cdot$mol), the ratio of $\Delta C/\gamma T_c$ is estimated to be larger than 2. Independently, high-resolution specific heat measurements on a single crystal in magnetic fields of up to 50 kOe revealed a jump of ΔC at T_c (Fig. 5.54). In the measurements, the data for the magnetic field of 50 kOe applied perpendicular to the b-c plane are taken as background consisting of the contributions from lattice and normal electrons. Using $\gamma = 34$ mJ/ (K$^2 \cdot$mol) derived from Pauli's paramagnetism [5.173], $\Delta C/\gamma T_c$ is evaluated to be 1.5 ± 0.15 and becomes rather consistent for the two measurements. Here we note that $\Delta C/\gamma T_c$ was evaluated to be 1.62 ± 0.2 at $T_c = 3.4$ K with $\gamma = (1.89 \pm 1.5)$ mJ/(K$^2 \cdot$mol) for κ-(ET)$_2$I$_3$ [5.174].

Energy Gap

With regard to the superconductivity energy gap, tunneling spectroscopy provided $2\Delta/k_B T_c$ up to 4.5 (the BCS value is equal to 3.53) [5.175, 176]. On the other hand, far-infrared absorption studies which covered the fre-

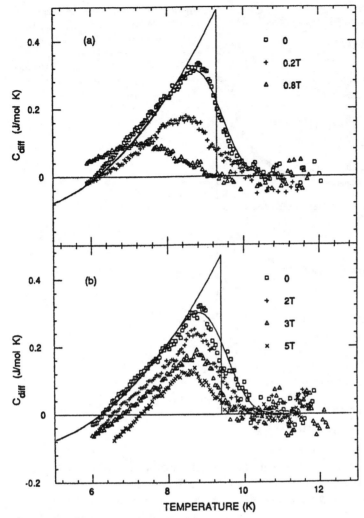

Fig. 5.54. Specific heat difference $\Delta C = C(H_\perp) - C(H_\perp = 5\,T)$ **(a)** for $H_\perp = 0, 0.2$ and 0.5 T, where $H_\perp$ is the field applied perpendicular to the b-c plane, **(b)** $\Delta C = C(H_\parallel) - C(H_\perp = 5\,T)$ for $H_\parallel = 0, 2, 3,$ and 5 T lying in the b-c plane. From [5.172]

quency range from 10 to 380 cm^{-1} could not find a conventional BCS gap. The measured absorption spectra did show neither a superconductivity gap nor interesting phonon features [5.177]. Resonant Raman spectra neither indicate a superconductivity gap opening [5.178].

Penetration Depth

Measurements of the **surface impedance** in the millimeter-wave frequency range $(1 \div 3$ cm$^{-1})$ revealed that the material behaves like a metal with a

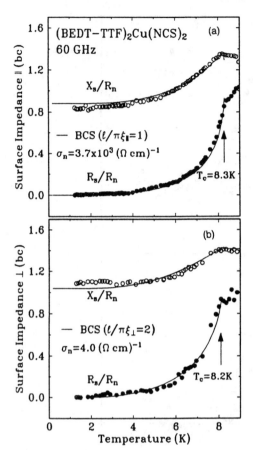

Fig. 5.55. The surface impedance of κ-$(ET)_2Cu(NCS)_2$ at 60 GHz as a function of temperature. R_S and X_s are the surface resistance and reactance, respectively, and R_n is the surface resistance at 9 K. (a) $E_{ac} \parallel c$, (b) $E_{ac} \parallel a$. From [5.179]

scattering rate of $\approx 6 \cdot 10^{11}$ s^{-1} in the normal state; the electrodynamics is in good agreement with calculations based on a BCS ground state [5.179]. The penetration depth and the coherence length are anisotropic, but the superconducting energy gap shows no indication of line nodes. Figure 5.55 depicts the resistance R_s and reactance X_s parts of the surface impedance. The sudden drop of R_s below T_c is expected for a superconductor at frequencies below the single-particle gap 2Δ. From the fit of the temperature dependence of R_s and X_s, $\omega\tau_\parallel = 0.15$ at 60 GH is deduced. The reactance is directly proportional to the microwave-penetration depth. The measured penetration depth $\lambda(T)$ is displayed together with other reported data in Fig. 5.56, which indicates a flat behavior for low temperatures. This is in agreement with the BCS prediction and conventional *s-wave pairing*. The results are consistent with those obtained by **muon-spin rotation** measurements by *Harshman* et al. [5.181], but controversial deviations from the BCS behavior below 1.5 K were reported by *Le* et al. [5.182]. In relation to this we should also note the controversy between **low-frequency magnetic**

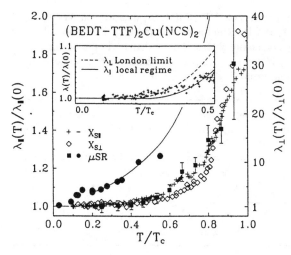

Fig. 5.56. Temperature dependence of penetration depth measured at 35 GHz (pluses) and 60 GHz (*open diamonds*). *Solid lines* are by *A chkir* et al. [5.180]. *Solid boxes* and *solid circles* are from μSR experiments by *Harshman* et al. [5.181] and by *Le* et al. [5.182]. The experiments probe $\lambda_{\parallel}$, except for the *open diamonds* where $\lambda_{\perp}$ corresponds to right axis. The *inset* demonstrates the comparison with BCS theory where the *dashed line* represents the London regime and the *solid line* the local regime. From [5.179]

measurements, exhibiting a strong temperature dependence of $\lambda(T)$ which suggests *unusual pairing* [5.183] and reversible magnetization that provides evidence of conventional Cooper pairing [5.184]. For low-field AC screening, *Hebard* et al. [5.185] claimed that an extrinsic temperature dependence is caused by existing defects through the nucleation of vortex pair, resulting in suppression of the screening. Table 5.11 lists the electrodynamical parameters determined by the millimeter wave measurements of *Dressel* et al.

Proton NMR Study

Measurements of NMR relaxation in the transition region towards superconductivity are interesting to understand the nature of the superconductor under investigation. However, the proton NMR relaxation rate $1/T_1$ exhibits an enormous increase below T_c, forming a sharp peak around 4 K, which then drops rapidly at lower temperatures [5.186]. The peak value of $1/T_1$ is 30 times as large as that at T_c in a 3-kOe field. The enhancement tends to be depressed in a magnetic field (Fig. 5.57). In BCS superconductors the enhancement, typically by a factor of 2, occurs just below T_c: it reaches a maximum around 0.9 T_c and then decreasing exponentially due to a finite gap opening. The present results differ from this in many respects. First, the peak of $1/T_1$ appears well below T_c. At higher fields the enhance-

Table 5.11. Electrodynamical properties of $(ET)_2 Cu(NCS)_2$ measured at 60 GHz, and $(ET)_2 Cu[N(CN)_2]Br$ measured at 35 GHz with the current flowing either parallel ($\parallel$) or perpendicular ($\perp$) to the highly conductiong plane (σ_n: resistivity in the normal state, δ: skin depth, ℓ: mean-free path, $1/(2\pi\tau)$: single-particle relaxation rate in the vicinity of the superconductivity transition. From [5.179]

κ-(ET)$_2$X X		T_c [K]	σ_n (T > T$_c$) [S/cm]	δ [μm]	λ [μm]	ℓ [nm]	ξ [nm]	$1/(2\pi\tau)$ [1/cm]	2Δ [1/cm]
Cu(NCS)$_2$	$\parallel$	8.3	$3.7\cdot10^3$	3.4	1.4	15	7	1.5	21
	$\perp$	8.2	4.0	100	40	3	0.5	5	21
Cu[N(CN)$_2$]Br	$\parallel$	11.3	$4\cdot10^3$	4.2	1.5	29	3.7	30	40
	$\perp$	11.3	6.4	110	38	5	0.6	5	40

ment of $1/T_1$ is reduced and the peak position shifts towards a lower temperature while at lower temperatures $1/T_1$ is less field-dependent. Since the sample used was polycrystalline, T_c(H) should differ from crystal to crystal due to the anisotropy of H_{c2}. However, at 3.28 kOe, for example, the maximum and the minimum T_c are expected at 10 and 8 K, respectively, and the shift of the peak cannot be explained in terms of different super-

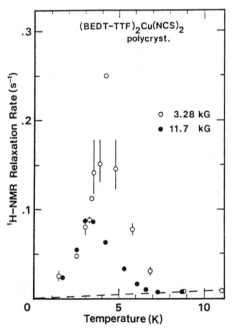

Fig. 5.57. Temperature dependence of proton NMR $1/T_1$ of κ-(ET)$_2$ Cu(NCS)$_2$ for two applied magnetic fields. From [5.186]

conducting states for each crystal. Furthermore, the size of the peak is too large.

One should note that a similar peak is observed in β-(ET)$_2$I$_3$, as described in Sect. 5.2.5, and also in κ-(ET)$_2$Cu[N(CN)$_2$]Br. As a possible origin of the anomalous peak the consequence of the field fluctuations generated by the **vortex motion** is considered. The relaxation comes from the Brownian motion of diffusive defects is a two-dimensional vortex system. The peak temperature corresponds to the melting of the vortex solid [5.187].

Vortex Pinning

With respect to vortex pinning in the superconducting state, the dynamical properties were studied by means of magnetization relaxation (creeping) for a slow process and the response to vibrations for a rapid process. The logarithmic relaxation rate of the magnetic moment in the superconducting state is observed and explained in terms of a thermally activated vortex flow. From the temperature dependence the average energy of a vortex-pinning center, U, is determined to be $7.2 \cdot 10^{-3}$ eV [5.188]. At very low temperatures, below about 0.5 K, the decay rate becomes temperature-independent, being understood that the process involves mainly the quantum tunneling process [5.189].

The depinning of a vortex has also been studied through the response to mechanical vibrations. Using the vibrating reed technique [5.190] or the vibrating superconductor method [5.191], measurements of damping and resonance-frequency shift for different geometries have been carried out as a function of temperature T and magnetic field H to find thermally activated depinning. The dissipation becomes maximum at T_D when $\omega\tau \approx 1$, ω being the angular frequency of the oscillation. The factor τ is proportional to $\exp(U/k_B T)$, where U is an activation energy determined by T and H. The measurements were carried out by keeping ω almost constant (around the mechanical resonance frequency). T_D depends on H (applied along the a axis) as represented by $dT_D/d\ln(H) \approx -1.2$ K in the field range of 1 to 10 kOe [5.190], whereas $dT_D/d\ln(H) \approx -0.76$ K below 1 kOe [5.191]. With an increase of H, the pinning is weakened and a vortex is driven more easily. Vortex-pinning characteristics can also be observed by ultrasonic attenuation [5.192] and mutual conversion between electromagnetic and ultrasonic waves [5.193].

With an increase of temperature, thermally activated depinning leads to a finite resistivity, which increases with the magnetic field. The resistivity shows approximately an exponential temperature dependence. It is superimposed upon the resistivity realized under thermal fluctuation; the overall resistivity is determined by thermal fluctuation and depinning. On cooling, the finite resistivity caused by fluctuations is suppressed by the pinning of vortices, resulting in zero resistance at low temperature.

Josephson Vortex

The structure of a two-dimensional superconductor is represented by an alternate stack of superconducting and non-superconducting sheets, as modeled by *Lawrence* and *Doniach* [5.194] (Appendix B). In this case, the response to a magnetic field is quite different depending on whether it is applied parallel or perpendicular to the sheets. For the parallel case, the core of a vortex fits between the sheets and its presence does not require an additional suppression of superconductivity. If the perpendicular component of the applied field is too weak to create normal cores in a sheet, the vortices are considered to be locked between the sheets but move freely along the sheets. This results in a **vortex lock-in state**, which can be observed by AC susceptibility measurements in a DC field [5.195]. The DC field applied parallel to the conducting planes causes a rapid decrease in the screening of the AC (2.5kHz) field. The pinning-force constant $k_p^{\parallel}$ for vortices moving parallel to the sheets (Josephson vortex) is extremely weak; its ratio to the pinning force for perpendicular vortices $k_p^{\perp}$ is evaluated to be $\approx 2 \cdot 10^{-3}$. The penetration depth associated with interlayer currents is found to be $\lambda_{\perp}$ (5K) ≈ 200 nm. Combined with $\lambda_{\parallel} \approx 0.6 \div 1.2$ μm from other measurements, this gives an anisotropy of $\gamma \approx 160 \div 330$. Further evidence of the weak coupling between superconducting sheets by the Josephson mechanism has been provided by the observation of microwave emission on application of a DC voltage due to the AC **Josephson effect** (Fig.5.58).

Effect of Uniaxial Strain

Hydrostatic pressure decreases T_c of κ-$(ET)_2 Cu(NCS)_2$ very rapidly. Then, it is interesting to see the effect of expansion. This is possible by elongating a crystal uniaxially. Furthermore, it is important to find the effects due to the change in intermolecular spacing in different crystalline orientations with respect to the pronounced anisotropy of the material. The application of uniaxial tensile stress was tried by fixing two ends of a single crystal to a substrate of copper or fused quartz and utilizing the differential thermal contraction between the crystal and the substrate; the thermal contraction of the latter is rather negligible compared to the former. When the assembly is cooled, an estimated tensile strain of $\Delta b/b \approx 9.5 \cdot 10^{-3}$ is applied and T_c was raised by (0.6 ± 0.5) K. On the other hand, on simultaneous application of elongational strain in the c-direction ($\Delta c/c \approx 2.7 \cdot 10^{-2}$) by fixing the planar surface of a thin crystal, the shift was represented by (0.0 ± 1.0) K, in contradiction to the expectation of an increase in T_c by ≈ 4 K according

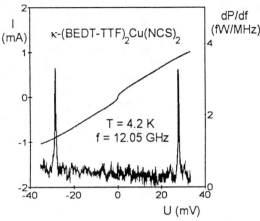

Fig. 5.58. V-I characteristics and microwave emission of a κ-$(ET)_2 Cu(NCS)_2$ single crystal. The peak of emission corresponds to a series of ≈ 1100 Josephson junctions. From [5.196]

to the BCS formula with an evaluated density of states that is based on the Hückel calculation [5.197]. The remarkable discrepancy implies the importance of the interlayer interactions, which are neglected in the calculation and in the model for the superconducting mechanism. Later work showed that the application of uniaxial compressional stress in the a-direction produces a decrease of T_c by $dT_c/dP_a = -2$ K/kbar [5.198]. It is probable that the enhancement of T_c due to the elongation of the bc plane is canceled by the associated compression in the a-direction appearing with the Poisson ratio.

On the other hand, the uniaxial-stress dependence of T_c has been obtained from the jump in the **thermal expansion** using Ehrenfest's relation as $\partial T_c/\partial P_a = -3.20$ K/kbar, $\partial T_c/\partial P_c = 1.46$ K/kbar and $\partial T_c/\partial P_b \simeq 0$ by *Kund* et al. [5.199]. Similar work carried out by *Lang* et al. gave $\partial T_c/\partial P_a = -4.8$ K/kbar and $\partial T_c/\partial P_c = -1.1$ K/kbar [5.200]. The origin of the difference is not clear. However, to explain the positive sign for the c-direction which means an increase of T_c with a decrease in the length along the c-axis, the contribution of a complex phonon spectrum is considered.

5.3.2 κ-$(ET)_2 Cu[N(CN)_2]Br$

The κ-type $(ET)_2 X$ superconductors with T_c higher than 10 K have been developed by adopting polymeric anions such as $Cu[N(CN)_2]Br^-$, $Cu \cdot [N(CN)_2]Cl^-$ and $Cu(CN)[N(CN)_2]^-$ with T_c of 11.8 K, 12.8 K and 11.2 K, respectively, determined by a resistive measurement. Among the salts, the κ-$(ET)_2 Cu[N(CN)_2]Br$ crystal can be prepared rather easily and exhibit su-

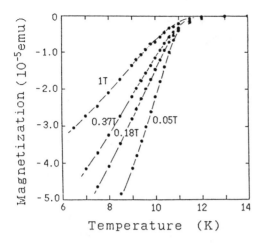

Fig. 5.59. Temperature dependence of DC magnetization of κ-(ET)$_2$Cu·[N(CN)$_2$]Br in the reversible region of magnetic field applied perpendicular to two-dimensional plane. The calculated temperature dependence of DC magnetization is represented by lines. The used parameters $\xi_\perp = 0.58$ nm, $\xi_\parallel = 2.3$ nm and $T_{c0} = 10.8$ K (Appendix B). From [5.162]

perconductivity under ambient pressure. The superconductivity characteristics of this kind of material shows the features of two-dimensionality.

The temperature dependence of the resistivity in the normal state is similar to that of κ-(ET)$_2$Cu(NCS)$_2$, yielding a broad maximum near 90 K, which is suppressed by weak hydrostatic pressure. The band structure and the associated Fermi surface are also similar, except for the energy gap between the open Fermi surface and the closed Fermi surface (Fig. 5.14). According to the crystal symmetry of κ-(ET)$_2$Cu[N(CN)$_2$]Br, a gap does not exist like in κ-(ET)$_2$I$_3$. However, it is noteworthy that an X-ray analysis has revealed the appearance of a superstructure with $(0, 0, c^*/2)$ although its origin and exact consequence have not yet been clarified [5.201].

Magnetic Transition to Superconductivity

An analysis of the transition behavior was performed on the temperature dependence of the magnetization in a magnetic field applied perpendicularly to the two-dimensional plane (Fig. 5.59). It was based on the **fluctuation**-renormalized two-dimensional Ginzburg-Landau functional (Appendix B). Accordingly, T_c in the absence of the magnetic field is determined to be 10.9 K, in contrast to the value of 11.8 K evaluated from the resistive transition characteristics. A similar treatment has been applied to salts with deuterated ET and ^{13}C substituted ET at CH$_2$ (Fig. 5.60). The T_c values in the absence of a magnetic field are listed in Table 5.10 together with the GL coherence lengths at 0 K in the two-dimensional plane $\xi_\parallel(0)$ and that in the

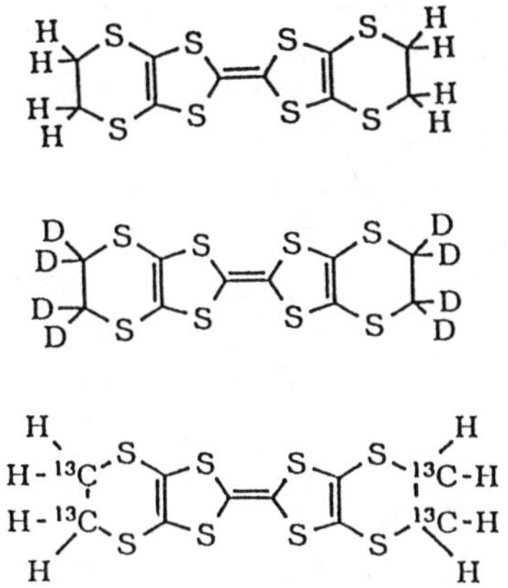

h_8-BEDT-TTF

d_8-BEDT-TTF

^{13}C-BEDT-TTF

Fig. 5.60. Isotope-substituted ET molecules

interlayer direction $\xi_{\perp}(0)$. It is noteworthy that $\xi_{\perp}(0)$ of κ-$(ET)_2Cu\cdot[N(CN)_2]Br$ is 0.58 nm, which is longer than 0.31 nm of κ-$(ET)_2Cu(NCS)_2$ but still shorter than the interlayer distance of 1.5 nm. This demonstrates that for these salts the superconductivity of the salts is confined to the ET sheet at a low temperature. The similar two-dimensional treatment has been carried out for the resistive transition behavior in a magnetic field. However, the fluctuation-renormalized treatment cannot provide a good fit in this case. The reason is ascribed to the percolative conductance due to inhomogeneity.

The dimensionality has further been investigated on the scaling behavior of the magnetization near the transition region (Appendix B). According to *Ullah* and *Dorsey* [5.202] the variable

$$t_G = \frac{T - T_c(H)}{(TH)^n} \tag{5.13}$$

exhibits scaling characteristics with $n = 2/3$ for the three-dimensional case whereas $n = 1/2$ for the two-dimensional case. By choosing the values of $T_c(H)$ to be those derived from the above-mentioned fit, the results illustrated in Figs. 5.61a and b were obtained for the two-dimensional and three-dimensional cases for κ-$(ET)_2Cu[N(CN)_2]Br$. Similar fits for κ-$(ET)_2\cdot Cu(NCS)_2$, as depicted in Fig. 5.61c and d. The figures confirm that a two-dimensional treatment is more appropriate for κ-$(ET)_2Cu(NCS)_2$.

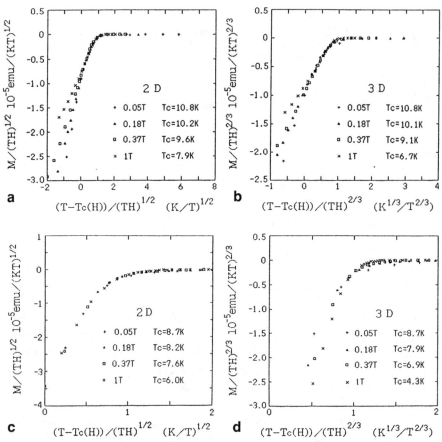

Fig. 5.61. Field-scaling fit of magnetization M around the transition region in κ-(ET)$_2$Cu[N(CN)$_2$]Br for two-dimensional scaling (**a**), for three-dimensional scaling (**b**), in κ-(ET)$_2$Cu(NCS)$_2$ for two-dimensional scaling (**c**), and for three-dimensional scaling (**d**)

^{13}C NMR Study

The studies on the superconducting state of κ-(ET)$_2$Cu[N(CN)$_2$]Br by ^{13}C NMR have brought about remarkable results [5.169, 203, 204]. When ^{13}C is substituted for ^{12}C in the central part of ET molecule within κ-(ET)$_2$Cu[N(CN)$_2$]Br crystal, it is adapted to probe the superconducting properties, since ^{13}C lies in the core of the conducting layer. This contrasts to ^{1}H existing at the outer end of an ET molecule, which serves to probe vortex states, producing a sharp peak of the relaxation rate (Fig. 5.57), as described in Sect. 5.3.1. When a magnetic field is applied perpendicularly to the superconducting sheets, the ^{13}C NMR relaxation rate $1/T_1$ is dominated by the normal-state electronic excitations in the vortex cores. Meanwhile, in the parallel-field orientation, relaxation by superconducting excitation

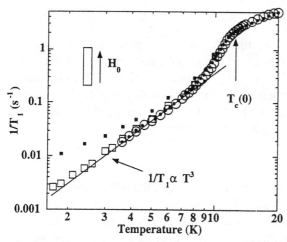

Fig. 5.62. ^{13}C NMR relaxation rate $1/T_1$ in parallel fields of 5.6 T ($\bigcirc$), 7.8 T ($\bullet$) and 7 T ($\square$). *Black squares* ($\blacksquare$) correspond to a field of 7 T with a slight misalignment (about $2°$). From [5.169]

within the sheet is expected. The temperature dependence of $1/T_1$ is obtained, as displayed in Fig. 5.62, $1/T_1$ being proportional to T^3 at a low temperature; no **Hebel-Slichter peak** was observed [5.169]. This rules out the possibility of the BCS **s-wave state** accompanying an isotropic energy gap and thus the weak-coupling electron-phonon mechanism as the source of superconductivity. It supports an **unconventional pairing** state with possible nodes in the gap function. On the other hand, the Knight shift gives the spin susceptibility in the superconducting state. Below T_c, the Knight shift decreases more quickly than the Korringa relation due to the normal state, and tends towards zero at low temperatures [5.169]. This implies that the spin pairing is singlet: in the case of a triplet state there should be no difference between the normal and superconducting spin susceptibilities [5.205]. Consequently, the pairing is considered to be due to **d-wave symmetry**. ^{13}C NMR studies have also shown that the normal state behavior is not that of a simple metal but highly enhanced, suggesting that antiferromagnetic fluctuations and spin-gap behavior may be present [5.206, 207]. This is supported by an ESR study yielding an abnormal temperature dependence: the intensity decreases below 50 K in connection with a linewidth decrease [5.208].

As to the superconducting gap symmetry, there has been a controversy around the experimental results on muon spin relaxation, magnetization, AC susceptibility and electromagnetic response. Some believed in an s-type symmetry [5.181, 184, 209] whereas other in an unconventional symmetry [5.182, 183], as touched upon in the previous section.

Isotope Substitution

With regard to the superconducting mechanism, the effect of deuteration of ET molecules have been investigated. As may be inferred from Table 5.10, T_c decreases from 10.9 to 10.6 K by deuteration of the ET molecule although the GL coherence lengths are not changed appreciably. However, it is noteworthy that T_c of κ-$(d_8$-ET$)_2$Cu[N(CN)$_2$]Br is somewhat cooling-process dependent and can be raised by very slow cooling.

AC Susceptibility

For the characterization of the superconductivity, the AC susceptibility has been measured as well as DC magnetization. In the case of two-dimensional superconductors, care should be taken concerning both the direction of the AC field with respect to the direction of the superconducting sheets and the AC frequency. The behavior has been investigated systematically using κ-(ET)$_2$Cu[N(CN)$_2$]Br crystals [5.210]. Figures 5.63a-c demonstrate the frequency dependences of the real part of the AC susceptibility χ' under DC magnetic field; both of the fields are applied perpendicularly to the two-dimensional conducting plane. For frequencies up to 1 kHz the onset temperature of χ' decreases due to the superconductivity shifts toward lower side without changing the transition behavior, but for 3.2 MHz the onset of the transition is not remarkably changed but the transition behavior is broadened. The results are explained in terms of a self-consistent linear response theory on **vortex dynamics** [5.211]. The theory describes the magnetic susceptibility of superconductors by self-consistently taking into account the electromagnetic dynamics of the vortices and the skin effect caused by quasi-particles (normal carriers). When the effective **penetration depth** of the AC field is less than the sample dimension, the diamagnetic response appears. The effective penetration depth is given by the real part of the complex penetration depth $\tilde{\lambda}$, determined by the vortex penetration depth, the skin depth due to quasi particles, and the London penetration depth. The temperature dependence of the length in a magnetic field of 2 kOe applied perpendicularly to the superconducting sheet is illustrated in Fig.5.64. At 17 Hz the effective penetration depth becomes comparable to the sample size of 0.5 mm at 7.2 K (Fig.5.64a), below which a diamagnetic signal appears (Fig.5.63a). For 3.2 MHz the penetration depth becomes comparable to the sample size just below T_c which is observed in the absence of a magnetic field and the diamagnetic signal starts to appear at that temperature (Fig.5.63c).

When the AC field is set parallel to the conducting plane on the application of a DC field in the perpendicular direction, the measured AC susceptibility behaves differently (Fig.5.65). To interpret the DC field dependence of χ', we have to take into account the effect of the interplane conduc-

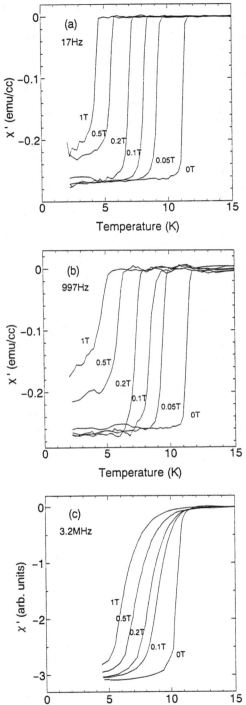

Fig. 5.63. Temperature dependences of the real part of AC susceptibility χ' for κ-(ET)$_2$ Cu-[N(CN)$_2$]Br under DC magnetic field applied perpendicular to the two-dimensional conducting plane. The direction of the AC field was set perpendicularly to the conducting plane. (**a**) 17 Hz, (**b**) 997 kHz and (**c**) 3.2 MHz. From [5.210]

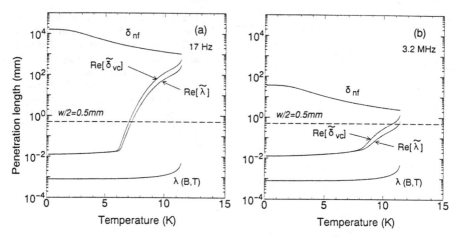

Fig. 5.64. Temperature dependences of the real part of effective complex penetration depth Re{$\tilde{\lambda}$}, the real part of the vortex penetration depth Re{$\tilde{\delta}_{vc}$}, the skin depth δ_{nf} due to the quasi particle flow and the London penetration depth $\lambda(B, T)$ at frequencies of (**a**) 17 Hz, and (**b**) 3.2 MHz under magnetic field of 2 kOe. The *broken line* represents the sample dimension. From [5.210]

tance, since the AC screening current has to flow to penetrate the two-dimensional superconducting plane. In this case the normal skin-depth effect is not so effective as in the case of an AC field applied perpendicularly to the conducting plane, because the skin depth in this case becomes much longer than the sample size. The transition behavior is affected by the structure of the vortices which are decoupled to be pancake-like and confined within the sheet due to superconductivity phase fluctuation between layers [5.212].

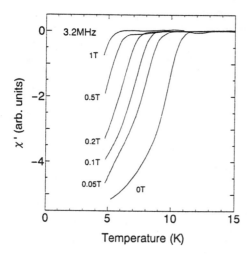

Fig. 5.65. Temperature dependences of the real part of AC susceptibility χ' of κ-(ET)$_2$Cu-[N(CN)$_2$]Br in a DC field applied perpendicular to the conducting plane. The direction of the AC field was set parallel to the conducting plane. From [5.210]

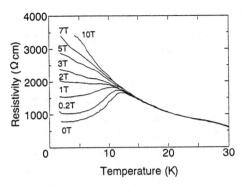

Fig. 5.66. Temperature dependence of the resistivity of κ-(ET)$_2$ Cu[N(CN)$_2$]Cl at ambient pressure. With increase of magnetic field, the resistive decrease below 13 K is suppressed. From [5.76]

5.3.3 κ-(ET)$_2$ Cu[N(CN)$_2$]Cl

The temperature dependence of the resistivity of this salt is depicted in Fig. 5.29 for the current directions parallel and perpendicular to the ET sheet as a function of hydrostatic pressure. At ambient pressure the in-plane resistivity exhibits a non-metallic temperature dependence but shows a slight decrease below 13 K. When a magnetic field is applied, the decrease is suppressed (Fig. 5.66), indicating that the decrease is due to superconductivity. With an increase of pressure, the resistivity below 13 K decreases steadily and exhibits zero resistance above 0.3 kbar (30 MPa). Meanwhile under ambient pressure the resistivity in the direction perpendicular to the sheet shows no decrease below 13 K. (The saturating behavior found in Fig. 5.29b below 20 K is due to the limitation of the input impedance of the voltmeter in use). However, it shows the corresponding decrease above 0.2 kbar and zero resistance above 0.3 kbar.

When the sheet resistance, i.e. the resistivity of each ET sheet, just above the superconductivity transition is evaluated, it decreases toward the universal value of h/e^2 or $h/4e^2$ (≈ 6.5 kΩ) per sheet. This indicates that the finite resistance in the superconductivity temperature region, observed above 0.3 kbar, can be related to the superconducting or normal electron localization, specific to two-dimensional system with disorder. The localized superconductivity decreases the average resistivity of the sheet. Since the salt is characterized by two-dimensionality, the zero resistance above 0.3 kbar in the direction perpendicular to the sheet is ascribed to Josephson coupling. When the **localization** is not fully developed within a sheet, a pair of superconducting regions in adjacent sheets can be Josephson-coupled, resulting in the resistance decrease in the out-of-plane direction between 0.2 and 0.3 kbar. Below 0.2 kbar, the localization prevails so as to make difficult to find Josephson-coupled pairs. The phase diagram shown in Fig. 5.30 represents the region of superconductivity yielding zero resistance as a complete-superonductivity state (C-SC), while the incomplete superconductivity yielding a finite resistance exhibits a crossover between absence and presence of superconductivity contributions due to Josephson coupling, designated by I-SC-1 and I-SC-2, respectively.

It is remarkable that superconductivity appearing at T_{c1} is suppressed below $T_{c2} = 6 \div 7$ K, resulting in a **reentrant resistive state**, i.e., the existence of superconductivity only in a finite temperature interval between T_{c1} and T_{c2} [5.213,214], as shown in Fig.5.29. T_{c2} depends on the cooling or heating procedure. The resistive state is enhanced even by a magnetic field [5.215]. The reentrant transition into a resistive state is of first order, which is thought to be accompanied by a subtle change in lattice structure being extremely sensitive to pressure, as evidenced by the variety of phases realized with weak pressure of less than 0.7 kbar (70 MPa), as displayed by Fig.5.30.

The decrease of T_c with pressure for κ-$(ET)_2 Cu[N(CN)_2]Cl$ [5.216] seems to be consistent with the influence of intermolecular distance to T_c, as discussed in relation to Fig.5.45. However, with regard to the relationship between isostructural κ-$(ET)_2 Cu[N(CN)_2]Cl$ and κ-$(ET)_2 Cu[N(CN)_2] \cdot$ Br, the latter yields stabilized superconductivity with lower T_c although the lattice is slightly expanded in the latter case (Table 5.2). Furthermore, the analogue crystals with I instead of Cl or Br is an insulator at low temperature even under a pressure of up to 5 kbar [5.77]. It is worthy to note that ET molecules are present with an eclipsed dimer conformation in the superconducting salt but, in the case of the iodine salt, no complete ordering occurs [5.217].

With regard to the alloyed anions of ET salts, that of β-type ET salts causes either a considerable reduction of the T_c value or a complete suppression of the superconducting transition, as described in Sect.5.2.7. However, the resistive measurements of κ-$(ET)_2 Cu[N(CN)_2]Cl_{1-x}Br_x$ revealed the T_c values of 11.3, 11.5 and 10.6 K for x = 0.5, 0.75 and 0.85, respectively, under ambient pressure [5.218], although the salt with x = 0.2 is insulating [5.219]. The structural study of the salt with x = 0.5 showed that disorder is not detected in the anion layer, implying that an ordered structure with alternating $Cu[N(CN)_2]Cl$ and $Cu[N(CN)_2]Br$ layers may be formed [5.220]. It was also found that the salt of κ-$(ET)_2 Cu \cdot [N(CN)_2]Br_{0.9}I_{0.1}$ shows a nonmetallic behavior under ambient pressure but becomes superconducting under an pressure of 3 kbar with $T_c = 5.9$ K [5.221].

5.4 Other ET Superconductors

5.4.1 (ET)$_2$ReO$_4$

The first ET superconductor developed was (ET)$_2$ReO$_4$ [5.222]. Its structure and physical properties are rather similar to those of the (TMTSF)$_2$X superconductors. However, because other, more promising (ET)$_2$X salts were discovered soon thereafter, further investigations of (ET)$_2$ReO$_4$ have been lagging.

In (ET)$_2$ReO$_4$, ET molecules are "zigzagged" along the a-axis (the stacking direction) providing cages for the ReO$_4$ anions. The unit cell consists of two formula units: (ET)$_4$(ReO$_4$)$_2$. The structure belongs to the P$\bar{1}$ space group and, as mentioned, is rather similar to that of the (TMTSF)$_2$X salts. The only differences are that the ET molecules are not planar and at room temperature the ReO$_4$ anions are ordered with a wave vector corresponding to $(0,\frac{1}{2},\frac{1}{2})$ [5.5].

Under ambient pressure this salt shows metallic behavior from room temperature to 81 K. There it undergoes a metal-insulator transition and the resistance abruptly increases by two orders of magnitude, demonstrating that the transition is predominantly first order. The room-temperature conductivities in the two-dimensional plane are $\sigma_a \approx 200$ S/cm and $\sigma_b \approx 10$ S/cm. The pressure phase diagram is displayed in Fig. 5.67. Between 4 and 6 kbar the resistivity has been reported to show considerable hysteresis,

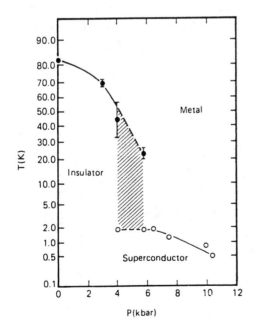

Fig. 5.67. Pressure phase diagram of (ET)$_2$ReO$_4$. From [5.222]

203

thus precluding the determination of a precise critical pressure. The resistive onset of superconductivity occurs near 2 K in the region of the critical pressure, as shown in the phase diagram. Above ≈ 7 kbar, the metal-insulator transition is suppressed completely and a transition to superconductivity appears.

An ESR study under ambient pressure has revealed that both the spin susceptibility and the linewidth decreases abruptly near T_{MI} [5.223]. Above T_{MI}, the linewidth varies linearly with temperature. These characteristics suggest that the transition is not due to a magnetic ordering, but is associated with the structural reordering, which is very likely due to a displasive modulation of the ET molecule and ReO_4 anion sublattices [5.224].

5.4.2 $(ET)_4Hg_3Cl_8$ and $(ET)_4Hg_3Br_8$

ET radical-cation salts with either a chloromercurate anion $[Hg_3Cl_8]^{2-}$ or a bromomercurate anion $[Hg_3Br_8]^{2-}$ show superconductivity, but they exhibit transport properties that are very different compared to other ET salts. This is due to a larger separation between cation and anion layers. Crystallographic data on the cation sublattice in $(ET)_4Hg_3Cl_8$ are a = 11.062(3) Å, b = 8.754(2) Å, c = 35.92(14) Å, β = 91.01(3)°, Z = 2 and V = 3478 Å^3. The space group is I2/c. The anion sublattice is apparently of I2 symmetry. Figure 5.68 illustrates a projection of the structure along the a-axis. A characteristic feature is that the ET radical-cation layers oriented parallel to the ab-plane are made up of parallel ET pairs with a pronounced overlap within a pair and with a rather short distance between adjacent ET planes. The dihedral angle between median planes of these molecules in the adjacent planes is about 84°. Chains of Hg_3Cl_8 anions lie in channels parallel to the a-axis between the cation layers. Measurements of the anisotropy in the

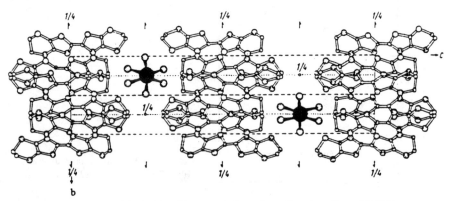

Fig. 5.68. Projection of $(ET)_4Hg_3Cl_8$ crystal along the a-direction. From [5.225]

conductivity show that it is not significant in the ab-plane (no larger than 2) while in the perpendicular plane it is very large ($\approx 5 \cdot 10^4$).

The results of an X-ray structural analysis provided a more accurate composition of this salt with the formula $(ET)_4 Hg_{2.89} Br_8$. The main packing scheme is the alternation of two-dimensional layers of ET molecules and layers of anions, as is typically the case for most of the ET salts. The distinguishing feature is described by two incommensurate monoclinic lattices with the lattice constants (I): a = 11.219(3) Å, b = 8.706(2) Å, c = 37.105 (7) Å, β = 90.97(4)°, space group I2/c; (II): a = 3.877(1) Å, b = 8.706(2) Å, c = 37.141(7) Å, β = 87.30(4)°, space group I2. An analysis of the two corresponding, independent sets of experimental data from one crystal indicated that the lattice I is comprised of ET and Br atoms and the lattice II consists of Hg atoms, and that the periodicity of Hg atoms differs from that of ET and Br atoms. Consequently, the Hg-Br bond length in a bromomercurate anion and the spacing between ET and Hg atoms are not constants [5.226].

The in-plane resistivity $\rho_{\parallel}$ of the crystal decreases only slightly upon cooling it from room temperature to liquid-helium temperature. For various crystals, the resistance ratio $R_{300}/R_6 \approx 5 \div 10$. Such a behavior of the resistivity is apparently related to the presence of an **incommensurate lattice** and a specific packing of the ET molecules in the cation layer. The out-of-

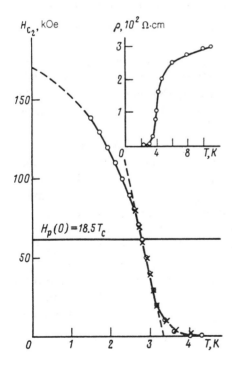

Fig. 5.69. Temperature dependence of the upper critical fields H_c^a and $H_c^{b'}$, parallel to the ab plane, of $(ET)_4 Hg_{2.89} Br_8$ ($\bigcirc$: $\mathbf{H} \parallel \mathbf{a}$ and $\times$: $\mathbf{H} \parallel \mathbf{b'}$). The *inset* shows the temperature dependence of the resistivity ρ_a of the same sample in the absence of the magnetic field. From [5.227]

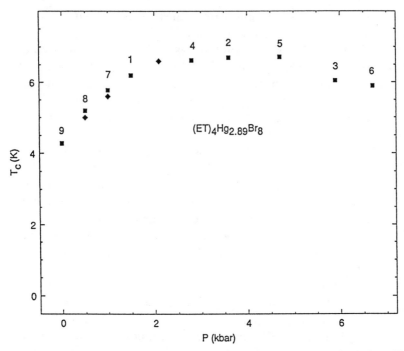

Fig. 5.70. Superconductivity transition versus pressure for $(ET)_4 Hg_{2.89} Br_8$. The numbered points are for the crystal with $\approx 45\%$ Meissner effect. The other points are for the crystal with $\approx 1\%$ Meissner signal. The numbering denotes the order in which points were taken. From [5.228]

plane resistivity of the crystal $(\rho_\perp)$, which increases slightly upon lowering the temperature, has a broad peak at $40 \div 60$ K. Below 20 K, both $\rho_\parallel$ and $\rho_\perp$ decrease rapidly and at 3.8 K the resistivity vanishes in both directions. The superconductivity transition begins at about 5.0 K and its midpoint lies at $T_c = 4.3$ K.

The critical field H_{c2} was determined from the midpoint of the superconducting transition on the resistivity curve in fields applied parallel and perpendicular to the conducting planes. The resultant $H_{c2\,\parallel}(T)$ curve is shown in Fig. 5.69. It is remarkable that $H_{c2\,\parallel}$ exceeds by far the **paramagnetic limit**. Although the Pauli limit in Fig. 5.69 was calculated using the T_c value of 3.3 K estimated by a linear extrapolation, the value remains 79 kOe based upon a free-electron model even if an onset value of 4.3 K is adopted. $\xi^\parallel(0)$ and $\xi^\perp(0)$ were evaluated to be 17 nm and 0.8 nm, respectively. The pressure dependence of T_c is displayed in Fig. 5.70. It reveals that a large positive derivative of more than 1 K/kbar results in a value of 6.7 K for T_c near 3.5 kbar and a change in the slope of the T_c-vs-pressure curve near 4 kbar.

5.4.3 $(ET)_2 M(CF_3)_4$ (Solvent)

An approach to control the intermolecular spacing of κ-type superconductors has been done by an electrocrystallization of large discrete anions such as $M(CF_3)_4^-$ (M: Cu, Ag and Au) with ET [5.229]. Two kinds of superconducting phases have been found. One phase typically crystallizes with a plate-like morphology and has a T_c value below 6 K; it was designated the κ_L-**phase** (L signifies the lower T_c phase). A second phase, which crystallizes in a filamentary needle-like morphology, has superconducting temperatures in the range of $7 \div 11$ K and was named the κ_H-**phase** (H signifies the higher T_c phase). Both of these phases contain molecules of the crystallization solvent in the solid structure. Thus, not only can these two phases be modified by an exchange of the anion's metal atom, but also by replacement of the crystallization solvent. So far more than 20 distinct organic superconductors have been found to yield superconductors, as compiled in Table 1.1.

Figure 5.71 exhibits an AC susceptibility in the superconductivity transition region for the κ_L and κ_H phases of $(ET)_2 Ag(CF_3)_4 TCE$. The pressure dependence of κ_H-$(ET)_2 Cu(CF_3)_4 (TCE)_x$ with $T_c \approx 9$ K under ambient pressure is plotted in Fig. 5.72. In the case of the $(ET)_2 Cu(CF_3)_4 TCE$ salt, the κ_L and κ_H phases are chemically similar, the only apparent difference is a smaller, and locally variable, content of solvent molecules in the κ_H salt so that it can be represented by $(ET)_2 Cu(CF_3)_4 (TCE)_x$ ($x < 1$). The κ_L-phase salt has a definite κ-phase crystallographic structure (Fig. 5.73), whereas the κ_H-phase salt has an as yet unknown structure. It is speculated that the notable difference in T_c between the two salts may be due to a

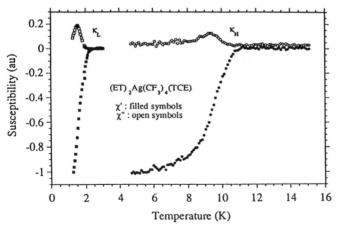

Fig. 5.71. AC susceptibility as a function of temperature for κ_L and κ_H phases of $(ET)_2 Ag(CF_3)_4 TCE$. From [5.229]

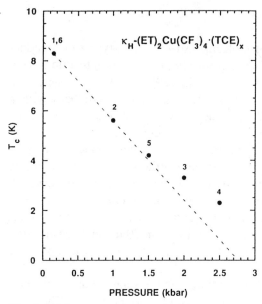

Fig. 5.72. Pressure dependence of the superconductivity onset T_c for a crystal of κ_H-$(ET)_2 Cu(CF_3)_4 (TCE)_x$. The numbers denote the sequence of determinations and the dashed line indicates the initial slope. From [5.230]

change from a *disordered crystallographic structure* of the $[Cu(CF_3)_4]^-$ anion in κ_L-type salt to an ordered structure in the κ_H-type salt. The anion ideally adopts D_{2d} symmetry (in the absence of distortions of the central CuC_4 core) with the Cu-C bonds located on the mirror planes. Any other conformation of the CF_3 groups leads to unacceptably short repulsive F ...

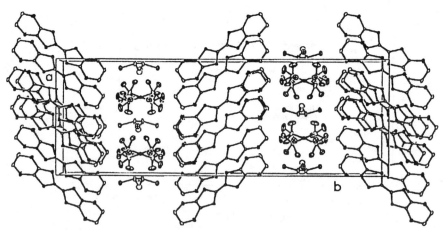

Fig. 5.73. Perspective projection of the unit cell of κ_L-$(ET)_2 Cu(CF_3)_4 TCE$ along the c-axis. From [5.231]

F contacts. In κ_L-phase crystals, however, the crystallographic mirror plane bisects the C-Cu-C angles, so that the anion is necessarily disordered. A disordered configuration has also been found in the solvent molecules. With regard to the κ_H-phase, it is conceivable that the removal of some of the solvent molecules allows for the rotation of the anion, so that the anion adopts an ordered arrangement, wherein the crystallographic symmetry elements are compatible with the internal symmetry of the anion [5.230].

5.4.4 α-(ET)$_2$ MHg(SCN)$_4$

The highly two-dimensional α-(ET)$_2$ NH$_4$ Hg(SCN)$_4$ exhibits a superconductivity transition at ≈ 1.8 K [5.60, 232]. The structure and normal electronic properties are described in Sect.5.1.5. The values of the in-plane (current parallel to the conducting layer) and out-of-plane (current perpendicular to the conducting layer) conductivities are $(2 \div 5) \cdot 10^2$ S/cm and $(3 \div 5) \cdot 10^{-4}$ S/cm at room temperature and increase to $10^4 \div 10^5$ S/cm and $(3 \div 5) \cdot 10^{-2}$ S/cm at 3 K, respectively. The anisotropy ratio of the in-plane and out-of-plane value is $10^5 \div 10^6$ [5.233]. The profile of the superconductivity transition is qualitatively different (Fig.5.74). The in-plane resistivity starts to decrease around 2.5 K and vanishes around 1.0 K, while the out-of-plane resistivity exhibits a comparatively sharp transition around 1.0 K. This indicates that, in the temperature range from 2.4 K to 1.0 K, the superconductivity coherence starts to grow gradually within layers but does not between layers.

The AC complex susceptibility, $\chi = \chi' - i\chi''$, measured for AC field applied perpendicular (in-plane susceptibility) and parallel (out-of-plane susceptibility) to the conducting layers is displayed in Fig.5.75. The in-

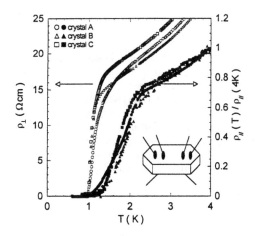

Fig. 5.74. In-plane ($\rho_{\parallel}$) and out-of-plane ($\rho_{\perp}$) resistivities of α-(ET)$_2$ · NH$_4$ Hg(SCN)$_4$ crystal measured with six-terminal method for three crystals. $\rho_{\perp}$ is represented with absolute values while $\rho_{\parallel}$ is normalized to the value at 4 K. From [5.232]

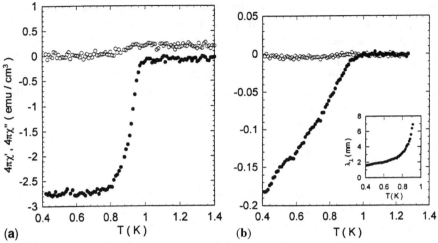

Fig. 5.75. AC susceptibility measured (**a**) in an AC-field of 47 Hz and 0.16 Oe (in amplitude) perpendicular to layer and (**b**) in an AC field of 1007 Hz and 0.16 Oe parallel to the layer. Inset shows the out-of-plane penetration depth deduced from the susceptibility data. From [5.232]

plane susceptibility χ' shows a sharp transition with an onset at 0.95 K and a saturation at 0.8 K. On the other hand, the out-of-plane susceptibility χ' changes gradually and reaches only 20 % of the perfect diamagnetization at 0.4 K. This implies that the **penetration depth** is comparable with the sample size. The estimated penetration at low temperature is 1.4 mm for the out-of-plane case, while it does not exceed the order of 1 μm for the in-plane case.

Figure 5.76 depicts the temperature dependence of the difference between C_p/T (C_p: specific heat) in the superconducting and normal states,

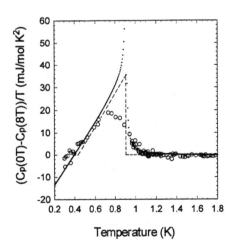

Fig. 5.76. Temperature dependence of the C_p/T difference between superconducting and normal states around the transition point for α-(ET)$_2$NH$_4$Hg(SCN)$_4$. The *dashed curve* represents the mean-field peak of BCS theory with $\gamma_{eff} = 25$ mJ/(mol·K^2) and the *dotted curve* denotes an analysis based on the two-dimensional Gaussian fluctuation model with $T_c = 0.91$ K. From [5.233]

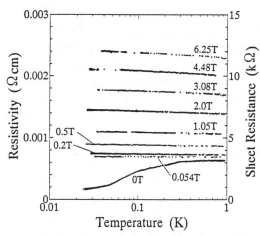

Fig. 5.77. Temperature dependence of the in-plane resistivity of α-(ET)$_2$KHg(SCN)$_4$ below 1 K in a magnetic field applied perpendicular to the conducting plane. From [5.61]

$C_p(0T)/T - C_p(8T)/T$, around the superconductivity transition. It is shown that the anomaly around 0.9 K due to the superconductivity transition is broadened and superconductivity **fluctuations** seem to exist up to $\simeq 1.3$ K. The dashed line in the figure represents the mean-field profile of BCS theory for the specific-heat jump $\Delta C_p/T_c = 1.43\gamma_{eff}$ with $\gamma_{eff} = 25$ mJ/ (mol$\cdot$K^2) at $T_c = 0.91$ K. The high-temperature tail of the peak is compared to a two-dimensional Gaussian fluctuation model. With the fit, there remains a serious discrepancy in the peak shape, as can be inferred from the dotted curve in the figure.

In contrast to α-(ET)$_2$NH$_4$Hg(SCN)$_4$ salts, α-(ET)$_2$MHg(SCN)$_4$ (M: K, Rb, Tl) undergoes a magnetic transition near 8 K giving a hump in resistivity, as described in Sect. 5.1.5. It had been assumed that these salts are considered not to possess a superconducting ground state due to the magnetic transition. However, by extending the measurement toward low temperatures, in κ-(ET)$_2$KHg(SCN)$_4$ a decrease in the resistivity has been found below 0.3 K in the absence of a magnetic field (Fig. 5.77). It is worthy to point out that the decrease is suppressed by a magnetic field of 500 Oe, indicating that the decrease is ascribable to superconductivity. The decrease is rather gradual and zero resistance is not attained even at 20 mK. A similar low-temperature resistive change has been found for α-(ET)$_2 \cdot$ RbHg(SCN)$_4$ and α-(ET)$_2$TlHg(SCN)$_4$ below 0.5 and 0.1 K, respectively. As the origin of such a low-temperature resistivity decrease, a **localization effect** specific to two-dimensional superconductors is considered with its sheet resistance exceeding the universal value of h/e^2 or $h/4e^2$ (≈ 6.5 kΩ), as described in the case of κ-(ET)$_2$Cu[N(CN)$_2$]Cl. Assuming that the in-plane resistivity is determined by a stack of two-dimensional sheets, the sheet resistance was evaluated to be 4 ± 2 kΩ for α-(ET)$_2$KHg(SCN)$_4$. The uncertainty is rather large since the shape of the sample is not regular and, in addition, its size is small. Taking into account the experimental uncertainty and the influence of the interlayer coupling on the actual threshold value, the estimated values are close to the critical sheet resistance. A similar situation has been found for α-(ET)$_2$MHg(SCN)$_4$ (M: Rb, Tl). In these cases, however, the effect of disorder-decreasing T_c [5.149] is not ruled out.

On the other hand, by applying **uniaxial compressive stress** along the cross-planar (b-axis) direction, T_c and H_c of α-(ET)$_2$NH$_4$Hg(SCN)$_4$ were increased by more than a factor of 3 over their ambient pressure values with a uniaxial stress of ≈ 4 kbar. In the case of α-(ET)$_2$KHg(SCN)$_4$, the onset of the resistance decrease implying the appearance of superconductivity was raised up to 2.5 K by the uniaxial stress of ≈ 3 kbar [5.234]. As a cause of the superconductivity enhancement, an increase in density of state is noted due to the expansion in the ac plane by the compression along the b-axis through the Poisson effect (an effect to elongate the transverse direction against a longitudinal uniaxial compressional stress), resulting in a decrease of the intermolecular transfer within the conducting layer. However, a simultaneous measurement of Shubnikov-de Haas effect revealed an enhancement of the effective mass from a zero-stress value of $(1.4 \pm 0.1)m_0$ to $(1.7 \pm 0.1)m_0$ by 1.6 kbar for α-(ET)$_2$KHg(SCN)$_4$ implying a role of the many-body effects. Furthermore, the density wave state was found to be suppressed probably violating the nesting condition.

Another experiment in which uniaxial stress has been applied in the cross-plane direction in α-(ET)$_2$KHg(SCN)$_4$ showed that the resistivity starts to decrease below 1 K and gradually reaches zero resistance below 0.1 K under 1.06 kbar [5.235]. The V-I characteristics were represented by $V \propto I^\alpha$, α varying from 1 at 1 K to 2 at 34 mK. The results are interpreted in terms of the Kosterlitz-Thouless transition with T_{KT} of 70 mK. Similar transition behavior was reported for α-(ET)$_2$NH$_4$Hg(SCN)$_4$ [5.236].

5.4.5 (ET)$_3$Cl$_2$(H$_2$O)$_2$

The ET molecule in an average oxidation state of +2/3 occurs in the salt (ET)$_3$Cl$_2$(H$_2$O)$_2$, which has also been seen to be superconducting [5.237]. The crystal strcutre of (ET)$_3$Cl$_2$(H$_2$O)$_2$ is triclinic and belongs to the P1 space group with a = 13.905(2) Å, b = 15.929(2) Å, c = 11.228(1) Å, α = 109.22(1)°, β = 97.08(1)°, γ = 94.67(1)°, and V = 2310.5(5) Å^3. Like in other ET complexes, the donors form a conducting sheet parallel to the ac-plane, and these sheets are separated from each other by anion sheets composed of Cl$_4$(H$_2$O)$_4$. The unit cell has Z = 2. Since there are six electron-donor molecules in the HOMO's in the unit cell, and the bands are 2/3

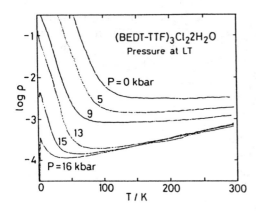

Fig. 5.78. Temperature dependence of the resistivity of $(ET)_3 Cl_2 (H_2 O)_2$ at several pressures. From [5.237]

filled, the four electrons are remaining in HOMOs. An extended Hückel calculation yields a semimetallic band structure with electron- and hole-pockets [5.237].

The room-temperature conductivity reaches 500 S/cm under ambient pressure, which is one of the highest obtained for ET salts; ET complexes usually exhibit a comparatively low conductivity (< 100 S/cm). The temperature dependence of the conductivity is displayed in Fig.5.78 as a function of the applied pressure. The clamp-cell method was used; the stated pressures were measured at low temperatures.

At ambient pressure, the resistivity is weakly metallic with a minimum near 160 K and a steep increase below 100 K. By application of pressure, the metallic region is extended toward lower temperatures. At 15 kbar, the resistivity has a minimum around 55 K, followed by a 30-fold increase to a maximum at 2 K and another rapid decrease. This last decrease is attributable to a transition to superconductivity whose midpoint of T_c is 1.1 K. Application of a magnetic field recovers the normal resistance observed at 16 kbar where the T_c onset is at 2.5 K. The critical field H_{c2} is axis-dependent reflecting the two-dimensional nature of this compound: $dH_c/dT = -16.6$ kOe/K for $H \| c$ and $dH_c/dT = -2.1$ kOe/K for $H \| b^*$. The anisotropy ratio is rather small compared to the other ET salts, which is consistent with a slightly smaller separation between the conducting layers.

5.4.6 $(ET)_4 M(CN)_4 (H_2 O)$

Superconductivity has been found in ET salts with a **divalent anion**, namely $(ET)_4 M(CN)_4 (H_2 O)$ (M: Pt, Pd). Figure 5.79 shows the crystal structure and donor arrangement of $(ET)_4 Pt(CN)_4 (H_2 O)$. The temperature dependence of the resistivity ρ is depicted in Fig.5.80 for various pressures. The energy-band structure is illustrated in Fig.5.81. At ambient pressure the salt

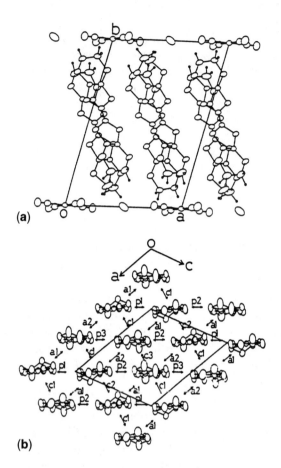

Fig. 5.79. Crystal structure of $(ET)_4 Pt(CN)_4 (H_2O)$ (**a**) and its donor arrangement (**b**). From [5.238]

is non-metallic below 60 K. The nonmetallic behavior is suppressed by pressure and undergoes a superconductivity transition, e.g., at 2 K under 6.5 kbar. A sister salt $(ET)_4 Pd(CN)_4 (H_2O)$ becomes superconducting at 1.2 K under 7 kbar [5.239]. Salts without H_2O, $(ET)_4 M(CN)_4$, have a structure

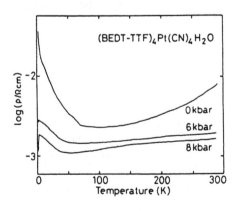

Fig. 5.80. Temperature dependence of resistivity of $(ET)_4 Pt(CN)_4 (H_2O)$ under various pressure. From [5.238]

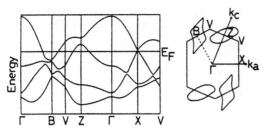

Fig. 5.81. Energy band structure of $(ET)_4 Pt(CN)_4 (H_2 O)$ calculated by tight-binding band approximation. From [5.238]

similar to the above-mentioned salts but do not undergo superconductivity transitions. The role of $H_2 O$ in the superconducting salt is open for future study.

5.5 Salts with ET Derivatives

In order to expand or improve the ET-type molecule yielding superconductors, partial substitution of sulfur atoms with other atoms has been carried out. To promote intermolecular interaction, a selenium atom with larger orbitals was introduced. Among the molecules bis(ethylenedithio)tetraselenafulvalene (abbreviated by BEDT-TSF or BETS) with Se in the TTF skeleton (Fig.1.1) has been found to provide one type of superconductor λ-$(BETS)_2 GaCl_4$. On the other hand, bis(ethylenedioxy)tetrathiafulvalene (abbreviated by BEDO-TTF, BEDO, or BO) (Fig.1.1) was synthesized by replacing S in outer-rings with oxygen atom. Although the rise in T_c was expected due to the lightness of the principal molecule in a superconductor according to the isotope effect provided that BCS theory is applicable, the T_c value remained below 3.5 K.

5.5.1 BETS Salts

The introduction of Se atoms in the ET skeleton was expected to increase the stability of the metallic state. Among the complexes based on BETS, plate crystals of MX_4 salts (M: Fe, Ga, In; X: Cl, Br) are isostructural to each other and have κ-type structures, while needle crystals belong to the triclinic system of $P_{\bar{1}}$ [5.240, 241] and its structure is called λ-**type**. All of the κ-type salts retain metallic conductance over a wide temperature range, as illustrated in Fig.5.82. It is interesting that, for $FeBr_4$ and $GaBr_4$ salts,

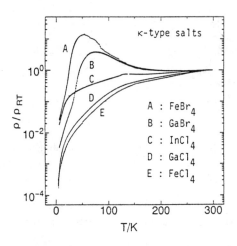

Fig. 5.82. Temperature dependences of κ-(BETS)$_2$ MX$_4$. From [5.242]

abnormal resistivity maxima were found at ≈ 50 and ≈ 70 K below which the resistivity decreases rapidly. This behavior is reminiscent of that of κ-type ET salts, as described in Sect. 5.1.5.

(BETS)$_2$ GaCl$_4$ with the λ-type structure exhibits a superconductivity transition at 8 K under ambient pressure (Fig. 5.83). More interestingly, λ-**(BETS)$_2$ FeCl$_4$** transforms to the insulating state around the same temperature. The resistivity of both salts has a maximum around 90 K [5.242]. Figure 5.84 displays the crystal structure of λ-**(BETS)$_2$ GaCl$_4$**. BETS molecules are apparently arranged along the [100] direction with a fourfold "quasi-stacking structure" with Se ... Se, Se ... S, S ... S, contacts shorter than the van der Waals distance along [001].

With regard to λ-(BETS)$_2$ FeCl$_4$, ESR measurement demontrated the antiferromagnetic interaction between Fe^{3+} ions, and the metal-insulator transition is accompanied by a transition of the magnetic state of the anion

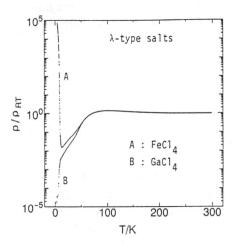

Fig. 5.83. Superconductivity transition of λ-(BETS)$_2$ GaCl$_4$ and insulator transition of λ-(BETS)$_2$ FeCl$_4$. From [5.242]

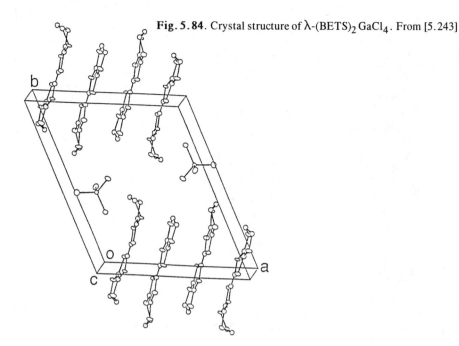

Fig. 5.84. Crystal structure of λ-(BETS)$_2$ GaCl$_4$. From [5.243]

[5.242]. The sharp metal-insulator transition occurring at 8 K under ambient pressure is suppressed when a magnetic field exceeding 100 kOe is applied. Figure 5.85 displays the magnetic-field dependence of the resistance at 4.2 K, measured with a current flowing along the [001] direction for two directions of the field applied to [100] ($\theta = 0°$) or to [010] ($\theta = 70°$). For $\theta = 70°$ and above 120 kOe, the high-field resistance follows an almost qua-

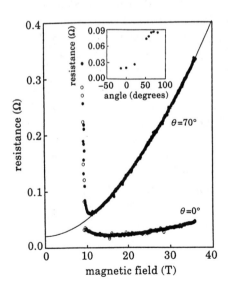

Fig. 5.85. Magnetic field dependence of the resistance of λ-(BETS)$_2$ FeCl$_4$ at 4.2 K in the two directions of the field applied parallel to [100] ($\theta = 0°$) ($\bigcirc$)) or to [010] ($\theta = 70°$) ($\bullet$). The *continuous curve* is a fit to an almost quadratic power law H$^\alpha$ with $\alpha = 1.85$. *Inset*: anisotropy of the resistance in the (001) plane at 15 T and 4.2 K. From [5.244]

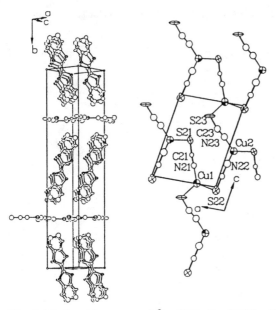

Fig. 5.86. Crystal structure of β_m-$(BO)_3\,Cu_2\,(NCS)_3$. From [5.247]

dratic H^α power law, with $\alpha \approx 1.85$. Above 200 kOe, a similar relationship is observed for $\theta = 0°$. It is worth pointing out that the minimum value of the resistance in a magnetic field for $\theta = 0°$ equals the values obtained either by extrapolating to zero field the previous fit for $\theta = 70°$ or by extrapolating down to 4.2 K, the metallic resistance measured without an applied magnetic field. It means that, whatever the orientation of the applied field perpendicularly to the [001] direction, the low-field insulating state is suppressed above 90 kOe. The field-restored highly conducting state is also obtained, above the same field value, in longitudinal-magnetoresistance measurements performed for fields applied along the axis of the [001] direction. This is not an orbital but a spin-dependent effect [5.244]. It has been observed that the sharp metal-insulator transition occurring at 8 K is also easily suppressed by applying pressure exceeding 4 kbar [5.245]. Applying pressure or a magnetic field is believed to respectively increase or decrease the dimensionality of the electron gas. Therefore, the suppression of the insulating state by such parameters with a mutually opposite influence, constitutes a striking feature. With regard to the effect of the magnetic field, it is believed that a ferromagnetic ordering of Fe^{3+} ions is induced to suppress the antiferromagnetic modulation of the localized spins of Fe^{3+} causing approximately $2k_F$ exchange potential to the conduction electrons.

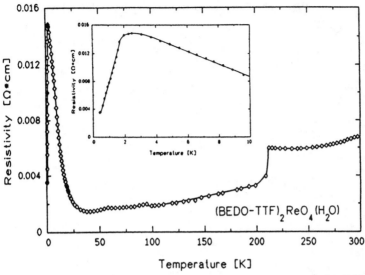

Fig. 5.87. Temperature dependence of resistivity parallel to c-axis for $(BO)_2 ReO_4 (H_2 O)$. The inset shows the resistivity in the temperature range 0.5 to 10 K. From [5.248]

5.5.2 BO Salts

The BO molecule acts as a donor and forms a stacking pattern because of the strong intermolecular C-H...O interactions. This donor has provided not only a number of metallic conductors forming complex salts with inorganic anions, but it also exhibits a strong tendency to form organic metals with organic acceptor molecules [5.246]. The conformation of BO is tub-shaped with a rather long inter-ring bond of 1.357(6) Å, particularly compared to that of ET (1.319Å), although all other bond lengths and angles appear normal. Among the BO complexes, two salts have been found to be superconducting; β_m-$(BO)_3 Cu_2 (NCS)_3$ [5.247] and $(BO)_2 ReO_4 (H_2 O)$ [5.248].

The crystal structure of the former salt is displayed in Fig.5.86. The donor layers are separated by anions that are constrained to lie on a crystallographic mirror plane. The $Cu_2 (NCS)_3$ anion shown on the right-hand side of Fig.5.86 is an infinitely connected two-dimensional polymer. Superconductivity was detected by an AC shielding experiment with an onset at 1.06 ± 0.02 K. With regard to the band structure, the dispersion relation of the highest occupied bands are largely represented by HOMO of BO. With the formal oxidation of $(BO)_3^+$ in the crystal with a unit cell containing three molecules, the highest occupied band is half-filled. The Fermi surface associated with this band consists of ellipse-like hole pockets.

Superconductivity was observed in $(BO)_2 ReO_4 (H_2 O)$ with a resistive onset at 2.5 K but with an AC-susceptibility onset at 0.9 K. The tempera-

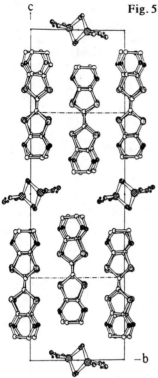

Fig. 5.88. Crystal structure of $(BO)_2 ReO_4 (H_2 O)$. From [5.249]

ture dependence of the resistivity is displayed in Fig. 5.87. The resistivity at 214 K is associated with the change in thermopower and ReO_4^--anion ordering. Two further transitions are found at $80 \div 90$ K and ≈ 35 K. These transitions might be caused by order-disorder phenomena either in anionic or/and cationic sublattices. Disorder could lead to an increase in resistivity below 35 K. By improving the quality of crystals, superconductivity was found to start at ≈ 3.5 K [5.249]. The metal-to-metal transition appearing at 213 K under ambient pressure decreases with pressure. Figure 5.88 illustrates the crystal structure.

6. Superconductors with Hybrid Molecules

In order to expand the basic molecular components, atomic substitutions of ET molecule have been performed: Replacement of a part of sulfur atoms with oxygen or selenium atoms has brought about BO or BETS molecules (Fig.1.1). On the other hand, a hybrid of TMTSF and ET molecule, DMET in which each half is conjugated at its central C=C bond (Fig.1.1) were found to form superconductors. Similar hybrid molecules with which superconductors are produced, like MDT-TTF, DMBEDT-TTF, DMET-TSF and DTEDT (Fig.1.1) have been synthesized. In this chapter superconductors with these non-centrosymmetric molecules are reviewed.

6.1 DMET Molecule and Its Salts

The non-centrosymmetric molecule dimethyl(ethylenedithio)diselenadithiafulvalene (DMET) is formed by conjugating half of a TMTSF and half of an ET molecule; its central TTF framework is depicted in Fig.1.1. As a result, DMET may have chemical properties intermediate between TMTSF and ET. DMET forms compounds in a 2:1 ratio, i.e., $(DMET)_2 X$, with a variety of monovalent anions which can bond to either the TMTSF or the ET end. At the same time its unsymmetry leads to a property unique among charge transfer crystals: Because the TMTSF end and the ET end have different thickness, due to the difference in the van der Waals radii of the constituting atoms to produce face-to-face stacks of DMET, successive molecules must be arranged with the DMETs rotated by 180° with respect to one another. It is interesting that in these cases the S and Se atoms face each other within a column in an alternating sequence.

The redox potential of DMET is almost the same as that of ET. Seven salts of $(DMET)_2 X$, prepared by electrocrystallization of DMET with TBA·X (tetra n-butylammonium) in chlorobenzene, exhibit superconductivity with T_c less than 2 K. Their counter anion is a linear one, I_3 (10.14Å in length), AuI_2 (9.42Å), IBr_2 (9.30Å), $Au(CN)_2$ (9.2Å), $AuBr_2$ (8.70Å) and $AuCl_2$ (8.14Å). The other anions with tetrahedral or octahedral structure do not yield superconductors.

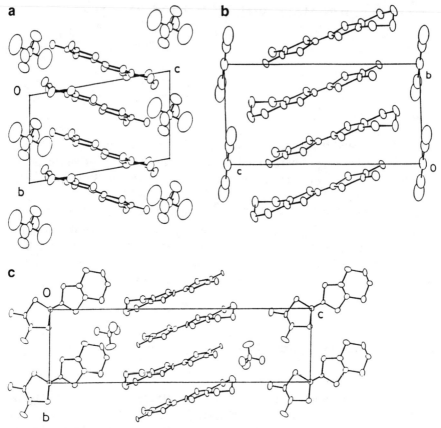

Fig. 6.1. Crystal structures of $(DMET)_2 PF_6$ (**a**), $(DMET)_2 Au(CN)_2$ (**b**) and $(DMET)_2 BF_4$ (**c**). Courtesy of *K. Kikuchi*. From [6.1]

In Fig. 6.1 the crystal structures of $(DMET)_2 PF_6$, $(DMET)_2 Au(CN)_2$, and $(DMET)_2 BF_4$ are shown as typical examples of these salts. The DMET molecules are stacked in alternating orientations and run in one direction both for $(DMET)_2 PF_6$ and for $(DMET)_2 Au(CN)_2$. As a result the stacking array makes two-dimensional conducting layers separated by anion layers. In $(DMET)_2 BF_4$, however, the stacks in the neighboring layers run in two nearly perpendicular directions but they also produce two-dimensional conducting layers.

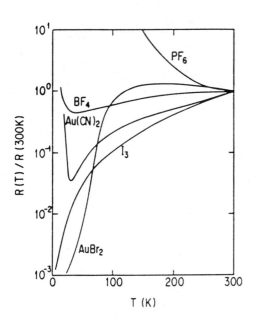

Fig. 6.2. Temperature dependence of the resistivity of $(DMET)_2 X$, for different X. Resistivities are normalized to the room-temperature value. Courtesy of *K. Kikuchi*. From [6.1, 2]

6.1.1 Electronic Properties

The DMET salts, $(DMET)_2 X$, show a rich variety of electron transport properties, ranging from insulator to superconductor, which are classified into five groups [6.1-3] as follows. The typical temperature dependence of the conductivity is exhibited in Fig. 6.2.

Group 1: $(DMET)_2 X$ salts with octahedral anions, X: PF_6, AsF_6.
These salts exhibit semiconducting behavior. Typical room-temperature conductivities of $(DMET)_2 PF_6$ and $(DMET)_2 AsF_6$ under ambient pressure are 300 and 200 S/cm, respectively. The resistivity increases monotonically with decreasing temperature, as depicted in Fig. 6.2 for $(DMET)_2 PF_6$, and hence they are semiconducting at room temperature.

For $(DMET)_2 PF_6$, Electron Spin Resonance (ESR) measurements exhibit a drop in the susceptibility below 25 K (Fig. 6.3a). The room-temperature value of the spin susceptibility is $\chi_{spin} = 2.0 \cdot 10^{-4}$ emu/mol, which decreases gradually to 25 K. Since the salt is insulating, the argument using the concept of Pauli paramagnetism is irrelevant to this temperature dependence. To account for the associated increase in linewidth, it is possible that some *magnetic orderings*, such as antiferromagnetic or spin-Peierls orderings, are present.

Group 2: $(DMET)_2 X$ salts with tetrahedral anions, X: BF_4, ClO_4, ReO_4.
Typical room-temperature conductivities are 130, 260 and 40 S/cm for $(DMET)_2 BF_4$ $(DMET)_2 ClO_4$ and $(DMET)_2 ReO_4$, respectively. The tem-

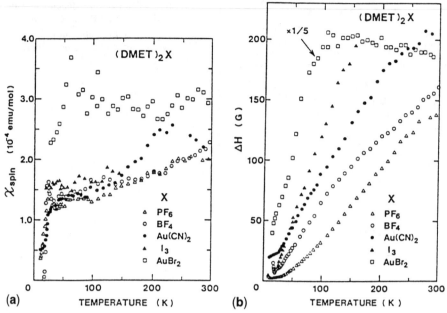

Fig. 6.3. Temperature dependences of the spin susceptibility χ_{spin} (**a**) and the peak-to-peak width ΔH (**b**) of the ESR absorption derivative for $(DMET)_2 X$ [X: PF_6, BF_4, $Au(CN)_2$, I_3 and $AuBr_2$]. The absolute value of χ_{spin} is calibrated for $(TMTSF)_2 PF_6$, $2.3 \cdot 10^{-4}$ emu/mol at room temperature. From [6.4]

perature dependence reveals that the metal-insulator transition occurs at a low temperature, e.g., at 40 K for $(DMET)_2 BF_4$ [6.2].

The susceptibility measured by ESR for $(DMET)_2 BF_4$ gives a value of $2.2 \cdot 10^{-4}$ emu/mol at room temperature. At lower temperatures, the observed ESR signal is of the Dysonian-type reflecting the metallic phase. The spin susceptibility drops sharply at 20 K and is accompanied by a linewidth broadening. These characteristics are typical for the appearance of **antiferromagnetic ordering**. From the proximity of the two temperatures for the change in conductivity and magnetization, the metal-insulator transition can be explained by the emergence of an SDW, which is reminiscent of the situation in $(TMTSF)_2 X$ salts.

Group 3: $(DMET)_2 X$ salts with gold dihalide and other anions, e.g., X: $AuCl_2$, AuI_2, $Au(CN)_2$.

Under ambient pressure $(DMET)_2 AuCl_2$ has a room-temperature conductivity of 230 S/cm and exhibits metallic behavior down to low temperatures. After showing a weak increase in the resistance below 3 K, it undergoes a superconducting transition, as illustrated by Fig. 6.4. The presence of superconductivity is verified by applying a magnetic field, after which the normal resistance is recovered [6.5].

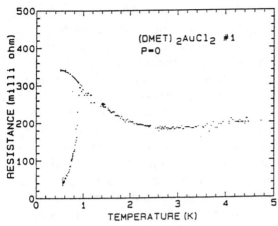

Fig. 6.4. Superconducting transition of $(DMET)_2 AuCl_2$ at ambient pressure shown by the recovery of the resistance after a magnetic field of 0.5 T has been applied. From [6.5]

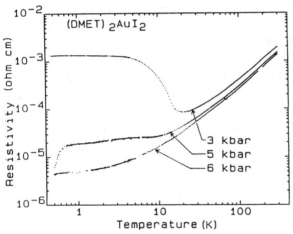

Fig. 6.5. Temperature dependence of the resistivity of $(DMET)_2 AuI_2$ at pressures of 3, 5 and 6 kbar. From [6.2]

For $(DMET)_2 AuI_2$, starting from a room-temperature conductivity of 300 S/cm the crystal continues to exhibit metallic behavior down to 20 K, below which an increase in the resistance is observed. Under a pressure of 3 kbar the transition temperature decreases. The resistivity still shows an increase, but reaches a saturating value, as displayed in Fig. 6.5. With further increase in the pressure, e.g. to 5 kbar, the increase is suppressed and superconductivity appears. Surprisingly, however, at 6 kbar the superconductivity disappears, although over the entire temperature range the dependence reflects a metallic phase.

Chronologically speaking, the first superconductor found among (DMET)$_2$X salts was **(DMET)$_2$Au(CN)$_2$** [6.6]. Under ambient pressure the room-temperature conductivity is 230 S/cm and increases monotonically with temperature decreasing to 28 K, where the resistance increase begins to appear, as depicted in Fig.6.2. At a pressure of 5 kbar the insulating phase is suppressed and the superconducting transition takes place.

The ESR measurements on (DMET)$_2$Au(CN)$_2$ are somewhat sample-dependent at lower temperatures. Figure 6.3 shows one example. The spin susceptibility is $2.3 \cdot 10^{-4}$ emu/mol at room temperature and has a broad peak around 230 K, below which it decreases with temperature. At 25 K, the rate of temperature dependence suddenly changes indicating the occurrence of a phase transition, which based on proton NMR measurements is thought to be a transition to an SDW phase [6.3]. This reveals the close similarity with the (TMTSF)$_2$X family.

Group 4: (DMET)$_2$X salts with linear and other halides, for example, X: I_3, I_2Br, IBr_2, SCN and AuBr$_2$.

Typical room-temperature conductivites of (DMET)$_2$I$_3$, (DMET)$_2$I$_2$Br, (DMET)$_2$IBr$_2$ and (DMET)$_2$SCN are 170, 320, 210 and 80 S/cm, respectively. Among these, **(DMET)$_2$I$_3$** and **(DMET)$_2$IBr$_2$** exhibit superconductivity under ambient pressure with T_c = 0.47 and 0.58 K, respectively [6.7]. On the other hand, **(DMET)$_2$I$_2$Br** and **(DMET)$_2$SCN** do not exhibit superconductivity, presumably due to the random distribution of unsymmetrical anions, as was found for the ET salts (Sect.5.2.7). In fact, at low temperatures the residual resistance of (DMET)$_2$I$_2$Br is higher than that of (DMET)$_2$I$_3$ (Fig.6.6), indicating that electron scattering is enhanced by the random potentials due to the anion distribution.

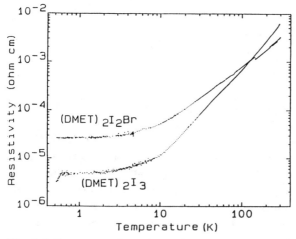

Fig. 6.6. Temperature dependence of the resistivity of (DMET)$_2$X. From [6.7]

The ESR measurements on $(DMET)_2 I_3$ [6.3] revealed that it is similar to salts of other groups. The temperature dependence of χ_{spin} is a weakly increasing function of temperature above 30 K. However, the linewidth exhibits a linear temperature dependence and is more pronounced than that for the salts of the groups 1-3. This, together with the absence of an instability in the Fermi surface, suggests that this compound has a higher dimensionality than the other $(DMET)_2 X$ salts. Thus, it is supposed that the electronic state of $(DMET)_2 I_3$ is similar to that of the $(ET)_2 X$ salts.

$(DMET)_2 AuBr_2$ has at least two different crystal structure with $Z = 1$ and 2, where Z is the number of $(DMET)_2 AuBr_2$ units in the unit cell. The $Z = 1$ crystal is assigned to group-4. Under ambient pressure the resistivity shows metallic behavior down to 0.5 K, as illustrated in Fig. 6.7 with the sample labelled p. The resistivity decreases monotonically with temperature but it does not exhibit a transistion to superconductivity.

Group 5: $(DMET)_2 X$ salts with X: $AuBr_2$.

$(DMET)_2 AuBr_2$ with a two-compound unit ($Z = 2$) belongs to this group; its crystal structure is depicted in Fig. 6.8. The centrosymmetric dimers are rotated by $\approx 80°$ to each other. At ambient pressure, the room-temperature conductivity of 14 S/cm decreases steadily down to $180 \div 120$ K, below which the semiconducting behavior changes into a metallic one down to low temperatures, as shown in Fig. 6.7 with data coded with r1 and r2 [6.2]. The low-temperature behavior is sample dependent; a certain crystal undergoes a superconducting transistion at 1.9 K under ambient pressure, as represented by the plots labelled r2. Another crystal labelled r1, does not show superconductivity under ambient pressure, but by applying a pressure of 1.5

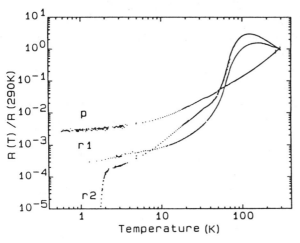

Fig. 6.7. Temperature dependences of the resistivity of $(DMET)_2 AuBr_2$ under ambient pressure. p denotes a sample belonging to Group-4, rl and r2 to Group-5. From [6.7]

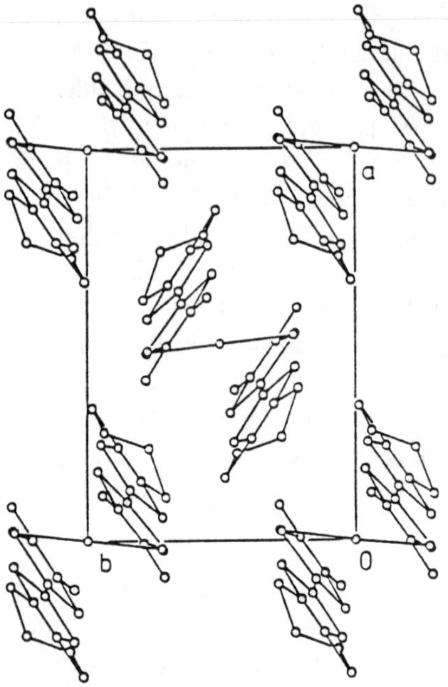

Fig. 6.8. Structure of the $(DMET)_2 AuBr_2$ crystal denoted by r2 in Fig. 6.7. From [6.7]

kbar, the resistance peak is shifted to about 190 K, and a resistance drop due to superconductivity appears at 1.6 K. With further increase in pressure, the resistivity peak is suppressed and the superconductivity transition temperature is decreased.

ESR measurements reveal yet another interesting feature in these salts. Within an experimental error, at least above 50 K, $\chi_{spin} = (2.90 \pm 0.25) \times 10^{-4}$ emu/mol is insensitive to temperature just as in the case of the Pauli paramagnetism. A sudden decrease below 50 K is assigned to the skin effect that is manifested by the appearance of the Dysonian-like line shape. The magnitude and the temperature dependence of ΔH are anomalous. At room temperature ΔH is as large as 930 G. This value is extraordinarily large for an organic conductor. With decreasing temperature, ΔH gradually increases reaching a maximum around 150 K, which is then followed by a rapid decrease. This anomaly is consistent with the behavior observed for the resistivity.

It is worthy to note that the **de Haas-van Alphen oscillations** have clearly been observed. This evidences the high quality of the grown crystals. The observation yielded two cross-sectional areas of $103 \div 104\%$ and 21% of the calculated first Brillouin-zone area. The corresponding effective masses are $6.0m_e$ for the large pocket and $3.8m_e$ for the small pocket [6.8].

Table 6.1. Superconducting Properties of $(DMET)_2 X$

X	σ_{RT} [S/cm]	T_c[a] [K]	P_c [kbar]	V [Å]
$Au(CN)_2$	230	1.1 (3.5kbar)	2.5	766.0
$AuCl_2$	230	0.83	0	761.5
AuI_2	300	0.55 (5kbar)	5.0	799.5
I_3	170	0.47	0	794.8
IBr_2	210	0.58	0	791.7
$AuBr_2$[b]		1.9	0	1537
	14	1.6 (1.5kbar)	1.5	1518

[a] Numbers in parentheses show the pressure under which the sample was placed during the T_c measurement.

[b] One sample exhibited superconductivity at ambient pressure but another only under pressure. See text.

6.1.2 Superconducting Properties

Typical characteristics of $(DMET)_2 X$ superconductors are summarized in Table 6.1. Four salts exhibit superconductivity under ambient pressure, although all have transition temperatures of less than 2 K.

In the case of pressure-induced superconductivity, T_c does not vary monotonically with pressure. An example of this is found for $(DMET)_2 \cdot Au(CN)_2$ [6.2]. Below 2 kbar, a transition to an insulating state is observed below 22 K, due to the emergence of an SDW phase. The superconducting transition exhibits a maximum T_c near 3.5 kbar, the superconductivity is suppressed above 7 kbar. The proximity of the superconductivity to the SDW phase is reminiscent of the phase diagram for $(TMTSF)_2 X$.

The conditions under which superconductivity in $(DMET)AuBr_2$ emerges are complicated, as mentioned in the previous subsection and shown in Fig.6.7. It occurs only within narrow temperature and pressure ranges. In addition, not all samples of this salt behave similarly. Figure 6.7 compared the behavior of a group-4 sample (curve p) and two group-5 samples (curves r1 and r2) all at ambient pressure. Note that only sample r2 displays a transition to superconductivity at $T_c = 1.9$ K; the highest critical temperature for any of the $(DMET)_2 X$ salts. Sample r1 requires 1.5 kbar pressure before it can become superconducting at 1.6 K.

Above 120 K, both Group-5 samples show an intriguing increase in resistivity with decreasing temperature. It should be noted that the broad maximum around $120 \div 180$ K corresponds to an insulator-metal-like transition in κ-$(ET)_2 Cu(NCS)_2$. It is also interesting that its stacking structure is similar to the structures of κ-$(ET)_2 I_3$ and κ-$(ET)_2 Cu(NCS)_2$, namely the checkered stacking depicted in Fig.6.8. It is supposed that the resistivity peak may be due to a precursor to a small polaron system with some *Mott-Hubbard insulator* character that is formed by the association of large electron-phonon interactions [6.9]. It was found that for sample r1 the position of the maximum shifts, depending on whether the temperature is being swept up or down, and also with increasing pressure. Crystallographic studies reveal no distinct differences in the two samples: both belong to the $P2_1/a$ space group and the difference in the lattice parameters is $<1\%$ [6.10]. The Meissner effect and other characteristics of the superconducting behavior have not yet been measured in detail.

6.2 MDT-TTF Salt

A methylenedithio-TTF (MDT-TTF) molecule is a hybrid of TTF and ET molecules (Fig.1.1). Its redox potentials indicate that its donor ability is comparable to that of ET. κ-$(MDT\text{-}TTF)_2 AuI_2$ has revealed superconducting properties [6.11]. The crystal structure is depicted in Fig.6.9. It is orthorhombic, Pbnm, with a $= 10.797$ Å, b $= 7.789$ Å, c $= 28.991$ Å, V $=$

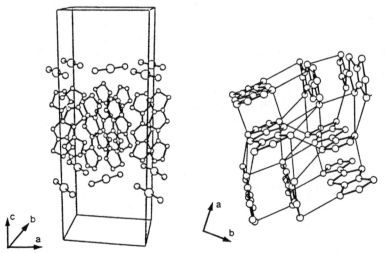

Fig. 6.9. Projections of the unit cell of $(MDT\text{-}TTF)_2 AuI_2$. From [6.11]

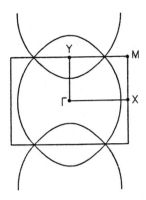

Fig. 6.10. Calculated Fermi surface of $(MDT-TTF)_2 AuI_2$. From [6.12]

2438.1 Å^3, Z = 4. Centrosymmetric donor dimers are rotated by 79.3° to each other to form a κ-type donor layer which is sandwiched by the anion layers of AuI_2. The next donor layer is determined by a mirror-reflection operation. As a result, the unit cell contains two layers of donor molecules. The calculated Fermi surface is depicted in Fig. 6.10.

The room-temperature conductivity is $20 \div 36$ S/cm. No semiconductor-like resistivity anomaly is observed down to low temperatures. Superconductivity with $T_c = 4.1 \div 4.3$ K is found under ambient pressure [6.13, 14]. T_c is suppressed by a pressure with the rate of $dT_c/dP = -0.92$ (± 0.10) K/kbar.

The ESR spin susceptibility at room temperature is $6 \cdot 10^{-4}$ emu/mole and shows a weak temperature dependence with a small minimum at about 20 K. The room-temperature value is rather high and this implies a strong electron-electron correlation in this salt. The ESR linewidth increases mon-

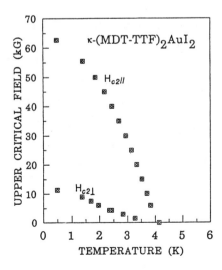

Fig. 6.11. Temperature dependence of the upper critical value of the field parallel or perpendicular to the conducting plane. From [6.14]

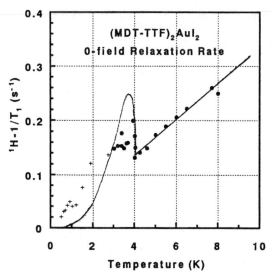

Fig. 6.12. Temperature dependence of $1/T_1$ in $(MDT\text{-}TTF)_2 AuI_2$. The solid line depicts the calculated $1/T_1$ for s-type symmetry. Data represented with filled circles were derived from an exponential-decay curve, whereas those with crosses were determined from the initial part of a non-exponential decay. From [6.17]

otonically down to 80 K and increases rapidly below it, whereas the g-value remains constant down to about 70 K [6.13]. Around 50 K, the resonance line is lost by unknown reasons.

The lower critical magnetic field H_{c1} is 50 Oe, whereas the temperature dependence of the upper critical field H_{c2} is plotted in Fig.6.11. The parallel and perpendicular coherence lengths are calculated as 160 and 20 Å, respectively. The interlayer-interaction or superconductivity dimensionality is evaluated via the magnetic penetration depth. The interlayer superconducting coupling is estimated to be stronger in this salt than in $\kappa\text{-}(ET)_2 \cdot Cu(NCS)_2$ and $\kappa\text{-}(ET)_2 Cu[N(CN)_2]Br$ [6.15]. The *^{1}H-NMR relaxation rate* measured via a field-cycling technique showed a clear enhancement of $1/T_1$, the *Hebel-Slichter coherence* peak, just below T_c (Fig.6.12). The maximum value of $1/T_1$ appears at $\approx 0.8 T_c$ and is 1.4 times as large as the Korringa value just above T_c. The result indicates that this material has a superconductivity gap with an s-wave symmetry [6.16]. The decay of $1/T_1$ below 3 K cannot be interpreted by an exponential temperature dependence. The origin has been ascribed to extra local fields exerted by trapped vortices [6.17].

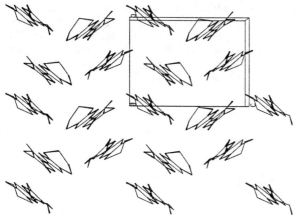

Fig. 6.13. Projection of DMBEDT-TTF molecules along the a axis of κ-(DMBEDT-TTF)$_2$ ClO$_4$. From [6.18]

6.3 DMBEDT-TTF Salt

(S,S)-dimethyl-bis-ethylenedithio-tetrathiafulvalene (DMBEDT-TTF) (Fig. 1.1) is a chiral molecule. It forms a superconductor of the κ-type donor packing: **κ-(DMBEDT-TTF)$_2$ ClO$_4$** whose molecular arrangement in the crystal is depicted in Fig. 6.13. This phase is minor product of the electro-crystallization and the major one is a semiconductive salt of the α'-type. The superconducting salt is monoclinic, P2$_1$, a = 34.868 Å, b = 8.509 Å, c = 11.910 Å, β = 93.37°, V = 3527.5 Å^3, Z = 4. The ClO$_4$ anions are disordered. The molecules of the dimer are related to each other with a pseudo-inversion center. The two five-membered rings of the TTF majority in a molecule are twisted relative to each other [6.18].

The salt is not highly conductive at room temperature ($\sigma_{\parallel}$ = 0.05 S/cm). The anisotropy of the conductivity, $\sigma_{\parallel}/\sigma_{\perp}$, is higher than 10^3. It exhibits a metal-insulator transition at 75 K. At a pressure of 5.8 kbar a sharp drop of the resistivity is observed below 3 K (Fig. 6.14).

6.4 DTEDT Salt

A molecule fused between TTF and vinylogus TTF, 2-(1, 3-dithiol-2-ylidene)-5-(2-ethanediylidene-1, 3-dithiole)-1, 3, 4, 6-tetrathiapentalene, in shorthand DTEDT, illustrated in Fig. 1.1, has been found to form a super-

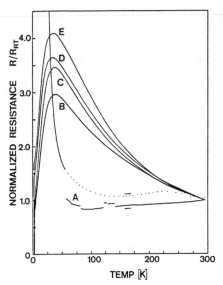

Fig. 6.14. Normalized resistivity of κ-(DMBEDT-TTF)$_2$ClO$_4$ as a function of temperature, at (A): 0 kbar, (B): 5.8 kbar, (C): 6.1 and 6.4 kbar, (D): 7.1 kbar, (E): 10 kbar. Resistivity jumps are due to microcracking in the crystal. From [6.18]

conductor such as **(DTEDT)$_3$Au(CN)$_2$**. The cyclic voltammogram of this molecule exhibits four pairs of single-electron redox waves. The first redox potential is comparable to that of TTF. The ΔE value (0.13V), ΔE being the difference between the first and second redox potentials, is considerably smaller than that of TTF (0.42V), indicating a small electron-electron correlation in the dication. The phases of HOMO of the sulfur atoms in the terminal vinylogous 1, 3-dithiole ring are reversed with respect to the other sulfur atoms (Fig. 6.15).

As expected for a system having both the extended π-conjugation and the *reduced on-site Coulomb repulsion*, this donor molecule affords a number of stable metals regardless of the counter anion. Most of them are metallic down to 1.4 K. These salts are prepared by electrochemical oxidation with TBA·X (tetra n-butylammonium) in chlorobenzene or TCE (1, 1, 2-trichloroethane) at 50°C. Among them, the Au(CN)$_2$ salt shows superconductivity with an onset T$_c$ of 4 K, but it does not reveal zero resistance even at 2 K (Fig. 6.16). Nevertheless, magnetic susceptibility measurements exhibit that a diamagnetic shielding at 2 K amounts to 10% of the perfect diamagnetism. It is noteworthy that the conductivity enhancement from room

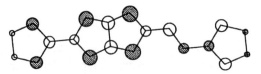

Fig. 6.15. HOMO of DTEDT. From [6.19]

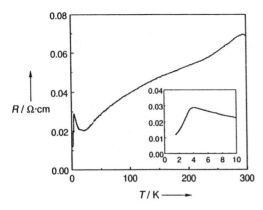

Fig. 6.16. Temperature dependence of the resistivity of (DTEDT)$_3$ · Au(CN)$_2$. The inset shows the resistivity in the range of $1.4 \div 10$ K. From [6.19]

temperature (15 S/cm) to just above T$_c$ (33 S/cm) is quite flat, i.e., an increase by only $2 \div 4$ times.

The stoichiometry of the DTEDT salts were estimated by Energy Dispersion Spectroscopy (EDS). The composition of the Au(CN)$_2$ salt was determined both by EDS and X-ray structural analysis as 1:0.39 and 1:0.4, respectively. Later, it was corrected to 3:1. It is triclinic, P1, a = 6.322 Å, b = 17.211 Å, c = 3.857 Å, α = 95.52, β = 95.52, γ = 93.75°, V = 414.8 Å, Z = 1. The donor molecules stack face-to-face with ring-over-bond type fashion along the c-axis (Fig. 6.17). It is worth to note that unsymmetrical DTEDT molecules are stacked unidirectionally, resulting in the absence of the crystal's centrosymmetry. Consequently the salt can be *electronically polarizable*. The donor packing pattern resembles that of β-(ET)$_2$I$_3$, although it does not show any indication of dimerization which β-(ET)$_2$I$_3$ has. The anion molecules lie along the c-axis. The calculated bandwidth is large (0.9eV). Though the interstack overlap is about one third of the interstack interaction, a closed Fermi surface is expected.

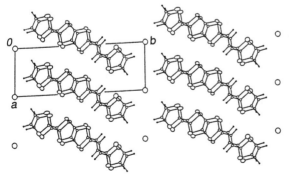

Fig. 6.17. Crystal structure of (DTEDT)$_3$ Au(CN)$_2$. Projection onto the ab-plane. From [6.19]

7. DMIT Salts: Anion Conductors

A new class of molecules, bis(4, 5-dimercapto-1, 3-dithiole-2-thione)-M (M: Ni or Pd), abbreviated by M(dmit)$_2$ and shown in Fig. 1.1, has been found to form highly conducting compounds. It is similar to a TTF-derivative, namely, it is isolobal to the TTF-derivative in the sense that M(dmit)$_2$ is given by replacing the central two carbons, C=C, in the TTF-derivative by a metal atom M. The d-orbital of M extends in the same directions as the valence orbitals of C=C. On forming ion radical salts, these molecules act as acceptor rather than donor. The oxidation states are not simple and change with the counter donor molecules. Furthermore, HOMO and LUMO of M(dmit)$_2$ are close in energy. This leads to a mixed contribution of HOMO and LUMO to the conductance.

7.1 TTF[M(dmit)$_2$]$_2$

The donor-acceptor compound of **α-TTF[Ni(dmit)$_2$]$_2$** has been found to show high electrical conductivity. At ambient pressure this reaches 300 S/cm at 300 K and increases to about 10^5 S/cm at 4.2 K [7.1]. It reveals a metallic temperature dependence under ambient pressure with an approximate $T^{1.65}$ power law between 300 and 10 K with a slight hump near 40 K. With a weak pressure of 500 bar it exhibits a nonmetallic behavior at low temperatures with a minimum near 50 K. With increase of pressure, the minimum temperature is lowered to ≈ 13 K at 5.5 kbar. It undergoes a clear superconductivity transition at 5.75 kbar. Figure 7.1 displays superconducting transitions at 9.5 and 12.3 kbar [7.2]. With an increase of pressure, T_c increases slightly with pressure and reached 2 K at 14 kbar, in contrast to the pressure dependence of T_c for both TMTSF and ET salts.

The crystal consists of stacks of Ni(dmit)$_2$ and TTF. It has a centered monoclinic structure with the space group C2/c having a = 46.22 Å, b = 3.732 Å, c = 22.86 Å and β = 119.19°. The molecules of TTF and Ni·(dmit)$_2$ stack in distinct columns parallel to the b-axis, with the needle axis as shown in Fig. 7.2, where only Ni(dmit)$_2$ molecules are depicted, for simplicity. These columns make alternating sheets of TTF and Ni(dmit)$_2$ paral-

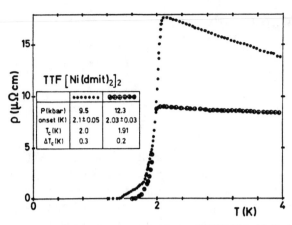

Fig. 7.1. Superconducting transitions of TTF[Ni(dmit)$_2$]$_2$ at 9.5 and 12.5 kbar. From [7.2]

lel to the bc-plane (Fig. 7.3) The complete structural data of TTF [Ni·(dmit)$_2$]$_2$ were presented in [7.3].

With regard to the electronic structure, it had been proposed that three-dimensional electronic properties are promoted because the S...S distances are shorter than twice the van der Waals distance of 3.70 Å between neighboring Ni(dmit)$_2$ groups or between Ni(dmit)$_2$ and TTF groups [7.1]. However, based on extended-Hückel band calculations *Kobayashi* et al. [7.4] demonstrated that the transverse interaction is not strong enough to make the system multi-dimensional. They claimed that the system consists essentially of an ensemble of *one-dimensional bands* and the resultant "*multi-Fermi surfaces*" ensure that the salt remains metallic even at low temperatures.

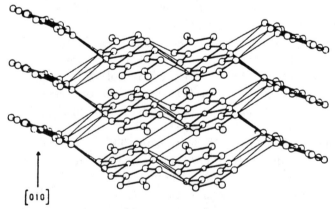

Fig. 7.2. End-on view of the Ni(dmit)$_2$ molecules in the bc-plane showing the two-dimensional arrangement of these units. From [7.1]

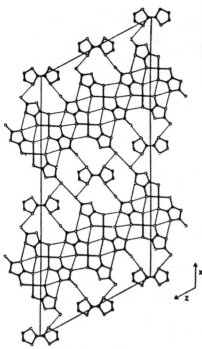

Fig. 7.3. A view of the crystal structure of TTF·[Ni(dmit)$_2$]$_2$ parallel to the b axis. The TTF and Ni(dmit)$_2$ molecules are repeated along the b axis at the unit cell distance of 3.73 Å. The four Ni(dmit)$_2$ molecules along z (or c) shown lie at different levels along the b axis. From [7.1]

Similarly to other planar π-cation molecules, Ni(dmit)$_2$ molecules tend to form columns in face-to-face stacks, which are connected by many inter-column S...S contacts. In order to form a two-dimensional band, the inter-molecular side-by-side interactions must be large. However, the amplitude of the LUMO of Ni(dmit)$_2$ on the outer sulfur atom of the five-membered hetero-ring is very small compared to that on the inner sulfur. Moreover, owing to the b_{2g} symmetry of the LUMO, the intermolecular transverse overlap integrals are almost canceled.

Figure 7.4 schematically describes HOMO and LUMO of an ideal M(dmit)$_2$ unit. They are built from in-phase and out-of-phase combinations of the same π-orbital of the dmit ligand. The three symmetry planes of the ideal dmit ligand do not allow the metal d orbitals to mix into HOMO. The nodal properties of the ligand orbital produce only a modest overlap, and hence relatively small mixing of the d_{xz} orbital into LUMO. Consequently the *HOMO-LUMO splitting* is not large [7.5].

The calculated band structure for the Ni(dmit)$_2$ slabs in α-TTF·[Ni(dmit)$_2$]$_2$ is depicted in Fig. 7.5. In this crystal Ni(dmit)$_2$ molecules are stacked uniformly within the column with every molecule slipped with re-spect to each other (Fig.7.2). In this case, the transfer integrals along the chain for both HOMO and LUMO have the same sign, and the two types of bands run parallel. However, the LUMO-based bands overlap appreciably

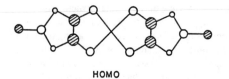

HOMO

LUMO

Fig. 7.4. LUMO (*upper*) and HOMO (*lower*) of an ideal M(dmit)$_2$ molecule. Hatched and non-hatched lobes refer to positive and negative contributions, respectively. From [7.5]

with the HOMO-based ones. It is noteworthy that the HOMO and LUMO bands of the acceptor are partially filled.

X-ray diffuse-scattering studies at ambient pressure reveal many sets of one-dimensional structural fluctuations associated with CDW instabilities. From a form-factor analysis, it is concluded that the periodic lattice distortions are only on Ni(dmit)$_2$ chains [7.6]. The slight hump in the temperature dependence of resistivity appearing near 40 K corresponds to the *CDW* with $q_1 = 0.4$ b*; two other weaker one-dimensional diffuse scatterings at $q_2 = 0.22$ b* and $q_3 = 0.18$ b* were detected. The reduced wave vectors are assigned to the intersection between the Fermi level and the bunches of the bands built on HOMO and LUMO of the Ni(dmit)$_2$ slab (Fig. 7.5). This accounts for the intrinsic one-dimensional multi-parallel

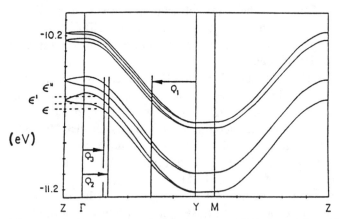

Fig. 7.5. Ambient-pressure band structure of Ni(dmit)$_2$ slabs of TTF[N(dmit)$_2$]$_2$ with marks of half the one-dimensional critical wave vectors measured by X-ray diffuse scattering. Γ, Y, Z and M refer to the wave vectors $(0, 0)$, $(b^*/2, 0)$, $(0, c^*/2)$ and $(b^*/2, c^*/2)$, respectively. The Fermi levels noted by ϵ_f, ϵ_f' and ϵ_f'' are those that are appropriate for charge transfers of 0, 0.5 and 1, respectively. From [7.5]

band structure, where LUMO of Ni(dmit)$_2$ overlaps not only with HOMO of TTF, but also with those of Ni(dmit)$_2$.

The central metal atom of Ni(dmit)$_2$ can be replaced with a transition-metal atom such as Pd and Pt with a large spatial expansion of the metal orbitals. The charge-transfer salt of **α'-TTF[Pd(dmit)$_2$]$_2$**, which is structurally isomorphous with α-TTF[Ni(dmit)$_2$]$_2$, shows superconductivity at T$_c$ as high as 6.5 K under 20 kbar [7.7]. Some crystals with a similar structure, at room temperature, hence named α-TTF[Pd(dmit)$_2$]$_2$, becomes superconducting with an onset T$_c$ of 1.7 K at a pressure of 22 kbar [7.8]. The α analogue shows an irreversible behavior associated with a monoclinic (α) to triclinic (β) phase transition near 240 K. By contrast the α' phase reveals a perfectly reversible behavior.

At ambient pressure the room-temperature conductivity of 770 S/cm decreases with temperature reflecting charge localization after reaching a maximum at T$_\rho$ = 245 ± 15 K. The conductivity peak shifts to a lower temperature with pressure, and superconductivity is observed above 16 kbar. The pressure-temperature phase diagram for α'-TTF[Pd(dmit)$_2$]$_2$ is depicted in Fig. 7.6. At 20.7 kbar the onset of the resistance drop appears at

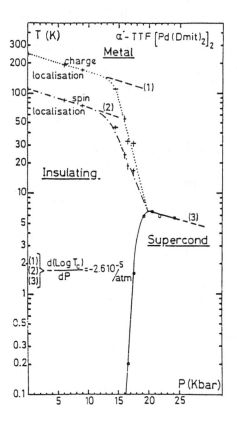

Fig. 7.6. Phase diagram of α'-TTF·[Pd(dmit)$_2$]$_2$. From [7.8]

6.4 K. Under ambient pressure the ESR measurement shows no anomaly near T_ρ but does so at 155 K. This suggests that the spin's degrees of freedom are not frozen around the resistivity minimum; only the charge is localizing around T_ρ. Below 155 K, a narrowing in the resonance signal implies spin localization. The temperatures of the charge and spin localizations decrease with increasing pressure (Fig.7.6). It is claimed that the same electron-phonon interaction is responsible for both the metal-insulator and the superconducting transitions, since the pressure dependeces $d\ln(T_{MI})/dP$ and $d\ln(T_c)/dP$ are close, and spin-fluctuation features are not found in the neighborhood of the superconducting phase. The low-pressure *CDW* and high-pressure superconducting ground states are competing in the intermediate pressure region.

X-ray scattering studies have provided structural evidence of a CDW instability. One-dimensional fluctuations observed as diffuse lines on the X-ray patterns condense into satellite reflections below about 40 K for Ni(dmit)$_2$ salt. The Pd derivative exhibits two kinds of scatterings which condense into satellite reflections at ≈ 150 and ≈ 105 K. By correlating the CDW wave vectors and the band structure for Ni(dmit)$_2$ and Pd(dmit)$_2$ salts, a charge transfer of 0.75e per TTF molecule is derived for TTF· [M(dmit)$_2$]$_2$ [7.8-10]. The absence of diffuse lines at $4k_F$ may indicate *weak electron correlations*.

The static magnetic susceptibility χ_s of α-TTF[Ni(dmit)$_2$]$_2$ and α'-TTF[Pd(dmit)$_2$]$_2$ is correlated to successive condensations of the CDW instabilities and to the partial density of states of multiple parallel bands arising from stacks. Both compounds are suggested to have small *on-site Coulomb repulsions* compared to the bandwidth [7.9]. The band structure of Pd(dmit)$_2$ slabs in α'-TTF[Pd(dmit)$_2$]$_2$ is similar to Ni(dmit)$_2$ slabs shown in Fig.7.5. Typical differences are in the bandwidth (0.6 and 0.8 eV for Ni(dmit)$_2$ and Pd(dmit)$_2$, respectively) and in the Fermi level correlates to the wave vectors of the one-dimensional super-lattices [7.5].

With regard to *^1H NMR relaxation* measured in α-TTF[Ni(dmit)$_2$]$_2$ under ambient pressure, deviations to the normal metal Korringa law were found indicating that TTF conducting chains are dominated by repulsive electron-electron interactions and one-dimensional correlation effects [7.11]. On the other hand, *^{13}C NMR* results for TTF[Ni(dmit)$_2$]$_2$ enriched by ^{13}C isotope in Ni(dmit)$_2$ provide evidence for an unusual multiband structure in which both the LUMO and HOMO bands of Ni(dmit)$_2$ acceptor stack cross the Fermi level. It has been suggested that CDW effects above 160 K are associated with the LUMO bands [7.12].

In TTF[Ni(dmit)$_2$]$_2$ CDW coexists with metal-like conductivity. Proton spin-lattice relaxation studies of the TTF stacks show that the metallic character is retained to 1 K. The ^{13}C NMR experiments give evidence that CDW involves the Ni(dmit)$_2$ stacks but the Ni(dmit)$_2$ subsystem is again metallic in the entire temperature region. The reason why CDW does not induce nonmetal transition on the Ni(dmit)$_2$ is ascribed to the multiband consisting of HOMO and LUMO bands. CDW could open a gap in only some of them so that the other metallic band will remain [7.12].

The two-band nature of TTF[M(dmit)$_2$]$_2$, in which two sets of Fermi surfaces for LUMO and HOMO bands exist, is considered to play an important role in the mechanism of superconductivity in them [7.13]. From the similarity of the M(dmit)$_2$ molecule with TTF in size and electronic structure, one is tempted to consider that the former has as large an on-site Coulomb energy U as the latter. Assuming an appreciable value of U, the M(dmit)$_2$ column can be modeled by a *two-chain Hubbard model*. In this modeling each dmit ligand is considered to be an electronic site. It is linked to the other dmit ligand in the same M(dmit)$_2$ molecule by a transverse transfer energy. The bonding and antibonding states made on the two sites in one molecule correspond to the HOMO and LUMO in the molecule, respectively. As described in Sect. 8.5.4, superconductivity is considered to occur due to a many-body effect if the band structure is favorable. Even if superconductivity is not induced by this mechanism alone, the detrimental effect, i.e., the effect of intraband matrix elements of the Coulomb interaction, is drastically reduced if the two-band superconductivity occurs with gap parameters in both bands having opposite signs. In TTF[Ni(dmit)$_2$]$_2$ the coupling of the LUMO electrons with the longitudinal acoustic phonons was found from a band-parameter calculations to be so strong that it may drive the above-mentioned two-band superconductivity [7.14]. On the other hand, coupling of the HOMO electrons with intramolecular vibrations in Ni·(dmit)$_2$ was observed to be modest [7.15].

7.2 (CH$_3$)$_4$N[Ni(dmit)$_2$]$_2$

(CH$_3$)$_4$N[Ni(dmit)$_2$]$_2$ is a salt which exhibits superconductivity with T$_c$ = 3.0 K at a pressure of 3.2 kbar. With further increase of the pressure up to 7 kbar, T$_c$ is raised to 5.0 K. The estimated in-plane coherence length at 0 K is 290 Å under 3.2 kbar [7.16, 17].

In (CH$_3$)$_4$N[Ni(dmit)$_2$]$_2$ the unique electron transferring capability of the Ni(dmit)$_2$ molecule is essential for the functioning of the salt as a conductor. This is because (CH$_3$)$_4$N$^+$ is a closed-shell cation which, in a first approximation, cannot contribute to electrical conduction. The salt is monoclinic, C2/c, with a = 13.856(4) Å, b = 6.498(2) Å, c = 36.053(11) Å, β = 93.83(3)°, V = 3239(2) Å and Z = 8 at room temperature [7.4]. The crystal is composed of alternating (and hence dimerizing) sheets of one-dimen-

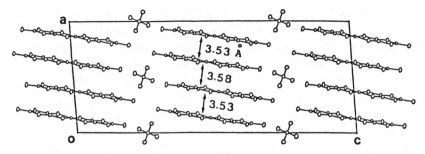

Fig. 7.7. Crystal structure of $(CH_3)_4 N[Ni(dmit)_2]_2$. From [7.16]

sional fourfold stacks of $Ni(dmit)_2$ with cations along the [001] axis (Fig. 7.7). The glide plane parallel to [001] causes the metal layers of $Ni(dmit)_2$ to stack along [110], and those along [1, − 1, 0] to alternate in the [001] direction.

The band structure for a $Ni(dmit)_2$ slab is shown in Fig. 7.8. Bands of HOMO and LUMO do not overlap in this case. The reason is ascribed to the dimerization along the stacks [7.6]. Because of the presence of two monomers per repeat unit, HOMO and LUMO generate two combinations: one bonding and one antibonding. If the transfer integrals t_{HOMO} and t_{LUMO} are larger than $\Delta/2$ (Δ: HOMO-LUMO splitting energy) the antibonding combination of HOMOs becomes higher than the bonding combination of LUMOs. According to the band structure, the partially filled band in $(CH_3)_4 N[Ni(dmit)_2]_2$ is mainly built from LUMO of the acceptor [7.18].

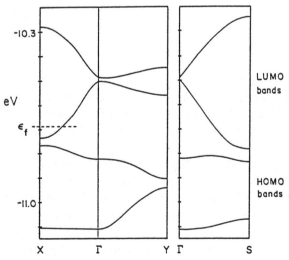

Fig. 7.8. Band structure of $(CH_3)_4 N[Ni(dmit)_2]_2$. From [7.5]

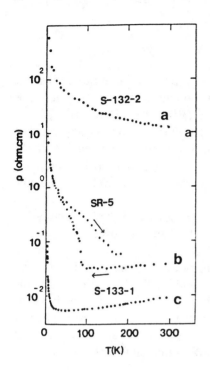

Fig. 7.9. The temperature dependence of the resistivity for three different crystals of $(CH_3)_4 N [Ni(dmit)_2]_2$. Curve a: the current is in the direction normal to the conducting plane. Curves b and c: the current is in the conducting plane. From [7.17]

Figure 7.9 depicts the temperature dependences of the resistivity for three samples [7.17]. Curve a represents the resistivity for the current direction *normal* to the conducting ab-plane. Curve b plots the behavior of most samples, where the resistivity in the ab-plane decreases slowly with temperature down to ≈ 100 K, where it shows a small jump. Below this temperature the resistivity increases slowly until a sharp rise occurs below ≈ 20 K. However, not all samples exhibit the jump near 100 K, as may be inferred from curve c of Fig. 7.9. This resistance jump is considered to be due to the freezing of the rotational motion of the methyl groups in $(CH_3)_4 N$, but it is not clear yet what discriminates the behavior of the types b and c. The rise of the resistivity below 20 K is thought to be due to the localization of electrons by the random potentials rather than a phase transition. An argument for this conjecture is based on an ESR measurement which depicts that ESR intensity increases monotonically with decreasing temperature, and shows no indication of superconductivity [7.17]. After applying hydrostatic pressure it was observed that both the resistance jump and the low-temperature resistance increase are suppressed and at a pressure of 3.2 kbar, a resistance drop due to the onset of superconductivity is observed at $T_c = 3.0$ K.

It should be noted that the sister compounds β-$(CH_3)_4 N[Pd(dmit)_2]_2$ and $(CH_3)_2 (C_2 H_5)_2 N[Pd(dmit)_2]_2$ exhibit superconductivity at 6.2 K under a pressure of 6.5 kbar and at 2.4 K under 4 kbar, respectively [7.19, 20].

7.3 α-(EDT-TTF)[Ni(dmit)₂]

A charge transfer salt formed between ethylenedithiotetrathiafulvalene (EDT-TTF) and N(dmit)₂ exhibits a superconductivity transition at 1.3 K under ambient pressure. The crystal of **α-(EDT-TTF)[Ni(dmit)₂]** has a segregated column structure with a solid crossing column, as illustrated in Fig.7.10a. The donors and acceptors form segregated sheets parallel to the crystallographic ab-plane with the molecular stacking along [110] and [010] in the EDT-TTF sheet and in the N(dmit)₂ sheet, respectively, as depicted in Fig.7.10b.

Figure 7.11 presents the temperature dependence of the in-plane resistance under various pressures. The resistivity decreases by cooling but increases slightly around 20 K and then decreases again exhibiting a hump around 14 K. The superconductivity transition takes place at 1.3 K (Fig. 7.12). *Tajima* et al. asserted that Ni(dmit)₂ and EDT-TTF chains contribute to the electrical conduction above 20 K, but only the Ni(dmit)₂ chain below 10 K, according to the symmetry coincidence of the angle-dependence of the magnetoresistance [7.23]. The increase of the resistivity on cooling toward 20 K is considered to be due to *SDW* formation [7.24].

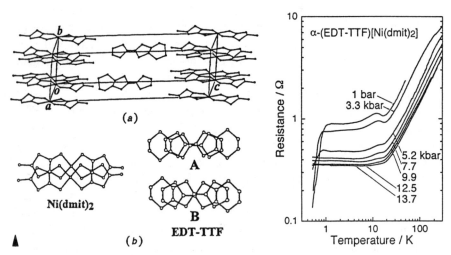

Fig. 7.10. Crystal structure of α-(EDT-TTF)[Ni(dmit)₂] (**a**) and stacking patterns of Ni(dmit)₂ and EDT-TTF molecules (**b**). Ni(dmit)₂ molecules and EDT-TTF molecules stack along the [010] and [110] directions, forming solid crossing columns. From [7.21]

Fig. 7.11. Temperature dependence of the resistivity of α-(EDT-TTF)[Ni(dmit)₂] under various pressures. The pressure value given for each curve is the one determined at room temperature. From [7.22]

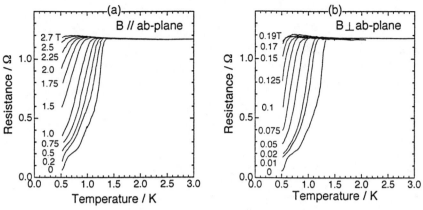

Fig. 7.12. Temperature dependence of the resistivity of α-(EDT-TTF)[Ni(dmit)$_2$] under various magnetic fields parallel (**a**) and perpendicular (**b**) to the ab-plane. From [7.22]

Increasing the pressure decreases T_c with $dT_c/dP \simeq -0.1$ K/kbar in contrast to the case for TTF[Ni(dmit)$_2$]$_2$. Studies of the Meissner and diamagnetic shielding effects reveal that the diamagnetic transition occurs at 1.39 K. The volume fraction was estimated to be $\approx 63\%$ (the diamagnetic shielding fraction reached $\approx 93\%$) [7.24]. The temperature dependence of resistance around the superconductivity transition in a magnetic field applied either parallel or perpendicular to the ab-plane is represented in Fig. 7.12. By regarding the resistance decrease of the residual resistance to 93% as a measure of the superconductivity onset, the temperature dependence of the upper critical field is obtained (Fig. 7.13). Assuming that the *coherence length* is isotropic within the ab-plane, the coherence length parallel and perpendicular to the ab-plane are given as $\xi_{\parallel}(0) \simeq 310$ Å and $\xi_{\perp}(0) = 24$ Å, $\xi_{\parallel}(0)$ is much longer than the lattice constant within the plane (a = 6.658Å, b = 7.627Å), whereas $\xi_{\perp}(0)$ is comparable to the lattice parameter, c = 27.385 Å.

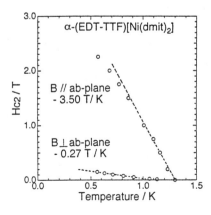

Fig. 7.13. Temperature dependence of the upper critical field H$_{c2}$ for the magnetic field parallel and perpendicular to the ab plane. From [7.22]

8. Mechanism of Superconductivity in Organic Materials

The mechanism of superconductivity is one of the most exciting areas in the study of organic superconductors. It is exotic and may be different from the BCS mechanism. The type of superconductivity, i.e., s-, p- or d-wave superconductivity, has been argued to be dependent on the mechanism. Moreover, it is interesting that some aspects resemble those in high-T_c oxide superconductors. Understanding the mechanism will clarify how the magnitude of T_c can reach, for example, 10 K for κ-$(ET)_2 Cu(NCS)_2$, and how it may be pushed higher, as well as why its pressure dependence is so large. These studies are still going on and firm conclusions might not always be available. In this chapter we review several models.

We begin with a brief description of the BCS (Bardeen-Cooper-Schrieffer) theory. Then, we discuss the types of waves, such as s-, p-, and d-waves, as well as the singlet and triplet pairings. We also review Little's model which gave the initial impetus to the search for organic superconductors. The second section is devoted to an exotic mechanism of superconductivity, namely that mediated by spin fluctuations. This mechanism is also being hotly debated for cuprate high-T_c superconductors. Next in Sect. 8.3, we analyze the mechanism due to intramolecular vibrations in the skeleton of the TTF-derivative molecules. We examine the contribution of the totally symmetric vibrational modes which strongly couple with HOMO of the molecules. Then in Sect. 8.4, we describe a potentially very important experiment, namely, the point-contact tunneling, and analyze the results in the framework of the strong coupling theory of superconductivity. The latter will be given an additional briefing in the following section. In the final Section 8.5 we touch on other theories, such as the g-ology, excitonic, bipolaron and two-band mechanisms. The possibility of overcoming the Pauli limit for H_{c2} in the case of triplet superconductivity is discussed. Finally, references to theories on likely polymer superconductors are presented.

8.1 BCS Theory of Superconductivity, and Little's Model

8.1.1 Attractive Interaction Due to Phonons

The Bardeen-Cooper-Schrieffer (BCS) theory [8.1] consists of two basic concepts: One is the attractive interaction between electrons, which occurs through the exchange of a phonon between two electrons; the other is the pairing wave function which maximizes the energy gain due to this attractive interaction, and leads to superconductivity. In this section we first derive the attractive interaction for an ordinary metal, and in the next section we will treat the pairing and the superconducting state.

The attractive interaction is caused by the electron-phonon interaction. The importance of this fact was first noticed by *Fröhlich* [8.2]. Discovery of the isotope effect on T_c [8.3,4] supported his assertion that the electron-phonon interaction plays an essential role in superconductivity. This is because the dependence of T_c on the mass of the constituting atom signifies that the lattice modes must be involved in superconductivity

The electron-phonon interaction of the deformation-potential type in ordinary superconductors [8.5,6] is given by the following Hamiltonian:

$$\mathcal{H} = \mathcal{H}_0 + \mathcal{H}' ,$$

$$\mathcal{H}_0 = \sum_{\mathbf{k}\sigma} \epsilon_{\mathbf{k}} c_{\mathbf{k}\sigma}^{\dagger} c_{\mathbf{k}\sigma} + \sum_{\mathbf{q}} \hbar\omega_{\mathbf{q}} b_{\mathbf{q}}^{\dagger} b_{\mathbf{q}} , \qquad (8.1)$$

$$\mathcal{H}' = i \sum_{\mathbf{k}\sigma} \sum_{\mathbf{q}} g_{\mathbf{q}} c_{\mathbf{k}+\mathbf{q}\sigma}^{\dagger} c_{\mathbf{k}\sigma} (b_{\mathbf{q}} - b_{-\mathbf{q}}^{\dagger}) ,$$

where $\mathcal{H}_0$ is the Hamiltonian of electrons and phonons without mutual interaction, and $\mathcal{H}'$ represents the coupling interaction; $c_{\mathbf{k}\sigma}^{\dagger}$ ($c_{\mathbf{k}\sigma}$) is the creation (annihilation) operator of an electron specified by the wave vector $\mathbf{k}$ and the spin σ, and $\epsilon_{\mathbf{k}}$ is the one-electron energy measured relative to the chemical potential, and ϵ_F the Fermi energy at absolute zero; $b_{\mathbf{q}}^{\dagger}$ ($b_{\mathbf{q}}$) are the creation (annihilation) operators of a phonon labeled by the wave vector $\mathbf{q}$; $\hbar\omega_{\mathbf{q}}$ is the eigenenergy of the phonon; and $g_{\mathbf{q}}$ is the coupling constant of the electron-phonon interaction.

In first order, $\mathcal{H}'$ causes electron-phonon scattering. In second order it leads to the above-mentioned exchange of a phonon between electrons, which brings about an indirect attraction between electrons. This process is illustrated in Fig.8.1. Here, the straight lines denote the electrons and the wavy line the phonons. This simply shows that one electron polarizes the

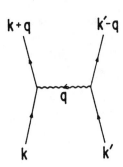

k+q   **k′-q**

q

k   **k′**

Fig. 8.1. Electron (*straight line*) emitting a phonon (*wavy line*) which is then absorbed by another electron. Vectors represent crystal momenta

lattice and the other electron interacts with the polarization. In order to get the expression of the resultant attractive interaction, one performs the following canonical transformation, which eliminates the first-order term in g_q:

$$\bar{\mathcal{H}} = e^{-iS}\mathcal{H}e^{iS}, \tag{8.2}$$

with

$$S = \sum_{k\sigma} \sum_{q} g_q c^\dagger_{k+q\sigma} c_{k\sigma} \left[\frac{b_q}{\epsilon_k - \epsilon_{k+q} + \hbar\omega_q} - \frac{b^\dagger_{-q}}{\epsilon_k - \epsilon_{k+q} - \hbar\omega_q} \right]. \tag{8.3}$$

Thus, S satisfies $i[\mathcal{H}_0, S] = -\mathcal{H}'$. Neglecting terms of the order of g_q^3, we obtain

$$\bar{\mathcal{H}} = \mathcal{H}_0 + \frac{i}{2}[\mathcal{H}', S] \tag{8.4}$$

$$= \mathcal{H}_0 + \sum_q |g_q|^2 \sum_{k\sigma} \sum_{k'\sigma'} c^\dagger_{k+q\sigma} c_{k\sigma} c^\dagger_{k'-q\sigma'} c_{k'\sigma'} \frac{\hbar\omega_q}{(\epsilon_{k'} - \epsilon_{k'-q})^2 - (\hbar\omega_q)^2}.$$

The second term means that the resulting indirect electron-electron interaction is attractive if $|\epsilon_{k'} - \epsilon_{k'-q}| < \hbar\omega_q$; the interaction constant is roughly equal to $-g_{\mathbf{q}}^2/\hbar\omega_{\mathbf{q}}$. Therefore, the above Hamiltonian can be further simplified to

$$\mathcal{H}_{SC} = \mathcal{H}_{el} + \mathcal{H}_{I} \, ,$$

$$\mathcal{H}_{el} = \sum_{\mathbf{k}\sigma} \epsilon_{\mathbf{k}} c_{\mathbf{k}\sigma}^\dagger c_{\mathbf{k}\sigma} \, , \qquad (8.5)$$

$$\mathcal{H}_{I} = - V_{BCS} \sum_{\mathbf{k},\mathbf{k'},\mathbf{q}} \eta(\mathbf{k'})\eta(\mathbf{k'}-\mathbf{q}) c_{\mathbf{k}+\mathbf{q}\uparrow}^\dagger c_{\mathbf{k'}-\mathbf{q}\downarrow}^\dagger c_{\mathbf{k'}\downarrow} c_{\mathbf{k}\uparrow} \, ,$$

where

$$V_{BCS} = 2\langle g_{\mathbf{k}-\mathbf{k'}}^2 / \hbar\omega_{\mathbf{k}-\mathbf{k'}} \rangle \, , \qquad (8.6)$$

and the average $\langle \cdots \rangle$ is taken over $\mathbf{k}$ and $\mathbf{k'}$ on the Fermi surface; $\eta_{\mathbf{k}} = \theta(\hbar\omega_D - |\epsilon_{\mathbf{k}}|)$, i.e. 1 for $|\epsilon_{\mathbf{k}}| < \hbar\omega_D$ and 0 otherwise; $\hbar\omega_D$ is the Debye energy, and the electron kinetic energy is measured relative to the chemical potential. We took into account that the phonon-state density becomes the maximum around the Debye energy $\hbar\omega_D$. The reduced potential V_{BCS} has the dimension of energy divided by the number density. Note that the interaction between electrons with the same spin is canceled.

8.1.2 BCS Theory of Superconductivity

For an isotropic metal with the attractive interaction described by (8.5), *Bardeen* et al. [8.1, 7] succeeded in obtaining the wave function for the ground state which is lower in energy than that of the normal state. If we have a perturbation term with negative off-diagonal matrix elements, the ground-state wave function is a linear combination ($\psi = \Sigma_j \alpha_j \phi_j$) of the original basic states ϕ_js with positive coefficients. For example, if the unperturbed ground states are degenerate, and each state is connected to n other states by the same matrix element $-V$, then a sum of the original set with equal coefficients makes the ground state to be lower in energy by $-nV$.

When we try to apply this principle to obtain the ground-state wave function, however, we encounter difficulties because of Fermi-Dirac statistics. In general, the matrix elements of $\mathcal{H}_I$ between states specified by occupation numbers may be of either sign. These states can alternatively be expressed by Slater determinants. We want to pick up a subset of those states between which the matrix elements of $\mathcal{H}_I$ are always of the same sign. This can be done by occupying the individual particle states in pairs, such that if one state in the pair is occupied, the other state in the pair is also occupied. For example, let us take a pair of electrons labeled by $\mathbf{k}\uparrow$ and $-\mathbf{k}\downarrow$ and another one labeled by $\mathbf{k}'\uparrow$ and $-\mathbf{k}'\downarrow$. Between the following two wave functions

$$\psi_1 = c_{\mathbf{k}\uparrow}^\dagger c_{-\mathbf{k}\downarrow}^\dagger \psi_0 \quad \text{and} \quad \psi_2 = c_{\mathbf{k}'\uparrow}^\dagger c_{-\mathbf{k}'\downarrow}^\dagger \psi_0 \, , \tag{8.7}$$

with ψ_0 being a suitable function, we always get a negative matrix element $-V_{BCS}$ for $\mathscr{H}_I$ in (8.5). The pairs should be chosen so that transitions between them are possible, i.e., they should all have the same total momentum. To form the ground state, the best choice is the pair $\mathbf{k}\uparrow$ and $-\mathbf{k}\downarrow$. The occupancy of these pairs may be specified by $1_{\mathbf{k}}$ or $0_{\mathbf{k}}$. Then, the best wave function will be the linear combination

$$\psi = \sum_{\cdots \mathbf{k}_1 \cdots \mathbf{k}_n \cdots} b(\cdots \mathbf{k}_1 \cdots \mathbf{k}_n \cdots) f(\cdots 1_{\mathbf{k}_1} \cdots 1_{\mathbf{k}_n} \cdots) \, . \tag{8.8}$$

where $f(\cdots 1_{\mathbf{k}_1} \cdots 1_{\mathbf{k}_n} \cdots)$ is the wave function for the state in which the pairs specified by $\cdots \mathbf{k}_1 \cdots \mathbf{k}_n \cdots$ are occupied, and $b(\cdots \mathbf{k}_1 \cdots \mathbf{k}_n \cdots)$ is the coefficient; the sum is over all possible configurations with the restriction that the total number of pairs is constant. Furthermore, *Bardeen* et al. approximated it in such a way that

$$b(\cdots \mathbf{k}_1 \cdots \mathbf{k}_n \cdots) = \cdots b(\mathbf{k}_1) \cdots b(\mathbf{k}_n) \cdots \, . \tag{8.9}$$

In addition, it is assumed that the factor $b(\mathbf{k})$ depends only on the energy $\epsilon_{\mathbf{k}}$. In the Fermi sea, i.e., in the ground level of the normal metallic state, all the states with $k < k_F$ are pairwise occupied and the states with $k > k_F$ are empty. We presume that due to the attractive interaction implied in (8.5), occupied pair states with $k > k_F$ and empty pair states with $k < k_F$ appear in the energy range of $-\hbar\omega_D < \epsilon_{\mathbf{k}} < \hbar\omega_D$.

Now, we turn to calculating the expectation value of $\mathscr{H}_I$, W_I, in the state given by (8.8). Non-vanishing matrix elements connect configurations which differ in only one of the occupied pairs. Matrix elements corresponding to $\mathbf{k} \rightarrow \mathbf{k}'$ are finite only if the state $\mathbf{k}$ is occupied as well as $\mathbf{k}'$ unoccupied in the initial configuration, and $\mathbf{k}'$ occupied and $\mathbf{k}$ unoccupied in the final configuration. The possibility that this occurs is

$$p(\epsilon_{\mathbf{k}})[1 - p(\epsilon_{\mathbf{k}'})]p(\epsilon_{\mathbf{k}'})[1 - p(\epsilon_{\mathbf{k}})] \, , \tag{8.10}$$

where $p(\epsilon_k)$ is the probability that a given state of the energy ϵ_k is occupied by a pair and is proportional to $|b(k)|^2$. Since matrix elements are proportional to the probability amplitudes rather than probabilities, the matrix element of $-V_{BCS}\, c^\dagger_{k'\uparrow} c^\dagger_{-k'\downarrow} c_{-k\downarrow} c_{k\uparrow}$ is given by

$$- V_{BCS}\sqrt{p(\epsilon_k)[1 - p(\epsilon_{k'})]p(\epsilon_{k'})[1 - p(\epsilon_k)]} \ . \tag{8.11}$$

Summing over k and k', we obtain the interaction energy

$$W_I \;=\; - [N(0)]^2 V_{BCS} \int_{-\hbar\omega_D}^{+\hbar\omega_D} \int_{-\hbar\omega_D}^{+\hbar\omega_D} \sqrt{p(\epsilon)[1 - p(\epsilon)]p(\epsilon')[1 - p(\epsilon')]}\; d\epsilon\, d\epsilon' \tag{8.12}$$

where $N(0)$ is the state density per spin at the Fermi energy. We assume that the state density is a slowly varying function of energy around the Fermi energy. The kinetic energy W_K coming from $\mathcal{H}_{el}$ in (8.5) is given by

$$W_K \;=\; 2N(0)\left[\int_{-\hbar\omega_D}^{+\hbar\omega_D} \epsilon\, p(\epsilon)\, d\epsilon - \int_{-\hbar\omega_D}^{0} \epsilon\, d\epsilon \right] \ . \tag{8.13}$$

By varying $W_K + W_I$ with respect to $p(\epsilon)$ to minimize it, we obtain the following equation:

$$\Delta \frac{1 - 2p(\epsilon)}{\sqrt{p(\epsilon)[1 - p(\epsilon)]}} \;=\; 2\epsilon \tag{8.14}$$

with

$$\Delta \;=\; N(0)V_{BCS} \int_{-\hbar\omega_D}^{+\hbar\omega_D} \sqrt{p(\epsilon)[1 - p(\epsilon)]}\; d\epsilon \ . \tag{8.15}$$

Equation (8.14) leads to

$$p(\epsilon) \;=\; \tfrac{1}{2}(1 - \epsilon/E) \tag{8.16}$$

with

$$E = (\epsilon^2 + \Delta^2)^{1/2} .\tag{8.17}$$

Substituting (8.16) into (8.15), we can obtain a relation which turns out to be the gap equation at absolute zero:

$$N(0)V_{BCS} \int_{-\hbar\omega_D}^{+\hbar\omega_D} \frac{1}{2E}d\epsilon = 1 .\tag{8.18}$$

Solving this, we get

$$\Delta = \frac{\hbar\omega_D}{\sinh[1/N(0)V_{BCS}]} \simeq 2\hbar\omega_D \exp\left[\frac{-1}{N(0)V_{BCS}}\right] ,\tag{8.19}$$

where the last equality is for the case of weak coupling ($N(0)V_{BCS} \ll 1$). The minimized total energy is given by

$$W = W_I + W_K = -N(0)\Delta^2 \left[1 + \sqrt{1+(\Delta/\hbar\omega_D)^2}\right]^{-1}$$

$$= -2N(0)\frac{(\hbar\omega_D)^2}{\exp[2/N(0)V_{BCS}] - 1} .\tag{8.20}$$

Thus, we have obtained a new ground-state wave function which is in energy lower by W than the normal state. BCS proved that this state has superconducting properties, but we will not derive them here [8.1]. Briefly, a similar pairing wave function made up of pairs such as $\mathbf{k}+\mathbf{q}\uparrow$ and $-\mathbf{k}\downarrow$ with a small but finite wave number $\mathbf{q}$ can form a metastable state with a finite current. Exitation energies in this state have a finite gap and, as a result, usual impurity scattering cannot cause the decay of the current, thereby, allowing resistance-free conduction.

If we lift the restriction that the total electron number be constant and perform the summation in (8.8), we get the famous BCS wave function

$$\psi = \prod_{\mathbf{k}} [1 + b(\mathbf{k})c^\dagger_{\mathbf{k}\uparrow}c^\dagger_{-\mathbf{k}\downarrow}]\psi_0 ,\tag{8.21}$$

where ψ_0 is the real vacuum wave function. Although this state does not have a constant pair number, it is known to have a very sharp amplitude maximum for one specified number of pairs, so that it can be taken as a very good approximation for the state with a fixed number of pairs. This non-number-conserving wave function is considered to be even more appropriate since it allows us to describe the Josephson effect [8.6] which is not discussed here. With this wave function, we can easily repeat the above calculation and get the same result if we set $p(\mathbf{k}) = |b(\mathbf{k})|^2/(1+|b(\mathbf{k})|^2)$. BCS extended their theory to finite temperatures, thus establishing the theory of superconductivity. In the next section, however, instead of following their original theory, we shall study a more easy-to-treat mean-field theory of superconductivity, mainly following *Richaysen* [8.7]. As we shall see, this also sheds light on another aspect of the mechanism of superconductivity.

8.1.3 Mean-Field Version of BCS Theory

Here we adopt the so-called **reduced Hamiltonian**

$$\mathscr{H}_{red} = \sum_{\mathbf{k},\sigma} \epsilon_{\mathbf{k}} c_{\mathbf{k}\sigma}^\dagger c_{\mathbf{k}\sigma} - V_{BCS} \sum_{\mathbf{k},\mathbf{k}'} \eta(\mathbf{k})\eta(\mathbf{k}') c_{\mathbf{k}\uparrow}^\dagger c_{-\mathbf{k}\downarrow}^\dagger c_{-\mathbf{k}'\downarrow} c_{\mathbf{k}'\uparrow}, \quad (8.22)$$

where $\eta(\mathbf{k}) = \theta(\hbar\omega_D - |\epsilon_{\mathbf{k}}|)$ is unity when $|\epsilon_{\mathbf{k}}| < \hbar\omega_D$, and zero otherwise. In the preceding subsection, to get W_I we used only a special part of $\mathscr{H}_I$ which transforms a pair to another pair. This part is extracted in (8.22). The rest of $\mathscr{H}_I$ can be treated perturbationally and is known not to be important. Since the wave function for the superconducting state as given in (8.21) yields the averages $\langle c_{\mathbf{k}\uparrow}^\dagger c_{-\mathbf{k}\downarrow}^\dagger \rangle$ and $\langle c_{\mathbf{k}\downarrow} c_{-\mathbf{k}\uparrow} \rangle$, we can further simplify (8.22) to the following average Hamiltonian in the mean-field sense, i.e., corresponding to (4.32):

$$\mathscr{H}_{av} = \sum_{\mathbf{k},\sigma} \epsilon_{\mathbf{k}} c_{\mathbf{k}\sigma}^\dagger c_{\mathbf{k}\sigma} - \sum_{\mathbf{k}} \eta(\mathbf{k})(\Delta c_{\mathbf{k}\uparrow}^\dagger c_{-\mathbf{k}\downarrow}^\dagger + \Delta^* c_{-\mathbf{k}\downarrow} c_{\mathbf{k}\uparrow}) + \frac{|\Delta|^2}{V_{BCS}} \quad (8.23)$$

with

$$\Delta = \sum_{\mathbf{k}} V_{BCS}\, \eta(\mathbf{k}) \langle c_{-\mathbf{k}\downarrow} c_{\mathbf{k}\uparrow} \rangle, \quad (8.24)$$

where the average $\langle \cdots \rangle$ is taken with $\mathscr{H}_{av}$, and Δ is the order parameter of the superconducting state. When we get a nonzero self-consistent solution of (8.24) at some temperature or for a set of material parameters, the system is superconducting for that condition. When we have only the trivial solution $\Delta = 0$, it cannot be superconducting.

Since (8.23) is quadratic in the creation and annihilation operators, we can diagonalize it by means of a canonical or Bogolubov transformation:

$$c_{k\uparrow} = u_k \alpha_{k1} + v_k^* \alpha_{k2}^\dagger ,$$

$$c_{-k\downarrow}^\dagger = -v_k \alpha_{k1} + u_k^* \alpha_{k2}^\dagger ,$$

$$\text{(8.25)}$$

where α_{ki} and $\alpha_{ki}^\dagger$ ($i = 1, 2$) are annihilation and creation operators satisfying the anticommutation relations. The following constraint is imposed on the coefficients u_k and v_k:

$$|u_k|^2 + |v_k|^2 = 1 . \tag{8.26}$$

Since the state $k\uparrow$ is linked only to $-k\downarrow$, the transformation among these two states is sufficient. Substituting (8.25) into (8.23) and demanding that the coefficients of $\alpha_{k1}\alpha_{k2}$ and $\alpha_{k2}^\dagger\alpha_{k1}^\dagger$ vanish, yield

$$2\epsilon_k u_k v_k + \Delta v_k^2 - \Delta^* u_k^2 = 0 . \tag{8.27}$$

We can solve the coupled equations (8.26, 27) as

$$|u_k|^2 = \tfrac{1}{2}(1 + \epsilon_k/E_k) , \quad |v_k|^2 = \tfrac{1}{2}(1 - \epsilon_k/E_k) , \tag{8.28}$$

with

$$E_k = \sqrt{\epsilon_k^2 + |\Delta|^2 \eta(k)} , \tag{8.29}$$

where $\eta(k) = \theta(\hbar\omega_D - |\epsilon_k|)$. One divides (8.27) by $u_k v_k$ and notices that it is the quadratic equation of $\Delta v_k/u_k$, which gives

$$\frac{\Delta v_k}{u_k} = -\epsilon_k \pm E_k . \tag{8.30}$$

We choose the plus sign, since it gives a positive excitation energy, and after using it in (8.26), one gets (8.28) except for the phase relation im-

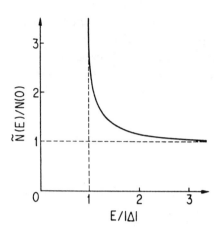

Fig. 8.2. State density $\widetilde{N}(E)$ of quasi-particle excitations in the superconducting level normalized by the state density in the normal state

posed by (8.30); there is no restriction on the phases of $u_{\mathbf{k}}$ and $v_{\mathbf{k}}$. The Hamiltonian of (8.23) is diagonalized to

$$\mathcal{H}_{\mathrm{av}} = \sum_{\mathbf{k}} E_{\mathbf{k}}(\alpha_{\mathbf{k}1}^{\dagger}\alpha_{\mathbf{k}1} + \alpha_{\mathbf{k}2}^{\dagger}\alpha_{\mathbf{k}2}) + \sum_{\mathbf{k}}(\epsilon_{\mathbf{k}} - E_{\mathbf{k}}) + \frac{|\Delta|^{2}}{V_{\mathrm{BCS}}} . \qquad (8.31)$$

This Hamiltonian shows that in the superconducting state there is a fermion-like quasi-particle excitation with the energy $E_{\mathbf{k}}$. This has the gap $|\Delta|$, as is seen in the state density $\widetilde{N}(E_{\mathbf{k}})$ of these quasi-particles in Fig. 8.2. With this, one can calculate the average in (8.24) by

$$\sum_{\mathbf{k}} \eta(\mathbf{k})\langle c_{-\mathbf{k}\downarrow}c_{\mathbf{k}\uparrow}\rangle$$

$$= \mathrm{Tr}\left\{\exp(-\beta\mathcal{H}_{\mathrm{av}}) \sum_{\mathbf{k}} \eta(\mathbf{k})c_{-\mathbf{k}\downarrow}c_{\mathbf{k}\uparrow}\right\} / \mathrm{Tr}\{\exp(-\beta\mathcal{H}_{\mathrm{av}})\} , \qquad (8.32)$$

where $\mathrm{Tr}\{...\}$ stands for the trace with respect to the eigenvectors, or functions, of $\mathcal{H}_{\mathrm{av}}$, and β is $1/k_{\mathrm{B}}T$ with T being the temperature. Using (8.23), one can rewrite (8.32) as

$$\sum_{\mathbf{k}} \eta(\mathbf{k})\langle c_{-\mathbf{k}\downarrow}c_{\mathbf{k}\uparrow}\rangle = \frac{1}{\beta}\frac{\partial}{\partial\Delta^{*}}\ln[\mathrm{Tr}\{\exp(-\beta\mathcal{H}_{\mathrm{av}})\}] + \frac{\Delta}{V_{\mathrm{BCS}}}$$

$$= \frac{1}{\beta} \frac{\partial}{\partial \Delta^*} \sum_{\mathbf{k}} \{2\ln[1 + \exp(-\beta E_{\mathbf{k}})] + \beta E_{\mathbf{k}}\} ,$$

$$= \sum_{\mathbf{k}} \frac{\Delta}{2E_{\mathbf{k}}} \tanh(\beta E_{\mathbf{k}}/2) . \tag{8.33}$$

From this relation together with (8.24), the self-consistency equation for Δ is completed in the form:

$$\sum_{\mathbf{k}} V_{BCS} \frac{\eta(\mathbf{k})}{2E_{\mathbf{k}}} \tanh(\beta E_{\mathbf{k}}/2) = 1 . \tag{8.34}$$

The limit of $T = 0$ gives the gap equation (8.18) obtained for the ground state. In the limit $\Delta \rightarrow 0$, the relation determining the superconducting transition temperature T_c is derived as

$$\frac{1}{N(0)V_{BCS}} = \int_0^{\hbar\omega_D} \frac{d\epsilon}{\epsilon} \tanh\left(\frac{\epsilon}{2k_B T_c}\right) , \tag{8.35}$$

$$\simeq \ln\left(\frac{\hbar\omega_D}{2k_B T_c}\right) - \int_0^{\infty} \left(\frac{\ln x}{\cosh^2 x}\right) dx , \tag{8.36}$$

where the second equality was obtained by partial integration over $1/\epsilon$, and we assumed that $\hbar\omega_D/2k_B T_c \gg 1$. The density $N(0)$ in the normal state is assumed constant throughout the band. The integral in the second line is equal to $-\ln(4\gamma/\pi)$ with $\gamma \simeq 1.78107$ being the Euler constant. Thus, we obtain the well-known BCS formula for T_c

$$k_B T_c = \frac{2\gamma\hbar\omega_D}{\pi} e^{-1/\lambda} \simeq 1.13 \, \hbar\omega_D \, e^{-1/\lambda} , \tag{8.37}$$

with the dimensionless electron-phonon interaction constant

$$\lambda = N(0)V_{BCS} . \tag{8.38}$$

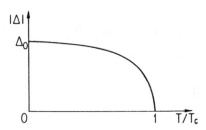

Fig. 8.3. Temperature dependence of the superconducting gap parameter

Since $V_{BCS} = 2\langle g_{\mathbf{q}}^2/\hbar\omega_{\mathbf{q}}\rangle$ does not depend on the mass M of the composing atoms, T_c is proportional to $\omega_D \propto M^{-1/2}$ in excellent agreement with the observations of the isotope effect. In the weak-coupling limit $N(0)V_{BCS} \to 0$, the well known BCS ratio of T_c to gap parameter Δ_0 at absolute zero that was found according to (8.19), is given by

$$\frac{2\Delta_0}{k_B T_c} = \frac{2\pi}{\gamma} \approx 3.53 . \tag{8.39}$$

At T = 0, only two-particle excitations are allowed, so that the real energy gap observed is $2\Delta_0$. The temperature dependence of Δ is obtained by solving (8.34) and displayed in Fig. 8.3.

Now, we reconsider the energy gain in the mean-field framework. At T = 0, the difference W between the total energies in both the superconducting and the normal states is derived from (8.31) as

$$W = \sum_{\mathbf{k}} (|\epsilon_{\mathbf{k}}| - E_{\mathbf{k}}) + \frac{|\Delta|^2}{V_{BCS}} . \tag{8.40}$$

This is in complete agreement with (8.20), if one substitutes the value of Δ at T = 0 derived from (8.34). From the results of (8.31, 40), we can see that the energy gain in the superconducting state relative to the normal state is provided by the lowering of the single-particle energies due to the superconducting gap at the Fermi energy; the gap can sustain itself by the attractive electron-electron interaction.

It should be ascertained that the system does have the properties of a superconductor when the order parameter Δ is finite. This has thoroughly been discussed in several well-known works [8.1, 9-11]. Therefore, we restrict ourselves to the problem of the mechanism and to T_c.

8.1.4 Effect of Coulomb Repulsion

In metals, besides phonon-mediated attractive interactions, Coulomb repulsive interaction between electrons is also important. The latter is much weakened due to screening by other electrons so that the interaction coefficient in the k-space is approximately given by

$$V_{Coul} = \frac{4\pi e^2}{N k_s^2} ,$$

(8.41)

where N is the number of atoms, and k_s is the inverse of the Thomas-Fermi screening length [Ref.8.5, pp.105-106] defined by

$$k_s^2 = 8\pi N(0) e^2 ,$$

(8.42)

being on the order of k_F. This repulsive interaction opposes the attractive interaction in the phase space around the Fermi surface where $|\epsilon_k| < \hbar\omega_D$. It also has an effect outside this phase space but in that region it enhances rather than suppresses superconductivity to our surprise. In such a system, the order parameter Δ_k is actually wave-number dependent and has the form

$$\Delta_k = \Delta_0 \eta(k) + \Delta_\infty [1 - \eta(k)] ;$$

(8.43)

Δ_∞ takes on the sign opposite to that of Δ_0. As a result, the repulsive Coulomb interaction outside the attractive shell region contributes to sustaining the order parameter Δ_0 within the shell region. We can easily check this by adding the Coulomb repulsive interaction with the constant coefficient given in (8.41) to the reduced Hamiltonian of (8.22). By performing the mean-field procedure in the same way as above with the exception that the k dependence of the gap parameter Δ_k is taken into account, we obtain T_c as follows:

$$k_B T_c = \frac{2\gamma\hbar\omega_D}{\pi} \exp\left[\frac{-1}{\lambda - \mu^*}\right]$$

(8.44)

with the so-called **Coulomb pseudo-potential**

$$\mu^* = \frac{N(0)V_{Coul}}{1 + N(0)V_{Coul} \ln(D/\hbar\omega_D)} ,$$

(8.45)

where D is the geometrical mean of the band half-widths which extend both upward and downward from the Fermi energy. Note that in the exponent, V_{Coul} divided by the denominator in (8.45), rather than V_{Coul}, is subtracted from V_{BCS}. This decreasing of the repulsive term is due to the above-stated constructive effect of the Coulomb interaction outside the attractive shell. Equation (8.44) can also be obtained by a more elaborate approach taking account of the retardation of the electron-phonon interaction [8.10]. Therefore, the condition for superconductivity can be expressed as

$$V_{BCS} > \frac{V_{Coul}}{1 + N(0)V_{Coul}\ln(D/\hbar\omega_D)} . \tag{8.46}$$

This means that even if V_{BCS} is smaller than V_{Coul}, i.e., even if there is no net attractive interaction in any part of the phase space, superconductivity can occur if the above condition is satisfied.

8.1.5 s-, p-, and d-Pairings

The BCS wave function defined by (8.8,9) can be rewritten as

$$\psi = \sum_{P} (-1)^P P\{\phi(\mathbf{r}_1-\mathbf{r}_2;\sigma_1\sigma_2)\phi(\mathbf{r}_3-\mathbf{r}_4;\sigma_3\sigma_4)...\phi(\mathbf{r}_{2\nu-1}-\mathbf{r}_{2\nu};\sigma_{2\nu-1}\sigma_{2\nu})\} ,$$

$$\tag{8.47}$$

where

$$\phi(\mathbf{r};\sigma_1\sigma_2) = \sum_{\mathbf{k}} \frac{b(\mathbf{k})}{\sqrt{2}} e^{i\mathbf{k}\cdot\mathbf{r}} [\alpha(\sigma_1)\beta(\sigma_2) - \beta(\sigma_1)\alpha(\sigma_2)] ; \tag{8.48}$$

P stands for the permutation of the electron labels, ν is the total number of pairs, and $\alpha(\sigma)$ and $\beta(\sigma)$ are the eigenfunctions for the up and down spin states, respectively.

The wave function $\phi(\mathbf{r}_1-\mathbf{r}_2;\sigma_1\sigma_2)$ designates a state of two electrons which make up the singlet spin state, and have specific coherent orbital motions with each other. The wave function ψ expresses a state in which ν pairs of electrons are degenerate in one state as if the pairs were Bose particles. These pairs are called **Cooper pairs**. The name comes from a real bound state of two electrons, which was found by *Cooper* [8.13] in treating a model where two electrons have a BCS-like attractive interaction but are only allowed to move in the energy region above the Fermi level. In (8.48), the volume of space where $\phi(\mathbf{r}; \sigma_1 \sigma_2)$ has a large amplitude is limited to the so-called **Pippard's coherence length** $\xi_0 = \hbar v_F / \Delta_0$ [8.1, 11, 14]. However, two electrons are not in a real bound state, since the wave function does not correspond to a discrete eigenenergy, and millions of other pairs also have this center of mass in that region.

In the BCS model, $b(\mathbf{k}) = b(k)$ means independence of the direction of $\mathbf{k}$, so that the $\mathbf{r}$-dependence of the pair function $\phi(\mathbf{r}; \sigma_1 \sigma_2)$ is s-wave-like. This is the reason why the BCS state is also known as the **s-wave superconducting state**. The spin state of ϕ is singlet, leading to the term of **singlet superconductivity**.

When the on-site Coulomb energy U is very large but there still is a finite-range attractive interaction, pair functions with non-zero angular momentum may become better at producing a superconducting state. This is because such functions vanish at $\mathbf{r} = 0$ and do not suffer energy loss from large U. If the angular momentum of such a pair function is equal to $1\hbar$ or $2\hbar$, this state is called a **p-pairing** or **d-pairing state**, respectively. In the p-pairing function, the orbital function is an odd function of $\mathbf{r}_1 - \mathbf{r}_2$, therefore the spin function must be even, i.e., the spin state must be triplet, since $\phi(\mathbf{r}_1 - \mathbf{r}_2; \sigma_1 \sigma_2)$ must be odd with respect to the permutation of the labels 1 and 2. Thus, p-pairing produces a triplet state. In this case the minus sign in the square bracket of (8.48) is changed to the plus sign. Since the d-pair function is even with respect to the permutation of orbital variables, it is singlet with respect to the spin part.

Superconductivity having a pair function with nonzero angular momentum $\ell\hbar$ is called *anisotropic*. The gap parameter in such a state is proportional to $b(\mathbf{k})$ and its $\mathbf{k}$-dependence is given by spherical harmonics with the same angular momentum. The gap parameter vanishes at points or along lines on the Fermi surface, giving rise to one-particle excitations without a finite gap, in contrast to s-pairing.

p-pairing is known to occur in superfluid ^{3}He at 3 mK under ambient pressure [8.15]. This is because the hard core of the ^{3}He atom prevents s-pairing. The possibility of d-pairing has been suggested for high-T_c cuprate superconductors [8.16] and some heavy-fermion systems [8.17, 18].

A magnetic field breaks the singlet superconductivity, since it increases the energy of one electron and decreases the energy of the other. For this type of superconductor there is an upper bound on the field in which superconductivity can survive, called the **Pauli limit** H_{Pauli} or **Clogston limit** [8.19, 20]. This nomenclature arises because, whereas the singlet superconducting state cannot gain any energy, the normal state gains an amount equal to $\frac{1}{2}\chi_{Pauli}H^2$. Hence, the normal state is recovered in the field beyond H_{Pauli} obtained from

$$\frac{1}{2}N(0)\Delta_0^2 = \frac{1}{2}\chi_{Pauli}H^2_{Pauli},\qquad(8.49)$$

where the left-hand side comes from (8.20). H_{Pauli} is rewritten as (3.13). This upper bound for the singlet superconductivity is considered to be valid with the allowance of a factor of 2, since (8.49) does not take account of the orbital diamagnetic susceptibility χ_{orb}. For a free-electron gas we have $\chi_{orb} = -\chi_{spin}/3$. χ_{orb} can be enhanced in materials with a strong spin-orbit interaction such as semimetallic Bi.

On the other hand, the triplet superconductor can exceed the Pauli limit. H_{c2}, which is the upper critical field, is bounded only by the limit coming from the orbital motion. It is boundless when the orbital destructive effect is negligible. Organics have the potential to realize this fascinating feature (Sect. 8.5.5).

8.1.6 Little's Model

In 1964, *Little* [8.21] proposed a new type of superconductor in which excitons, i.e., intramolecular or intraatomic electronic excitations, play the role of phonons. According to the BCS theory, in ordinary superconductors phonons mediate the attractive electron-electron interaction. *Little* was inspired by a question raised by *London* [8.22] whether a superfluid-like state might occur in certain macromolecules which play an important role in biochemical reactions. In superconductors due to such an exciton mechanism, T_c might possibly be elevated up to 2000 K. This is because the Debye energy $\hbar\omega_D$ for T_c in (8.37) is replaced by the excitation energy $\hbar\omega_{ex}$ of the excitons, which is larger than $\hbar\omega_D$ by about two orders of magnitude. However, the interaction between electrons and excitons has to be sufficiently strong, i.e., $N(0)V^{(ex)}_{BCS}$ must be of order unity, $V^{(ex)}_{BCS}$ being the coefficient of the attractive interaction owing to excitons. It is the counterpart of V_{BSC} described by (8.6) for the excitonic superconductor and given by

$$V^{(ex)}_{BCS} = \langle \frac{2|g_q^{(ex)}|^2}{\hbar\omega_{ex}(\mathbf{q})} \rangle,\qquad(8.50)$$

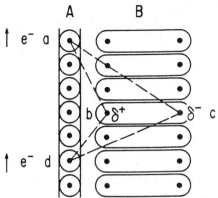

Fig. 8.4. Electron-exciton interaction in an excitonic superconductor. From [8.23]

where $g_\mathbf{q}^{(ex)}$ is the electron-exciton interaction coefficient. Here we take account of the extended nature of excitons with the wave number $\mathbf{q}$; $\langle ...\rangle$ means the average over the Fermi surface.

In Fig. 8.4 we present a one-dimensional structure of an organic, which is *Little's* model for an excitonic superconductor. The section labeled A is a conducting polymer, the "spine" of the conductor, like the carbon chain in polyene. The section labeled B is a series of polarizable side chains or "arms" attached to the spine. The exciton mechanism comes about as follows: Let us assume an electron at the site a is moving along the spine. As it moves past the polarizable side chains B, it repels part of the electronic charge from the end b to the other end c on each side chain. These charges set up a Coulomb potential attractive to another electron at the site d. These charges result from the virtual electronic, i.e., excitonic excitations in groups of the side chains. *Little* showed an example of a side chain made of a dye molecule, which is a resonating hybrid of the two limiting structures illustrated in Fig. 8.5; the positive charge resonates between the two nitrogen sites.

This idea was extended by *Ginzburg* [8.24] to a surface superconductor. In his model, conducting electrons are in surface levels which interact with a dielectric film or a polarizable monomolecular layer deposited on the surface.

These models triggered more arguments on the possibility of attaining a finite value of T_c in one- and two-dimensional superconductors. In such systems thermal fluctuations are known to push T_c down to zero [8.25, 26]. However, the critical temperature can be made finite by aggregating low-dimensional superconductors so that they interact with each other through Josephson coupling, forming a three-dimensional superconductor.

Since then, many attempts have been made to produce an excitonic superconductor in both types of systems. No success has yet been reported,

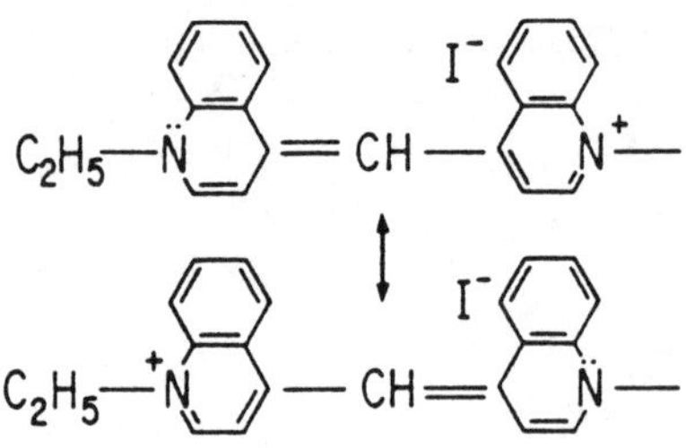

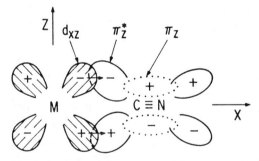

$C_2H_5-\ddot{N}$ ⬡ $=CH-$ ⬡ $\overset{+}{N}-$ I⁻

Fig.8.5. Resonance between two electronically polarized states of a dye group in a side chain. From [8.21]

$C_2H_5-\overset{+}{N}$ ⬡ $-CH=$ ⬡ $\ddot{N}-$ I⁻

however. Recently, *Little* recognized that the electron-exciton interaction must be made stronger by arranging the excitonic system in as close a contact with the conducting chain as possible. In his new model [8.23], *Little* considered a compound like KCP, i.e. $K_2 Pt(CN)_4 Br_{0.30} \cdot xH_2O$ with x being indefinite, in which the conduction electrons move along a transition-metal-atom chain in the d_{z^2}-p_z orbital band. The d_{xz} and d_{yz} orbitals, which are orthogonal to all the states in this band, bond with the orbitals of ligands (Fig.8.6). If the ligand is part of a dye group such as the resonance-stabilized hybrid depicted in Fig.8.5, then, as the charge moves back and forth in this group, it will also move partially on and off the metal. The very close proximity of the d_{xz} level and the d_{z^2}-p_z band should induce strong coupling.

Little emphasized another important point, that is, in order for the coupling to be best enhanced, the energy level of the metal d_{xz} orbital should be close to that of the ligand and within the magnitude of the transfer energy between the two orbitals.

Fig.8.6. Overlap between a transition metal d_{xz}-orbital and π-orbitals of a ligand. From [8.23]

8.2 Superconductivity Due to Spin Fluctuations

8.2.1 Theory

Some research groups have asserted that since the SuperConducting (SC) and SDW phases are neighboring in the phase diagram for the $(TMTSF)_2 X$ salts, the superconductivity can be mediated by spin fluctuations [8.27-29]. This mechanism works in the case of p-wave superfluidity in liquid ^{3}He [8.15] and was argued to drive d-wave superconductivity in heavy-Fermion systems [8.17, 18].

A general argument [8.29] shows that ferromagnetic fluctuations enhance triplet pairing but suppress singlet pairing. Superfluidity in liquid ^{3}He is believed to be due to ferromagnetic fluctuations. On the other hand, the same argument leads to the conclusion that in nearly antiferromagnetic uniform systems both kinds of pairing are suppressed [8.29]. It has been suggested that this treatment is not necessarily correct when there is a spatial non-uniformity. *Scalapino* et al. [8.28] carried out a heuristic argument in which they obtained an attractive interaction owing to spin fluctuations and leading to d-wave pairing. They introduced Fermi-surface-averaged spectral weights of the effective interaction $V(\mathbf{k}',\mathbf{k},\omega)$

$$F(\omega) = -\pi^{-1} \langle \operatorname{Im}\{V(\mathbf{k}',\mathbf{k},\omega)\} \rangle , \qquad (8.51)$$

where $\operatorname{Im}\{\cdot\cdot\}$ means the imaginary part of the function. The Fermi-surface average depends on the form of pairing, to be discussed below. From this function we get the dimensionless coupling constant

$$\lambda = 2 \int_0^\infty \frac{d\omega}{\omega} F(\omega) , \qquad (8.52)$$

which characterizes the strength of the pairing interaction. Substituting (8.51) into this equation and using the Kramers-Kronig relation, we find

$$\lambda = - \langle \operatorname{Re}\{V(\mathbf{k}',\mathbf{k},0)\} \rangle , \qquad (8.53)$$

where $\operatorname{Re}\{\cdot\cdot\}$ denotes the real part of the function. Thus, we only need the zero-frequency limit of the effective interaction.

To model the pairing mechanism mediated by SDW spin fluctuations, we consider a three-dimensional Hubbard model with a repulsive on-site

Coulomb energy U. The treatment of the spin susceptibility $\chi(\mathbf{Q})$ in the so-called **Random-Phase Approximation** (RPA) leads to the condition

$$U\chi_0(\mathbf{Q}) = 1 \tag{8.54}$$

for the SDW instability. Here, $\chi_0(\mathbf{Q})$ is the zeroth-order spin susceptibility given by

$$\chi_0(\mathbf{Q}) = \frac{1}{N} \sum_{\mathbf{k}} \frac{f(\epsilon_{\mathbf{k}+\mathbf{Q}}) - f(\epsilon_{\mathbf{k}})}{\epsilon_{\mathbf{k}} - \epsilon_{\mathbf{k}+\mathbf{Q}}} \tag{8.55}$$

with the band energy

$$\epsilon_{\mathbf{k}} = -2t(\cos k_x + \cos k_y + \cos k_z) - \mu \tag{8.56}$$

for a cubic, tight-binding band; the lattice constant is chosen to be unity; μ is the chemical potential; N is the number of electronic sites. For a half-filled band ($\mu = 0$), the Fermi surface exhibits perfect nesting with the wave vector $\mathbf{Q}_0 = (\pi,\pi,\pi)$. At this wave vector $\chi_0(\mathbf{Q}_0)$ grows logarithmically at low temperatures varying as $\ln(t/T)$, so that (8.54) is always satisfied at some low T for any positive value of U. For $\mu \neq 0$, $\chi_0(\mathbf{Q}_0)$ has a finite maximum at zero temperature. Therefore, for a fixed value of U, we have a critical value of the chemical potential μ_c so that

$$U\chi_0(\mathbf{Q}_0, \mu_c, T=0) = 1 \; ; \tag{8.57}$$

we choose a negative value of μ_c. Here, we are interested in studying the pairing interaction which is due to the exchange of SDW-type spin fluctuations for $\mu < \mu_c < 0$. We will assume that the superconducting transition temperature T_c is sufficiently low so that $\chi_0(\mathbf{Q}_0)$ is practically equal to its $T = 0$ value.

The basic RPA diagrams for the even-parity singlet pairing potential

$$V_s(\mathbf{k},\mathbf{k}') = U + \frac{U^3 \chi_0^2(\mathbf{k}'-\mathbf{k})}{1 - U^2 \chi_0^2(\mathbf{k}'-\mathbf{k})} + \frac{U^2 \chi_0(\mathbf{k}'+\mathbf{k})}{1 - U\chi_0(\mathbf{k}'+\mathbf{k})} \tag{8.58}$$

are illustrated in Fig. 8.7. Since $V_s(\mathbf{k},\mathbf{k}')$ mainly consists of the zero-frequency limit of the spin fluctuation propagator, this potential describes the interaction owing to spin fluctuations.

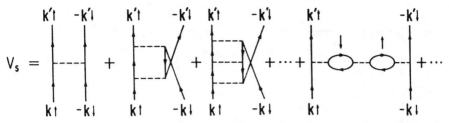

Fig. 8.7. Diagrams giving spin-fluctuation contributions to the even-parity singlet-pairing interaction V_s. *Straight lines* show electrons propagating and *dashed lines* present the on-site Coulomb repulsion U. From [8.28]

If we take the projection of the $Y_{\ell m}$ part of the interaction as a function of $(\mathbf{k} \cdot \mathbf{k}')/kk'$ for the continuum case, we conclude, as mentioned above, that the antiferromagnetic spin fluctuations suppress the both even- and odd-parity pairings. Here, we take account of the discreteness of the electronic sites and in order to decompose V_s, we use tight-binding cubic harmonics:

$$g_0(\mathbf{k}) = 1 \quad \text{(s-wave-like)},$$

$$g_{s^*} = A + \cos(k_x) + \cos(k_y) + \cos(k_z) \quad \text{(extended s(s^*)-wave-like)},$$

$$g_x = \sin(k_x) \quad \text{(p-wave-like)} \tag{8.59a-c}$$

along with g_y and g_z; and

$$g_{x^2-y^2} = \cos(k_x) - \cos(k_y) \quad \text{(E_g d-wave-like)},$$

$$g_{3z^2-r^2} = 2\cos(k_z) - \cos(k_x) - \cos(k_y) \quad \text{(E_g d-wave-like)},$$

$$g_{xy} = \sin(k_x)\sin(k_y) \quad \text{(T_{2g} d-wave-like)} \tag{8.59d-f}$$

along with g_{xz} and g_{yz}. Utilizing these functions to weight the averages, the constants for the various types of pairings are defined by

$$\bar{\lambda}_\alpha = -\int \frac{d^2k}{|v_\mathbf{k}|} \int \frac{d^2k'}{(2\pi)^3 |v_{\mathbf{k}'}|} g_\alpha(\mathbf{k}') V_s(\mathbf{k}',\mathbf{k}) g_\alpha(\mathbf{k}) \bigg/ \int \frac{d^2k}{|v_\mathbf{k}|} g_\alpha^2(\mathbf{k}), \tag{8.60}$$

where $v_\mathbf{k}$ is the velocity of the electron, and the integration is performed over the Fermi surface. This expression is for even-parity pairing. (For odd-parity pairing, see the original literature [8.28]). Another coupling constant

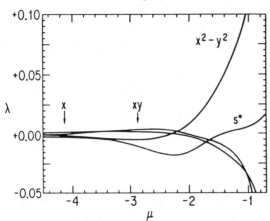

Fig.8.8. Effective coupling λ_{s*}, λ_x, $\lambda_{x^2-y^2}$, and λ_x versus μ for U = 4. The unit of energy is the transfer energy between the nearest neighbor sites t. From [8.28]

λ_z which gives a measure of the effective-mass renormalization is defined by

$$\lambda_z = \int \frac{d^2\mathbf{k}}{|v_\mathbf{k}|} \int \frac{d^2\mathbf{k}'}{(2\pi)^3|v_{\mathbf{k}'}|} \left[\frac{U^3\chi_0^2(\mathbf{k}'-\mathbf{k})}{1-U\chi_0(\mathbf{k}'-\mathbf{k})} + \frac{U^2\chi_0(\mathbf{k}'-\mathbf{k})}{1-U^2\chi_0^2(\mathbf{k}'-\mathbf{k})}\right] \Bigg/ \int \frac{d^2\mathbf{k}}{|v_\mathbf{k}|}.$$

(8.61)

The effective coupling constant for α-wave pairing is defined by

$$\lambda_\alpha = \frac{\overline{\lambda}_\alpha}{1+\lambda_z}.$$

(8.62)

A rough estimate of T_c^α is given by $k_B T_c^\alpha = \hbar\omega_c \exp(-1/\lambda_\alpha)$ with ω_c being the cutoff frequency given by the spectral function $F(\omega)$ of (8.51).

Figure 8.8 presents the results of a numerical evaluation of λ_α versus μ for U = 4. The energy is in units of t. As μ approaches the SDW-instability limit, or $\mu_c = -0.71$, $\lambda_{x^2-y^2}$ rapidly increases. The s^*-wave coupling λ_{s*} also becomes attractive in this region. As μ decreases, the strength of these couplings also decreases. At $\mu = -2.2$, the λ_{xy} coupling becomes the most attractive channel, although λ_{xy} is small.

When U is increased, μ_c becomes more negative. For U = 6 and 8, the qualitative behavior of the couplings is similar to the U =4 case, although

$\lambda_{x^2-y^2}$ becomes smaller, and λ_{xy} increases. When U exceeds 8, λ_{xy} becomes dominant near the instability. To better understand this behavior *Scalapino* et al. [8.28] used the Fourier transform of $V_s(\mathbf{k}', \mathbf{k})$

$$V_s(\boldsymbol{\ell}) = \sum_{\mathbf{q}} e^{i\mathbf{q}\cdot\boldsymbol{\ell}} V_s'(\mathbf{q}) . \qquad (8.63)$$

Here $V_s'(\mathbf{q})$ is presumably a modified form of $V_s(\mathbf{k}', \mathbf{k})$ that is obtained by changing the sign in front of $\mathbf{k}$ in the third term of (8.58) so that this function becomes a function of $\mathbf{q} = \mathbf{k}'-\mathbf{k}$. Even if we employ the modified function in (8.58), the result remains the same. Results for U = 4 and $\mu = -1$ and -2.5 are depicted in Fig.8.9. Here, the strength of the interaction at the origin, i.e., on site, is normalized to unity. Since $\mu_c = -0.71$ for U = 4, Fig.8.9a corresponds to a band filling, which is close to the SDW instability. In this case, the pairing interaction is repulsive on site, as one would expect, but attractive for pairs that are placed on the nearest-neighbor sites. It is repulsive for the next-nearest-neighbor sites, etc. As μ approaches μ_c, this sign pattern persists, and the range of the interaction increases as $[1-U\chi_0(\mathbf{Q}_0,\mu)]^{-1/2}$, whereas its strength goes as $[1-U\chi_0(\mathbf{Q}_0,\mu)]^{-1}$. As μ decreases, both strength and range decrease. In addition, as the wave vectors spanning the Fermi surface decrease with decreasing μ, the scale for the oscillations increases, leading to the $V_s(\boldsymbol{\ell})$ structure at $\mu = -2.5$ (Fig.8.9b). Here, the nearest neighbor pairing potential is repulsive, while the next-nearest-neighbor pairing potential is attractive. From Figs.8.9a,b it would appear that when U = 4, the $(d_{x^2-y^2}, d_{3z^2-r^2})$-like states are favored for $\mu = -1$, while the (d_{xy}, d_{yz}, d_{xz})-like states are favored for $\mu = -2.5$.

The mechanism of superconductivity under discussion is essentially the same as that proposed by *Kohn* and *Luttinger* [8.30]. These researchers pointed out that the screened Coulomb interaction between two electrons

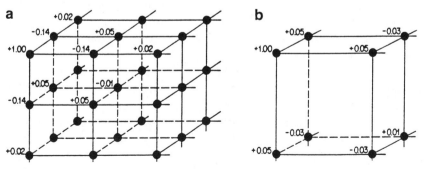

Fig.8.9. Even-parity pairing interaction V_s in real space for U = 4 and (**a**) $\mu = -1$ and (**b**) $\mu = -2.5$, normalized such that the on-site interaction has unit magnitude. From [8.28]

spatially decays with an undulation known as the **Friedel oscillation** [8.31], so that there is a spatial region where this potential is negative. A suitable pairing function can make superconductivity appear by utilizing the negative part of the screened Coulomb interaction.

This mechanism makes use of the non-uniformity of the space, i.e., the spatially restricted nature of the electronic sites linked by transfer energies, and picks up a small amount of negative potential. Therefore, the effective coupling is relatively weak. It is remarkably enhanced for μ near μ_c, but here the RPA approximation overestimates the effective coupling, because the enhanced spin fluctuations compete with the superconductivity. Further work is needed to obtain reliable estimates of T_c.

Miyake et al. [8.32] independently developed a similar line of reasoning to show an attractive interaction mediated by spin fluctuations in heavy-Fermion systems. *Bourbonnais* et al. [8.33] gave an argument which is based somewhat on a one-dimensional model.

The superconductivity in $(TMTSF)_2 X$ salts has been attributed to this mechanism by the aforementioned researchers [8.27-29, 33], since the superconducting phase neighbors the SDW phase in the P-T phase diagram due to good Fermi-surface nesting. When SDW disappears under pressure due to a deterioration of the nesting, spin fluctuations with wave vectors close to the optimal Q_0 should be excited easily. In such a picture, the fast decrease of T_c under pressure is caused by a reduction of these spin fluctuations due to the deterioration of nesting. *Scalapino* et al. also pointed out that a rapid decrease of T_c by normal impurities and defects in these salts becomes understandable, since d-wave superconductivity should be as sensitive to those imperfections as p-wave superconductivity.

The applicability of this mechanism to real systems has not yet been established conclusively, at least for organic superconductors. In the context of high-temperature cuprate superconductors, *Shimahara* and *Takada* [8.34] treated a two-dimensional isotropic model with nearly half-filled bands. They found numerically that superconductivity produced by mediation of spin fluctuations in the RPA scheme yields very high values of T_c, on the order of the highest SDW transition temperature T_{SDW}. When they take account of the self-energy correction, however, T_c decreases, and the superconducting region is very much restricted.

Shimahara applied the same treatment to a quasi one-dimensional Hubbard system modeling the $(TMTSF)_2 X$ and $(DMET)_2 X$ salts [8.35]. This time he obtained the maximum value of $T_c \approx 1$ K close to the observed maximum value of 1.4 K for Bechgaard's salts. However, since he assumed $T_{SDW} = 25$ K for the case of good nesting, T_c obtained still looks slightly too low in view of $T_{SDW} \leq 12$ K for these salts.

Hirsch [8.36] found a tendency of pairing in the two-dimensional (2-D) Hubbard model in quantum-Monte-Carlo calculations. However, *Hirsch* et al. [8.37] and others [8.38] concluded from more extensive numerical studies that this tendency is not strong enough to establish a superconducting ground state in the model [8.16, 39]. They suggested that the RPA theory of mediation by spin fluctuations must also take careful account of their effects to the self-energy and the vertex correction, which may be very bad for the superconductivity.

Two-dimensional models of superconductivity of Coulombic origin aroused much interest in the context of high-T_c cuprate superconductors as well as organic and heavy-fermion superconductors. In contrast, indications of the occurrence of d-wave superconductivity in the 2-D t-J model were obtained by *Dagotto* et al. [8.40] and by *Ohta* et al [8.41]. The t-J model is derived from the Hubbard model in the limit of large U by a transformation in which the doubly occupied states are removed [8.42]. In this model the electron hops from an electronic site to neighboring ones with the transfer energy t as in the Hubbard model, but only in the case where the destination site is empty; each site accommodates, at most, one electron so that it has a net spin when it has an electron; the spins on the neighboring sites have the exchange interaction with the coupling constant J. The exchange interaction exerts power to form a spin-singlet pair of electrons on neighboring sites. This is considered in favor of driving superconductivity. The above-mentioned transformation is exact in the limit of large U but for the intermediate value of U the neglect of higher-order terms in the derived Hamiltonian should be cautiously examined. Among those terms the most important ones are the so-called **three-site terms**. *Li* et al. carried out a variational Monte-Carlo calculation taking account of these terms and employing larger sizes of systems with up to 226 electrons [8.43]. They found that the d-wave and s+id wave SC states are stabilized with an electron density slightly smaller than unity per site. This work suggests that similar SC states would be obtained if one succeeds in treating the 2-D Hubbard model with a similar size of the system.

The 2-D Hubbard model was studied by means of the quantum Monte Carlo method. This method treats a system of small size cut out of the infinite system and extrapolates the results to infinity. A difficulty, called the *negative-sign problem*, has prevented this method from increasing the system size and the on-site Coulomb energy and decreasing the temperature. This may be the reason for the failure to get direct evidence of the occurrence of superconductivity in those models. Recently, a verification of the occurrence of superconductivity in this model has attracted much attention again. Two groups presented positive indications of the occurrence of d-wave superconductivity in the ground state with slight hole doping by calculating superconducting correlation functions via the projection Monte-Carlo method [8.44, 45]; one group obtained the indications for a model taking account of the next-nearest-neighbor transfer energy t' in such a band situation where the energy level of the van Hove singularity is close to the Fermi energy [8.44]. These calculations were done for U up to about 4t due to the restriction coming from the method. This model was also studied by means of the variational Monte-Carlo method. A precursory work suggested that it requires d-wave superconductivity as its ground state for slight hole doping with U = 10t, although with large error bars for the energy gain and the gap amplitude obtained [8.46]. *Nakanishi* and co-workers achieved a definite minimum of the total energy as a function of the d-wave gap function by the same method with a wide range of U [8.47]. The energy gain of the superconducting state was found to quickly increase with increasing U up to U = 8t, to saturate and then to turn to decrease with further increase of U above 12t. Based on the above-mentioned results the 2-D Hubbard model is considered to have a parameter region for which the d-wave superconducting ground state occurs. This must be related to the superconductivity of high-T_c cuprates and possibly to that of organic superconductors.

Normally, electron-phonon interactions help the s-wave superconductivity but not the d-wave one. This is because in the gap equation, the contribution to the electron-phonon pairing potential from the **k**-space region with a positive gap parameter is almost completely canceled by that from the region with a negative parameter. Thus, the above mechanism competes with the phonon-mediated mechanism.

8.2.2 NMR Relaxation Rate

Measurements of the proton NMR relaxation rate $1/T_1$ in the superconducting state of $(TMTSF)_2 ClO_4$ revealed an anomalous temperature dependence around T_c [8.48]. As depicted in Fig.3.24, the observed values $1/T_1$ decrease for temperatures below T_c but do not exhibit the enhancement just below T_c. This is a prominent feature, known as **coherence peak**, of the BCS s-pairing state. *Hasegawa* et al. [8.49] asserted that this feature strongly suggests that in salts under consideration the superconducting state is d-wave pairing, presumably owing to spin fluctuations.

Hasegawa et al. have calculated $1/T_1$ as a function of the temperature T below T_c for the following four types of superconducting gap parameters which the $(TMTSF)_2 X$ salts might take:

$$s1 : \quad \Delta(\mathbf{k}) = \Delta_{s1}[1 + C_{s1}\cos(bk_y)] , \qquad (8.64a)$$

$$s2 : \quad \Delta(\mathbf{k}) = \Delta_{s2}[\cos(bk_y) + C_{s2}] , \qquad (8.64b)$$

$$t1 : \quad \Delta(\mathbf{k}) = \Delta_{t1}\,\mathrm{sgn}(k_x)[1 + C_{t1}\cos(bk_y)] , \qquad (8.64c)$$

$$t2 : \quad \Delta(\mathbf{k}) = \Delta_{t2}\sin(bk_y) , \qquad (8.64d)$$

where the constants $|C_{s1}|$, $|C_{s2}|$ and $|C_{t1}|$ are smaller than unity. The NMR relaxation rate $1/T_{1s}$ in the superconducting state with the gap parameter given by (8.64) is expressed by

$$
\frac{T_{1n}}{T_{1s}} = \int_0^\infty \frac{dE}{2k_B T \cosh^2(E/2k_B T)}
$$

$$
\times \left[\left\{ \int_0^{2\pi} \frac{d(bk_y)}{2\pi} \mathrm{Im} \left\{ \frac{E}{[\Delta^2(k) - E^2]^{1/2}} \right\} \right\}^2 \right.
$$

$$
\left. + \left\{ \int_0^{2\pi} \frac{d(bk_y)}{2\pi} \mathrm{Im} \left\{ \frac{\Delta(k)}{[\Delta^2(k) - E^2]^{1/2}} \right\} \right\}^2 \right] , \qquad (8.65)
$$

where $1/T_{1s}$ is renormalized by the relaxation rate $1/T_{1n}$ in the normal state. The contribution to the integration over bk_y comes from the region where $|\Delta(k)| \leq E$.

Without a suitable choice of parameter values (presumably $C_{s2} = 0$), the calculated results are illustrated in Fig. 8.10. As is seen, the enhance-

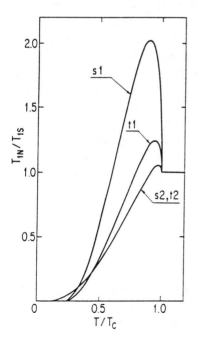

Fig. 8.10. Temperature dependence of the NMR relaxation rate normalized to the value in the normal state. From [8.49]

275

ment of $1/T_{1s}$ below T_c is much reduced in the cases of s2 and t2. This is because in those cases the gap parameters vanish at some values of k_y on the Fermi surface. Consequently, the unusually large peak in the density of states for $E \geq \Delta$, seen in the s1 case, becomes only moderate.

Hasegawa et al. argued that the absence of an enhancement for $T \leq T_c$ can be understood to be due to the gaplessness of the state density in the s2- or t2-type state, and that the observed temperature dependence, proportional to T^3, is approximately reproduced in the calculated one for the lower temperature region. Therefore, they presumed that the superconductivity in $(TMTSF)_2 ClO_4$ is of the d-pairing type owing to spin fluctuations, as stated in Sect. 8.2.1. A similar T dependence of $1/T_1$ has been observed for heavy-fermion systems which are also considered to have d-pairing superconductivity [8.18].

In the κ-$(ET)_2 Cu(NCS)_2$ the proton relaxation rate $1/T_1$ was reported to exhibit a smooth temperature dependence from the high-temperature side of T_c to the low-temperature side [8.50]; incidentally, it increases with further decreasing the temperature; this produces a giant peak around 4 K on which we do not touch. The ^{13}C NMR relaxation $1/T_1$ in κ-$(ET)_2 Cu \cdot [N(CN)_2]Br$ was shown to give a temperature dependence around T_c similar to that in $(TMTSF)_2 ClO_4$, i.e., the absence of the coherence peak and rapid decrease of $1/T_1$ proportional to T^3 below T_c [8.51-53]. It should be noticed, however, that the proton relaxation rate $1/T_1$ in κ-$(MDT-TTF)_2 AuI_2$ reveals a coherence peak below T_c (Fig. 6.12) [8.54].

Disagreements on data of an identical material are seen among different groups for the temperature dependence of the penetration depth. With an μSR experiment the Bell-Labs group reported a temperature dependence at low temperatures consistent with the BCS theory for κ-$(ET)_2 Cu(NCS)_2$ [8.55]. A similar dependence was obtained by the UCLA group with the help of a resonant microwave-cavity technique [8.56]. However, complex susceptibility measurements provided a T^2 dependence of the penetration depth [8.57] which suggests an unconventional pairing. Both were for the same material. By μSR measurements the Uemura group presented a linear temperature dependence at low temperatures, indicating anisotropic superconducting pairings, for κ-$(ET)_2 Cu(NCS)_2$ and κ-$(ET)_2 Cu[N(CN)_2]Br$ [8.58].

8.3 Electron-Molecular-Vibration Interaction

There are two types of electron-phonon interactions in molecular crystals. One interaction is intramolecular in which the molecular distortions due to phonons shift the molecular orbital level up and down, working like perturbations of the potential energy. This type of interaction has been known as the **Electron-Molecular-Vibration** (EMV) **coupling**. The other is intermolecular coupling in which the phonons modulate the interrelation between two neighboring molecules and so change the transfer energy between them. The latter interaction is weaker in $(TMTSF)_2 X$ salts than in most organic conductors. It drives a Peierls transition in the temperature range of $50 \div 200$ K in TCNQ salts. The $(TMTSF)_2 X$ salts have no Peierls transition down to 12 K in spite of the good Fermi-surface nesting. Good nesting is manifested by the SDW transition around that temperature in most TMTSF salts. A precursory lattice softening was reported for $T > 12$ K in $(TMTSF)_2 PF_6$ [8.59,60], but the tendency corresponding to a CDW transition is hindered by another Fermi-surface instability with respect to the SDW state. On these grounds *Weger* et al. [8.61] obtained the upper bound for the dimensionless electron-acoustic phonon interaction λ_{ac} in $(TMTSF)_2 X$ as

$$\lambda_{ac} < 0.16 \ . \tag{8.66}$$

If one substitutes this upper bound and an upper bound for the Debye energy $(\hbar \omega_D \lesssim 200 k_B)$ into the BCS expression (8.37) for T_c without taking account of the repulsive Coulomb interaction, one gets $T_c < 0.4$ K.

The smallness of λ_{ac} in TTF stacks (from which TMTSF is derived) is known from an extended-Hückel-type band calculation of TTF $\cdot$TCNQ by *Berlinsky* et al. [8.62]. If only the longitudinal phonons are taken into account, they yield $\lambda_{ph} = 0.05$ for the TTF column and $\lambda_{ph} = 0.11$ for the TCNQ column. This is roughly in qualitative agreement with the observation that the TCNQ column makes a Peierls transition at 54 K and the TTF at the lower temperature of 49 K [8.63]. The smallness of intermolecular electron-phonon interactions in columns consisting of TTF-derivative molecules may be due to the larger spatial extent of p- and d-orbitals of sulfur atoms, and their analogues Se and Te.

On the other hand, theory indicates that the intramolecular electron-phonon interactions, i.e. EMV interactions, are much stronger in TTF than in TCNQ [8.64, 65]. Mathematically, only totally symmetric (abbreviated by a_g) intramolecular vibrations couple linearly with non-degenerate Molecular Orbital (MO) levels. Two important a_g modes of TTF, ν_6 and ν_3, are illustrated in Fig. 8.11. Such an MO level is shifted up or down in energy by an amount linearly proportional to the change of the normal coordinate corresponding to the a_g-mode vibration. These interactions turn into electron-phonon interactions and mediate attractions between electrons when the molecules are aggregated into a bulk metal. There, after one hole passes by and lowers the potential energy at one molecule, another hole coming to the same site enjoys the lowered potential energy. In this way, two electrons

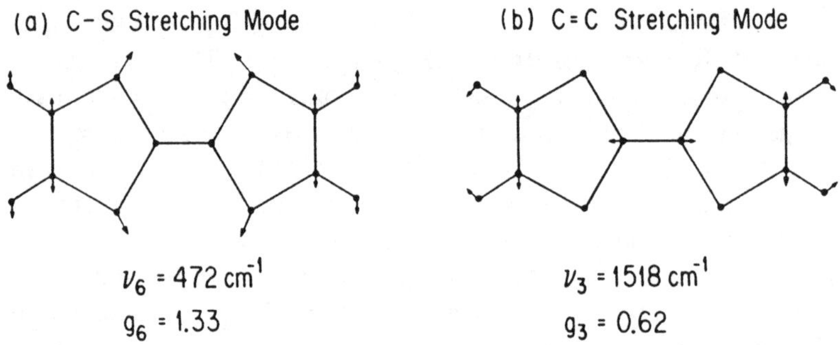

(a) C-S Stretching Mode   (b) C=C Stretching Mode

$$\nu_6 = 472 \text{ cm}^{-1}$$
$$g_6 = 1.33$$

$$\nu_3 = 1518 \text{ cm}^{-1}$$
$$g_3 = 0.62$$

Fig. 8.11a,b. Two important totally symmetric (a_g) intramolecular vibrations of the TTF molecule; ν_i and g_i (i = 3, 6) are the frequency and the coefficient of intraction with HOMO, respectively. From [8.64], see also [8.66]

can have an attractive interaction. This is very similar to the situation in ordinary superconductors with deformation-potential-type interactions.

The dimensionless electron-phonon interaction constant λ_6 for the central C-S stretching mode ν_6 of TTF (Fig. 8.11a) is estimated to 0.3, which through (8.37) gives $T_c = 24$ K if one sets $\nu_6 = 472$ cm^{-1} for $\hbar\omega_D$. The details of this calculation will be given in Sects. 8.3.1 and 3. For the central C=C stretching mode ν_3 which is illustrated in Fig. 8.11b, $\lambda_3 = 0.2$ and $T_c = 17$ K without taking into account other interactions. The sum of the λ_is for all a_g modes in TTF is about 0.6. Thus, EMV interactions are probably giving quite important contributions to the BCS-type attractive interaction driving the superconductivity.

One might think that such appreciable electron-phonon interactions must lead to a Peierls transition at a higher temperature, but this is not the case. It is known that CDW caused by intramolecular electron-phonon interactions is opposed by the on-site Coulomb interaction [8.67] while that due to the intermolecular electron-phonon interaction is not [8.68, 69]. In the CDW state the electronic charge density changes periodically in space causing an excess or deficiency of charge density at each site. This costs much on-site Coulomb energy [8.70]; if the on-site Coulomb energy U is big enough, CDW disappears. CDW caused by intermolecular interactions is not a real CDW but rather a bond ordering, in which the interrelation between the neighboring electronic sites is periodically modulated in space as in polyacetylene. Since it does not cause a charge accumulation or deficiency, U does not oppose it, at least when it is not large.

In this section the possibility of superconductivity mainly driven by the EMV interactions is investigated [8.71]. The on-site Coulomb interaction and the electron-acoustic-phonon interactions are also taken into account. The superconductivity which appears in this scheme is of the s-wave type.

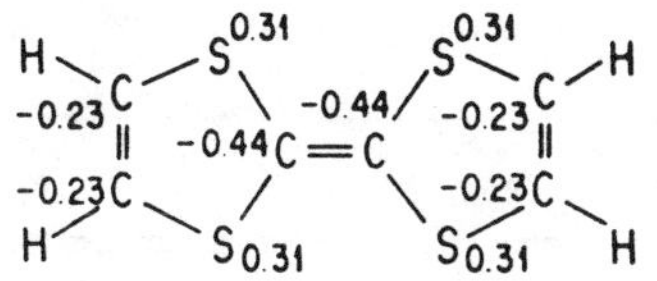

TTF

$$H \diagdown_C\diagup^{S^{0.31}}\diagdown \diagup^{S^{0.31}}_C\diagdown H$$

Fig. 8.12. Coefficients of p_z-orbitals of constituent atoms in HOMO of TTF. From [8.64]

8.3.1 EMV Interactions in TTF Molecules

In the TTF molecule, when it deforms according to the vibration patterns of the two important a_g-modes, shown in Fig. 8.11, exceptionally strong shifts in the energy of the Highest Occupied Molecular Orbital (HOMO) occur. This orbital constitutes the conduction band in salts made of TTF molecules. Therefore, the Electron-Molecular-Vibration (EMV) couplings are an important source of electron-phonon interaction in these salts.

Figure 8.12 gives coefficients for the atomic p_z-orbitals constituting HOMO which is a π-orbital with b_{1u} symmetry. The z-axis is perpendicular to the molecular plane. The d-orbitals of sulfur are neglected. Since the coefficients for the p_z-orbitals of sulfur and the central carbon have different signs, HOMO is antibonding between C and S, and bonding for the central C=C bond. Since the C and S coefficients are much larger than the others, charge is concentrated in the central part of the molecule. The C-S stretching mode changes the distances between the central carbon and sulfur atoms, along the directions of the normal-mode displacements depicted by the arrows in Fig. 8.11, which lowers the energy of the HOMO level E_γ. Meanwhile, the displacements due to the C=C stretching mode, also indicated in Fig. 8.11, lift the energy level.

This linear coupling of the HOMO level E_γ with the dimensionless normal coordinate Q_i (i being the index of the mode) is expressed by

$$\frac{\partial E_\gamma}{\partial Q_i} Q_i = g_i \hbar\omega_i (b_i + b_i^\dagger) . \tag{8.67}$$

Here, $\partial E_\gamma / \partial Q_i$ is equal to $g_i \hbar\omega_i \sqrt{2}$ by definition, where g_i is a non-dimensional coupling constant, and $\hbar\omega_i$ is the quantum energy; $b_i^\dagger$ (b_i) is the creation (annihilation) operator of a quantum of the intramolecular vibration of the ith a_g-mode, and $Q_i = (b_i + b_i^\dagger)/\sqrt{2}$. The values of $\hbar\omega_i$ for all a_g modes of TTF were calculated by *Bozio* et al. [8.66] and those of g_i were obtained by *Lipari* et al. [8.64, 65] and are listed in Table 8.1. They were found to be in reasonable agreement with experiments [8.72, 73]. A similar calculation for TCNQ reveals that its coupling constants are appreciably smaller [8.64].

HOMOs in TTF-derivative molecules such as ET [8.74] and TMTSF [8.75] are very similar to those of TTF (Fig. 8.13). The charge distribution

Table 8.1. Characteristics of the a_g modes ν_i (i = 1, 2, ..., 7) of a TTF molecule

Mode	ν_1	ν_2	ν_3	ν_4	ν_5	ν_6	ν_7
Frequency[a] ν_i [cm^{-1}]	3099	1559	1518	1077	740	472	253
Interaction constant[a,b] g_i	0.03	0.23	0.62	0.16	0.49	1.33	0.16
Relaxation energy[c] $E_{r,i}$ [meV]	0.3	10.2	72.4	3.4	22.0	103.5	0.8
Electron-phonon coupling constant[d] λ_i	0.001	0.028	0.200	0.010	0.061	0.287	0.002

[a] From [8.64, 65]
[b] Defined in (8.67)
[c] Defined in (8.83)
[d] Dimensionless, equal to $N_F V_i$ with $t_{eff} = 0.1625$ eV

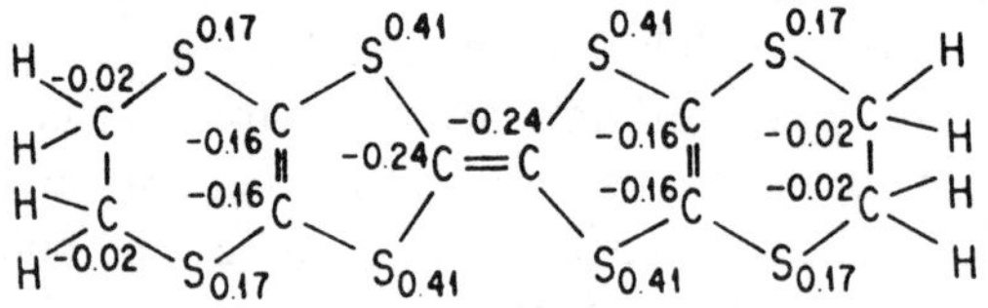

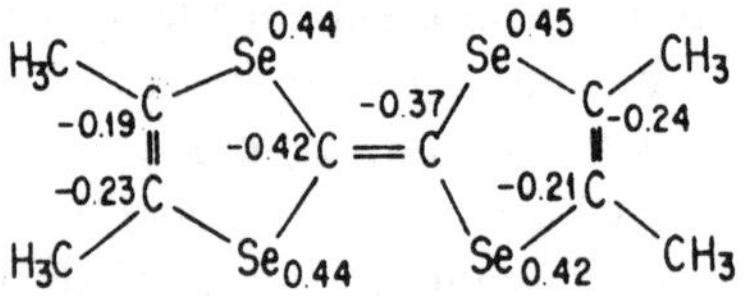

Fig. 8.13. Coefficients of p_z orbitals of constituent atoms in HOMOs of ET and TMTSF. From [8.74, 75]

is concentrated near the center of the molecule. These molecules have similar C-S (or C-Se) and C=C stretching a_g-modes with almost the same values of the quantum energy $\hbar\omega_i$, although the energy of the C-Se stretching mode is lowered to about 285 cm^{-1} due to the heavier Se atom. Therefore, one can expect very similar strong linear couplings in organic conductors composed of these and other TTF-derivatives such as DMET and MDT-TTF. The characteristics of HOMO of TTF and ET, i.e., the concentration of charge in the central part and the b_{1u} pattern of the coefficients for the atomic p_z-orbitals, were also confirmed for DMET [8.76], MDT-TTF [8.77, 78], EDT-TTF and EDT-DSDTF [8.78].

8.3.2 Model and T_c

We can write down the electron-phonon Hamiltonian for a TTF-derivative molecule according to *Lipari* et al. [8.64] as

$$\mathcal{H}_{mol} = E_\gamma \sum_\sigma c_\sigma^\dagger c_\sigma + \sum_i \hbar\omega_i (b_i^\dagger b_i + \tfrac{1}{2}) + \sum_{i\sigma} g_i \hbar\omega_i c_\sigma^\dagger c_\sigma (b_i + b_i^\dagger) , \quad (8.68)$$

where E_γ is the HOMO energy level, c_σ ($c_\sigma^\dagger$) is the annihilation (creation) operator of a hole in the HOMO state with the spin σ; $\hbar\omega_i$ is the quantum energy of the ith a_g-mode vibration, $b_i (b_i^\dagger)$ is the annihilation (creation) operator of the phonon of the same mode, and $g_i \hbar\omega_i$ is the coupling constant introduced by (8.67). The HOMO orbital is assumed to be non-degenerate. For an organic conductor made up of a uniform arrangement of TTF-like molecules, one gets the following Hamiltonian for the whole system

$$\mathcal{H} = \sum_{k\sigma} \xi_k c_{k\sigma}^\dagger c_{k\sigma} + \sum_{q,i} \hbar\omega_i (b_{qi}^\dagger b_{qi} + \tfrac{1}{2})$$

$$+ \sum_{k\sigma qi} \frac{g_i \hbar\omega_i}{\sqrt{N}} c_{k+q\sigma}^\dagger c_{k\sigma} (b_{qi} + b_{-qi}^\dagger) , \quad (8.69)$$

where $c_{k\sigma}$ is the annihilation operator of the hole with the wave vector $\mathbf{k}$ and the spin σ in the tight-binding band ϵ_k composed of HOMOs of the constituent molecules; $\xi_k = \epsilon_k - \mu$ with μ being the chemical potential; $b_{qi} = \Sigma_\ell b_{\ell i} \exp(i\mathbf{q}\cdot\mathbf{R}_\ell)N^{-1/2}$ is the annihilation operator of the phonon with the wave vector $\mathbf{q}$ that originates from the ith a_g-mode vibration of the molecule located at $\mathbf{R}_\ell$; N is the number of TTF-derivative molecules.

Following the approximation employed by BCS [8.1] in deriving (8.5), one gets the following reduced Hamiltonian for superconductivity:

$$\mathcal{H}_{BCS} = \sum_{k\sigma} \xi_k c_{k\sigma}^\dagger c_{k\sigma} - \sum_{i=1}^{\nu} \sum_{kk'} V_i \eta_i(k)\eta_i(k') c_{k\uparrow}^\dagger c_{-k\downarrow}^\dagger c_{-k'\downarrow} c_{k'\uparrow} \quad (8.70)$$

with

$$V_i = 2g_i^2 \hbar\omega_i/N , \quad (8.71)$$

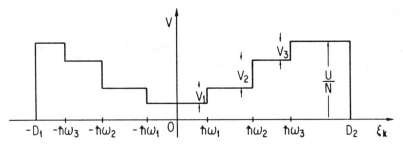

Fig. 8.14. Over-all interaction coefficient $V = -\sum_i \sum_{\mathbf{k},\mathbf{k}'} V_i \eta(\mathbf{k}) \eta(\mathbf{k}')$ versus $\xi_{\mathbf{k}}$ in the special case where $\xi_{\mathbf{k}} = \xi_{\mathbf{k}'}$. For simplicity, we assume $\nu = 4$

where we have defined $\eta_i(\mathbf{k}) = 1$ when $|\xi_{\mathbf{k}}| < \hbar\omega_i$, and $\eta_i(\mathbf{k}) = 0$ otherwise. In (8.70) we included the attractive interaction owing to acoustic phonons by setting $\omega_1 = \omega_D$ and also the Coulomb interaction by setting $\omega_\nu = \infty$ and $V_\nu = -U/N$. The number ν is equal to the number of a_g-modes plus 2.

The $\mathbf{k}$ dependence of the coupling constant in (8.70) is schematically illustrated in Fig. 8.14 for the case of $\xi_{\mathbf{k}} = \xi_{\mathbf{k}'}$ and $\nu = 4$, $-D_1$ and D_2 being the lower and upper bounds of the band, respectively. The superconducting gap parameter is defined by

$$\Delta(\mathbf{k}) = \sum_i V_i \eta_i(\mathbf{k}) \sum_{\mathbf{k}'} \eta_i(\mathbf{k}') \langle c_{-\mathbf{k}'\downarrow} c_{\mathbf{k}'\uparrow} \rangle . \tag{8.72}$$

We relabel the phonon modes so that $\omega_1 \leq \omega_2 \leq \cdots \leq \omega_\nu$. After the standard procedure one gets the gap equation

$$\Delta(\mathbf{k}) = \sum_i V_i \eta_i(\mathbf{k}) \sum_{\mathbf{k}'} \eta_i(\mathbf{k}') \frac{\Delta(\mathbf{k}')}{2E_{\mathbf{k}'}} \tanh\left[\frac{E_{\mathbf{k}'}}{2k_B T}\right] , \tag{8.73}$$

where $E_{\mathbf{k}} = \sqrt{\xi_{\mathbf{k}}^2 + |\Delta(\mathbf{k})|^2}$. This can be rewritten as

$$\Delta(\mathbf{k}) = \sum_i \eta_i(\mathbf{k}) \widetilde{\Delta}_i , \tag{8.74}$$

where

$$\widetilde{\Delta}_i = V_i \sum_{\mathbf{k}} \eta_i(\mathbf{k}) \frac{\Delta(\mathbf{k})}{2E_{\mathbf{k}}} \tanh\left(\frac{E_{\mathbf{k}}}{2k_B T}\right). \tag{8.75}$$

Since $\Delta(\mathbf{k}) = \Sigma_i \widetilde{\Delta}_i$ for $\mathbf{k}$ on the Fermi surface, i.e., for $\xi_{\mathbf{k}} = 0$, this model gives s-wave superconductivity. If one assumes, for simplicity, a constant state density per spin N_F throughout the band from $-D_1$ to D_2 one obtains

$$\widetilde{\Delta}_i = V_i N_F \int d\xi \theta(\hbar\omega_i - |\xi|) \frac{\Delta(\mathbf{k})}{2E_{\mathbf{k}}} \tanh\left(\frac{E_{\mathbf{k}}}{2k_B T}\right), \tag{8.76}$$

where $\theta(x)$ is the step function defined as 1 for $x \geq 0$, and as 0 otherwise. The integral can be carried out for successive energy regions defined in terms of $\omega_1 \leq \omega_2 \leq \dots \leq \omega_\nu$.

For $T = T_c$, i.e., in the limit of $\Delta(\mathbf{k}) \to 0$, (8.74) reduces to the following set of equations:

$$\Delta_i = \sum_{j=i}^{\nu} \lambda_j \sum_{\ell=1}^{j} \Delta_\ell x_\ell, \quad i = 1,\dots,\nu, \tag{8.77}$$

where $\Delta_i = \Sigma_{j=i}^{\nu} \widetilde{\Delta}_j$, $\lambda_j = N_F V_j$ and $x_1 = \ln(2\gamma\hbar\omega_1/\pi k_B T_c)$; γ is the Euler constant 1.78107 and $2\gamma/\pi = 1.13$;

$$x_i = \frac{1}{2}\ln \frac{\min(\hbar\omega_i, D_1) \cdot \min(\hbar\omega_i, D_2)}{\min(\hbar\omega_{i-1}, D_1) \cdot \min(\hbar\omega_{i-1}, D_2)} \quad (i \geq 2) \tag{8.78}$$

with $\min(x, y)$ denoting the smaller value of x and y. When $\hbar\omega_i$ and $\hbar\omega_{i-1}$ are smaller than D_1 and D_2, $x_i = \ln(\omega_i/\omega_{i-1})$. Since (8.77) is a coupled linear equation for Δ_i, $i = 1,..,\nu$, the determinant of the matrix composed of coefficients of Δ_j must vanish. Consequently one can obtain the expression of the unknown parameter x_1, and then T_c, as follows:

$$k_B T_c = \frac{2\gamma\hbar\omega_1}{\pi} \exp\left[-\cfrac{1}{\lambda_1 - \cfrac{1}{x_2 - \cfrac{1}{\lambda_2 - \cfrac{1}{x_3 - \cdots - \cfrac{1}{x_\nu - 1/\lambda_\nu}}}}}\right]. \qquad (8.79)$$

This is a reasonable result. If we set $\lambda_i = 0$ except for λ_1 and λ_ν with the νth term denoting the Coulomb interaction, (8.79) is reduced to

$$k_B T_c = \frac{2\gamma\hbar\omega_1}{\pi} \exp\left[\frac{-1}{\lambda_1 - \mu^*}\right], \qquad (8.80)$$

which is the formula for T_c of weak-coupling superconductors including the so-called **Coulomb pseudo-potential** [8.12]

$$\mu^* = \frac{\mu}{1 + \mu \ln(\sqrt{D_1 D_2} / \hbar\omega_1)}, \qquad (8.81)$$

where $\mu = -N_F V_\nu = N_F U/N$. As we can see from (8.80), superconductivity can appear even if the coupling coefficient is not negative in any part of the phase space, i.e. $V_1 < U/N$, only if $\lambda_1 - \mu^*$ is negative. This is considered to be due to the phenomenon of retardation.

When $\hbar\omega_{\nu-1} > D_1$ and D_2, the $V_{\nu-1}$ term in (8.80) is combined with $V_\nu = -U/N$ into $-(U/N - V_{\nu-1})$. Therefore, the effect of the a_g-modes with $\hbar\omega_i > D_1$ and D_2 is reduced to a static screening of the on-site Coulomb interaction. Note that this direct way of reducing U through mediation of very-high-frequency modes is not very efficient in raising T_c.

EMV coupling to the vertex correction was suggested to enhance T_c by *Pietronero* and *Strässler* [8.79].

8.3.3 Evaluation of T_c for β-(ET)$_2$I$_3$

Now we apply (8.79) to obtain the value of T_c for β-(ET)$_2$I$_3$. In the present model the effects resulting from the constituent molecules are included through $\hbar\omega_i$ and g_i for each a_g-mode. Those of the crystal are accounted for in the band parameters N_F, D_1, D_2, the coupling constant with the acoustic phonons, $\lambda_1 = \lambda_{ac}$, and the Debye energy $\hbar\omega_D$. The on-site Coulomb energy U is a complicated quantity determined by the nature of the individual molecules, the polarization of the surrounding molecules and the screening effect of surrounding electrons. *Lipari* et al. [8.64, 65] obtained the values of $\hbar\omega_i$ and g_i for the a_g-modes of TTF, as compiled in Table 8.1. The labeling in the table is after those researchers with $h\nu_i = \hbar\omega_i$ in descending order. In the ET molecule [8.80], the central C-S stretching mode and the central C=C stretching mode have frequencies very close to those of the corresponding modes in TTF; they have strong EMV couplings. As shown in Figs. 8.12 and 13, HOMOs on both molecules are very similar in that the coefficients of the atomic p_z-orbitals on the central carbon and on the neighboring sulfurs are large, and are of opposite signs. Therefore, for ET we tentatively employ the a_g-mode characteristics of TTF (Table 8.1).

A reported value of the transfer energy $t_{\parallel}$ along the TTF stack in TTF·TCNQ is $0.65/4 = 0.1625$ eV [8.81]. For β-$(ET)_2 I_3$ the Fermi surface is closed in the k_x-k_y plane [8.82]. Plasma-frequency data give $t_{\parallel} = 0.19$ eV and $t_{\perp} = 0.08$ eV assuming a tight-binding band on a rectangular lattice [8.83]. When we numerically calculate the state density N_F at the Fermi energy ϵ_F and express it by (as for a quasi one-dimensional conductor with a 1/4-filled band having $t_{\parallel} = t_{eff}$)

$$N_F = \frac{N}{\pi t_{eff}\sqrt{2}} , \qquad (8.82)$$

$t_{eff} = 0.178$ eV is obtained. Another analysis showing the effect on optical data by interband transitions across the significant dimerization gap estimates that the difference between the top of the band and the Fermi energy is 0.11 eV, assuming a quadratic **k** dependence in the k_x-k_y plane [8.84]. This gives $t_{eff} = 0.104$ eV. A total bandwidth of 0.5 eV was theoretically obtained in [8.82]. Although the 0.178 eV value is considered to give a better global band feature, it takes no account of the dimerization gap. As a compromise we tentatively employ the above-mentioned value $t_{eff} = 0.1625$ eV coming from the TTF system when estimating λ_i (Table 8.1) and also for obtaining T_c of β-$(ET)_2 I_3$. The upper and lower bounds are given by $D_1 = N/4N_F = \pi t_{eff}/2\sqrt{2}$ and $D_2 = 3D_1$, respectively, from the assumption of a constant density of states.

For U we use the value 0.4875 eV estimated for TTF in TTF·TCNQ [8.81], because not enough quantitative data are available for ET. The coupling constant $\lambda_{ac} = N_F V_1$ for an attractive interaction due to acoustic phonons is expected to be small. For the TTF chain in TTF·TCNQ, λ_{ac} is estimated to be on the order of 0.05 [8.62]. *Weger* et al. [8.61] argued that λ_{ac} of $(TMTSF)_2 X$ salts is less than 0.16 based on the fact that CDW does not show up for $T > 20$ K. We therefore set, for the moment, $\lambda_{ac} = 0.15$. The Debye energy $\hbar\omega_D$ is equal to 52 K, which is the value one gets if one assumes that TTF is a rigid object having no internal freedom of motion [8.71].

Substituting all these values into (8.79) we derive $T_c = 4.2$ K. If we decrease U to 0.409 eV, keeping the other parameters unchanged, we get $T_c = 8$ K. If we decrease t_{eff} to 0.1285 eV, or if we increase λ_{ac} to 0.2653, the same 8 K value is obtained. On the other hand, $t_\parallel = 0.30$ eV, or U = 0.66 eV, or $\lambda_{ac} = 0.02$ gives $T_c = 1$ K. Although $\mathcal{H}_{BCS}$ in (8.70) is a crude first-step approximation, these estimates illustrate that $T_c = 8$ K is found in the present scheme when the main contribution to superconductivity comes from the attractive interaction mediated by the a_g-mode vibrations.

In the following, the plausibility of the employed values for the coupling constants g_i is discussed. As is known from the work of *Rice* [8.85, 86], a molecular dimer absorbs the infrared radiation via coupling of the charge transfer between the two molecules with a_g-mode intramolecular vibrations. The peaks of these absorptions are slightly shifted to lower frequencies from the a_g-mode frequencies of the separated molecules, or those observed by the Raman effect. From the magnitudes of these shifts the values of g_i can be determined. Table 8.2 lists the results of such an analysis for TTF. The calculated g_i values are in reasonable agreement with the measurements. The value of g_6 for the most important C-S stretching mode is slightly too large, and that of g_2 for the central C=C stretching is slightly smaller.

Table 8.2. Experimental and calculated values of the coupling constant g_n in TTF·Br and TTF·CA (chloranil)

Mode	ω_n [cm^{-1}]	TTF·Br g_n[a]	TTF·CA g_n[b]	Calculated g_n[c]
ν_2	1505	0.23	0.09	0.23
ν_3	1420	0.76	0.65	0.62
ν_4	1073		0.08	0.16
ν_5	758	0.52	0.40	0.49
ν_6	501	1.08	1.14	1.33
ν_7	264		0.31	0.16

[a] From [8.73]
[b] From [8.72]
[c] From [8.64, 65]

Another set of experiments suggests larger values of the coupling constants. The relaxation energy E_{rel} accompanying the removal of an electron from a neutral molecule is given by

$$E_{rel} = \sum_i E_{r,i} = \sum_i g_i^2 \hbar\omega_i , \qquad (8.83)$$

where the summation is over all a_g-modes. This energy represents the difference between the vertical and the adiabatic photoionization energies [8.87]. Its value is tabulated for TTF-derivative molecules in Table 8.3 after *Shaik* et al. [8.88]. As can be seen, the calculated value of E_{rel} for TTF is smaller than the observed values for TTF, ET and TMTSF. Part of the discrepancy may come from the quadratic EMV coupling [8.89]. These two data sets suggest, however, that the calculated values of g_i for TTF are in the correct range.

Kozlov et al. performed the calculation of g_i for the ET molecule [8.90]. They first adjusted the force-field constants so that the frequencies of the a_g normal modes of ET$^+$ are reproduced, and thereafter calculated g_i. The central C=C stretching mode has the frequency $\nu_3 = 1427$ cm^{-1} with $g_3 = 0.53$; the central-part C-S stretching mode has $\nu_9 = 508$ cm^{-1} with $g_9 = 0.34$; g_3 and g_9 are divided by $\sqrt{2}$ so that they conform to the present definition. The frequencies are very close to the corresponding ones in TTF. The coupling constant g(C=C) for the C=C stretch is slightly larger than the corresponding g_3 for TTF (Table 8.2); g(C-S) for the C-S stretch is appreciably smaller than the corresponding g_9 of TTF. *Kozlov* et al. suggested that the coupling constants are sensitive to small molecular field changes. *Faulhaber* et al. advanced still smaller coupling constants for both modes by means of an abinitio density-functional calculation [8.91]. A more recent calculation of ET gave g(C=C) = 0.73 and g(C-S) = 1.21 [8.92]. Raman-scattering measurements produced values of the dimen-

Table 8.3. Observed and calculated values of the relaxation energy $E_{rel} = \Sigma_i\, g_i^2 \omega_i$

	TTF	BEDT-TTF	TMTSF
$(E_{rel})_{obs}$[a] [eV]	0.52	0.29	0.31
$(E_{rel})_{calc}$[b] [eV]	0.21		

[a] From [8.88]
[b] From [8.64, 65]

sionless coupling constants λ_i for ET in κ-$(ET)_2Cu(NCS)_2$ as follows: $\lambda(C=C) = 0.14$ (0.17) and $\lambda(C-S) = 0.23$ (0.30) for the b(c) direction of the electic field [8.93]. An appreciable coupling of the C-S stretching mode was also observed in a point-contact tunneling experiment [8.93]. Recently, *Yartsev* obtained g(C-S) ≈ 0.7 by fitting of the phase-phonon theory to the optical conductivities of ET compounds [8.95]. Therefore, our employment of the coupling constants of TTF for ET is not unreasonable, although it does not have a solid quantitative basis.

Kübler et al. [8.96] presented band calculations for β-$(ET)_2I_3$. In contrast to the result of *Mori* et al. [8.74], they suggested that several molecular orbital levels of ET exist in the close vicinity of the HOMO level and that consequently a few bands cross the Fermi energy. Reports on the Shubnikov-de Haas effect of β-$(ET)_2IBr_2$ [8.97] and on the high-T_c state of β-$(ET)_2I_3$ [8.98] clearly show the existence of a cylindrical Fermi surface with a cross-sectional area equal to half of the Brillouin zone in the conducting plane, in agreement with the extended Hückel result [8.74]. *Kübler* et al. supported it with a recalculated result [8.99], changing their opinion. Other observations suggest the existence of a different size of the cross-sectional area [8.100, 101] and a small spherical Fermi surface [8.102]. The reason for them is not yet clarified. However, our assumption of a conduction band that is composed of a single b_{1u}-type HOMO has been given firm grounds for ET superconductors. Even if the above-mentioned small spherical pocket exists, the present scheme of superconductivity is not seriously affected, since the contribution of extra carriers is smaller.

The model for SDW in $(TMTSF)_2X$ (Chap. 4) gave a value of U = 0.34 eV from the condition that T_c = 11.5 K. Another model for SDW which takes account of not only U but also the BCS-like attractive interaction suppressing SDW was found to yield U = 0.37 eV [8.103], when T_c is set to 1 K and T_{SDW} = 12 K. Since the volume of a TMTSF molecule is close to that of ET, the value of U is expected to be similar for both species. Therefore, an appropriate value for ET should be smaller than 0.4875 eV employed in the above estimates of T_c. An analysis of infrared optical data for κ-$(ET)_2Cu(NCS)_2$ [8.104] led to a value of the state density, which gives a spin susceptibility close to the observed one [8.105] even if the enhancement factor is not taken into account. The values of the effective mass m^* = 5m [8.97] and 3.7m [8.98] obtained from magnetotransport data for β-$(ET)_2X$ also produce similar values for the spin susceptibility. These suggest that U is not so large in ET superconductors. The smallness of U is considered to be due to the screening of the surrounding current carriers, as discussed in Sects. 4.4.1 and 3.

8.3.4 Other $(ET)_2 X$ Salts, Pressure Effects, and the Isotope Effect

Many $(ET)_2 X$ superconductors with X: I_3, IBr_2, AuI_2, $Cu(NCS)_2$ etc. are known, having crystal structures such as the types labeled α, β, θ and κ (Table 1.1). In addition, the superconductor $(ET)_3 Cl_2 (H_2 O)_2$ was reported, which has a hole concentration per ET of 2/3 [8.106]. The T_c values range from 0 to 13 K. An extensive comparative study of these superconductors is expected to reveal essential information on the mechanism of superconductivity in organic superconductors. Since the model presented in this section is independent of the arrangement of molecules, it is applicable to all these superconductors.

The assumption that the conduction band consists of b_{1u}-type HOMOs gets more support from κ-$(ET)_2 Cu(NCS)_2$ which provides a clear evidence for it from an excellent agreement with the Shubnikov-de Haas data and band calculations [8.107]. The observation of an STM current image that indicates the presence of a carrier concentration around the central TTF skelton is in agreement of our picture for HOMO of ET [8.108].

The state density is in good agreement with the spin susceptibility in ET superconductors, as described at the end of Sect. 8.3.3. The spin susceptibility of the κ-type $Cu(NCS)_2$ salt is reported to be $4.2 \cdot 10^{-4}$ emu/mol [8.105]. It is larger than $3.0 \cdot 10^{-4}$ emu/mol [8.109] for the β-type AuI_2 salt having $T_c = 4.8$ K. Both data hold for low temperatures. The spin susceptibility of the β-type I_3 salt at low temperatures has been reported to be $3.2 \cdot 10^{-4}$ emu/mol in [8.110] and $4.6 \cdot 10^{-4}$ emu/mol in [8.111]. Although all these numbers are not precise, it is sure that the state density of the above-mentioned ET superconductors is three times larger than that of the Bechgaard salts. This is considered to be an important factor that leads to higher values of T_c for the ET superconductors.

If we use the total state density $0.76 \cdot 10^{34}$ $erg^{-1} \cdot cm^{-3}$ reported for κ-$(ET)_2 Cu(NCS)_2$ in [8.104], which gives $t_{eff} = 0.088$ eV in (8.82), and employ the same values for the other parameters as in the previous estimates (Sect. 8.3.3), we get $T_c = 18.6$ K from (8.79). This prediction convincingly supports that the EMV mechanism gives T_c sufficiently high to explain the observation. At least it provides a substantial contribution to the BCS-type attractive interaction.

Concerning the state density, the effective mass and so on of κ-$(ET)_2 Cu(NCS)_2$ several sets of values have been reported from various measurements and band calculations. Since a consistent interpretation of these values have not yet been established, they are juxtaposed here so that the reader can see the involved but interesting situation at present. The above-mentioned value of the total state density is equal to the state density $N_F = 3.25/(eV \cdot ET \cdot spin)$ where ET means one ET molecule. It gives the bare spin susceptibity $\chi_0 = 3.2 \cdot 10^{-4}$ emu/mol, which is 1/1.3 times the observed

χ_{spin}. It is deduced from the optically obtained plasma frequencies, yielding $m_b = 5.5m$ and $m_c = 4.1m$ for the b and c directions; m being the free electron mass. The band calculation of *Mori* produced the form of the Fermi surface but not the state density [8.107]. A recent band calculation gave $N_F = 1.6/(eV \cdot ET \cdot spin)$, $m_\alpha^* = 1.72m$ and $m_\beta^* = 3.05m$, which are for the smaller α- and β-orbitals, respectively [8.112]. Shubnikov-de Haas oscillations gave $m_\alpha^* = 3.5m$ and $m_\beta^* = 6.5m$ [8.113] which lead to $N_F \approx 2.8/(eV \cdot ET \cdot spin)$. m_α^* was shown to decrease under pressure in correlation with a decrease in T_c. The γ value of the specific heat measured in an applied magnetic field was reported to be $25mJ/(K^2 \cdot mol)$, which strangely is a common value for both the $NH_4Hg(SCN)_4$ salt with $T_c = 1$ K and the $Cu \cdot [N(CN)_2]Br$ salt with $T_c = 12$ K [8.114, 115]. This value yields $N_F = 1.33/(eV \cdot ET \cdot spin)$ which is much smaller than the other values of N_F. Clearly, it is necessary to decide on the bare band mass or state density, and the nature of other values so that we can determine the theoretical value of T_c. In other words, the organic superconductors represent a good testing ground for the many-body theory as well as the magneto-oscillatory effects.

The β-type $(ET)_2X$ superconductors with X: I_3, AuI_2 and IBr_2 are known to display certain systematics. The value of T_c for β-$(ET)_2I_3$ in its high-T_c state decreases very rapidly under pressure (Fig. 5.40). The pressure dependence of T_c for κ-$(ET)_2Cu(NCS)_2$ is still larger, as seen in Fig. 5.46. T_c of the AuI_2 or IBr_2 salts under ambient pressure is close to that of the I_3 salt at a pressure under which the stacking periods are uniform [8.116]. The low-T_c value of the I_3 salt with the incommensurate lattice modulation at ambient pressure is conjectured to come partly from the shorter stacking period due to the disruption of the uniform linear arrangement of the I_3^- anions. The shorter stacking works as an effective pressure. The shortening of the stacking period due to a lattice modulation at T ≈ 175 K is clearly seen in Fig. 8.15. The magnitude of the shortening corresponds to a pressure of 2 kbar. A part of the decrease of T_c must come from the randomness introduced by the incommensurate lattice distortion.

A correlation was found between T_c of all known ET superconductors and the effective volume V' available per hole, which includes the space occupied by the BEDT-TTF molecules but excludes that of anions [8.118, 119]. As seen in Fig. 8.16, T_c increases as V' increases. This tendency is in good agreement with the rapid decrease for T_c of these superconductors under pressure.

The superconductors κ-$(ET)_2Cu[N(CN)_2]X$ (X: Br, Cl) discovered later presumably lie on the extention of the correlation line between T_c and V' in Fig. 8.16. The Br salt is an ambient-pressure superconductor with $T_c = 11.6$ K [8.120], and the Cl salt becomes superconducting under a pressure above 300 bar with the maximum $T_c = 12.8$ K [8.121]. Without pressure the latter is basically a localized spin antiferromagnet with a canted ferromagnetic

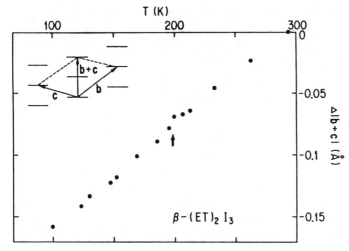

Fig. 8.15. Temperature dependence of the shrink of the lattice constant in the stacking direction of the ET column. The notation of lattice vectors follows the original data. From [8.117]

component [8.122]. This suggests that in the limit of a large effective volume V' per ET the system becomes a Mott insulator. This tendency is observed in the precursive enhancement of the resistivity around 100 K in κ-type superconductors with $T_c \geq 10$ K (Fig. 5.31); when this tendency is suppressed at lower temperatures, superconductivity appears at about 10 K. This is seen in the T-dependence of the resistivity of κ-$(ET)_2$Cu· $[N(CN)_2]$Cl under pressure [8.123] and is clearly indicated in the common T-dependence of ^{13}C NMR $(T_1 T)^{-1}$ of κ-$(ET)_2$X with X = $Cu(NCS)_2$, $Cu[N(CN)_2]$Br and $Cu[N(CN)_2]$Cl down to 50 K [8.124].

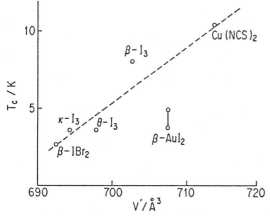

Fig. 8.16. Dependence of T_c on V', the effective volume per ET molecule excluding the volume of the anion, in $(ET)_2$X superconductors. From [8.118, 119]

From this discussion it seems that one plausible mechanism for T_c to decrease under pressure or in a compressed space is through the decrease of the coupling constants g_i. If we assume that, instead of $\partial E_\gamma / \partial Q_i$ in (8.67), $\partial E_\gamma / \partial q_i$ is independent of pressure, since $q_i = Q_i (\hbar / \omega_i)^{1/2}$ is the conventional normal coordinate for the ith a_g-mode independent of ω_i, then, we can rewrite the dimensionless coupling contant λ_i as

$$\lambda_i = \frac{N_F}{\omega^2} \left[\frac{E_\gamma}{\partial q_i} \right]^2 . \tag{8.84}$$

As usual, with the squeezed unit-cell volume, ω_i is expected to increase since the a_g-mode vibration is accompanied by a volume change. This leads to a decrease of λ_i, resulting in a decrease of T_c, although this must still be checked via measurements of the decrease of ω_i. Since a major part of the attractive interaction that comes from the mediation of a_g-modes is cancelled by the opposing on-site Coulomb interaction, a slight decrease of the former might sharply decrease T_c, as noted in the above-given estimate.

Another important factor which directly decreases T_c is the state density N_F. It decreases under pressure or with a reduction of the effective volume V' per molecule. This is because the decreased intermolecular separation increases the intermolecular transfer energies and thus the bandwidth. However, an increase of N_F in the high-T_c state of the I_3 salt over that in the low-T_c one was reported to be less than 10 % [8.125].

Another possibility is that the sensitive pressure dependence comes from the pressure dependence of the second band that consists of another type of molecular orbitals at the Fermi energy. The existence is, however, getting less likely, as described in Sect. 8.3.4. Such a band would more effectively screen the on-site Coulomb energy U in the first band. This can also enhance T_c through a two-band mechanism which will be discussed in Sect. 8.5.4.

In order to test the EMV mechanism of superconductivity the Orsay group started to substitute ^{13}C for ^{12}C on the carbon sites of the central double bond of the ET molecule. They reported that $T_c = 7.8$ K of the undeuterated β_H-$(ET)_2 I_3$ decreased by 0.2 K (± 0.1K), i.e., $\Delta T_c / T_c = -2.5\%$ [8.126]. They regarded it too large since the reduction of the average C=C stretching-mode frequency was $\Delta \omega / \omega = -1.8\%$. They ascribed the large decrease of T_c to inelastic scattering which remains still appreciable at low temperatures since a T-dependence of the resistivity shows up. This leads, as they demonstrated, to an enhanced isotope effect. The Argonne group carried out intensive measurements of the isotope effect. They reported that the ^{13}C substituted samples of κ-$(ET)_2 Cu(NCS)_2$ and κ-$(ET)_2 Cu[N(CN)_2]Br$ did not give measurable shifts of T_c within an uncertainty of ± 0.1 K [8.127]. In β_H-$(ET)_2I_3$ they did not detect a systematic decrease of T_c [8.128] in contrast to the Orsay report. They also replaced all the eight sulphur atoms, mostly ^{32}S, of ET with ^{34}S and statistically found the decrease of T_c to be 0.08 K in β-$(ET)_2Cu[N(CN)_2]_2Br$, i.e., $\Delta T_c/T_c = -0.7\%$ [8.129]. Since the frequencies of the two a_g modes associated with C-S stretching decrease by 2.7% on this substitution, they concluded that the C-S stretching modes of ET is not a dominant mediator of interelectron attractions. Rather they argued that the decrease is consistent with a small BCS isotope shift due to the change of intermolecular modes through the entire mass of ET. They substituted all S by ^{34}S and peripheral C by ^{13}C in ET of κ-$(ET)_2Cu(NCS)_2$ and obtained $\Delta T_c = -0.12 \pm 0.05$ K and $\alpha = 0.26 \pm 0.11$ [8.130].

These results suggest that intermolecular vibrations may be more important. EMV coupling may not be so effective, hampered by the appreciable on-site Coulomb energy. g(C-S) may not be so large, being much smaller in ET, TMTSF and so on than in TTF. Another problem is the vertex correction. Since the relevant phonon frequencies are not much smaller than the bandwidth in the 10-K-class organic superconductors, the vertex correction for the electron-phonon interaction is basically not negligible. However, in a similar case of the doped fullerene superconductors, the vertex correction was indicated to strongly enhance T_c [8.79]. There are other works suggesting that vertex corrections are negligible [8.131]. Therefore, this may not be a cause for the apparent ineffectiveness of the MVE mechanism.

Incidentally, the replacement of hydrogens on the peripheral part by deuterons was found to increase T_c of β_L-$(ET)_2I_3$, κ-$(ET)_2Cu(NCS)_2$, κ-$(ET)_2Cu[N(CN)_2]Cl$ and κ-$(ET)_2Cu(CN)[N(CN)_2]$ by about 1 K [8.132]. Although this is still an open question, it is in accord with the increase of the effective volume per ET, which correlates with the increase of T_c (Fig. 8.16).

The Padova goup treated a model in which both EMV couplings and the interaction with intermolecular phonons are taken into account, and discussed the isotope effect. It is thus possible to reconcile the model with the observed weak isotope effect [8.133].

8.3.5 Other TTF-Derivative Complexes

It is natural to expect that EMV interactions play a major role in driving superconductivity in the other organic superconductors that are composed of TTF-derivative molecules, such as $(TMTSF)_2X$, $(DMET)_2X$, and $(MDT-TTF)_2X$ (Chaps. 3 and 6). As already stated in Sect. 8.3.1, the TMTSF molecule has a HOMO orbital similar to that of TTF with the charge distribution concentrated in the central part consisting of C and Se. TMTSF has important a_g-modes, the C=C stretch and the C-Se stretch which are similar to the C=C and C-S stretching modes in TTF, except that the C-Se stretching frequency is 40% lower than that of the C-S stretch due to the heavier mass of the Se atom [8.134]. This is a desirable feature since, in principle, the lower value of ω_{C-Se} decreases the Coulomb pseudo-potential μ^*. If one sets to zero all the coupling constants, except those for the C-Se mode and the Coulomb interaction, one obtains T_c from (8.80) with $\mu^* = \mu/[1 + \mu \times \ln(\sqrt{(D_1 D_2)}/\hbar\omega_{C-Se})]$, and ω_1 being replaced by ω_{C-Se} The decrease of μ^* compensates more than the decrease of the prefactor $\hbar\omega_{C-Se}$ as long as $\hbar\omega_{C-Se}$ is more than several times larger than $k_B T_c$.

In TMTSF salts there is one complicating factor, i.e., the good nesting property of the Fermi surface which lead to the SDW instability. The nesting property can also be considered to enhance the repulsive Coulomb interaction in the vicinity of the Fermi surface, hampering the rise of T_c for s-wave superconductivity in the $(TMTSF)_2 X$ salts. It has been argued that spin fluctuations contribute to d-wave superconductivity [8.27-29]. As discussed in Sect.8.2, this kind of argument has some merits and some difficulties, since the nesting property is restricted to the $(TMTSF)_2 X$ salts. It is more reasonable to presume a mechanism common to all superconductors that are composed of TTF-derivative molecules.

The HOMO orbitals of the asymmetrical TTF-derivative molecules are also very likely to be essentially the same as those of TTF in the symmetry and in the concentration of charge in the central part. This has been found to be true in the case of DMET [8.76] and holds for MDT-TTF, too [8.77, 78]. The $(MDT-TTF)_2 AuI_2$ superconductor was reported [8.136] to have a band structure very similar to that of κ-$(ET)_2 Cu(NCS)_2$, in addition to the similar packing pattern. Since these molecules are asymmetric, they must have a higher number of a_g-modes. However, the most strongly coupling modes, i.e., the C=C and C-S (or C-Se) stretching modes, are likely to be very similar to those in the symmetric species. Therefore, we can expect that EMV interactions also play a major role in driving superconductivity in the compounds of these asymmetric molecules.

At low temperature the spin susceptibility of $(DMET)_2 X$ salts ranges in $(1 \div 3) \cdot 10^{-4}$ emu/mol [8.137], i.e., between the values of $(TMTSF)_2 X$ and $(ET)_2 X$ salts. The values of T_c are also confined to the intermediate range, however with a stronger concentration on the low-temperature side. The latter fact may be due to some remaining disorders in the systems. The highest value of T_c is 1.9 K for κ-type-like $(DMET)_2 AuBr_2$, which has also the largest value of the spin susceptibility. The correlation between T_c and the spin susceptibility indicates the important role of the state density in agreement with the EMV mechanism.

Another interesting feature of κ-type-like $(DMET)_2 AuBr_2$ is the hump of the resistivity-vs-temperature curve in the intermediate temperature range [8.137]. A similar feature is shared by some samples of κ-$(ET)_2 \cdot Cu(NCS)_2$ [8.138] which has a higher value of T_c. This may be a precursory effect to form a localized small polaron system due to strong EMV couplings and the narrowness of the electron band or to form a Mott insulator state. The small values of conductivity on the order of 10 S/cm might come from the same origin.

As described in Sect. 7.1, since $(CH_3)_4 N[Ni(dmit)_2]_2$ becomes superconducting, the $Ni(dmit)_2$ system itself is considered to drive supercunductivity, although the role of the TTF column in α-$TTF[M(dmit)_2]_2$, with M: Ni and Pd, is not yet clear. The EMV interactions in $M(dmit)_2$ have not yet been studied extensively, but are expected to be strong. This is reasonable since an extended-Hückel calculation demonstrates that the charge of the Lowest Unoccupied Molecular Orbital (LUMO) and HOMO are concentrated in the central part of the $M(dmit)_2$ molecule [8.139]. As seen in Fig. 1.1, its structure in the central part of the molecule is similar to that of TTF, except for the replacement of C=C by M.

8.3.6 Fullerene Superconductors

In the fullerence superconductors $A_3 C_{60}$, with A being an alkali metal atom, the LUMO levels were demonstrated to be strongly coupled with intramolecular vibrations [8.140-143]. Since the LUMO levels are degenerate, not only a_g but also h_g modes can mediate the attractive interaction between electrons. Experimental evidence that phonons of these modes are actually strongly coupled with doped carriers came from the observation of the broad widths of these modes by Raman scattering [8.144] and inelastic neutron scattering [8.145]. In this family the isotope effect due to a replacement of ^{12}C by ^{13}C was shown to give a very large isotope effect, for example, with $\alpha \approx 0.37$ in the case of A: Rb [8.146]. For this reason the mechanism of superconductivity mainly mediated by the intramolecular vibrations is almost completely established. More details on experimental studies of the mechanism will be found in Sect. 10.3.3.

8.4 Intermolecular Electron-Phonon Interactions

8.4.1 Electron-Phonon Coupling Coefficient

The combination $\alpha^2(\omega) F(\omega)$ of the electron-phonon interaction constant $\alpha(\omega)$ and the phonon state density $F(\omega)$ is well known to be obtained from the tunneling characteristics of strong-coupling superconductors in which the electron-phonon interaction is strong. It is called the **Eliashberg function**. The Eliashberg equation [8.10] with a suitable choice of the Eliashberg function yields values of T_c in good agreement with the observed values. However, the usual tunneling experiment on strong-coupling superconductors is not the only way to obtain this function. For superconductors of any coupling strength, a measurement in the normal state of the current (I) vs voltage (U) characteristics of a junction with a point contact of a very small cross-sectional area reflects the scattering which the current carriers

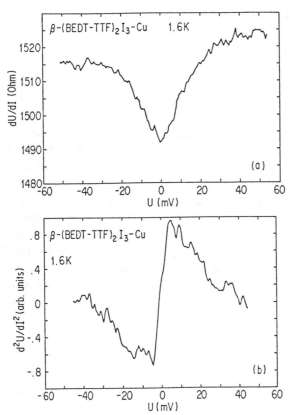

Fig. 8.17. (a) dU/dI and (b) $d^2 U/dI^2$ characteristics of a high-resistance contact between β-(ET)$_2$ I$_3$ and Cu. From [8.149]

undergo when they pass through the point contact [8.147, 148]. The second derivative $d^2 U/dI^2$ is proportional to $\alpha^2(\omega)F(\omega)$. This method was applied to β-(ET)$_2$I$_3$ [8.149, 150], and the data have been analyzed within the framework of the strong-coupling theory of superconductivity [8.151]. Because this experimental method provides direct information on the coupling strength of specified electron-phonon interactions, we present the results of both experiments and analysis in detail.

A single crystal of β-(ET)$_2$I$_3$ was pressed against a copper plate by a screw device to produce a junction [8.149]. The results are very sensitive to the contact resistance. For the proportionality $d^2 U/dI^2 \propto \alpha^2(\omega)F(\omega)$ with ω = eU to be valid, the contact area πa^2 must be such that the effective radius a of the point contact is much smaller than the energy-dependent mean-free path $\ell(E)$ of the current carriers. The results are plotted in Fig. 8.17.

In a second experiment, two platelet single crystals of β-(ET)$_2$I$_3$ were pressed against each other at their edges with a spring [8.150]. The force on the two crystals is much weaker than in the previous setup. The surfaces of the crystals were parallel to the conducting plane. The I-U dependence is depicted in Fig. 8.18. The I-U characteristics are much more stable than in the preceding setup.

The scale in Fig. 8.18c is expressed by the relation

$$\frac{d(dU/dI)}{dU} = \frac{m\alpha^2 F}{nea\hbar} , \tag{8.85}$$

where the radius a of the point-contact is determined from the zero-voltage-contact resistance given by

$$R_s = \frac{4\hbar k_F}{3ne^2 \pi a^2} . \tag{8.86}$$

This system has R_s = 72 Ω, which leads to a = 78 Å. The validity of these numbers should be taken with caution because this is a two-dimensional, highly anisotropic system, whereas (8.85, 86) have been derived for the usual isotropic metals. The $\alpha^2(\omega)F(\omega)$ curve in Fig. 8.18c are calculated by averaging the positive and negative bias parts of $d^2 U/dI^2$. Due to slight asymmetry, a dip appears at U = 2.5 meV, which makes the low-energy peak look very pronounced. Here, the contributions due to voltages above 25 meV are completely neglected.

The absolute value of $\alpha^2 F$ was determined by assuming ideal experimental conditions, for instance, an undamaged surface, a perfect contact, and an isotropic electronic band. The absolute scale of $\alpha^2 F$ is difficult to

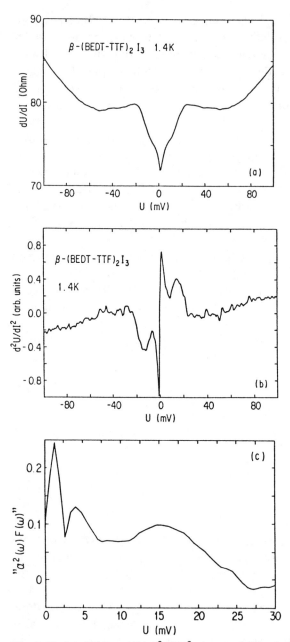

Fig. 8.18. (a) dU/dI and (b) d^2U/dI^2 characteristics of a low-ohmic contact between two samples of β-(ET)$_2$I$_3$. (c) Eliashberg function $\alpha^2(\omega)F(\omega)$, neglecting contributions from the higher voltage region. The error in the scale is within a factor of 2. From [8.150]

pinpoint. If one assumes that it is reliable within a factor of 2, then the dimensionless electron-phonon coupling constant λ defined by

$$\lambda = 2 \int \alpha^2(\omega) \frac{F(\omega)}{\omega} d\omega \qquad (8.87)$$

is found to be nearly equal to 1. The width of the observable voltage region is expected to mainly correspond to the intermolecular phonon energies, according to the data for other molecular crystals [8.152].

8.4.2 Calculation of T_c

A semiquantitative value of T_c can be deduced from (8.37) in terms of λ, (8.87), as will be discussed in the next subsection. Quantitatively, it changes with the ω-dependence of the Eliashberg function $\alpha^2(\omega)F(\omega)$. With the data derived from Figs.8.17 and 18, *Nowack* et al. [8.150] calculated T_c as a function of λ by numerically solving the Eliashberg gap equation in the strong-coupling region using the Bergmann-Rainer computer program [8.151]. They varied the scale of $\alpha^2 F$, which is proportional to the value of λ. Figure 8.19 displays the curves T_c vs λ for two values of the Coulomb pseudo-potential $\mu^* = 0$ and 0.1 for each form of $\alpha^2 F$ deduced from Figs.8.17 and 18.

Since the estimate $\lambda \approx 1$ is believed to be good within a factor of 2, the curves in Fig.8.19 show that $T_c = 1$ to 8 K should be accessible on the basis of the $\alpha^2 F$-vs-ω data. As can be seen, the values of T_c for given values of λ are very different. This reflects the fact that soft phonon modes with $\hbar\omega \approx 1$ meV have a large effect on λ but not on T_c. It has been argued that in the first setup the screw device generates a high pressure in the sample, which shifts the frequency of the soft phonon upwards. It was suggested that this is the reason for the difference between the forms of the $\alpha^2 F$-vs-ω curves and also the reason for the high value of $T_c = 8$ K in the sample under pressure and the low value of $T_c = 1$ K at ambient pressure.

Although this point-contact experiment looks very powerful, the above data still have to be considered preliminary. Although finite values of $d^2 U/dI^2$ at U higher than 60 mV (close to the eigenenergy of the a_g-type C-S stretching mode) are obtained, the data are noisy. Another remarkable feature is that $d^2 U/dI^2$ is slightly negative in the region from 25 to 60 meV. This is not due to noise but a systematic tendency, since dU/dI clearly has a negative slope in this region. This behavior has not been observed in other metals, e.g., Pb up to 75 mV [8.147]. Since $d^2 U/dI^2$ must be proportional

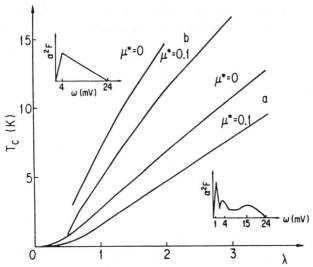

Fig. 8.19. T_c calculated from the strong-coupling theory vs. the dimensionless electron-phonon coupling constant λ. The λ values depend on the scale of $\alpha^2(\omega)F(\omega)$. Curves *a* have been calculated using the $\alpha^2(\omega)F(\omega)$ function of Fig. 8.18(c) (*lower inset*) and curves *b* have been obtained from Fig. 8.17b (*upper inset*), for two values of the Coulomb pseudo-potential μ^*. From [8.150]

to $\alpha^2 F$ which is always positive. It suggests some problem peculiar to this experimental situation or the low-dimensional band structure. If this is the case, the contribution to d^2U/dI^2 for $U \geq 60$ mV becomes appreciable. In fact, an Ukrainian group carried out this type of experiments on β-$(ET)_2 I_3$ and reported the result of d^2U/dI^2 in which $\alpha^2(\omega)F(\omega)$ has sizable contributions in the range of $50 \div 80$ meV [8.93], as expected from the strong EMV coupling. Further measurements on such systems are necessary. One of the interesting systems to investigate and to compare with this result is $(TMTSF)_2 ClO_4$. The effects of intramolecular vibrations could be evaluated since the eigenenergy of the analogous C-Se stretching is changed to 300 cm^{-1}. In addition, the influence of spin-fluctuation modes on the superconductivity, as discussed in Sect. 8.2.1, could be investigated.

It has also been suggested [8.149] that librational modes (a restricted rotation) have a large influence on T_c. The eigenenergies of these modes are usually rather small, lower than $200k_B$. Since the amplitudes of these rotations should be sharply reduced by a decrease in the volume around the molecule, their effect on T_c should be strongly pressure-dependent.

Recently, the frequency of transverse acoustic phonons of the κ-$(ET)_2 Cu(NCS)_2$ crystal were measured by ineleastic neutron scattering. Significant frequency changes were observed for phonons, the energy of which is close to the superconducting energy gap [8.153]. This suggests an appreciable contribution of those phonons to the pairing.

8.4.3 Supplement to the Strong-Coupling Theory of Superconductivity

The expression for T_c starts to deviate from the BCS form given in (8.37) or (8.44) with an increase in the strength of the electron-phonon interaction, i.e., as λ approaches unity. A compact expression for T_c in strong-coupling superconductors was obtained by *McMillan* [8.154]. For such systems *Eliashberg* [8.155] and *Nambu* [8.9, 156] derived a set of equations for the gap parameter $\Delta(\omega)$ as a function of frequency ω. *McMillan* started from a Green's function expression for the electrons and phonons in a uniform three-dimensional superconductor. They derived a coupled equation for $\Delta(\omega)$ at any temperature in terms of the Eliashberg function $\alpha^2(\omega)F(\omega)$. Taking a trial function of the form

$$\Delta(\omega) = \Delta_0 \theta(\omega_0 - |\omega|) + \Delta_\infty \theta(|\omega| - \omega_0) \tag{8.88}$$

with ω_0 being the maximum phonon frequency, *McMillan* calculated Δ_0 and Δ_∞ as consistently as possible and obtained

$$k_B T_c = \hbar \omega_0 \exp\left[\frac{-(1+\lambda)}{\lambda - \mu^* - (\langle\omega\rangle/\omega_0)\lambda\mu^*}\right], \tag{8.89}$$

where the dimensionless electron-phonon interaction constant λ is defined by (8.87), the Coulomb pseudo-potential μ^* is the same as (8.45) except that ω_D is replaced by ω_0. The average phonon frequency $\langle\omega\rangle$ is defined by

$$\langle\omega\rangle = \int_0^{\omega_0} d\omega\, \alpha^2(\omega)F(\omega) \Big/ \int_0^{\omega_0} d\omega\, \alpha^2(\omega)F(\omega)/\omega . \tag{8.90}$$

Equation (8.89) is a slight modification of the expression for T_c, (8.44), for weak-coupling superconductors.

McMillan numerically solved the Eliashberg equations for niobium. The phonon density of states, $F(\omega)$, was approximated by neglecting states for which $\hbar\omega < 100k_B$; α was assumed to be constant. The values obtained are well approximated by

$$T_c = \frac{\theta}{1.45}\exp\left[\frac{-1.04(1+\lambda)}{\lambda - \mu^*(1+0.62\lambda)}\right], \tag{8.91}$$

where Θ is the Debye temperature and can as well be replaced by $\hbar\omega_0/k_B$ or $\hbar\langle\omega\rangle/k_B$: for niobium, $\Theta = 277$ K, $\hbar\omega_0 = 330$ k_B and $\hbar\langle\omega\rangle = 230$ k_B.

The tunneling experiment on strong-coupling superconductors provides the ω-dependence of $\Delta(\omega)$. The $\Delta(\omega)$ obtained reveals remarkable structure reflecting the peak in $\alpha^2(\omega)F(\omega)$. This demonstrates that conversely $\alpha^2(\omega)F(\omega)$ can be deduced from the tunneling data for $\Delta(\omega)$, as well.

Equation (8.91) suggests that the maximum value of T_c could be around 40 K even if λ is increased, since λ is observed to be inversely proportional to $\langle\omega^2\rangle$. However, this result depends on the assumed form of $\alpha^2 F$. *Allen* et al. [8.157] actually demonstrated that T_c increases proportionally to $\lambda^{1/2}$ in the limit of large λ in a different model.

8.5 Other Theories of Superconductivity

8.5.1 The g-ology

Soon after the discovery of superconductivity in $(TMTSF)_2 PF_6$ under pressure, *Barišić* et al. [8.158] tried to apply the g-ology scheme (Sect.2.3.8) to explain this phenomenon, because superconductivity was believed to be essentially one dimensional. They discussed a mechanism which is based on acoustic phonons, intramolecular vibrations such a$_g$-modes discussed in Sect.8.3, and the Coulomb interaction. They concluded that because of the forward Coulomb scattering that results in a small momentum exchange, the well screened g_2 should decrease to nearly half the value of g_1 for the backward scattering which is not well screened because of a large momentum exchange. Then, the (g_1,g_2) point in the g-ology diagram is located near the boundary between the superconducting and insulating phases. Here, the pressure was assumed to move the system from the SDW region into the superconducting region.

Later *Barišić* [8.159] concluded that the Coulomb interaction alone is not sufficient to induce superconductivity and that phonon interactions must be included. Thus, their picture approaches that of *Horovitz* et al. [8.160, 161] described in Chap.4.

Here, one should recall that in $(TMTSF)_2 X$ salts superconductivity appears only when the SDW phase vanishes due to a deterioration of the Fermi-surface nesting, as concluded in Chap.4. Therefore, superconductivity needs a two- or three-dimensional effect to overcome the essentially one-dimensional Fermi-surface instabilities, i.e., CDW and SDW instabilities. Unfortunately, no evidence of systems where the superconducting phase is the most stable phase in the one-dimensional g-ology phase diagram has been reported.

8.5.2 Excitonic Model

The relevance of an excitonic mechanism contributing to the superconductivity of β- $(ET)_2 I_3$ has been discussed by *Nakajima* [8.162]. Alternating stacks of conducting layers and insulating sheets of highly polarizable molecules are taken as the model for the salt. The conduction band in each conducting layer is simplified to a two-dimensional free-electron model in the xy-plane. The electron hops in the z-direction between the two-dimensional layers, generating a tight-binding band with a transfer energy.

Interlayer as well as intralayer attractive interactions are due to the polarization of the electrons localized in the insulating sheet, as depicted in Fig. 8.20. Here, a solid line represents the Green's function of an electron in a conducting layer, a dashed line the Green's function of an electron in an insulating sheet, and a wavy line the relevant matrix elements of the Coulomb interaction. Then, the effective attractive interaction to be used in the mean-field superconducting-gap equation takes the form

$$V(\mathbf{k}, \mathbf{k}') = g_0 + 2g_1 \sin(k_z)\sin(k_z') + 2g_2 \cos(k_z)\cos(k_z') , \qquad (8.92)$$

where

$$g_0 = 4[(v_d^2 + v_{ex}^2)/E_G] - V_0 , \qquad (8.93a)$$

$$g_1 = 2[(v_d^2 - v_{ex}^2)/E_G] - V_1 , \qquad (8.93b)$$

$$g_2 = 2[(v_d^2 + v_{ex}^2)/E_G] - V_1 . \qquad (8.93c)$$

Here, in defining the wave-vector component k_z, we take the interlayer spacing as the unit length; E_G is the gap of the electronic excitation in the insulating molecule; v_α and v_{ex} are the Coulomb matrix elements defined in Fig. 8.20; V_0 and V_1 are the intralayer and interlayer direct Coulomb interaction, respectively.

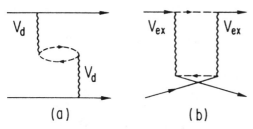

Fig. 8.20a,b. Two processes giving the excitonic interaction between conduction electrons. *Solid lines*: electron Green's function in the conducting layers; *dashed lines*: for the insulating layer; *wavy lines*: Coulomb interaction. From [8.162]

The respective gap parameter can be written in an abbreviated form as

$$\Delta = \Delta_0 , \quad \Delta_1 \sin(k_z) , \quad \text{or} \quad \Delta_2 \cos(k_z) , \tag{8.94}$$

corresponding to the maxima of g_0, g_1 or g_2 with

$$k_B T_c \approx \Theta \exp\left[\frac{-1}{\rho \cdot \max(g_0, g_1, g_2)}\right] , \tag{8.95}$$

where Θ is the smaller of the gap E_G and the Fermi energy, and ρ is the state density in the conducting layer.

From the signs in front of v_{ex} in g_1 and g_2 of (8.93b, c), g_2 is always larger than g_1. Since $V_0 > V_1$, $g_2 > g_0$ might be possible with g_2 being positive. It has been suggested that the high and low T_c states of β-(ET)$_2$I$_3$ correspond to the above-mentioned isotropic and anisotropic states. This interpretation has the advantage that the difference between $T_c = 1$ and 8 K corresponds to a small difference in the term $\max(g_1, g_2, g_3)$ if $\Theta \approx 10^3$ K.

8.5.3. Bipolaron Model, and a Treatment of the Polaron Effect

Another mechanism of superconductivity for organics has been suggested by *Mazumdar* [8.163] for the extreme limit of electron localization due to strong Coulomb repulsion. In this model, the screening of the Coulomb interaction is very bad for the special value of the charge transfer ratio $\rho = 1/2$ [8.164]. It is assumed that in each layer of (TMTSF)$_2$X and (ET)$_2$X, the electron sites make a rectangular lattice, and that holes are localized at every other lattice point, forming a face-centered rectangular sublattice. When an extra hole is added to a vacant lattice point, a high Coulomb energy between nearest neighbors is required to move any hole by one lattice constant. However, when a pair of holes on nearest neighbors is moved by the same amount with the help of some means, the new configuration has the same total Coulomb energy as before. This idea led him to the concept of **bipolarons**, pairs of current carriers, which can condense into a superfluid state. The same might also be true for oxide superconductors.

This picture has flaws, however, because it neglects the normal metallic properties of conducting salts and also relies too heavily on geometrical features. The actual arrangement in these compounds allows a hole to move in oblique directions without loss of Coulomb energy. Furthermore, there exists an ET superconductor, namely (ET)$_3$Cl$_2$(H$_2$O)$_2$, in which $\rho = 2/3$ [8.106].

Nasu [8.165] has developed an involved treatment for the phase diagram of the Hubbard-Peierls system in one and two dimensions, taking account of the polaron effect, i.e., the effect of phonon clouds that accompany the moving and also the localized electrons. He included in his model the transfer energy between electronic sites t, the on-site Coulomb energy U, and the electron-phonon interaction S. He obtained the phase diagrams for superconducting, CDW, SDW, and bipolaron insulator phases in the t, U, S and ω, phonon frequency, parameter space. He also found the transition from large polarons to small polarons with decreasing t/ω.

8.5.4 Two-Band Mechanism

The possibility of a two-band mechanism for organic superconductors has been pointed out by *Emery* as an alternative possibility to a paramagnon-mediated mechanism [8.27]. A band calculation for β-$(ET)_2 X$ has suggested the existence of several bands within 0.5 eV from the Fermi level [8.96]. Recently, experimental data have become available which indicate two closed Fermi surfaces in these salts [8.100-102]. Calculations using the extended-Hückel method revealed that in the ET molecule the next-highest band lies 0.6 eV below the b_{1u} HOMO [8.166]. Therefore, there is the possibility of a two-band mechanism to work, although not so large in actual systems. At least the effect of a second band on the parameters of the main band must be investigated. On the other hand, in $(TMTSF)_2 X$ salts the next highest MO lies 2 eV below ϵ_F and LUMO lies 1.5 eV above ϵ_F so that their influence on the conductivity should be negligible.

There are two kinds of interband interactions through which a Cooper pair can pass from one band to another. One is the BCS-type interaction mediated by phonons through, presumably, non-adiabatic electron-phonon interactions with a negative coefficient as in (8.5) [8.167]. In this case the phase space available for Cooper pairs and, consequently, T_c are increased. The other is of Coulombic origin [8.168]. In the present case the relevant two bands are tight-binding bands originating in the two molecular orbitals $\phi_1(\mathbf{r})$ and $\phi_2(\mathbf{r})$. The Coulomb interaction in the second-quantization formulation is expressed in terms of two-center integrals involving these molecular orbitals. Among these integrals, there are the on-site Coulomb energy U for each band, the exchange integral, and the so-called **exchange-like integral** defined by

$$K = \int d\mathbf{r} \int d\mathbf{r}' \phi_1{}^*(\mathbf{r}) \phi_1{}^*(\mathbf{r}') \frac{e^2}{|\mathbf{r} - \mathbf{r}'|} \phi_2(\mathbf{r}') \phi_2(\mathbf{r}) . \tag{8.96}$$

With it one obtains the following interband interaction:

$$\mathcal{H}_{int} = \frac{1}{N} \sum_{\mathbf{k},\mathbf{k}'} K(c^{\dagger}_{1\mathbf{k}\uparrow}c^{\dagger}_{1-\mathbf{k}\downarrow}c_{2-\mathbf{k}'\downarrow}c_{2\mathbf{k}'\uparrow} + \text{H.c.}) , \qquad (8.97)$$

where $c_{i\mathbf{k}\sigma}$ ($i = 1,2$) refers to the ith band formed by the ith orbital in the tight-binding scheme, the sum over $\mathbf{k}$ has no BCS-like restriction on the energy range, and H.c. denotes the Hermite conjugate term. When $\phi_i(\mathbf{r})$'s, $i = 1,2$, are real functions, K is positive. This interband BCS-form interaction, nevertheless, can increase T_c if one chooses antiphase order parameters for the two bands. Roughly speaking, the effect of K is to diminish the suppressive effect of U in the main superconducting band [8.169].

Although the overlapping bands in β-(ET)$_2$X are getting less likely, there is the possibility that besides the main band that consists of the b_{1u} molecular orbitals of the ET molecules, the second band forms a small three-dimensional spherical carrier pocket. Since no analysis of the magnitudes of the interaction coefficients is available, it is difficult to tell how important these interactions are in the existing organics. The two-band effect works in cooperation with the various phonon mechanisms in the formation of s-pairing superconductivity. It might also contribute to the extraordinarily large pressure dependence of T_c, since the second band seems quite pressure-sensitive, if it exists.

If two bands have a good interband nesting, a new type of divergence coming from an interband ladder diagram emerges [8.170]. As a result, the effect of the exchange-like interaction is divergently enhanced and elevates T_c.

Canadell et al. reported on new band calculations for TTF[M(dmit)$_2$]$_2$ (M: Ni, Pd) in which two bands consisting of LUMO and HOMO orbitals on the M(dmit)$_2$ column cross the Fermi energy [8.171], as described at the end of Sect. 7.1.

The two-chain Hubbard model consists of two chains of the 1-D Hubbard model, which are linked by transfer energy between neighboring electronic sites on both chains. In the t-J ladder model in which two t-J chains are linked by both transfer energy and exchange interaction between both chains. These models are two-band models of superconductivity and have attracted strong interest since they are simple and expected to fill the gap between the 1-D and 2-D models. Like the case of two dimensions the slightly doped t-J ladder gives evidence for the occurrence of superconductivity [8.172] while the Hubbard ladder leads to a clear enhancement of superconducting fluctuations in the slightly doped case [8.173-175] but the superconducting ground state is not yet firmly established for it. Interestingly, both models have a parameter region in which a spin-gap state, or spin-liquid state, occurs [8.174-176]. The SC region lies in it as a part.

These models may be relevant to the superconductivity in dmit superconductors, especially TTF[Ni(dmit)$_2$]$_2$ superconductors described in Sect. 7.1. In this superconductor the planar Ni(dmit)$_2$ molecule is stacked in a column and the dmit group can be regarded as an electronic site. Then, the network of the dmit group makes a ladder system. As described in Sec. 7.1, this material has two bands, i.e., HOMO and LUMO bands, crossing the Fermi level, which can be regarded as coming from the bonding and anti-bonding orbitals on a ring made of two dmit groups, respectively. As discussed in the present section, such a two-band system has a possibility to drive superconductivity by the on-site Coulomb interaction alone if the band situation is favorable. Although the mechanism in TTF[Ni(dmit)$_2$]$_2$ is still open, this possibility poses a very interesting problem. At least this two-band nature can drastically reduce the effect of the intraband Coulomb repulsion which suppresses superconductivity [8.177]. Incidentally, in other Pd(dmit)$_2$ superconductors, although the band situation is more complicated, tendencies were noticed that the superconductivity occurs in such a situation where HOMO and LUMO are both around the Fermi level [8.178].

8.5.5 Possibilities of Triplet Pairing and High-Field Superconductivity

The possibility of triplet pairing was raised even at the time of the initial discovery of organic superconductors. As is seen in Sect. 8.2, the problem of whether superconductivity is due to s-, d- or p-wave pairing has not yet been settled. The p-pairing has been suspected because the superconductivity is very quickly destroyed by defects and impurities. The triplet-pairing superconducting phase neighbors the SDW phase in the simple version of the g-ology diagram. This suggested the possibility of the triplet nature of superconductivity in (TMTSF)$_2$X salts. The anisotropic band structure, especially in the Bechgaard salts, facilitates anisotropic electronic interactions.

The observation that they show the type-II nature despite sample cleanness may be related to the type of superconductivity in Bechgaard salts. It may also be connected with the moderate anisotropy of $H_{c2}^{(i)}$ (i = a,b,c). *Gor'kov's* recipe to calculate $H_{c2}^{(i)}$ gives much larger anisotropy than observed from the established quasi one-dimensional band structure of (TMTSF)$_2$X [8.179-182].

If a salt is superconducting above the Pauli limit of the magnetic field H_{Pauli}, (3.13), then an exotic pairing mechanism is implied. The maximum value of the observed H_{c2} looks limited by the Pauli limit in the case of both

TMTSF and ET superconductors [8.183, 184], although *Laukin* et al. [8.185] asserted that in the high-T_c state of β-$(ET)_2 I_3$ it exceeds H_{Pauli}.

Lebed et al. [8.186, 187] pointed out an interesting consequence if the Bechgaard salts have triplet superconductivity. The magnetic field is detrimental to superconductivity in two ways: through increasing the kinetic energy of the orbital motion of the electron, and exerting a force through the spin. If one applies a magnetic field to $(TMTSF)_2 X$ along the b'-direction, the semiclassical path in the c*-direction is confined to the width of $2ct_c/ev_F H$. In a field higher than t_c/μ_B, or about 20 T, this becomes smaller than the lattice constant $\tilde{c}$ in the c-direction. In this situation the electron motion is restricted to one ab-sheet so that the magnetic field no more gives a detrimental effect through the orbital freedom. In the limit of 2D, the magnetic field has no effect to the orbital motion in this configuration. Furthermore, since in the triplet superconducting state the field does not break superconductivity by exerting opposite forces into the pairing electrons, either. Therefore, in this configuration the triplet superconductivity is expected to revive in an extremely high field. Using the mixed representation of the Green's function $G_{++}(\omega_n, \mathbf{k}_\perp; x, x')$, which is employed in the theory of a magnetic-field-induced SDW in Sect. 9.2.1, *Lebed* et al. obtained the mean-field superconducting gap equation

$$\frac{\Delta(x)}{|g|} = \int_{|x-y|>d} dy 2\pi k_B T \frac{\Delta(y) \cdot J_0(2\lambda \sin^2[(x-y)/x_H] \sin^2[(x+y)/x_H])}{\hbar v_F^2 \sinh(2\pi k_B T |x-y|/\hbar v_F)},$$

(8.98)

where $x_H = 2c\hbar/eH\tilde{c}$, $\lambda = 4t_c c/ev_F H\tilde{c}$ with $\tilde{c}$ being the lattice constant in the c-direction, and d is a cut-off which is related to the coupling constant g on the left-hand side by $1/|g| = 2\ln(\hbar v_F/\pi k_B T_c d)$. This equation allows for a periodic solution, namely $\Delta(x+\pi x_H/2) = \Delta(x)$. Hence, a logarithmic singularity of the integral in the zero-temperature limit is present for any value of the magnetic field. For triplet pairing, the Zeeman-energy term plays no role in (8.98), in contrast to the case of singlet pairing. Sufficiently strong magnetic fields can noticeably diminish the orbital effects, especially in the presently chosen geometry. As a result, the above equation gives a solution that describes restoration of superconductivity in sufficiently strong fields. This new feature has to appear for $\lambda \approx 1$ or $H \approx 10$ T. *Burlachkov* et al. [8.187] calculated H_{c2} for $(TMTSF)_2 X$ when the magnetic field is applied precisely along the b'-direction. The expected phase diagram is shown in Fig. 8.21.

Dupuis et al. extended this argument [8.189-191]. They found that (8.98) has a more general solution $\Delta_Q(x) = e^{iQx}\tilde{\Delta}_Q(x)$ where $\tilde{\Delta}_Q(x+\pi x_H/2)$

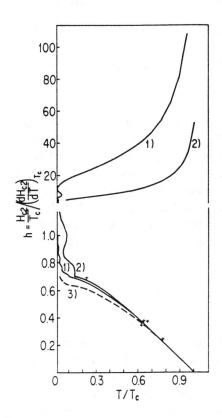

Fig. 8.21. Stability curves, $H_{c2}(T)$, for the triplet (curves *1* and *2*) and singlet (curve *3*) superconductivity in a quasi-one-dimensional superconductors such as $(TMTSF)_2 X$ $(H \| b')$. The experimental data (*crosses*) are for $(TMTSF)_2 AsF_6$ at $P = 11$ kbar [8.176]. Curves *1* and *3* were obtained for $t_c/k_B T_c = 7$; curve *2* for $t_c/k_B T_c = 4$. From [8.187]

$= \tilde{\Delta}_Q(x)$ and $-2/x_H < Q \leq 2/x_H$. This includes the above-mentioned solution which corresponds to $Q = 0$. Although degenerate in Q in a low field, in high fields $(eH/mc \gg k_B T)$ the highest $T_c(H)$ is obtained for $Q = 0$ or $Q = 2/x_H$ so that the superconducting subphases with $Q = 0$ and $Q = 2/x_H$ appear alternately with the first-order boundary with increasing field. They demonstrated that even in the singlet SC state, the SC state survives in very high fields although at much lower temperatures. They suggest that this state is a kind of Larkin-Ovchinnikov-Fulde-Ferrel (LOFF) state that has a spatial nonuniformity. A LOFF state of a 2-D system was also reported in [8.192].

Hasegawa et al. studied the case of anisotropic superconductivity under the assumumption of an attractive interaction between neighboring sites [8.193]. The results are basically similar to the s-wave singlet case.

Recently, (8.98) giving the upper critical field was shown to be able to be expanded to the quasi 2-D case [8.194]. This work pointed out that even in the case of singlet pairing the present effect expands the superconducting region in the H-T plane beyond the Ginzburg-Landau theory.

Fig. 8.22. Structure of undistorted polyacene

Naughton et al. investigated the superconductivity under very high fields on $(TMTSF)_2 ClO_4$ applying the field in the b'-direction [8.195-197]. They found that at very low temperatures the resistive state, turned away from the SC state by the magnetic field, shows a decrease of resistivity, a possible precursor effect to recover the superconductivity. They also found that the value of H_{c2} for the field parallel to b' becomes larger than that for the field parallel to a, although their definition of H_{c2} contains an arbitrariness [8.197]. Both values of H_{c2} exceed the Pauli limit of this superconductor.

8.5.6 Superconductivity in Polymers

There have not yet been any reports of superconductivity in organic polymers. Several groups have recently presented theoretical discussions of this possibility. *Kivelson* et al. [8.198] pointed out that in polyacene $(C_4 H_2)_n$ (Fig. 8.22) the Fermi surface lies at the edge of the Brillouin zone (Fig. 8.23). Nevertheless, an accidental degeneracy between valence and conduc-

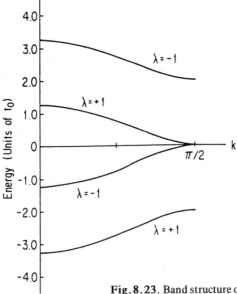

Fig. 8.23. Band structure of the undistorted polyacene. From [8.198]

tion bands makes it metallic. The dispersion relation is quadratic and the state density diverges at the Fermi level. *Kivelson* et al. suggested that this system may possibly contain instabilities including superconductivity. According to *Kertesz* et al. [8.199], the energy gain due to a periodic lattice distortion giving rise to an energy gap at the Fermi level is higher than second order in the CDW order parameter. This means that there should be practically no Peierls instability in polyacene. *Aono* et al. [8.200] studied polyacene modified by attached side chains which may work as excitonic systems in terms of the g-ology framework. They suggested that the interaction is too weak to induce superconductivity.

Yamabe et al. [8.201] found that another one-dimensional graphite, polyacenacene, also has a metallic band structure. This system consists of three conjugated carbon chains, instead of two as in the case of polyacene, which are linked by sp^2 hybrid bonds between neighboring carbons in the neighboring chains [8.202].

A non-adiabatic electron-lattice, i.e., vibronic, interaction was suggested by *Tachibana* et al. [8.203, 204] to induce an attractive interaction between electrons in conducting polymers. This may be yet another mechanism for superconductivity.

9. Field-Induced Spin Density Wave and Magnetic Oscillations

The magnetic-Field-Induced Spin Density Wave (FI-SDW) is one of the most delicious fruits that the growing field of organic conductors has produced. It has led to the development of a very interesting branch of research that encompasses a rich variety of phenomena. The concept of FI-SDW provides strong support for the nesting model of SDW and also for the denesting mechanism of the disappearance of SDW under pressure. In this chapter, we first introduce a variety of dramatic phenomena attributed to the magnetic-field-induced SDW. Then, we study the Gor'kov-Lebed theory and the related development that explains the appearance as due to the one-dimensionalization of the electron motion in the field. Next, we will discuss FI-SDW from the energetic viewpoint. This demonstrates that the quantization of the orbital motion of current carriers in the k-space in the vicinity of the SDW gap generates energy gain. This enables the SDW order to be restored, even if this order is destroyed by the deterioration of the Fermi-surface nesting. Some of the very interesting physical properties are obtained in the FI-SDW subphases, specified by the occupancy of the Landau subbands, in good agreement with observations.

In the later part of the theoretical discussion Green's function theories of FI-SDW are introduced. These enable us to treat the problem in the entire H-T region. We can see basic physical properties of organic conductors. Interestingly, the theory has a structure similar to that of the Green's function theory of BCS superconductivity. Recent experimental and theoretical results on FI-SDW are reported in Sect. 9.5

Independent of the large-period oscillation due to FI-SDW, rapid oscillations were observed and remain controversial. Angualr oscillations have stirred up interest in organic superconductors. The present status of the experimental and theoretical studies on these phenomena will be outlined in Sect. 9.6.

9.1 Initial Experiments

Soon after the discovery of superconductivity in $(TMTSF)_2 PF_6$, *Kwak* et al. [9.1] found Shubnikov-de Haas-like oscillation of the magnetoresistance, as illustrated by Fig.9.1. This was observed in the metallic state of the salt under a pressure above the critical value of $P_c \simeq 6$ kbar. Although practically no magnetoresistance was found for the magnetic field H perpendicular to the c-axis, that for the parallel field rose steeply. The oscillation is superimposed on the latter, which is clearly seen through the third derivative in Fig.9.2. The oscillation has a well-defined period in H^{-1}, $\Delta(1/H) = 0.013$ $(Tesla)^{-1}$, and has a $\cos\theta$ dependence, θ being the angle between the field direction and the normal to the conducting plane. This phenomenon was first interpreted in terms of the Shubnikov-de Haas effect, by postulating the existence of cylindrical carrier pockets parallel to the c-axis and with cross sections of 1% of the Brillouin-zone cross section in the conducting plane.

The Shubnikov-de Haas oscillation of the resistance is caused by successive crossings of the Fermi energy in Landau levels. The 1/H period which corresponds to the extremum cross section of the Fermi surface is given by [9.2]

$$\Delta(1/H) = \frac{2\pi e}{c\hbar A}, \tag{9.1}$$

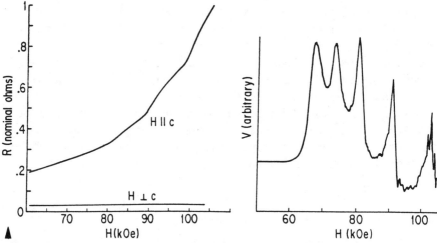

Fig. 9. 1. Resistance R along the a-axis of $(TMTSF)_2 PF_6$ as a function of the magnetic field H parallel and perpendicular to the crystal c-axis (T = 1.1 K, P = 7.4 kbar) [9.1]

Fig. 9. 2. Third-harmonic signal at 150 Hz with constant-amplitude 50-Hz field modulation versus magnetic field for $(TMTSF)_2 PF_6$ (T = 1.1 K, P = 6.9 kbar). From [9.1]

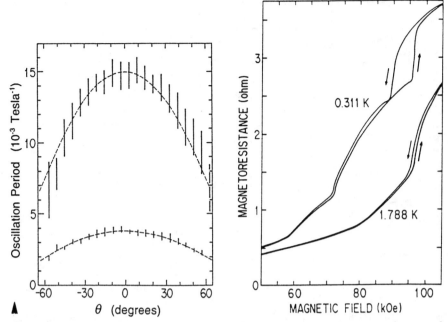

Fig. 9.3. Angular dependence of the oscillation periods of the magnetoresistance of $(TMTSF)_2 ClO_4$. *Dashed curves* represent fits given by $0.0150\cos\theta$ $(Tesla)^{-1}$ and $(0.0036\cos\theta + 0.0002)$ $(Tesla)^{-1}$, where θ is the angle between the magnetic field and the c^*-axis. From [9.4]

Fig. 9.4. Hysteresis of the magnetoresistance in the relaxed state of $(TMTSF)_2 ClO_4$ that is observed during upward and downward sweeps in the magnetic field. From [9.5]

where A is the area of the extremum cross section in the k-space. From the sharp rise of the magnetoresistance, *Kwak* et al. presumed that there are compensated electron and hole pockets.

The presumed carrier pockets are impossible to reconcile with the simple band model expressed by (4.12) with $t_a \approx 10t_b$, however. In contrast to the Shubnikov-de Haas effect, the oscillation was found to have a threshold field H_{th} above which the oscillation appears. This threshold goes up very quickly with an increase in pressure [9.3]. A similar oscillation of the magnetoresistance was found in $(TMTSF)_2 ClO_4$ by *Bando* et al. [9.4] with a period of $\Delta(1/H) = 0.015$ $(Tesla)^{-1}$. They also noticed an additional, more rapid overlapping oscillation with a shorter period of $\Delta(1/H) = 0.0036$ $(Tesla)^{-1}$. Both periods have the $\cos\theta$ dependence shown in Fig. 9.3, which reveals an orbital origin of the phenomena. Later, *Kajimura* et al. [9.5] noticed a hysteresis of the magnetoresistance in a slowly cooled sample of $(TMTSF)_2 ClO_4$ (Fig. 9.4). The oscillation has no constant periodicity. The results strongly suggest successive phase transitions that are accompanied by sharp changes in the resistance rather than a Shubnikov-de Haas pheno-

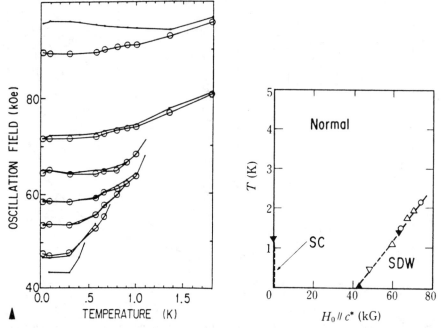

Fig. 9.5. Temperature dependence of the magnetic fields at which a sharp change of the resistance occurs. *Points* and *open circles* show the field values for upward and downward sweeps, respectively. From [9.5]

Fig. 9.6. Phase diagram for the relaxed state of $(TMTSF)_2 ClO_4$ derived from NMR data [9.7], specific-heat measurements ($\blacktriangledown$) [9.6], and magnetoresistance data: Δ [9.3], ∇ [9.4], $\blacktriangle$ [9.5]

menon. The phase diagram is depicted in Fig. 9.5. Incidentally, they also observed that the oscillation becomes less dramatic when the sample is rapidly cooled. They did not find the small-period oscillation. This suggests that the latter effect is sample dependent.

A maximum in the specific heat of the relaxed $(TMTSF)_2 ClO_4$ was also recorded at the threshold field [9.6]. Threshold field vs temperature data are shown in a phase diagram (Fig. 9.6); the large-period oscillation appears only in the semimetallic region above the threshold field. The magnetic, or SDW, nature of this phase was established by the disappearance of a ^{77}Se NMR signal [9.7, 8]. Thus, it was concluded that this kind of oscillation takes place in the SDW phase which is induced by high magnetic fields. An increase of the temperature of the transition to the field-induced SDW state with an increasing field was also reported for $(TMTSF)_2 PF_6$ at a fixed pressure above the critical one P_c [9.9].

A similar, large-period oscillation was noted in $(TMTSF)_2 ReO_4$ at pressures between ≈ 9.5 and ≈ 11 kbar [9.10]. The observed features are

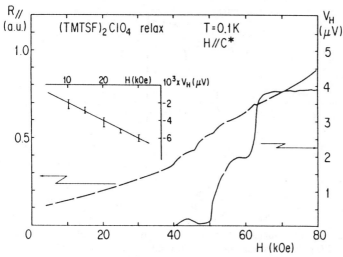

Fig. 9.7. High-magnetic-field Hall voltage (*continuous curve, right-hand axis*) of the related state of $(TMTSF)_2 ClO_4$ at $T = 0.1$ K [9.11]. *Inset* shows the low-field variation. The transverse magnetoresistance is also displayed (*dashed curve, left-hand axis*)

similar to those of the $(TMTSF)_2 PF_6$ under pressure and of the relaxed $(TMTSF)_2 ClO_4$. The ReO_4 system also exhibited a small-period oscillation in the magnetoresistance with $\Delta(1/H) \simeq 0.0030$ (Tesla)$^{-1}$.

Further amazing properties of the field-induced SDW state were discovered. As in the quantum Hall effect [9.11], the Hall voltage V_H shows a step-wise field dependence (Fig. 9.7), although the values of V_H do not yield simple ratios as expected [9.12-14]. When samples were very slowly cooled so that they are well relaxed, *Ribault* [9.15] found sign reversals of V_H in the field region of the well developed FI-SDW phase that is displayed in Fig. 9.8.

Another surprising result is the magnetization of the relaxed $(TMTSF)_2 ClO_4$ in the c*-direction seen by *Naughton* et al. [9.16]. It has a sharp jump at points supposed to be phase transitions (Fig. 9.9). These jumps in the magnetization correspond to the discontinuities in V_H. This is not due to the de Haas-van Alphen effect because the direction of the magnetization jumps is opposite to that expected. Specific heat data also clearly show these phase transitions inside the FI-SDW state [9.17, 18] (Fig. 9.10). Other anomalies believed to be associated with phase transitions, for example, the oscillatory thermopower, have been observed [9.19].

We would like to add a final remark about the small-period, or rapid, oscillation in the magnetoresistance. This type of oscillation is also found in the SDW state in $(TMTSF)_2 PF_6$ at ambient pressure with $\Delta(1/H) = 0.0043$ (Tesla)$^{-1}$ [9.20]. Appearing below about 10 K, it becomes stronger upon cooling, then turns weaker and suddenly disappears below 4 K. Another

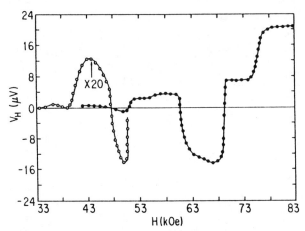

Fig. 9.8. Hall voltage V_H versus magnetic field at $T = 0.1\,K$ for an extremely slowly cooled sample of $(TMTSF)_2\,ClO_4$ (70 hours from 30 to 4.2 K). From [9.15]

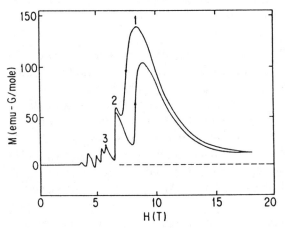

Fig. 9.9. Magnetization versus magnetic field at $T = 60$ mK for relaxed $(TMTSF)_2\,ClO_4$. From [9.16]

very interesting feature of this oscillation in $(TMTSF)_2\,ClO_4$ is the observation of two series of antiphase oscillations in the resistivity which have a different field dependence (Fig. 9.11) [9.21-23]. Recent experimental and theoretical results are reviewed in Sect. 9.6.1.

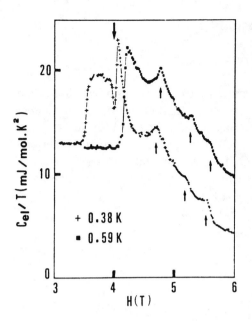

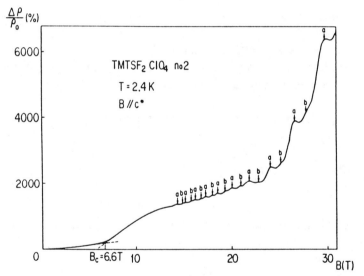

Fig. 9.10. Specific-heat C_{el}/T versus magnetic field for $(TMTSF)_2 ClO_4$. From [9.18]

Fig. 9.11. Double series of oscillations labeled *a* and *b* in the transverse magnetoresistance of $(TMTSF)_2 ClO_4$ at T = 2.4 K. B denotes the magnetic field. From [9.21]

9.2 Theory of the FI-SDW Instability

9.2.1 The Gor'kov-Lebed Theory

Why is SDW restored by high magnetic fields in systems in which it is suppressed under pressure or due to structural conditions? *Gor'kov* and *Lebed* were the first to give the answer to this question [9.24]. They succeeded in calculating the spin susceptibility $\chi(\mathbf{Q})$ in response to a spatially periodic magnetic field of the wave vector $\mathbf{Q}$. They demonstrated that $\chi(\mathbf{Q})$ becomes divergent in a magnetic field; it is finite in the absence of the field due to deteriorated nesting of the Fermi surface.

The model band is similar to (4.52) and given by

$$\epsilon_{\mathbf{k}} = v_F \hbar(|k_x| - k_F) + \epsilon_{\perp}(\mathbf{k}_{\perp}) , \qquad (9.2)$$

where

$$\epsilon_{\perp}(\mathbf{k}_{\perp}) = 2t_{b*}\cos(bk_y) + 2t_{c*}\cos(\widetilde{c}k_z) + 2t'_{b*}\cos(2bk_y)$$
$$= + 2t'_{c*}\cos(2\widetilde{c}k_z) . \qquad (9.3)$$

Here $\widetilde{c}$ is the lattice constant in the c-direction. In the presence of the magnetic field H parallel to the z-axis, one obtains the Schrödinger equation for the one-electron state, employing the Landau gauge $\mathbf{A} = (0,Hx,0)$, as

$$[v_F \hbar(\pm\hat{k}_x - k_F) + \epsilon_{\perp}(\hat{k}_y - eHx/c\hbar, \hat{k}_z)]\phi(\mathbf{r}) = E\phi(\mathbf{r}) , \qquad (9.4)$$

with [9.25]

$$\hat{k}_x = -i\partial/\partial x, \ \hat{k}_y = -i\partial/\partial y \text{ and } \hat{k}_z = -i\partial/\partial z ;$$

c is the light velocity; $\pm$ refers to the positive and negative component of k_x, respectively. Thanks to the linearity with respect to k_x, one can easily determine the following eigenvalue and eigenfunction for the newly introduced wave vector $\mathbf{k}$:

$$E_{\mathbf{k}} = v_F \hbar(|k_x| - k_F) , \qquad (9.5)$$

$$\phi_{\mathbf{k}} = \exp\left\{i\left[k_x x + \mathbf{k}_{\perp} \cdot \mathbf{r}_{\perp} \mp \frac{1}{\hbar v_F} \int^x du \, \epsilon_{\perp}(k_y - eHu/c\hbar, k_z)\right]\right\} , \qquad (9.6)$$

where $\mathbf{k}_{\perp} = (0,k_y,k_z)$ and $\mathbf{r}_{\perp} = (0,y,z)$.

Using this solution and a mixed representation of the Green's function, *Gor'kov* and *Lebed'* obtained the spin susceptibility $\chi(Q_0)$ for the optimum wave vector $Q_0 = (2k_F, \pi/b, \pi/\tilde{c})$ in the following way: The field operator $\psi_\sigma(r)$ for the σ spin component is divided into positive and negative wave-number components:

$$\psi_\sigma(r) = \psi_{+\sigma}(r) + \psi_{-\sigma}(r) . \tag{9.7}$$

Then, for the positive component, the Green's function $G_{++}(r\tau, r'\tau')$ is defined by

$$G_{++}(r\tau, r'\tau') = -\langle \hat{T}\psi_{+\sigma}(r,\tau)\psi_{+\sigma}^\dagger(r',\tau')\rangle , \tag{9.8}$$

where $\hat{T}$ is the chronological operator which arranges the field operators in an order with respect to the imaginary times τ and τ'. One can Fourier-transform it with respect to $\tau-\tau'$ [9.26] and the perpendicular space coordinate $(r-r')_\perp$, obtaining the mixed representation $G_{++}(i\omega_n, k_\perp; x, x')$. This can further be rewritten in terms of the slowly varying part g_{++} defined by

$$G_{++}(i\omega_n, k_\perp; x, x') = \exp[ik_F(x-x')]g_{++}(i\omega_n, k_\perp; x, x') . \tag{9.9}$$

In the magnetic field H specified by the vector potential $A = (0, Hx, 0)$ and in the absence of the Coulomb interaction, g_{++} satisfies the following equation

$$[i\hbar\omega_n + i\hbar v_F \partial/\partial x - \epsilon_\perp(k_y - eHx/\hbar c, k_z)]g_{++}(i\omega_n, k_\perp; x, x') = \delta(x-x') . \tag{9.10}$$

Integration yields

$$g_{++} = \frac{\text{sgn}(\omega_n)}{i\hbar v_F}\theta[\omega_n(x-x')]$$

$$\times\exp\left[-\frac{\omega_n(x-x')}{v_F} - \frac{i}{\hbar v_F}\int_{x'}^{x}\epsilon_\perp(k_y - eHu/\hbar c, k_z)du\right] , \tag{9.11}$$

where sgn(..) and $\theta[..]$ are the sign and step functions, respectively. In the same way we can obtain G_{--}.

The spin susceptibility for the wave vector $\mathbf{Q}$ is obtained by a summation of the standard ladder diagram as

$$\chi(\mathbf{Q}) = \frac{\chi_0(\mathbf{Q})}{1 - U\chi_0(\mathbf{Q})} , \tag{9.12}$$

where $\chi_0(\mathbf{Q})$ is the zeroth-order spin susceptibilitzy. It is given in terms of G_{++} and G_{--} by

$$\chi_0(\mathbf{Q}) = T \sum_{\omega_n, \mathbf{k}_\perp} \int dx' \exp[-iQ_x(x-x')]$$
$$\times G_{++}(i\omega_n, \mathbf{k}_\perp; x, x') G_{--}(i\omega_n, \mathbf{k}_\perp - \mathbf{Q}_\perp; x', x) , \tag{9.13}$$

where Q_x and $\mathbf{Q}_\perp$ are the x- and perpendicular components of $\mathbf{Q}$. Using g_{++} and g_{--}, one gets

$$\chi_0(\mathbf{Q}) = \frac{1}{2}N(0) \int_0^{2\pi} \frac{du}{2\pi} \int_d^\infty \frac{dx}{x_T \sinh(x/x_T)} \left[e^{iq_\| x} \right.$$
$$\times \exp\left[i \frac{T_\perp(u-\kappa x) + T_\perp(u-bQ_y-\kappa x) - T_\perp(u) - T_\perp(u-bQ_y)}{\hbar\kappa v_F} \right]$$
$$\left. + \exp\{x \to -x\} \right] , \tag{9.14}$$

where $\exp\{x \to -x\}$ means a term similar to the preceding one with x replaced by $-x$; $N(0) = Na_s/2\pi\hbar v_F$ is the state density, with a_s being the molecular spacing in the a-direction, $x_T = \hbar v_F/2\pi k_B T$ with T being the temperature and k_B the Boltzmann constant,

$$q_\| = Q_x - 2k_F , \quad \kappa = \frac{beH}{\hbar c} , \tag{9.15, 16}$$

$$T_\perp(u) = \int_0^u \epsilon_\perp(u'/b, k_z) du' . \tag{9.17}$$

The lower limit d of the integral over x represents the cutoff defined below. In the present case of $q_\parallel = 0$ the above integral is reduced to

$$\chi_0(\mathbf{Q}_0) = N(0) \int_d^\infty J_0\left[\left(\frac{4t'_{b*}}{\hbar\kappa v_F}\right)\sin(\kappa x)\right] J_0\left(\frac{4t'_{c*}x}{\hbar v_F}\right) \frac{dx}{x_T \sinh(x/x_T)} , \qquad (9.18)$$

where J_0 is the Bessel function. The cutoff d is determined from the divergence of (9.12), or

$$N(0)U \cdot \ln \frac{\hbar v_F}{\pi k_B T_0 d} = 1 , \qquad (9.19)$$

for $t'_{b*} = t'_{c*} = 0$ in the absence of a field, with T_0 being the SDW transition temperature for this condition. If $|t'_{c*}| \ll |t'_{b*}|$, one can approximate $J_0(4t'_{c*}x/\hbar v_F) \simeq 1$ in (9.18).

When $H = 0$, the other Bessel function in (9.18) is reduced to $J_0(4t'_{b*}x/\hbar v_F)$, which becomes smaller as x increases, and lets disappear the $\ln(1/T)$ singularity of the integral. For a finite value of H, the argument of this Bessel function in (9.18) vanishes periodically as a function of x so that the $\ln(1/T)$ singularity is restored. Averaging over x and using the relation

$$\frac{1}{\pi} \int_0^\pi J_0(z\sin\phi)d\phi = J_0^2(\tfrac{1}{2}z) , \qquad (9.20)$$

one can reduce (9.18) to

$$\chi_0(\mathbf{Q}_0) \simeq N(0)J_0^2\left[\frac{2t'_{b*}}{\hbar\kappa v_F}\right]\ln\left(\frac{\hbar v_F}{\pi k_B Td}\right) , \qquad (9.21)$$

where $\mathbf{Q}_0$ is the optimal wave vector $(2k_F, \pi/b, \pi/c)$. From the condition of the divergence of (9.12), or $1 - U\chi_0(\mathbf{Q}_0) = 0$, one can get the finite SDW transition temperature T_{SDW} except for the values of H which make J_0 vanish. The H dependence of T_{SDW} is sketched in Fig.9.12. This result means that the SDW ordering is restored in a field.

In a magnetic field the electron motion in the b-direction is, in the semiclassical picture, restricted to a narrow region of width on the order of $(c\hbar/eH)(t_b/\hbar v_F)$, as shown in Fig.2.14. This one-dimensionalization of the electron motion is interpreted as being the reason for the restoration of the SDW instability characteristic of the one-dimensional electron gas.

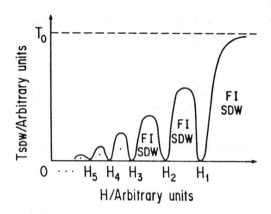

Fig. 9.12. Infinite series of phase transitions occurring in the presence of the magnetic field H [9.24]

This viewpoint was also taken by *Chaikin* [9.27]. He proposed that the transition temperature T_{SDW} variation as a function of field H is described by

$$T_{SDW} = T_0 \exp(-H_A/H) \tag{9.22}$$

with constants T_0 and H_A. This gives a fair fitting to the observed $T_{SDW}-H$ data.

9.2.2 FI-SDW with a Shifted Wave Vector

The Spin-Density-Wave (SDW) that occurs in the boundary-parameter region is of the transient type having an SDW wave vector $\mathbf{Q}$ slightly shifted from the optimum wave vector $\mathbf{Q}_0$, as discussed in Sect.4.5.1. Then, in a FI-SDW restored by an applied field it is also plausible to have a shifted wave vector. The Orsay group [9.28-31] tested this idea by calculating the spin susceptibility $\chi(\mathbf{Q})$ given in (9.12) for the shifted wave vector $\mathbf{Q} = (2k_F + q_{\parallel}, Q_{\perp})$. Assuming, for simplicity, $t'_{c*} = 0$, we can reduce the susceptibility $\chi_0(\mathbf{Q})$ of the unperturbed system from (9.14) to be

$$\chi_0(\mathbf{Q}) = N(0) \int_{\kappa d/2}^{\infty} \frac{dy}{r \sinh(y/r)}$$

$$\times \int_0^{2\pi} \frac{du}{2\pi} \cos(2\nu y - 2z \sin y \cos u - 2z' \sin 2y \cos 2u) , \tag{9.23}$$

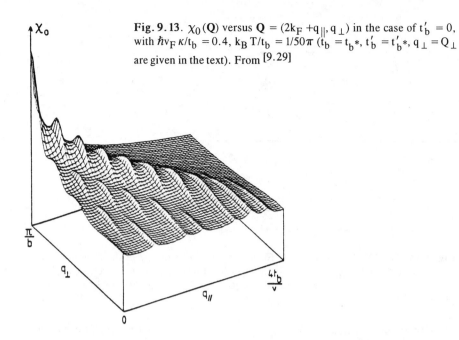

Fig. 9.13. $\chi_0(\mathbf{Q})$ versus $\mathbf{Q} = (2k_F + q_\parallel, q_\perp)$ in the case of $t'_b = 0$, with $\hbar v_F \kappa/t_b = 0.4$, $k_B T/t_b = 1/50\pi$ ($t_b = t_{b}*$, $t'_b = t'_{b}*$, $q_\perp = Q_\perp$ are given in the text). From [9.29]

where $\nu = q_\parallel/\kappa$, $z = (4t_{b*}/\hbar\kappa v_F)\cos(bQ_\perp/2)$, $z' = (2t'_{b*}/\hbar\kappa v_F)\cos(bQ_\perp)$ and $r = \hbar\kappa v_F/4\pi k_B T$; κ has been defined by (9.16). When $\sin(y)$ and $\sin(2y)$ in the argument of the cosine function vanish for $y = \pi \times$integer and if $2\nu y = 2\pi \times$integer, the cosine factor under the integer sign becomes unity. Therefore, when ν is an integer or

$$q_\parallel = Q_x - 2k_F = \kappa \times \text{integer} , \tag{9.24}$$

the integration with respect to y picks up positive contributions and becomes proportional to $\int_{\kappa d/2}^{\infty} dy/[r \cdot \sin(y/r)] \propto \ln(2r/\kappa d) \propto \ln(1/T)$, i.e., it recovers the $\ln T$ singularity.

When we set $t'_{b*} = 0$, (9.23) reduces to

$$\chi_0(\mathbf{Q}) = N(0) \int_{\kappa d/2}^{\infty} \frac{dy}{r\sinh(y/r)} \cos(2\nu y) J_0(2z\sin y) . \tag{9.25}$$

The result of numerically integrating this function is depicted in Fig. 9.13. As expected from the above discussion, $\chi_0(\mathbf{Q})$ shows local maxima when $q_\parallel = \kappa \cdot$integer. The absolute maximum is always found at $\mathbf{Q} = \mathbf{Q}_0 = (2k_F, \pi/b)$. The peaks are located along a curve drawn in the $q_\parallel$ vs $Q_\perp$ plane that satisfies the following geometrical condition: if one moves a piece of

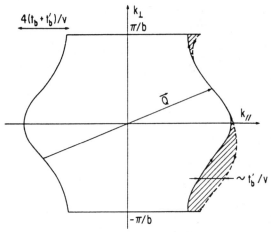

Fig. 9.14. The SDW wave vector **Q** for which the left-hand side of the Fermi surface is translated so that it becomes tangent to the right-hand side. In a magnetic field, $\chi_0(\mathbf{Q})$ has a local maximum when the area of the hatched part is quantized in terms of $2\pi eH/c\hbar$. From [9.29]

the Fermi surface near $k_x = -k_F$ by **Q**, the two pieces of the Fermi surface touch tangentially (Fig. 9.14).

When we take a finite value of t'_{b*}, the situation changes drastically. As in the preceding case, local maxima are located at $q_{\parallel} = n\kappa$ but the absolute maximum is now displaced to a finite value of $q_{\parallel} = n_0\kappa$ (Fig. 9.15 with $n_0 = 2$). With increasing H, the value of n_0 which specifies the position of the maximum decreases and finally tends to zero (Fig. 9.16). Each maximum is logarithmically divergent at low temperatures. Therefore, for any value of H, $\chi(\mathbf{Q})$ given by (9.12) becomes divergent at low temperatures with n equal to a finite integer; only for a very high field we have $n_0 = 0$. Figure 9.16 also suggests that with increasing H the value of n_0 for the most stable state changes in sequential steps.

Incidentally, in the absence of a field $\chi(\mathbf{Q})$ is also obtained from (9.23) as its limit as $H \to 0$. The result is illustrated in Fig. 9.17. The position of the absolute maximum of $\chi_0(\mathbf{Q})$ at $\mathbf{Q}_1$ is slightly shifted from $\mathbf{Q}_0$ at absolute zero. This reveals that in the absence of the field, the SDW instability, occurring because of the divergence of $\chi(\mathbf{Q})$, starts at $\mathbf{Q}_1$. This means that at temperatures near the SDW transition, SDW **Q** may be shifted from the optimal $\mathbf{Q}_0$ as in the case of the transient type (Sect. 4.5.1).

Now when we introduce a finite value for t'_{c*}, the logarithmic divergence of each maximum is arrested. Thus, a finite value of H above the threshold H_{th} becomes necessary for $\chi(\mathbf{Q})$ to be divergent. The scattering of electrons by defects and impurities is considered to raise the value of H_{th}.

$(DMtTSF)_2 X$ salts have neither a superconducting [9.32] nor a FI-SDW phase [9.33, 34], but they remain metallic down to very low temperatures,

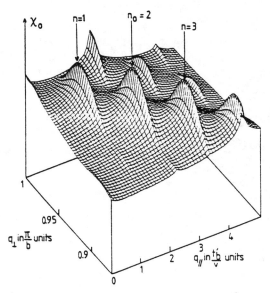

Fig. 9.15. $\chi_0(\mathbf{Q})$ versus $\mathbf{Q}$ in the case where $t'_b/t_b = 0.1$, $\hbar v_F \kappa/t'_b = 1.158$, $k_B T/t'_b = 1/40\pi$. The main series of peaks are labeled by the quantum number n; the absolute maximum is at $n = n_0 = 2$. From [9.29]

and have a band structure similar to that of $(TMTSF)_2X$. Although the asymmetric dimethyltrimethylene-tetraselenafulvalene molecules are crystallographically well ordered in the same way as DMET, the observed negative magnetoresistance when X is BF_4, ClO_4 and ReO_4 [9.33, 34] suggests that weak disorder may remain in the DMtTSF arrangement since the de-

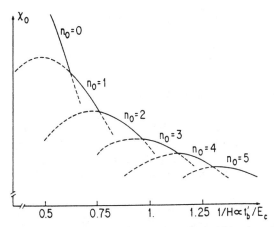

Fig. 9.16. Absolute maximum of $\chi_0(\mathbf{Q})$ as a function of 1/H in units of $t'_b/\hbar v_F \kappa$, shown as a succession of segments with an increasing quantum number n_0 ($t'_b/t_b = 0.1$, $k_B T/t'_b = 1/40\pi$). From [9.29]

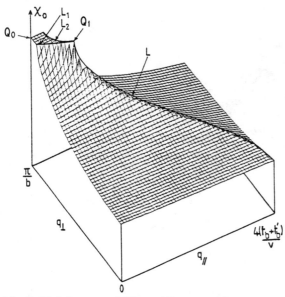

Fig. 9.17. Spin susceptibility $\chi_0(\mathbf{Q})$ versus $\mathbf{Q}$ in the absence of a magnetic field at $T = 0\,\mathrm{K}$, $t_b/t_a = t_b'/t_b = 0.1$. The edge curves L_1, L_2, and L correspond to $\mathbf{Q}$ defined by the point at which the two sections of the Fermi surface are tangent as in Fig. 9.14. From [9.30]

gree of asymmetry of DMtTSF is weaker than in DMET and MDT-TTF. Therefore, a likely possibility is that such disorder has raised H_{th} inaccessibly and suppressed the superconductivity as well. In organics superconductivity is very quickly suppressed by disorders, as described in Sects. 3.2.4 and 5.2.7. Both phenomena are not yet fully understood theoretically.

9.3 Energetics of FI-SDW

9.3.1 Energy Gain of the SDW State in Magnetic Fields

Reasonings on the origin of the energy gain of the SDW state in applied magnetic fields provide a more intuitive picture of the behavior in this state. This is possible because the one-particle eigenvalues in the magnetic field, derived from full quantum-mechanical calculations, can be seen to well correspond to semiclassical counterparts.

The band model in (4.52) taking account of the multiple transverse transfer energies describes the real mechanism of denesting. Instead, however, we use the simple t_a-t_b model in this section. This is because it was originally studied extensively and provides the identical feature of FI-SDW.

As is discussed in Sect.4.4.1, in the context of SDW with the optimum wave vector Q_0 in the absence of fields, the energy gain $E_n - E_{SDW}$ of the SDW phase decreases as t_b increases (Fig.4.9). For $t_b > t_{b,cr}$ SDW with Q_0 is lost. In a narrow range of $t_b \geq t_{b,cr}$, a transient type SDW, the energy of which is depicted by the dashed curve, appears. Above this region there is no SDW phase which has an appreciable energy gain or an appreciable value of T_{SDW}. In such a parameter region, for SDW to be restored by an applied magnetic field, the field must provide an energy gain to the SDW state. The existence of such an extra energy gain is indicated by the enhancement of the magnetic susceptibility χ of $(TMTSF)_2AsF_6$ in the SDW state for the field parallel to the c^*-axis (Fig.3.13) [9.35]. This figure provides evidence that SDW has its spin polarized in the b'-direction in the insulator state of $(TMTSF)_2X$. The increase of χ_{c*} for $T < 12$ K is unexpected, since in the normal antiferromagnetic state this must be constant [9.36]. The increase in χ at lower temperatures leads to an energy gain over the total energy of the normal state (of $\frac{1}{2}\Delta\chi H^2$). By the definition of χ the energy change in the applied magnetic field is given by $-\frac{1}{2}\chi H^2$. Judging from the condensation energy of the SDW state, the estimate of $\frac{1}{2}\Delta\chi H^2$ is in the right order of magnitude.

From the angle dependence $1/\cos\theta$ of the field for a stepwise change of the magetoresistance and other properties, the energy gain should be closely related to the orbital motion in the ab-plane. In fact, closed orbits in the SDW states do occur and their quantization with respect to the magnetic field yields the energy gain [9.37-40]. First, we consider the orbital quantum levels semiclassically [9.37, 38] and evaluate the change of the total energy due to orbital quantization. We restrict ourselves to the case of the so-called **optimal SDW wave vector** $Q_0 = (2k_F, \pi/b)$. The value of the gap parameter M is fixed to the value obtained at absolute zero for $t_b < t_{b,cr}$ in the band model of (4.26). We use the set of parameter values employed previously [9.41], i.e., $t_a = 3690k_B$, $M = 20.46k_B$, $t_{b,cr} = 325.59k_B$ and $t_b = 326k_B$. With these values for M and Q, the top of the lower band and the bottom of the upper band, which are separated by the SDW gap, are depicted as functions of k_y in Fig.9.18 for the case of H = 0. Figure 9.19 represents the state density corresponding to such a state. The lower band is fully occupied and the upper one empty.

When the magnetic field is applied along the z-axis, the electron moves along an equi-energetic curve in the $k_x k_y$-plane, whose projection onto the Ek_y-plane is a horizontal line (Fig.9.18). When this line is cut by the extremum curve showing the top of the lower band, the equi-energetic curve must be a closed orbit such as the orbits labeled a and b in Figs.9.18 and 20. When this line is not cut, then it corresponds to an open orbit such as the orbit labeled e.

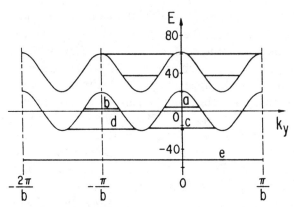

Fig.9.18. Projections of the semiclassical, closed orbits a, b, c, and d onto the E-k_y plane for H = 9.513 Tesla. The undulating curves represent the top of the lower band and the bottom of the upper band on either side of the SDW gap. The origin of the ordinate is arbitrary. For convenience, the figure is extended beyond of the Brillouin zone $|k_y| \leq \pi/b$. From [9.38]

According to the semiclassical theory, the area S_k of the closed orbits in the k-space must satisfy the quantization condition:

$$S_k = S_0 (n + \tfrac{1}{2}) \quad \text{with} \quad n = 0,1,2,\dots , \tag{9.26}$$

where $S_0 = 2\pi eH/c\hbar$ [9.2]. This comes from the Bohr-Sommerfeld quantization condition $\int pdq = \hbar(n+const)$ and leads to quantized energy levels. The degeneracy of each level corresponds to the area in the phase space S_0 as follows:

$$\zeta = \frac{NabS_0}{(2\pi)^2} = \frac{NabeH}{2\pi c\hbar} , \tag{9.27}$$

where N is the number density of the electronic sites. Examples of quantized levels are illustrated in Fig.9.18 for H = 9.513 Tesla, i.e., $\mu_B H = 6.4k_B$. This discretization into quantized levels drastically changes the state density in the energy region where there are closed orbits. On the other hand, in regions of open orbits such as orbit e, there is practically no change in the state density. Since the total number of states in the lower bands must be the same in both absence and presence of a field, the boundary between the closed-orbit and open-orbit regions employed in calculating the total energy can be defined by the orbit satisfying $S_k = nS_0$; here n is the number of the quantized closed orbits in the lower band (Fig.9.18), S_k is the area enclosed by two equi-energetic open orbits (orbit e in Fig.9.20) and the Brillouin-zone boundary.

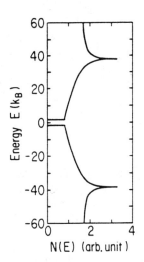

Fig. 9.19. State density N(E) versus energy E in the SDW state with $M = 20\,k_B$. From [9.38]

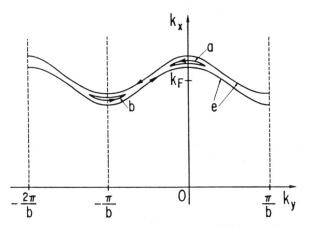

Fig. 9.20. Examples of closed and open orbits in the k_x-k_y space. The orbits a, b, and e correspond to the levels a, b, and e in Fig. 9.18, respectively. The figure is extended beyond the Brillouin zone in the k_y-direction. From [9.38]

Discretization of the eigenenergies in the range of closed orbits (Fig. 9.19) lowers $E_{SDW} - E_n$. The state density N(E) of the lower band for one k_x-branch per spin in the absence of the field is shown in Fig. 9.21. The semiclassical quantum level ϵ_q is defined by (9.26), which can be rewritten as

$$\int_{\epsilon_q}^{\epsilon_c} N(E)\,dE = (n + \tfrac{1}{2})\zeta \quad \text{with} \quad n = 0,1,2,\dots, \tag{9.28}$$

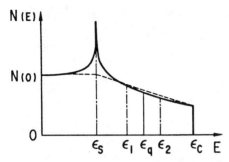

Fig. 9.21. The state density N(E) as a function of energy E for the SDW-reorganized lower band in the absence of a field. N(0) is the state density per spin per one k_x-branch in the normal state. This illustrates that all the states between the energies ϵ_1 and ϵ_2 converge to the level ϵ_q under the field, therby decreasing the energy. See text. From [9.38]

where ϵ_c is the upper bound of the lower band, and ζ is the degeneracy factor defined by (9.27). In a magnetic field all states between the levels ϵ_1 and ϵ_2 defined by

$$\int_{\epsilon_1}^{\epsilon_c} N(E)\,DE = (n+1)\zeta \quad \text{and} \quad \int_{\epsilon_2}^{\epsilon_c} N(E)\,DE = n\zeta \qquad (9.29)$$

are concentrated at the level ϵ_q. The energy change resulting from this re-arrangement between ϵ_1 and ϵ_2 is given by

$$\int_{\epsilon_1}^{\epsilon_2} dE(\epsilon_q - E)N(E) \simeq \frac{1}{24}(\epsilon_2 - \epsilon_1)^3 N'(\epsilon_q) < 0 \; ; \qquad (9.30)$$

the inequality results from the inequality $N'(\epsilon_q) < 0$ in the energy region between ϵ_s and ϵ_c defined in Fig. 9.21. If $N'(\epsilon_q) > 0$, we obtain an increase of the total energy as in the case of the free-electron gas, which gives the Landau diamagnetism [9.2]. This clearly demonstrates that the orbital quantization in the closed-orbit region of the lower band contributes to the energy gain of the SDW state.

In Fig. 9.21 the dotted curve illustrates an approximation for N(E) with $N(\epsilon_c) \simeq N(0)/2$. It is valid for $t_b \approx t_{b,cr}$. We assume that the integral of N(E) between ϵ_s and ϵ_c is just equal to ζ times an integer. We then get the following estimate by summing the above quantity:

$$E_{SDW} - E_n \simeq - \frac{\xi^2}{6N(0)} = -\frac{1}{2}\chi^{(0)}_{Pauli}\frac{H^2}{3}[2ma_s\,bt_a\sin(a_sk_F)]^2 \quad (9.31)$$

with

$$\chi^{(0)}_{Pauli} = 4N(0)\mu_B{}^2\,, \tag{9.32}$$

where m is the mass of the bare electron, and a_s the spacing between neighboring molecules in the TMTSF stack. The last factor on the right-hand expression in (9.31) turns out to be equal to 0.80 for our set of parameters. The difference (9.31) is plotted in Fig. 9.22. This figure also displays the result of a more elaborate semiclassical calculation [9.38], to which (9.31) is a good approximation. In the following subsection, we will find that these curves are in fair agreement with the result of the fully quantum calculation [9.38].

9.3.2 Full Quantum-Mechanical Calculation

Our band model is that given by (4.26) as in the preceding subsection. For convenience, we again treat holes as particles. The model is based on the two-dimensional tight-binding scheme for a rectangular lattice with the basic periodic vectors **a** and **b**. The transfer energy t_a between the nearest-neighbor electronic sites along the a-axis is about ten times larger than t_b

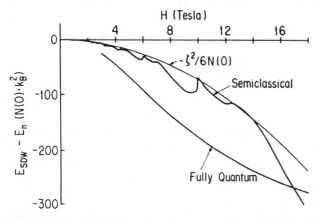

Fig. 9.22. Semiclassical and quantum-mechanical energy difference $E_{SDW} - E_n$ as a function of the magentic field H. The *thin curve* is the $-\xi^2/6N(0)$ approximation to the semiclassical result. The parameters are $t_a = 318$ meV, $t_b = 326\,k_B$ and $T_{SDW}{}^{(0)} = 11.5$ K; in this case $t_{b,cr} = 325.6\,k_B$, where $E_{SDW} - E_n = 0$ for $H = 0$. From [9.38]

along the b-axis. The x- and y-coordinates are set along the a- and b-axes, respectively. The magnetic field H is applied in the z-direction normal to the conducting sheet. We choose the Landau gauge, i.e., $\mathbf{A} = (0, Hx, 0)$. The on-site Coulomb interaction drives SDW in our system. Assuming that SDW in the magnetic field has the gap parameter M, the wave vector $\mathbf{Q}$, and a polarization pointing in the x-direction, we obtain the following effective Hamiltonian:

$$
\mathscr{H}_{eff} = \sum_{\mathbf{k},\sigma} c_{\mathbf{k}\sigma}^{\dagger} \left[-2t_{a}(\cos ak_{x} - \cos ak_{F}) - 2t_{b}\cos\left(bk_{y} - i\frac{beH}{c\hbar}\frac{\partial}{\partial k_{x}}\right) \right.
$$

$$
\left. - \mu_{B}H\sigma - \mu \right] c_{\mathbf{k}\sigma} + \sum_{\mathbf{k}} [M(c_{\mathbf{k}\uparrow}^{\dagger}c_{\mathbf{k}-\mathbf{Q}\downarrow} + c_{\mathbf{k}-\mathbf{Q}\uparrow}^{\dagger}c_{\mathbf{k}\downarrow}) + \text{H.c.}] , \quad (9.33)
$$

with c and e being the light velocity and the absolute value of the electronic charge, respectively; $\partial/\partial k_{x}$ operates on the subscript of the annihilation operator $c_{\mathbf{k}\sigma}$ and gives the effect of the magnetic field on the orbital motion. The other notation is as usual, with H.c. denoting the Hermite conjugate. The first term is a straightforward rewriting of the result of *Peierls* [9.25] for the tight-binding band in the magnetic field applied to the second-quantization scheme.

Following the derivation of the variational free energy and the gap equation in [9.42], we finally obtain the following variational total energy of the SDW state in a field at absolute-zero temperature as a function of M and $\mathbf{Q}$:

$$
E_{SDW} = \sum_{i} (E_{i} + \mu)f(E_{i}) + \frac{2M^{2}}{I} , \quad (9.34)
$$

where E_{i} is the eigenvalue of $\mathscr{H}_{eff}$, $I = U/N$ the coefficient of the Coulomb interaction; $f(E_{i})$ is the distribution function which is unity when E_{i} is less than zero and zero otherwise; the chemical potential μ is always adjusted so that the total electron number is constant, i.e., in the present case equal to half the number of the electronic sites N.

In the course of the derivation of (9.34) we took the expectation value of the Coulomb interaction term for the ground state of $\mathscr{H}_{eff}$. For this we assumed that the terms such as $\langle c_{\mathbf{k}\sigma}^{\dagger}c_{\mathbf{k}-\mathbf{Q}\bar{\sigma}}\rangle \langle c_{\mathbf{k}'\bar{\sigma}}^{\dagger}c_{\mathbf{k}'+\mathbf{Q}\sigma}\rangle$, $\bar{\sigma}$ denoting the inverse of σ, give the main contribution. We neglected terms containing the expectation values of two operators whose wave-vector difference is not equal to **Q**. The resulting variational energy is highly reasonable, since its variation with respect to M gives the same SDW gap equation that we obtain if we start from the definition of the gap parameter by

$$M = -I \sum_{\mathbf{k}} \langle c^{\dagger}_{\mathbf{k}-\mathbf{Q}\bar{\sigma}} c_{\mathbf{k}\sigma} \rangle = -I \sum_{\mathbf{k}} \langle c^{\dagger}_{\mathbf{k}\sigma} c_{\mathbf{k}-\mathbf{Q}\bar{\sigma}} \rangle, \quad \sigma = \uparrow \text{ and } \downarrow, \quad (9.35)$$

and look for the self-consistency equation by calculating the expectation values $\langle \cdots \rangle$ for the average Hamiltonian.

Since we have

$$2\cos b \left[k_y - i \frac{eH}{c\hbar} \frac{\partial}{\partial k_x} \right] c_{k_x, k_y, \sigma}$$

$$= \exp(ibk_y) c_{k_x + \kappa, k_y, \sigma} + \exp(-ibk_y) c_{k_x - \kappa, k_y \sigma}, \quad (9.36)$$

where κ is defined by (9.16), the operator $c_{k_x, k_y, \sigma}$ is coupled in the eigenvalue equation to $c_{k_x + n\kappa, k_y, \sigma}$ and $c_{k_x \pm Q_x + n\kappa, k_y \pm Q_y, \bar{\sigma}}$, where n is the integer, and Q_x and Q_y are the components of **Q**. We take account of the possibility that Q_y shifts from the optimal value $Q_y^{(0)}$, as first noticed in [9.41]. Then, the eigenvalue problem is reduced to that of a matrix composed of the coefficients of these operators, which can be rewritten as

$$\begin{pmatrix} \bullet & \bullet & \bullet & \bullet & \bullet & & & & \\ & \bullet & \bullet & \bullet & \bullet & \bullet & & & \\ & & C & & A_n & M & C & & \\ & & D & M & B_n & & D^* & & \\ & & & C & & A_{n-1} & M & C & \\ & & & D & M & B_{n-1} & & D^* & \\ & & & & \bullet & \bullet & \bullet & \bullet & \bullet \\ & & & & & \bullet & \bullet & \bullet & \bullet & \bullet \end{pmatrix} \quad (9.37)$$

where with $x = ak_x$

$$A_n = -2t_a[\cos(x+n\delta)-\cos(ak_F)] - \mu_B H\sigma - \mu$$

$$- \frac{M^2}{2t_a[\cos(x+n\delta)-\cos(x+aQ_x+n\delta)] + 2t_b[1-\cos(aQ_y)] + 2\mu_B H\sigma},$$

$$B_n = 2t_a[\cos(x-aQ_x+n\delta) - \cos(ak_F)] + \mu_B H\sigma - \mu$$

$$- \frac{M^2}{2t_a[\cos(x-aQ_x+n\delta)-\cos(x-2aQ_x+n\delta)] + 2t_b[\cos(aQ_y)-\cos(2aQ_y)] - 2\mu_B H\sigma},$$

$$C = -t_b, \quad D = -t_b e^{-ibQ_y}. \tag{9.38}$$

The above matrix is of infinite dimension with $-\infty < n < +\infty$. The k_y-dependence disappears after a unitary transformation. So does the phase attached to the gap parameter M, even if we give it a non-zero value. The eigenvalues are independent of the direction of the SDW polarization. To suppress the off-diagonal elements connecting the wave-number region, namely $-2k_F < k_x < 2k_F$ to the outer region, we have brought in the correction terms of the form $M^2/(\cdots)$. The eigenvalues of (9.37) around the Fermi energy were obtained by diagonalizing the matrix truncated to a suitable restricted size. With $x = ak_x \approx ak_F$ and $-N_t \leq n \leq N_t$, N_t being a sufficiently large integer, it was found that the eigenvalues in the energy region of closed orbits become independent of N_t. They are bunched together in narrow bands which very closely correspond to semiclassical levels. When H = 10 T, N_t = 50 is sufficient, resulting in a 202×202 matrix. As H is reduced, N_t must inverse-proportionally be increased. The parameters $|A_n|$ and $|B_n|$ with $n = \pm N_t$ have to be larger than $4t_b$.

The eigenvalues are periodic functions of k_x with the period κ defined by (9.16) and are independent of k_y. For the energy region of the lower band the eigenvalues are illustrated in Fig.9.23 as functions of $\Delta k_x = \mathrm{mod}(k_x-k_F,\kappa)$. We see that they make narrow bands (**Landau bands**) corresponding to the semiclassical quantized levels. The total accommodation number of each band is equal to the semiclassical degeneracy factor ζ defined by (9.27). The bandwidth increases as the energy level comes near the open-orbit energy region. It is remarkable that even in the open-orbit region there is an appreciable band gap near the closed-orbit region, although it decreases rapidly as the gap moves away. The position of the narrow gap is in good agreement with the value determined from the condition $S_k = nS_0$ with n being the number of the Landau bands formed above the gap in the lower band. S_k is the area of the open orbit defined in Sect.9.3.1.

In the energy region below a narrow gap, the state density is not appreciably modified by the effect of the field on the orbital motion. Therefore, we can neglect the field effect in the lower region and use the correct eigen-

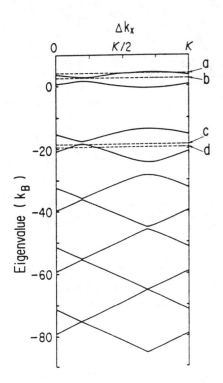

Fig. 9.23. Quantum eigenvalues of H_{eff} as functions of $\Delta k_x = \text{mod}(k_x - k_F, \kappa)$ in one period $0 \leq \Delta k_x \leq \kappa$ for the field $H = 9.513$ Tesla. The *dashed lines a, b, c,* and *d* are the semiclassical levels illustrated in Fig. 9.18. The chemical potential μ is added to the energy. From [9.38]

values above an appropriate narrow gap to compute $E_{SDW} - E_n$ by using (9.34), and obtain the fully quantum-mechanical curve in Fig. 9.22. In the region between 2 and 16 Tesla the resulting $E_{SDW} - E_n$ is twice lower than the semiclassical result. This is due to a small shift of the fully quantum-mechanical eigenvalues from the semiclassical levels. The oscillation observed on the semiclassical curve in Fig. 9.22 now turns out spurious.

Since we have kept the variational variables fixed in this calculation, the result for $E_{SDW} - E_n$ shown in Fig. 9.22 is the upper bound. Therefore, we can conclude that when the system has an SDW order in a magnetic field, it gains an energy at least on the order of $4N(0)(\mu_B H)^2$.

In the data of *Mortensen* et al. [9.35], the c^*-component of the susceptibility χ_{c^*} in the SDW state increases with lowering the temperature by about 40 % of the normal-state Pauli susceptibility. Our result demonstrates an increase by 200 % of the nonenhanced Pauli susceptibility. When we take account of the enhancement factor and also of the parameter dependence of $E_{SDW} - E_n$, our result is in fair agreement with the experiment. Of course, the decrease of the energy in our calculation depends almost exclusively on the z-component of the field in agreement with the observation. With the energy gain obtained in the magnetic field, the SDW phase becomes lower in energy than the normal state even if $t_b > t_{b,cr}$, as is illustrated in Fig. 9.24. This means that SDW is induced by the applied field.

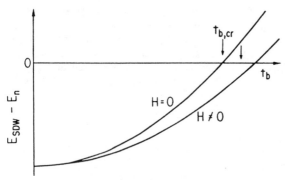

Fig. 9.24. Occurrence of FI-SDW due to a decrease of $E_{SDW} - E_n$ in the applied magnetic field: When the system has the t_b value indicated by the *unlabeled arrow*, SDW appears in the applied field because E_{SDW} is lower than E_n

In mean-field theory [9.36] the enhancement factor even in the SDW state is the same as that in the normal state. In the present calculation, the enhancement is neglected since it involves no essential change in the present scheme.

9.3.3 Successive Phase Transitions Among FI-SDW Subphases

Judging from the results on the transient SDW in Sect. 4.5.1, it is very probable that SDW restored by the energy gain in the magnetic field has a wave vector $\mathbf{Q}$ slightly shifted from the optimum $\mathbf{Q}_0$. With a changing field, $\mathbf{Q}$ and the SDW gap parameter M can adjust so that E_{SDW} always takes on a minimum. Therefore, to find the real solution in our scheme, we have to minimize the energy with respect to $\mathbf{Q}$ and M. In order to facilitate later comparison with experimental results on $(TMTSF)_2 PF_6$, the following parameter values for this salt are used [9.37]:

$$t_a = 245 \text{ meV} = 2843k_B \ , \quad T_{SDW}^{(0)} = 11.5 \text{ K} \ , \quad t_{b,cr} = 285.6k_B \ , \quad (9.39)$$

where again $t_{b,cr}$ is defined as the upper bound of t_b that is allowed by the stability condition for SDW with $\mathbf{Q} = \mathbf{Q}_0$, i.e., (4.44). Here we choose $t_b = 335k_B \gg t_{b,cr}$, since in this case interesting things appear at relatively high fields where less computing time is needed. With fixed $\mathbf{Q}_0$ and M_0, M_0 being the mean-field value at T = 0 K for $t_b < t_{b,cr}$, the energy difference $E_{SDW} - E_n$ is given by the smooth, unlabeled curve in Fig. 9.25a. As H tends to zero, the energy difference assumes quite a large positive value, since $t_b \gg t_{b,cr}$. As H increases, it decreases and finally becomes negative.

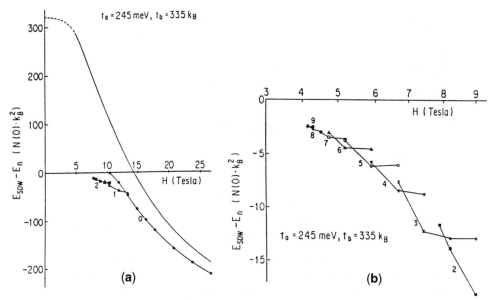

Fig. 9.25a,b. Local minima resulting from minimization of the energy difference $E_{SDW} - E_n$ with respect to the SDW parameters as a function of the magnetic field H for $t_b = 335$ k_B, $t_a = 245$ meV, $T_{SDW}{}^{(0)} = 11.5$ K, and $t_{b,cr} = 285.6$ k_B. The *smooth, unlabeled curve* in (a) is for the fixed parameter values $Q = Q_0$ and $M = M_0$. Other local minimum curves with changed Q and M are labeled by the integers 0, 1, 2, ... They will be shown to denote the number of filled magnetic quantum levels N_{QL} formed in the upper band above the SDW gap. (b) Same as (a), but with expanded scales. From [9.39]

When **Q** and M are varied, many well defined local minima, to be explained below, appear. Each of them is lowered and raised again as the field is swept. These local minima are plotted as functions of the field (the curves in Figs. 9.25a, b labeled by integers). Therefore, the real minimum is given by the envelope of the curves, which switches from one local minimum to another one at fields where the curves cross each other. Such crossovers entail first-order phase transitions repeated successively as the field is changed. This was first unambiguously demonstrated in [9.39].

We define the new variable N_{QL} by

$$Q_x = 2k_F - \kappa N_{QL} , \tag{9.40}$$

with κ given by (9.16). The energy of the system was actually found to have very sharp minima at the integer values of N_{QL}. When N_{QL} is an integer, the Fermi energy lies in the gap, i.e., in a Landau gap between adjacent Landau bands. In this situation the existence of the gap decreases the single-particle energy and leads to energy gain. The whole Brillouin Zone (BZ) below the SDW gap accommodates $2[Nab/(2\pi)^2]Q_x 2\pi/b$ electrons. Since it is

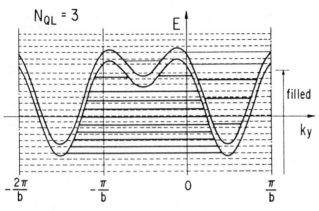

Fig. 9.26. Distribution of the magnetic quantum levels in the subphase with $N_{QL} = 3$. Double-peaked curves as a function of k_y represent the bottom of the upper band and the top of the lower band on either side of the SDW gap. The Brillouin zone is extended into the negative region of k_y to facilitate drawing of the levels of semiclassical quantized closed orbits indicated by the horizontal *straight lines*. The extremum curves are assymmetric about $k_y = 0$ due to the shift of Q_y from π/b. The horizontal *dashed lines* show the average levels of fully quantum magnetic bands or Landau bands. The *thick lines* are semiclassical levels corresponding to filled Landau bands and the *thin lines* are empty. As required for $N_{QL} = 3$, three Landau bands in the upper band are filled ($H = 5\,k_B/\mu_B$, $M = 5.8\,k_B$, $\phi_0 = 2.92°$). From [9.39]

full when $Q_x = 2k_F = \pi/2a$, the shrinkage of Q_x by $N_{QL}\kappa$ results in a replacement of $2\zeta N_{QL} = [2Nab/(2\pi)^2]N_{QL}\kappa(2\pi/b)$ electrons into the upper band; here ζ is the degeneracy factor of the Landau band and the factor 2 comes from the spin degree of freedom. This means that the N_{QL} Landau bands in the upper band above the SDW gap are filled. It is sufficient to know only such cases with integer values of N_{QL}. For each integer value we actually optimize the energy gain with respect to $Q_y = (\pi - 2\phi_0)/b$ and M. The integer labels in Fig. 9.25 correspond to N_{QL} values.

Figure 9.26 depicts the distribution of Magnetic Quantum Levels (MQL) in the state that gives the local minimum (curve *3* in Fig. 9.25b). Fully quantum-mechanical magnetic levels corresponding to closed orbits actually form narrow bands or **Landau bands**. The average level of Landau band is indicated by a dashed line, and solid lines represent now the semiclassical MQLs. The correspondence between the two is quite good. Only when multiple, closely spaced semiclassical levels are almost degenerate, do they deviate from their quantum-mechanical counterparts. Therefore, we can associate each fully quantum-mechanical level, or Landau band, to a semiclassical MQL. We can identify whether each MQL comes mainly from the lower band or the upper band. We see that all the Landau bands that are formed in the lower band are fully occupied. In addition, the Landau bands corresponding to the three semiclassical MQLs formed in the

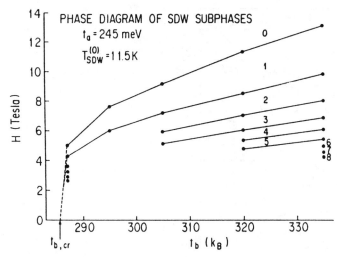

Fig. 9.27. Calculated phase diagram of the SDW subphases in the plane of a magnetic field H versus a weaker transfer energy t_b. The integers denote the values of N_{QL} for the lowest energy subphases in the specified region. The *dashed line* is a guide for the eye, since no crossover was found for $t_b < t_{b,cr}$. From [9.39]

upper band are filled. Therefore, in this situation $N_{QL} = 3$ and, thus, N_{QL} means the number of fully occupied Landau bands formed from the upper band. Since the Fermi energy lies in a broad gap, the one-electron energy is suppressed. This brings in the energy gain for the SDW state in the magnetic field. These features are common to all the other local-minimum states.

When the magnetic field increases, the highest occupied Landau band originating from the upper band is pushed above all the Landau bands coming from the lower band in Fig. 9.26. Then this becomes energetically less favorable than the situation in which this Landau band is empty, i.e., $N_{QL} = 2$. This is why the local minimum curves for $N_{QL} = 2$ and 3 cross over with increasing field. The crossover points are plotted in the H-t_b plane (Fig. 9.27). The integers denote the values of N_{QL} which specify the subphase between two crossover fields. The accompanying changes in Q_x, ϕ_0, and M are shown in Fig. 9.28. In Fig. 9.29 we plot the anisotropic part of the real magnetization in the z-direction that is given by

$$\mathscr{M} = -\frac{\partial E_{SDW}}{\partial H}.$$

(9.41)

In Fig. 9.27 we see successive phase transitions as the field is swept over all t_b values above $t_{b,cr}$. As the field is increased, we finally reach the subphase with $N_{QL} = 0$. In this state we find $\mathbf{Q} = \mathbf{Q}_0$ and the value of M ap-

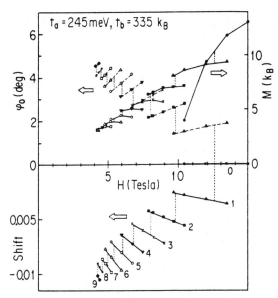

Fig. 9.28. Parameters of the SDW subphases at $t_b = 335\,k_B$ as a function of the field H. *Shift* is defined by $Q_x/2k_F - 1$, ϕ_0 determines Q_y via $Q_y = (\pi - 2\phi_0)/b$. The integers denote the values of N_{QL}. The vertical *dashed lines* display the jumps due to phase transitions. When $N_{QL} = 0$, *Shift* $= 0$ and $\phi_0 = 0$. From [9.39]

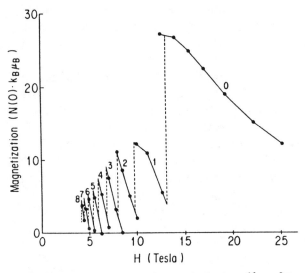

Fig. 9.29. Anisotropic part of the magnetization $\mathcal{M} = -\partial E_{SDW}/\partial H$ given by (9.41) in the z-direction as a function of the field H. The unit $N(0)k_B \mu_B$ of the ordinate is equal to 0.44 erg/Oe/mole. The vertical *dashed lines* depict the jumps due to phase transitions between subphases. From [9.39]

proaches M_0. Therefore, we note that even as $t_b > t_{b,cr}$ the high field stabilizes the simple SDW, as described in Chap.4.

In order to know whether a threshold H_{th} exists in the present model, we calculated curves for large N_{QL} values in lower magnetic fields, but the resulting energy gain became of the same order as the computational errors. Therefore, it seems that in this model the threshold field is also zero, as is the case with the Gor'kov and Lebed' model discussed in Sect.9.2.1.

9.3.4 Comparison of Theory and Experiment

The most direct check of calculations can be made by a comparison with magnetization measurements on $(TMTSF)_2 ClO_4$ [9.16]. Features of the calculated anisotropic magnetization, illustrated in Fig.9.29, are close to those observed in the following situation: with increasing field, the magnetization shows repeatedly an instantaneous increase and a subsequent smooth decrease; the rate of the increase grows larger and larger; the magnetization almost always has the same sign; after the final biggest jump it gradually decreases monotonically. The calculated magnetization is smaller than that observed by almost one order of magnitude. This is closely related to the fact that $|E_{SDW} - E_n|$ determined for $H = 15$ Tesla and $t_b = 335k_B$ is about 1/7 of the energy decrease obtained by integrating the observed magnetization using (9.41). The smallness of the calculated value is mainly due to the fact that $E_{SDW} - E_n$ is largely biased in the positive direction, since $t_b = 335k_B \gg t_{b,cr} = 285.6k_B$, as is seen in Fig.9.25a. Actually, in the case of $t_b = 287k_B$ close to $t_{b,cr}$ the energy gain at $H = 15$ Tesla is 1/3 of the observed value. Recent experiments [9.43] reveal that the magnetization becomes negative at high fields. This discrepancy will be discussed in Sect.9.5.1.

In the SDW subphases deduced in Sect.9.3.3, only electrons occupying the Landau bands that are formed from the upper band, are mobile because the filled Brillouin zone cannot contribute to transport properties. Therefore, the carrier density n is given by [9.28, 39]

$$n = 2 \zeta N_{QL} = \frac{1}{\widetilde{c}} \frac{eHN_{QL}}{\pi c} , \qquad (9.42)$$

where ζ is the degeneracy factor of the Landau level, $\widetilde{c}$ is the lattice constant in the z-direction, and the factor 2 comes from the spin degeneracy. Although the up-spin and down-spin pockets appear in different k-space locations as do the a- and b-orbits in Fig.9.20, they are always in pairs.

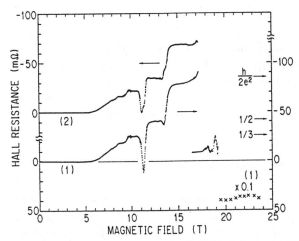

Fig. 9.30. The Hall resistance of $(TMTSF)_2 PF_6$ in the FI-SDW state under hydrostatic pressure at 0.5 K is plotted against the magnetic field along c*for two pairs (1), (2) of opposite contacts on a single crystal. The lower part is the Hall resistance of pair (1) on a reduced scale. The quantized values $h/2ne^2$ (12.9/n kΩ) per molecular layer are marked on the right-hand side for the pair (1). From [9.44]

Then, the Hall voltage V_H for the current J in the a-direction is plausibly given by

$$\frac{V_H}{J} = \frac{H}{nec} \frac{1}{L_z} = \frac{h}{2e^2 N_{QL}} \frac{\widetilde{c}}{L_z}, \qquad (9.43)$$

where L_z is the size of the sample in the z-direction, and h is Planck's constant. The sign of the current carriers is chosen to be positive, since they are actually holes in the normal state of $(TMTSF)_2 X$. If we assume that the mobility of the carriers in the a-direction is proportional to H^2, we find the resistance in this direction as

$$R(H) \simeq \frac{H^2}{n} \approx \frac{H}{N_{QL}}. \qquad (9.44)$$

These quantities depend on the magnetic field, as depicted in Fig. 9.30. The step-wise change of the Hall voltage V_H is well reproduced [9.12-14]. The heights of these steps for $(TMTSF)_2 ClO_4$, however, do not yield the ratio expected from (9.44). Later in Sects. 9.5.1, 2 we shall see that this is due to the superlattice potential of the orientationally ordered ClO_4 anions. The Hall voltage which results for $(TMTSF)_2 PF_6$ in the FI-SDW state under pressure, was found to produce the ideal voltage steps, given by (9.44), to

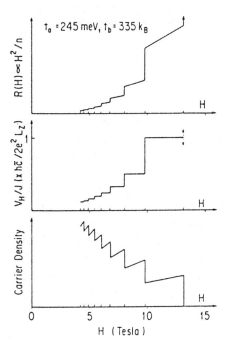

$t_a = 245\,meV$, $t_b = 335\,k_B$

$R(H) \propto H^2/n$

$V_H/J \,(\times \hbar\bar{c}/2e^2\,L_z)$

Carrier Density

H

H

H

0   5   10   15

H (Tesla)

Fig. 9.31. The magnetic-field dependence at $t_b = 335\,k_B$ of the carrier density, the absolute value of the Hall resistance V_H/J, and the magnetoresistance R(H). The *vertical marks* on the *abscissa* denote the phase transition points. From [9.39]

our satisfaction (Fig. 9.30) [9.43, 44]. The abrupt change of the sign of V_H for high fields [9.15, 46, 47], as seen in Figs. 9.8, 30 is of a different nature. The high-field features will be discussed in Sect. 9.5.1-4.

The behavior of the calculated magnetoresistance (Fig. 9.31) is also close to that observed [9.1, 5, 46]. The hysteresis of the resistivity of both perchlorate [9.5] and $(TMTSF)_2 PF_6$ under a 7-kbar pressure [9.46] can be interpreted on the basis of the first-order nature of the transition between the SDW subphases.

Finally, we examine the periodicity of the transitions. When one examines the inverse of the fields at which crossovers between phases occur, then the separation between them is nearly constant, similarly to the case of the Shubnikov-de Haas oscillation. The present results should be compared with data on $(TMTSF)_2 PF_6$, which is not distorted by anion ordering as in $(TMTSF)_2 ClO_4$. The calculated period $\Delta(1/H)$ for $t_b = 335 k_B$ is 1.45 times the period observed for $(TMTSF)_2 PF_6$ [9.1] and increases when t_b approaches $t_{b,cr}$. It becomes 2.85 times the period observed at $t_b = 287 k_B \lesssim t_{b,cr}$, but it is close to the period observed for $(TMTSF)_2 ClO_4$ [9.4, 5, 16]. It should be remarked that this calculated period is not uniquely determined by the area of the carrier pocket, which forms in the upper band. It is on the order of 0.3 % of the Brillouin zone S_{BZ} in the $k_x k_y$-plane. The size of the pocket itself decreases as the field increases. The observed periods correspond to 1% of S_{BZ} for $(TMTSF)_2 PF_6$ [9.1] and 0.7% for $(TMTSF)_2 ClO_4$ [9.4, 5], if we apply (9.1).

9.3.5 FI-SDW in a Refined Model

Since the main contribution to the deterioration of the nesting properties comes from the multipleness of the transverse transfer energies, we shall treat FI-SDW by the more refined model according to (4.49). Very similar results are obtained for FI-SDW, and important qualitative features of FI-SDW in the $(TMTSF)_2 X$ salts can be reproduced [9.40]. This model taking advantage of second harmonics is more realistic for $(TMTSF)_2 X$ than that expressed by (4.26). The parameter τ_{cos} assumes values close to those for the upper bound $\tau_{cos}^{(cr)} = (M_0 + \epsilon_0)/2$, determined by (4.57), for the stability of SDW with the optimum wave vector Q_0. This was concluded to be the most pressure-sensitive parameter; τ_{cos} increases under pressure or when the donor X is ClO_4 and violates the stability condition in (4.57).

Following the same procedure as in the previous subsection, we get the phase diagram of the FI-SDW subphases in the H-τ_{cos} plane. As shown in Fig.9.32, it is very similar to the previous diagram (Fig.9.27) in the H-t_b plane. The employed parameter values are stated in the figure. The value of M_0 specifies the coupling constant I through (4.40). As demonstrated in Chap.4, in a limited range of τ_{cos} above the upper bound of the stability $\tau_{cos}^{(cr)} = (M_0 + \epsilon_0)/2$, we have the transient type of SDW with $Q \neq Q_0$ even in

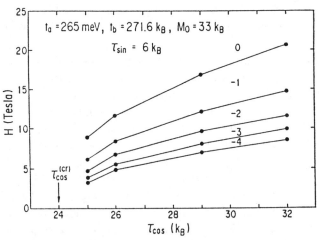

Fig.9.32. Phase diagram of the FI-SDW subphases in the H-τ_{cos} plane. The integers denote the value of N_{QL} for the subphase having the lowest energy. The upper bound of τ_{cos}, above which SDW with $Q = Q_0$ becomes unstable, is $(M_0 + \epsilon_0)/2$. From [9.40]

the absence of a field; SDW with $\mathbf{Q} = \mathbf{Q}_0$ is no more stable. With the application of a magnetic field perpendicular to the ab-plane, the latter SDW phase becomes more stable and extends in the parameter space. This phase consists of subphases where Q_x is quantized to the value of (9.40) but with $N_{QL} \leq 0$, in contrast to the case for $N_{QL} \geq 0$ discussed in the previous subsection.

This FI-SDW model qualitatively reproduces experimental features for an increasing magnetic field [9.40]. A few remarks are made here concerning the type of carriers in the FI-SDW state, and the magnetization. In the present case where $2\tau_{cos} - \epsilon_0 > M_0$, we see that the subphases have $N_{QL} \leq 0$, as mentioned. The phase space below the SDW gap, or the new first Brillouin zone, accommodates $\frac{1}{2}N - 2\{N_{QL}$ electrons. Since the negative value of N_{QL} makes this number bigger than $\frac{1}{2}N$, $|N_{QL}|$ completely empty Landau bands appear in a lower band below the SDW gap (Fig.9.33). As in the previous model, we find good correspondence between the semiclassical Landau levels and the averages of the fully quantum-mechanical magnetic Landau bands. The Fermi energy lies in the energy range, where the extremum curves are relatively flat, between the Landau levels formed above and below the SDW gap in this energy region. Since empty Landau levels are left in the prominent part of the extremum curve of $E_-^{(ex)}$ versus $\eta = bk_y + \phi_0$, the FI-SDW state with a negative N_{QL} has current carriers of the type opposite to those in the normal state. This result is consistent with the

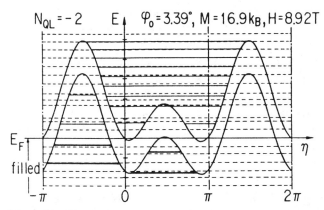

Fig. 9.33. Distribution of the magnetic quantum levels for $\tau_{cos} = 26\,k_B$ and $\tau_{sin} = 6\,k_B$. Curves with the double minimum as a function of $\eta = bk_y + \phi_0$ are the extremum curves neglecting the effect of the field to the orbital motion. Averages of fully quantum magnetic bands are designated by horizontal *dashed lines*. Semiclassical Landau levels are denoted by horizontal *solid lines*, those that are completely occupied quantum magnetic bands are shown by *thick lines*. Averages of the quantum levels and the semiclassical levels are in excellent agreement except when multiple close semiclassical levels exist. Note that two semiclassical levels below the SDW gap are empty, corresponding to $N_{QL} = -2$. From [9.40]

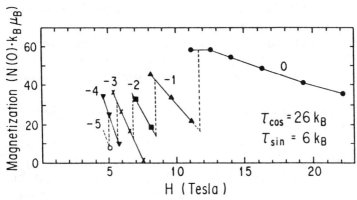

Fig. 9.34. Anisotropic part of the magnetization defined by (9.41) as a function of the field H, with $\tau_{cos} = 26\,k_B$, for the case shown in Fig. 9.31. From [9.40]

change of the sign of the Hall voltage upon the appearance of FI-SDW or the transient SDW (SDW2) [9.9, 12-14].

The difference between carrier types comes from the difference of the sign of $t'_b = \tau_{cos} - \tfrac{1}{2}\epsilon_0$ and, therefore, from the different features of the extremum curves (Fig. 4.12). The latter plays an important role in a semiclassical consideration. With $t'_b > 0$ we have the situation displayed in Figs. 4.12a-c. In Sect. 9.3.3 we considered Figs. 4.12d, e with $t'_b < 0$.

From the similarity of the extremum, the FI-SDW state can be interpreted, as stated before, to be the transient SDW (SDW2) state which is more stable. The region of stability is extended in the parameter space by an energy gain which comes from the bunching of one-electron energy levels due to orbital quantization in the magnetic field. This picture is also supported by the observation that in the PF_6 salt in the absence of a field the SDW transition temperature for $T \leq 2$ K at a pressure of 6 kbar is markedly increased when a magnetic field is applied [9.9].

The anisotropic part of the magnetization, given by (9.41) is shown as a function of H in Fig. 9.34 for $\tau_{cos} = 26k_B$ and $\tau_{sin} = 6k_B$. The unit $N(0) \times k_B \mu_B$ is equal to 0.409 erg/(Oe·mol) for $t_a = 0.265$ eV. The features are essentially the same as those observed for the simple model illustrated by Fig. 9.29.

The energy gain is monotonically increasing in the highest subphase with $N_{QL} = 0$ and saturates in the high-field limit. The saturation value can be calculated by the semiclassical treatment employed in Sect. 9.3.1, using a simplified trapezoid form of the state density for the closed-orbit energy range. The result is given by

$$(\text{energy gain})_{sat} = \frac{5}{6}N(0)(2\tau_{cos} - \epsilon_0)^2 \ . \tag{9.45}$$

Saturation is reached when the last single semiclassically-closed orbit comes to the lowest edge of the closed-orbit energy range below the SDW gap at

$$H_{sat} = \frac{3c}{2a_s \, bet_a} \sin(a_s k_F) |2\tau_{cos} - \epsilon_0|, \tag{9.46}$$

where $2\tau_{cos} - \epsilon_0$ is the amplitude of the undulation of the extremum curve in Fig.4.6 with $\mathbf{Q} = \mathbf{Q}_0$, $2(2\tau_{cos} - \epsilon_0)$ being the depth of the closed-orbit energy range where semiclassical Landau levels are formed; it should be nearly equal to $2M_0$. With the parameter values used here we find $H_{sat} \simeq 50$ Tesla. Both quantities given by (9.45, 46) are in rough agreement with those obtained by the full quantum-mechanical calculation. By choosing an appropriate value for $2\tau_{cos} - \epsilon_0$ at the observed value M_0, the calculated saturation value of the energy gain can be raised towards the observed saturation value at ≈ 15 Tesla [9.16]. In the high-field region, the calculated magnetization gradually decreases and never becomes negative. This contrasts with the recently observed rapid decrease [9.43] in the high-field region, to be discussed in Sect.9.5.1.

The refined model with transverse second harmonics is successful in determining the direction of the shift of Q_x and the type of current carriers in the FI-SDW and transient SDW states. It also improves the saturation value of the energy gain in the high field. Nevertheless, it does not completely reproduce more quantitative details of the FI-SDW subphases such as the average period between transitions, $\langle \Delta(1/H) \rangle$, and the field dependence of the magnetization. The theoretical values of $\langle \Delta(1/H) \rangle$ are fairly reasonable, however, suggesting that the theory is not far off. Although the new parameter τ_{sin} does not play such an important role as τ_{cos}, it has a large effect on the period $\langle 1/\Delta H \rangle$, the energy gain, etc. [9.40]. Since the contribution to the energy gain comes from a rather wide range of band energies, it may be necessary to take account of its $\mathbf{k}$-dependence, not only in the vicinity of the Fermi surface, but also in a wider energy range.

The type of current carrier changes between the normal and FI-SDW, states which is in good agreement with all observations [9.9, 12-14]. The signs of the carrier charges agree with those observed by *Ribault* et al. [9.12] and *Oshima* et al. [9.13], but are opposite to the results of *Chaikin* et al. [9.14] and *Kwak* et al. [9.9]. The difference may come from difficulties of the experimental techniques used.

In order to explain the small-period oscillation [9.4, 10, 20-23], direct nesting with the wave vector $\mathbf{Q} = (2k_F + \Delta Q_x, 0, 0)$ [9.48] or longitudinal nesting with $\mathbf{Q} = (2k_F, 0, \pi/c)$ [9.49] was suggested so that large carrier pockets appear. However, these types of SDW nesting vectors induce so weak a divergence of the wave number-dependent susceptibility that the coupling constants corresponding to these wave vectors could not possibly compensate for the weakness.

9.4 Green's Function Theory of FI-SDW

9.4.1 Basic Equations

Maki et al. [9.50-52] and *Poilblanc* et al. [9.53-56] developed essentially identical Green's function theories of FI-SDW which cover the temperature range from absolute zero to the transition point. This enables us not only to complete the phase diagram in the H-T plane but also to calculate thermodynamic and dynamical properties. An outline of the theory and its basic results are sketched in this section.

We follow the discussion of *Virosztek* et al. [9.51], although our notation is more self-explanatory. We employ the convention $\hbar = 1$. The Hamiltonian of the system is given in terms of the field operator $\psi_\sigma(\mathbf{r})$ for the σ spin component by

$$\hat{\mathcal{H}} = \sum_\sigma \int d\mathbf{r}\, \psi_\sigma^\dagger(\mathbf{r})\, \hat{h}_0\, \psi_\sigma(\mathbf{r}) + \frac{1}{2} \sum_{\sigma,\sigma'} \int d\mathbf{r}\, \psi_\sigma^\dagger(\mathbf{r})\, \psi_\sigma(\mathbf{r})\, \psi_{\sigma'}^\dagger(\mathbf{r})\, \psi_{\sigma'}(\mathbf{r}) \,,$$

where (9.47)

$$\hat{h}_0 = v_F(\pm\hat{k}_x - k_F) + \epsilon_\perp(\hat{k}_y - eHx/c, \hat{k}_z) + \sigma\mu_B H \,, \qquad (9.48)$$

with

$$\epsilon_\perp(k_y, k_z) = -2t_b \cos(bk_y) - 2t_b' \cos(2bk_y) - 2t_c \cos(\tilde{c}k_z) \,, \qquad (9.49)$$

and $\hat{k}_x = -i\partial/\partial x$, $\hat{k}_y = -i\partial/\partial y$, and $\hat{k}_z = -i\partial/\partial z$; $\pm$ denotes the positive and the negative wave-number components, respectively; H is the magnetic field parallel to the z-direction, and $\tilde{c}$ is the lattice constant in the c-direction. We divide the field operator into negative and positive wave number components

$$\psi_\sigma(\mathbf{r}) = \psi_{-\sigma}(\mathbf{r}) + \psi_{+\sigma}(\mathbf{r}) \,. \qquad (9.50)$$

Then, the Green's function is defined by

$$G_{+\sigma+\sigma}(\mathbf{r}\tau, \mathbf{r}'\tau') = -\langle \hat{\mathcal{T}} \psi_{+\sigma}(\mathbf{r}, \tau)\, \psi_{+\sigma}^\dagger(\mathbf{r}', \tau') \rangle \,, \qquad (9.51)$$

where $\hat{\mathcal{T}}$ is the chronological operator on the the imaginary times τ and τ' [9.26]. In the mean-field approximation the equation of motion for the Fourier component $G_{+\sigma+\sigma}(i\omega_n;\mathbf{r},\mathbf{r}')$ with the Matsubara frequency $\omega_n = \pi k_B T(2n+1)$ is decoupled as follows:

$$(i\omega_n - \hat{h}_0)G_{+\sigma+\sigma}(i\omega_n;\mathbf{r},\mathbf{r}') + \Delta_\sigma(x)F_{-\bar\sigma+\sigma}(i\omega_n;\mathbf{r},\mathbf{r}') = \delta(\mathbf{r}-\mathbf{r}') , \quad (9.52)$$

where another Green's function $F_{-\bar\sigma+\sigma}(i\omega_n;\mathbf{r},\mathbf{r}')$ is the Fourier component of

$$F_{-\bar\sigma+\sigma}(\mathbf{r}\tau,\mathbf{r}'\tau') = -\langle \hat{\mathcal{T}}\psi_{-\bar\sigma}(\mathbf{r},\tau)\psi^\dagger_{+\sigma}(\mathbf{r}',\tau')\rangle \exp(i\mathbf{Q}\cdot\mathbf{r}) \qquad (9.53)$$

and the SDW gap parameter $\Delta_\sigma(x)$ is defined by

$$\Delta^*_\sigma(x) = U\langle \psi^\dagger_{+\sigma}(\mathbf{r})\psi_{-\bar\sigma}(\mathbf{r})\rangle \exp(i\mathbf{Q}\cdot\mathbf{r})$$

$$= UT\sum_{\omega_n} F_{-\bar\sigma+\sigma}(i\omega_n;\mathbf{r},\mathbf{r}) ; \qquad (9.54)$$

here $\bar\sigma$ denotes the opposite of σ. We expect that SDW induced by the field has the mean value $U\langle\psi^\dagger_{+\sigma}(\mathbf{r})\psi_{-\bar\sigma}(\mathbf{r})\rangle$ with the main spatial Fourier component of the wave vector $-\mathbf{Q}$. We still allow the gap parameter $\Delta_\sigma(x)$ to be x-dependent. In a similar way we obtain the equation of motion for $F_{-\bar\sigma+\sigma}(i\omega_n;\mathbf{r},\mathbf{r}')$. Applying to both equations the partial Fourier transform $\int d\mathbf{r}_\perp \exp[-i\mathbf{k}_\perp\cdot(\mathbf{r}-\mathbf{r}')_\perp]$ yields the following equations for the Green's functions in the mixed representation:

$$[i\omega_n - v_F(-i\partial/\partial x - k_F) - \epsilon_\perp(\mathbf{k}_\perp - e\mathbf{A}_\perp/c) - \sigma\mu_B H]G_{+\sigma+\sigma}(i\omega_n,\mathbf{k}_\perp;x,x')$$

$$+ \Delta_\sigma(x)F_{-\bar\sigma+\sigma}(i\omega_n,\mathbf{k}_\perp - \mathbf{Q}_\perp;x,x') = \delta(x-x') ,$$

$$\qquad (9.55)$$

$$[i\omega_n - v_F(i\partial/\partial x - k_F) - \epsilon_\perp(\mathbf{k}_\perp - \mathbf{Q}_\perp - e\mathbf{A}_\perp/c) + \sigma\mu_B H]$$

$$\times F_{-\bar\sigma+\sigma}(i\omega_n, \mathbf{k}_\perp - \mathbf{Q}_\perp; x, x') + \Delta^*_\sigma(x)G_{+\sigma+\sigma}(i\omega_n, \mathbf{k}_\perp; x, x') = 0 ,$$

where $\mathbf{A} = (0, Hx, 0)$.

Performing the following phase transformation:

$$G_{+\sigma+\sigma}(i\omega_n, \mathbf{k}_\perp; x, x') = g(x, x')\exp\{i[\phi(x) - \phi(x')]\} ,$$

$$f_{-\bar\sigma+\sigma}(i\omega_n, \mathbf{k}_\perp - \mathbf{Q}_\perp; x, x') = f(x, x')\exp\{i[\phi'(x) - \phi(x')]\} ,$$

$$\qquad (9.56)$$

with

$$\phi(x) = \frac{1}{v_F} \int_0^x du [v_F k_F - \sigma\mu_B H - \epsilon_\perp(k_y - eHu/c, p_z)] \,,$$

(9.57)

$$\phi'(x) = \frac{1}{v_F} \int_0^x du [v_F (Q_x - k_F) - \sigma\mu_B H + \epsilon_\perp(k_y - eHu/c - Q_y, k_z - Q_z)] \,,$$

we obtain the simpler equations

$$(i\omega_n + iv_F \partial/\partial x) g(x, x') + \tilde{\Delta}(x) f(x, x') = \delta(x - x') \,,$$

$$(i\omega_n - iv_F \partial/\partial x) f(x, x') + \tilde{\Delta}^*(x) g(x, x') = 0 \,,$$

(9.58)

where

$$\tilde{\Delta}(x) = \Delta_\sigma(x) \exp\{i[\Phi(x) - \Phi(0)]\}$$

(9.59)

and

$$\Phi(x) = q_x(x - x_0) + \beta\cos[\kappa(x - x_0)] - \alpha\sin[2\kappa(x - x_0)] \,.$$

(9.60)

Here q_x and q_y are defined through $\mathbf{Q} = (2k_F + q_x, \pi/b + q_y, \pi/\tilde{c})$; $x_0 = (k_y - q_y/2)c/eH$ and $\kappa = ebH/c$;

$$\alpha = \frac{2t_b'}{v_F \kappa}\cos(bq_y) \,, \quad \beta = \frac{4t_b}{v_F \kappa}\sin(\tfrac{1}{2}bq_y) \,.$$

(9.61)

Equations (9.58) are no longer dependent on the Zeeman energy. They are self-consistent together with (9.54). We can obtain another set of similar equations when we start from $G_{-\sigma-\sigma}(i\omega_n; \mathbf{r}, \mathbf{r}')$.

9.4.2 Solution of the Green's Functions

The Green's functions $g(x,x')$ and $f(x,x')$ are constructed in terms of the eigenfunctions $[u_n(x), v_n(x)]$ by

$$g(x,x') = \sum_j \frac{u_j(x)\,u_j^*(x')}{i\omega_n - E_j},$$

$$f(x,x') = \sum_j \frac{v_j(x)\,u_j^*(x')}{i\omega_n - E_j},$$

$$(9.62)$$

where the eigenfunctions satisfy

$$\begin{bmatrix} E_j + iv_F\,\partial/\partial x & \tilde{\Delta}(x) \\ \\ \tilde{\Delta}^*(x) & E_j - iv_F\,\partial/\partial x \end{bmatrix} \begin{bmatrix} u_j(x) \\ \\ v_j(x) \end{bmatrix} = 0 \,. \tag{9.63}$$

The potential $\tilde{\Delta}(x)$ is expanded into

$$\tilde{\Delta}(x) = \Delta_\sigma(x)\exp[-i\phi(0) + iq_x(x-x_0)] \sum_{n=-\infty}^{\infty} I_n \exp[in\kappa(x-x_0)] \,, \quad (9.64)$$

where I_n is defined by

$$I_n = I_n(\alpha,\beta) = i^n \sum_{\ell=-\infty}^{\infty} J_\ell(\alpha)J_{n-2\ell}(\beta) \tag{9.65}$$

with $J_\ell(z)$ denoting a Bessel function.

Equation (9.63) is solved in terms of a plane wave $\exp(ikx)$. The off-diagonal term mixes it with $\exp[i(k+q_x+n\kappa)x]$. If we assume $\Delta_\sigma(x) = \text{const} = \Delta$, the eigenenergy E_j forms a series of bands, the **Landau bands** or **subbands**. Gaps, called **Landau gaps**, with the magnitudes $2|\Delta I_n|$ appear at $k = -(q_x+n\kappa)/2$.

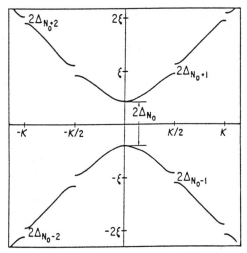

Fig. 9.35. Quasi-particle energy spectrum in the FI-SDW state as a kind of extended band scheme. The energy spectrum develops a series of energy gaps at $k_x = \pm Q_x/2 + n\kappa/2$, $n = 0, \pm 1, \pm 2, \ldots$. The Fermi energy lies in the gap at $Q_x/2 + n\kappa/2 = k_F$ or $n = N_{QL}$ defined by (9.40) due to the requirement of conservation of the electron number. In order to achieve maximum energy gain, it lies in the largest gap, i.e., that is specified by $n = n_0$, which also maximizes $\left| I_n \right|$. In the figure $\xi = v_F \kappa/2$. The abscissa is for κ. From [9.51]

In order for the SDW gaps to provide the maximum energy gain, the Fermi level, i.e. the zero level in the present case, must lie in the middle of an energy gap. Therefore, q_x must satisfy

$$q_x = -\kappa N_{QL} , \tag{9.66}$$

where N_{QL} is an integer. When $\left| I_{n_0} \right| \gg \left| I_{n_0 \pm 1)} \right|, \left| I_{n_0 \pm 2)} \right|, \cdots$, $N_{QL} = n_0$ is the most favorable one. If we take into account the largest gap exactly and the others perturbatively, the unperturbed energy is given by

$$E^\pm(k) = \pm (\xi^2 + \left| \Delta I_{N_{QL}} \right|^2)^{1/2} \quad \text{with} \quad \xi = v_F k . \tag{9.67}$$

For other perturbational gaps the eigenvalues are illustrated in Fig. 9.35 in an extended-zone scheme. This shows essentially the same eigenvalues obtained in Sect. 9.3.2 (Fig. 9.23) but in a more systematic way.

The Green's functions are given in the single-gap approximation by

$$g(x,x') = -\frac{1}{L}\sum_k \frac{i\omega_n + \xi}{\xi^2 + |\Delta I_{N_{QL}}|^2} e^{ik(x-x')} ,$$

(9.68)

$$f(x,x') = \frac{1}{L}\sum_k \frac{\Delta I_{N_{QL}}}{\xi^2 + |\Delta I_{N_{QL}}|^2} \exp\left[i\phi(0) - \frac{i\pi N_{QL}}{2}\right] e^{ik(x-x')} .$$

9.4.3 Basic Properties

By using (9.68), the gap equation is rewritten as

$$\frac{U}{b\tilde{c}}|I_{N_{QL}}|^2 T \sum_{\omega_n} \int \frac{dk}{2\pi}(\omega_n^2 + \xi^2 + |\Delta I_{N_{QL}}|^2)^{-1} = 1 ,$$

(9.69)

which is a BCS-type gap equation. Then, the SDW transition temperature T_{SDW} is given by

$$k_B T_{SDW} = 1.13\epsilon_c \exp(-1/\lambda_{N_{QL}}) ,$$

(9.70)

where $\lambda_{N_{QL}} = (U/2\pi v_F b\tilde{c})|I_{N_{QL}}|^2$ and $\epsilon_c = v_F/\gamma d$ with d being the cutoff defined in (9.19) and $\gamma = 1.78107$.

As the magnetic field decreases from a very high value, the maximum value of $|I_n|^2$ moves from $|I_0|^2$ to $|I_1|^2$, $|I_2|^2$, ... in sequence (Fig.9.36). With decreasing field this gives rise to a series of first-order transitions, closely related to the sequence of the maxima in χ_0 displayed in Fig.9.16. The phase diagram obtained by the present method is shown in Fig.9.37. The phase boundary between two adjacent subphases is independent of the temperature. The phase diagram in the H-T plane is thus completed [9.58], interpolating the results for T_{SDW} in Sect.9.2.2 and the phase diagram at T = 0 in Sects.9.3.3 and 5.

The single-gap approximation employed in deriving (9.68) is crude, especially when $|I_{N_{QL}\pm 1}| \approx |I_{N_{QL}}|$, so that (9.70) is of limited reliability. Actually, when the other gaps are included, the phase diagram changes quantitatively [9.52-50]. The phase boundaries between subphases become slightly inclined.

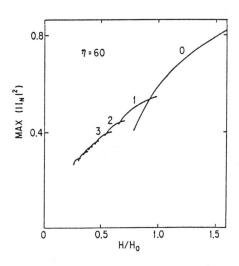

Fig. 9.36. Maximum value of $|I_n|^2$ as a function of the magnetic field. The *curve segments* are characterized by successive values of the quantum number n_0; H_0 is the scaling field defined by $H_0 = (t_b/2t_a \cos x_F)^2 c\hbar/ea_s b$. From [9.51]

This theory yields the temperature dependence of the gap parameter Δ. We also get the free-energy difference ΔF by the aid of

$$\Delta F = F_{SDW} - F_n = 2\int_0^\Delta \left[\frac{1}{\lambda} - g(\Delta')\right] \Delta' d\Delta',\qquad (9.71)$$

where $g(\Delta)$ is defined by the gap equation $1/\lambda = g(\Delta)$. The magnetization given by $\mathcal{M} = -\partial(\Delta F)/\partial H$ is found to be in good agreement with that obtained numerically at $T = 0$ K [9.51]. The specific heat can also be determined [9.52-54]. When plotted as a function of the field, the specific heat

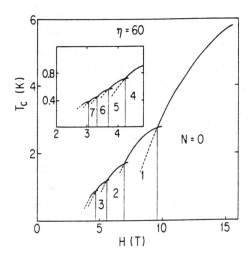

Fig. 9.37. Phase diagram of FI-SDW subphases. The *solid curve* shows the transition temperature T_{SDW}, vertical *thin lines* exhibit the boundaries between two adjacent SDW subphases. *Broken curves* are continuations of the second-order transition curves into the next subphases. From [9.51]

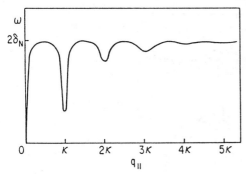

Fig. 9.38. Schematic collective mode spectrum showing minima of roton-like excitation versus $q_{\parallel}$; $\delta_N = |\Delta I_N|$. From [9.57]

at T_{SDW} was found to show large jumps at the fields of transition between FI-SDW subspaces [9.52, 56].

Dynamical properties have been studied by using the Green's function formalism. Figure 9.38 illustrates the spectrum of roton-like excitation in the FI-SDW state [9.57].

9.5 Developments of FI-SDW

9.5.1 Phase Diagram of $(TMTSF)_2ClO_4$

The FI-SDW theory described in the preceeding sections and called **standard theory**, has been found to be successful in giving essential features of the phase diagram of $(TMTSF)_2PF_6$ in the T-vs-H plane. However, the phase diagram of $(TMTSF)_2ClO_4$ is known to have delicate disagreements with calculations, presumably due to the modulation of the band by the superlattice potential of ClO_4 anions with the wave vector $Q = (0, \frac{1}{2}, 0)$. *Chaikin* suggested [9.58] that the reentrance to the normal state and also the H dependence of the magnetization are quite sensitive to the cooling rate, or the degree of anion ordering, of the $(TMTSF)_2ClO_4$ sample. The reentrance is quickly retarded to higher fields as the cooling becomes faster, and is finally lost. They also found that $(TMTSF)_2ClO_4$ shows agreement with the standard model when the anion ordering is suppressed by a pressure above ≈ 5 kbar [9.59].

Figure 9.5 plots the phase diagram in the H-vs-T plane for the low field. The solid curves separating the FI-SDW subphases, could be indexed by the Landau quantum number N_{QL} which could be specified by the Hall voltage of the plateau in each subphase according to (9.43). They should form a regular integer series with increasing field. As is seen in Fig. 9.7, the

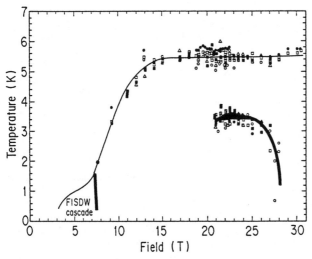

Fig. 9.39. T-H phase diagram for $(TMTSF)_2 ClO_4$ at high magnetic fields arising from the compilation of measurements shown in Fig. 9.40 and other similar measurements. The *thin curve* represent a second-order transition. The *heavy curves* correspond to first-order transitions. From [9.60]

Hall voltages do not give such a good series of ratios in contract to Fig. 9.30 for $(TMTSF)_2 PF_6$.

On the high-field side, although the standard theory predicts that the subphase with $N_{QL} = 0$ is realized here as in Fig. 9.36, the observed phase diagram of this part is complicated. Observability of the phase boundaries depends on the physical properties employed. The most recent result is shown in Fig. 9.39 based on measurements of ρ_{zz}, ρ_{xy}, magnetization, etc., examples of which are displayed in Fig. 9.40. In the range from 8 to 30 Tesla, the upper transition is of the second order and the lower-temperature transition in 21 Tesla $<$ H $<$ 28 Tesla is of the first order. Apparently the presence of the first-order transition is in discrepancy with the theory.

Before this phase diagram was published, the above-mentioned second-order phase boundary at high fields was considered to continue to the lower phase boundary [9.61]. No phase boundary was asserted to exist around a temperature of 5.5 K in a field above 25 Tesla. The phase region around T ≥ 2 K and in the field range H $> 25 \div 28$ Tesla was believed to be a normal state which continues to the normal state in the absence of the field. The negative magnetization, seen in Fig. 9.40b for T $= 2$ K suggests that the free energy at H ≈ 30 Tesla increases close to that of the normal state due to some unknown mechanism. This is consistent with the previous phase diagram which gives the reentrance to normal state. A transition to an insulating state was indicated to occur at T ≈ 2 K in the high-field region with

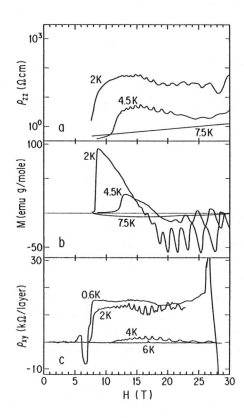

Fig. 9.40. Fixed-temperature field sweep of ρ_{zz}, ρ_{xy} and magnetization. Phase transition points are compiled into Fig. 9.39. From [9.60]

H > 25 Tesla [9.62]. The nature of this lower-temperature insulator phase has been controversial.

9.5.2 Advances in the Theory of FI-SDW

Motivated to resolve the discrepancy between theory and experiment, a few groups took account of the band modification due to the anion superlattice potential with $\mathbf{Q} = (0, \frac{1}{2}, 0)$. *Lebed* and *Bak* obtained a mean-field result suggesting that in the high-field limit T_{SDW} decreases in proportion to $H^{-1/2}$, oscillating with the frequency

$$\Delta\left(\frac{1}{H}\right) = \frac{\pi b e v_F}{4 c t_b} \tag{9.72}$$

as a function of H in the high-field limit [9.63]. Concerning the decrease of T_{SDW} they argued that the superlattice potential with $(0, \frac{1}{2}, 0)$ gives rise to two types of electron wave functions: one (+) has a maximum at the even

359

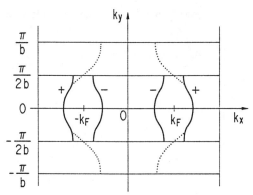

Fig. 9.41. *Thick curves* show Fermi surfaces in the reduced zone for $(TMTSF)_2 ClO_4$ in the anion-ordered state. The $\pm$ curves are the upper and the lower bands, respectively. The *dotted curves* represent the Fermi surface in the absence of anion ordering, or that of $(TMTSF)_2 PF_6$. From [9.64]

chains, and the other ($-$) at the odd ones (Fig.41). The interaction with the local SDW potential diminishes with increasing field because SDW with the optimal wave vector corresponds to pairing between the two different types of wave functions. Further detailed analyses were advanced by *Gor'kov* and *Lebed* [9.64].

A similar idea was developed by *Osada* et al. [9.65]. They calculated the q-dependence of the spin susceptibity $\chi_0(q)$ including the superlattice potential. As is seen in Fig.9.42, the obtained susceptibility reveales that local maxima with even Landau quantum number N are split into two (N^+ and N^-) peaks, which are relatively smaller than those with odd quantum numbers. Therefore, with an increasing field, the subphase with odd quantum numbers, i.e., ... 5, 3, 1, are considered to appear successively. This is in accordance with the phase diagram of $(TMTSF)_2 ClO_4$ because the Hall voltages below 25 Tesla basically yields the ratio ..., 1/3, 1 [9.12].

Further, *Osada* et al. showed that in the highest-field region the FI-SDW state with $N = 0$ takes T_{SDW} and oscillates with the period given in (9.72), which can intermittently exceeds T_{SDW} for the $N = 1$ FI-SDW. The respective phase diagram is illustrated in Fig.9.43.

McKerman et al. [9.60] who proposed the phase diagram in Fig.9.39 present an alternative interpretation in terms of Multi-Order Parameter (MOP). They suggest that the upper second-order transition is due to the SDW formation by electrons of one reduced Brillouin zone, which may also induce a small gap for electrons of the second reduced zone. For the lower first-order transition the gap in the second zone increases jump-wise. They interpreted the Hall voltage between the two transition temperature consistently. Further, *McKerman* et al. suggested no oscillation of the transition

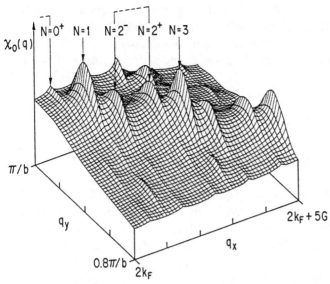

Fig. 9.42. The q dependence of the spin susceptibity $\chi_0(\mathbf{q})$ at a fixed magnetic field $(\hbar v_F \kappa/4t_b = 0.03)$ and temperature $(k_B T/t_b = 0.001)$ with $t'_b/t_b = 0.1$ and a subband splitting equal to $\hbar v_F \kappa/4$. From [9.65]

temperatures as a function of the field at variance with the result of [9.63, 64, 66].

Another puzzle was presented by the observation that, when one applies sufficient pressure to $(TMTSF)_2 NO_3$ so that its SDW disappears, SDW cannot be induced again even with very high magnetic fields [9.67]. *Osada* et al. succeeded in clarifying the reason [9.68]. $(TMTSF)_2 NO_3$ has a

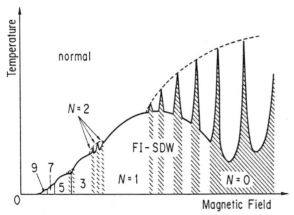

Fig. 9.43. Theoretical H-T phase diagram schematically given by *Osada* et al. [9.66]. N is the absolute magnitude of the Landau quantum number N_{LQ}

pair of electron and hole pockets with the superlattice potential having $\mathbf{Q} = (\frac{1}{4},0,0) = (2k_F,0,0)$. In the applied magnetic field, both pockets have Landau quantum subbands with the Fermi level always lying in the gap between the subbands. This situation was found stable against FI-SDW formation.

Theory and experiment have made much progress but there still remain gaps to be filled.

9.5.3 Jumps of the Specific Heat

Via precise measurements of the specific heat, *Pesty* et al. [9.69] found that the FI-SDW subphase is further divided by jumps of C and the isothermal coefficient of the magnetization $\alpha_T = (\partial M/\partial T)_H$ like in a first-order phase change, when they sweep the magnetic field at a fixed temperature. This division becomes more and more frequent with a decreasing temperature so that the new phase boundaries bifurcate to generate a tree-structure with decreasing temperature (Fig. 9.44).

This phenomenon was interpreted in terms of multiple components of the FI-SDW order parameter, independently by *Lebed* [9.70] and by *Machida* et al. [9.71-73]. In the theory presented in Sect. 9.3 a main gap parameter which corresponds to a Landau quantum number N_{QL}, or k_x-component $2k_F + N_{QL}\kappa$ in another word, is assumed to exist. However, in the mean-field treatment, other components for different numbers are found to be coupled since in the magnetic field κ is a new Bragg wave number, or a

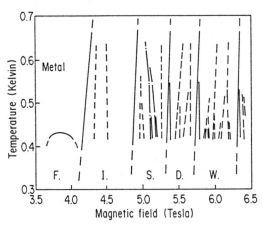

Fig. 9.44. Treelike phase diagram built up from simultaneous measurements of C_B and $(\partial M/\partial T)_B$. The *solid lines* represent the transitions between FI-SDW phases with different Landau quantum numbers. The *dashed curves* show the transitions indicated by the measurements inside the subphases. From [9.69]

new reciprocal lattice, brought in by the magnetic field. These subsidiary gap parameters make jump-wise changes even within a subphase when sweeping the field, and produce new phase boundaries. These jumps are considered to become more and more frequent with decreasing temperature. Such a picture allows us to understand the tree structure of the new phase boundaries.

9.5.4 Sign Reversal of the Hall Voltage

Another long-standing puzzle in the FI-SDW phase is the sign reversal of the Hall constant, or the **Ribault anomaly** (Fig. 9.8). It was observed for an extremely slowly cooled sample of $(TMTSF)_2 ClO_4$. Such an anomaly was also reported for the PF_6 salt [9.47].

Machida et al. interpreted this anomaly as being due to the multiple order-parameter components that have been explained in the previous section [9.74]. They gave examples in which the conductivity tensor σ_{xy} deviates from $2N_{QL} e^2/h$, a value the subphase specified by the Landau quantum number N_{QL} should take on. Although in the neighborhood of the normal FI-SDW transition the main component is overwhelmingly large so that $\sigma_{xy} = 2N_{QL} e^2/h$ is satisfied. Near the boundary between different subphases they found that various order-parameter components are comparable in magnitude, giving rise to a different value of σ_{xy} even with a different sign.

Another explanation was proposed by the Orsay group [9.75]. They calculated the spin susceptibility $\chi_0(q)$ for the band energy (4.52) with an additional term $2t_3 \cos(3bk_y)$ (they assume $t_{sin} = 0$ and $\phi = 0$). They found that the maxima of $\chi_0(q)$ appear for the even Landau quantum numbers N_{QL} even when $|t_3|$ is much smaller than $|t_b'|$ but that the maximum for $N_{QL} = 2m$ with the integer m is degenerate with that for $N_{QL} = -2m$ and that a finite small value of t_4, the coefficient of the fourth harmonic in the band energy, lifts the degeneracy. The maximum for the odd Landau quantum number $N_{QL} = 2m+1$ is not degenerate with $N_{QL} = -2m-1$ and the dominant one is decided by the sign of t_b', the coefficient of the second harmonic, as in the standard theory. As a consequence, they suggest that the sequence of plateaus with $N_{QL} = 1, 2, -2, 3, 4, -4, 5$ is possible. They depend on a delicate change of the band parameters caused by the pressure, etc. This feature is in accord with the fact that the sign change occurs in the subphases with even quantum numbers in the case of $(TMTSF)_2 PF_6$.

9.5.5 FI-SDW in (DMET-TSeF)$_2$X

FI-SDW was found in another family, namely (DMET-TSeF)$_2$X, with X: AuI$_2$, AuBr$_2$ and AuCl$_2$ [9.76]. This indicates that the phenomenon is not specific to the Bechgaard salts and is a generic nature of the quasi 1-D systems. Note that the AuI$_2$ salt becomes superconducting, the other two do not. Another salt with X: I$_3$ shows superconductivity, but no FI-SDW. Thus the occurrence of FI-SDW seems to be uncorrelated with the occurrence of superconductivity, in contrast to the assertion of *Yakovenko* [9.77, 78].

9.5.6 Other Topics Related with FI-SDW

Phase transitions among FI-SDW subphases were detected by specific heat [9.79], thermopower [9.62, 80] and sound-velocity [9.81] measurements.

The magnetoresistance of (TMTSF)$_2$ClO$_4$ in FI-SDW was demonstrated by *Osada* et al. [9.82] to be very nonlinear with respect to the applied electric field. It grows by a factor of 2 in a moderate electric field. This is attributed to a sliding of FI-SDW, but it is difficult to understand why the resistivity should increase when SDW moves.

Tüttö et al. [9.83] and *Suzumura* et al. [9.84] presented a mechanism of pinning SDW around an impurity in quasi 1-D conductors. They showed that the charge density makes a Friedel oscillation with the wave number $2k_F$ since both up- and down-spin electron density components of SDW tune their phases around the impurity potential to get an energy gain from it. *Chang* et al. [9.85] presented a theory demonstrating that conventional impurities in the SDW system are similar to the magnetic impurities in superconductors. Experimentally, the anion disorder was shown to move the

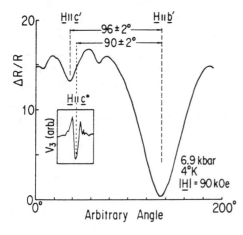

Fig. 9.45 Angular dependence of the relative transverse magnetoresistance along the a-axis in a (TMTSF)$_2$PF$_6$ crystal at 4.0 K under 6.9 kbar. The origin of the angle of the magnetic field is arbitrary. *Inset*: same sample at 1.1 K. From [9.3]

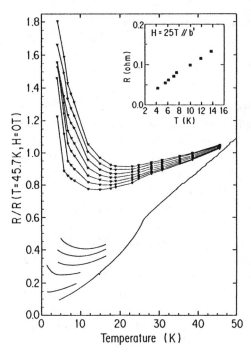

Fig. 9.46. Resistivity curves in the under field H∥c* for two samples. The *continuous curves* are the measurements performed at low fields 12.3, 10, 8, 6, 3 and 0 Tesla, respectively. The high-field points at 27, 25, 22.5, 20, 17.5, 15 and 12.5 Tesla were taken at fixed temperatures. The *inset* shows resistance vs temperature curves at H = 25 Tesla parallel to the b′ direction. From [9.89]

FI-SDW transition to lower temperatures and to higher magnetic fields [9.86].

The rotation diagram of the a-axis transverse magnetoresistance exhibits a dip at the magnetic field inclined by 30° from the c* axis to the b′ axis (Fig. 9.45). *Lebed'* [9.87] and *Chen* and *Maki* [9.88] have proposed theories of FI-SDW for a field inclined in the c*b′ plane and suggested that the SDW-transition temperature is reduced around this angle. Such considerations opened up a way to find a rich field of magneto-oscillations in organic conductors in tilted magnetic fields.

Recently, an old problem of giant transverse magnetoresistance was revived in a report of finding it in $(TMTSF)_2 ClO_4$ (Fig. 9.46). This phenomenon was first observed for $(TMTSF)_2 PF_6$ in the early days of organic superconductors [9.90-92]. It is considered to show the tendency of localizing electrons in a transverse magnetic field.

9.6 Other Oscillatory Effects in Magnetic Fields

9.6.1 Rapid Oscillations

The puzzle of the rapid, or small-period, oscillations displayed in Fig. 9.11 has attracted much interest among theoreticians and experimentalists. They made steady advances, although the puzzle has not yet been resolved. This phenomenon was observed in $(TMTSF)AsF_6$ [9.93, 94], as well as pressurized $(TMTSF)_2ReO_4$ [9.10] and $(TMTSF)_2NO_3$ [9.95], and, as stated in Sect. 9.1, in $(TMTSF)_2ClO_4$ and $(TMTSF)_2PF_6$.

Except for $(TMTSF)_2NO_3$ these have no closed orbits in the neighborhood of the Fermi level so that it can have neither Shubnikov-de Haas nor de Haas-van Alphen oscillations. The oscillation periods, $1/\Delta(1/H)$, of these are $200 \div 300$ Tesla. The oscillations were observed in the SDW and FI-SDW states. In the cases of $(TMTSF)_2ClO_4$ and pressurized $(TMTSF)_2 \cdot ReO_4$ they were also found in the normal state. Surprisingly, even after FI-SDW has been induced with increasing field, they keep the identical period and phase. The temperature dependence of the amplitude is strange. It increases with decreasing temperature but starts to decrease around several K, and vanishes at very low temperatures. The peak position seems to coincide with the temperature of the non-understood Takahasi anomaly inside the SDW phase described in Sect. 4.6.1. The oscillations are mostly observed only in transport properties like the resistivity but in the case of $(TMTSF)_2ClO_4$ also in thermodynamic properties such as the magnetization [9.43, 60], the specific heat [9.96] and the sound velocity [9.81]. In the normal state of $(TMTSF)_2ClO_4$ they were found only in the transport properties [9.97]. The oscillations in this case is considered to come from the superlattice potential with $Q = (0, \frac{1}{2}, 0)$ because they disappear when the anion order of ClO_4 is suppressed by applying pressure [9.98]. Recently, another period, about twice the previous one, of weaker resistivity oscillations was detected when the pressure exceeded a critical value above which the superlattice disappears [9.99, 100]. Observation of two sets of oscillations was also reported by *Howe* et al. [9.101]. $(TMTSF)_2ClO_4$ has a still further complication. As seen in Fig. 9.11, it has two series of oscillations called a and b at high fields. These series have different temperature-, anion ordering- and/or sample-dependences of the amplitudes [9.102, 103]. The abrupt change of the amplitude of the so-called **a-series** around 30 Tesla suggests a phase transition between FI-SDW and the possibly normal, very resistive state. The phase difference between the two series slightly depended on the anion-ordering [9.103].

In the normal state of $(TMTSF)_2NO_3$ there is a closed orbit, with the size corresponding to the oscillations, due to the anion ordering of NO_3

with $\mathbf{Q} = (\frac{1}{2}, 0, 0)$ so that the oscillations represent the normal Shubnikov-de Haas and de Haas-van Alphen effects. However, in its SDW state there is no more a closed orbit. The rapid oscillations have been observed in the SDW state where SDW exists either without magnetic field or induced by the field.

There is no successful theory which explains all aspects of the rapid oscillations comprehensively. The attempt to clarify the mechanism of the rapid oscillations started from an argument as follows: The one-electron part of the Hamiltonian in the magnetic field applied perpendicularly to the conducting plane is solved with the eigenenergy (9.5) and the eigen wave-function (9.6). Various matrix elements, in terms of such eigen wave-functions of, for example, the on-site Coulomb energy, are oscillatory as a function of the magnetic field [9.104]. Then the life time of the single electron gets an oscillatory component due to the Umklapp process of the Coulomb scattering. The oscillation period due to this mechanism is given by

$$\Delta\left[\frac{1}{H}\right] = \frac{\pi b e v_F}{4 c t_b (2 + 2|\cos\phi|)^{1/2}} \quad , \tag{9.73}$$

where ϕ is the band parameter defined by (4.20 and 22). The life time of the electron can be also oscillatory due to the electron-phonon scattering process with the period given by $\Delta(1/H) = \pi b e v_F / 4 c t_b$, identical to (9.72) [9.104, 105]. The result has the shortcoming of an activation-energy-type Boltzmann factor.

By taking account of the oscillatory matrix elements for the Umklapp process, *Yakovenko* calculated the second-order correction to the free energy in the normal state. From this he obtained oscillatory correction terms for the specific heat and magnetization [9.106]. The period is the same as (9.73), although he adopted a model with $\phi = 0$.

Lebed found that these perturbational terms are qualitatively changed when one includes the superlattice potential with $\mathbf{Q} = (0, \frac{1}{2}, 0)$ due to the anion ordering [9.107]. The oscillation amplitude is larger by orders of magnitude. The oscillation period is equal to that in (9.72). This is twice (9.73) if one assumed $\phi = 0$. He asserted that this produces the rapid oscillations in the normal state of $(TMTSF)_2 ClO_4$. He calculated also the correction terms for the magnetization and the spin susceptibility, which he found to be of 0.1 and $1 \div 10$ of the unperturbed quantities, respectively [9.108]. He also obtained the correction to the specific heat of the order of 1% [9.109]. *Gor'kov* and *Lebed* pointed out that the current response function has cyclotron-resonance-type singularities from the same origin [9.110]. *Lebed* argued that these oscillatory terms violates Landau's Fermi liquid theory and considered the system to be a non-Fermi liquid [9.108].

The matrix element of the Umklapp process also modifies the single-particle energy, which is basically described by the Landau bands (Figs. 9.23 and 35). This matrix element gives rise to additional minute gaps in the Landau bands [9.111]. Their magnitude oscillates with the period

$$\Delta\left(\frac{1}{H}\right) = \frac{\pi b e v_F}{4 c t_b \left|\sin(2\phi + b q_y /2)\right|} \ , \tag{9.74}$$

where q_y is the shift of the y-component Q_y of the SDW wave vector from the optimal one. There are two series of oscillations which are antiphase with each other. These two series may lead to the a- and b-series of resistivity peaking, if there remained still a small number of carriers in the SDW state maybe due to defects. The minute gaps would allow to understand the observed strong electric-field dependence of the amplitude of oscillations. This scheme could basically provide weak oscillations of thermodynamic properties. But it is not known yet if it is big enough to explain the observations or not.

The oscillatory matrix element also appeared in the theory of FI-SDW for $(TMTSF)_2 ClO_4$ with anion ordering (Sect. 9.5.2). It gave the oscillatory peaking of T_{SDW} with the period given by (9.72). Thus, the periodic SDW phase changes are a possible origin of the rapid oscillations at high fields in $(TMTSF)_2 ClO_4$.

Concerning the transport oscillations in the normal state of $(TMTSF)_2 ClO_4$ Yan et al. [9.112, 113] pointed out the possibility of the Stark quantum-interference effect (different from the electric-field effect in the case of an atom) which had been known in Mg [9.114]. With the superlattice potential, the Brillouin zone is divided into two, as shown in Fig. 9.41, and different from the one in Fig. 4.3. Then the semiclassical orbits formed in the field have an interference due to the magnetic breakthrough when the two orbits approach each other near the new zone boundary, giving rise to oscillations in transport properties as a function of the field. The oscillation period due to this effect is inversely proportional to the area enclosed by the two semiclassical orbits going in the same direction. It is equal to (9.72). The temperature dependence is argued to be in a fair agreement with experiment [9.97]. This oscillation appears basically in σ_{bb}. It is desirable for such a point to be checked.

There is another possible mechanism with rapid oscillations. This ascribes the oscillations to the quantization of the edge orbits [9.115]. In the semiclassical picture an orbit encountering the sample surface is reflected and starts another round, keeping the momentum parallel to the surface. This gives the so-called **skipping orbit**. This orbit is quantized because its spatial extent is restricted near the surface. This mechanism can give an explanation of the sample-dependence. However, a fully quantum-mechanical treatment of surface levels does not give an oscillatory term in the free energy [9.116].

Recently, *Uji* et al. [9.117] carried out extensive studies on the rapid oscillations. As stated above, they concluded that the effect in the normal state of $(TMTSF)_2 ClO_4$ is due to the Stark quantum-interference effect [9.97]. In the FI-SDW states of the same salt they recognized two mechanisms working; they extracted one component of the oscillations which give the temperature dependence of the amplitude according to the Lifschitz-Kosevich formula. The researchers ascribed it to the Shubnikov-de Haas oscillation due to closed orbits that are generated by the SDW potential, the anion superlattice potential and the magnetic breakthrough [9.117, 118]. The mechanism for the other components still remains puzzles. The aforementioned closed orbits were initially proposed by *Kishigi* and *Machida* for the rapid oscillation in $(TMTSF)_2 ClO_4$ and $(TMTSF)_2 NO_3$ [9.119]. *Uji* et al. further suggested that the rapid oscillation in the resistivity of $(TMTSF)_2 PF_6$ may be due to the closed orbits generated in a similar way from the reason of the temperature dependence of the amplitude being close to that given by the Lifschitz-Kosevich formula [9.120].

9.6.2 Lebed Resonance, Danner Oscillation, and Other Magnetic Oscillations

Lebed and *Bak* first suggested the possibility of a commensurability effect such that the scattering rate of electrons in the Bechgaard salts should show a peak-wise increase at special angles for such magnetic field for which the semiclassical electron path is periodic in k-space, when the field is tilted from the least conducting c-axis to the next conducting b-axis [9.121]. These angles φ are defined by

$$\tan\varphi = \frac{m}{n} b/\widetilde{c} \tag{9.75}$$

in the case of the orthorhombic lattice, where m and n are integers and b and $\widetilde{c}$ are lattice constants. Under the field the electron follows a semiclassical orbit in k-space defined by the intersection between the constant energy surface and the plane perpendicular to the field. In the quasi 1-D band these orbits are open orbits and extend one-dimensionally in the extended Brillouin zone. When folded into the reduced Brillouin zone all paths are aperiodic except the above-mentioned commensurate orbits. *Osada* et al. found by experiment that the resistivity in those directions of the field shows dips [9.122]. This effect is called the **Lebed resonance** or **commensurability**

resonance. They pointed out that when the magnetic field is parallel to $(0, mb, n\tilde{c})$, there is a single electron eigenvalue.

$$E_{k_x, k_y, k_z} = \frac{\hbar^2 k_x^2}{2m^*} - \sum_{m, n, G_{mn} = 0} t_{mn} \cos(mbk_y + n\tilde{c}k_z) , \qquad (9.76)$$

where $G_{mn} = 0$ denotes that the summation has to be carried out for the integers m and n satisfying $\tan\theta = mb/n\tilde{c}$ with θ being the tilt angle of the field; t_{mn} is the transfer energy between two electronic sites separated by $(0, mb, n\tilde{c})$. This means that there are single-particle states that have a finite group-velocity component along the field. Therefore, the conductivity component for this direction increases. They considered that this leads to the dips [9.123].

Maki derived an semiclassical expression of σ_{zz} as a function of the tilt angle of the field [9.124]. σ_{zz} was shown to have a series of resonance-like peaks for the special angles determined by (9.75). This is considered to be a mathematical expression of the above-mentioned intuitive picture.

Lebed gave another interpretation [9.125]. The Coulomb Umklapp scattering provides the background resistivity [9.121]. When the applied field direction is not commensurate i.e., not parallel to vector $(0, mb, n\tilde{c})$ for any integers m and n, then the single-electron energy is given by $\hbar^2 k_x^2 / 2m^*$. In such a case where the electron motion is one-dimensional, the resistivity due to the Coulomb Umklapp process is enhanced, while for the magic-angle case, part of the electrons take on two-dimensional motion so that their Coulomb Umklapp scattering rate is reduced [9.125]. He asserts that this gives dips which are sufficiently large.

Danner et al. found special angles where the resistance has a peak when the field is slightly titled from the a-axis, the most conducting axis, to the c-axis, the least conducting one, in the ac plane [9.126]. They noticed that in such an orientation the k_z range where the semiclassical electron orbit moves, becomes restricted in the extended zone scheme. In this situation the average value of the group-velocity component $\langle v_z \rangle$ parallel to the z-direction is proportional to $J_0(2\tilde{c}t_b B_x / \hbar v_F B_z)$ according to the semiclassical calculation, where $J_0(...)$ is a Bessel function, $\tilde{c}$ is the lattice constant, and B_i is the field component to the ith direction. The finite value of $\langle v_z \rangle$ leads to little magnetoresistance. However, there are special angles where the Bessel function vanishes so that the resistivity is enhanced especially at low temperatures. Actually they observed such peaks in $(TMTSF)_2 ClO_4$ and succeeded to estimate the value of t_b from the positions of the peaks for $(TMTSF)_2 ClO_4$.

Danner et al. did an analogous experiment on $(TMTSF)_2 PF_6$ and found that it gives similar peaking but it is quickly smeared out by a small field in the intermediate conducting b-axis [9.127]. They believed that this weak field makes the interlayer hopping weaker and the interplanar transport incoherent, thus leading to a confinement of the electron to one layer asserted as being due to strong correlation [9.128]. According to the model of *Strong* et al. [9.128] the conducting sheets of $(TMTSF)_2 PF_6$ are marginal Fermi liquids due to the interlayer coupling t_c. Introducing the b-direction magnetic field dephases the interplane tunneling and leads to an effective t_c which is below the threshold for Fermi-liquid behavior. *Danner* et al. argued that this occurs in the PF_6 salt because its band is half-filled and that it does not in the ClO_4 salt because its reduced zones are not half-filled.

Osada et al. pointed out another effect of the titled magnetic field [9.129]. When the field is perpendicular to the nearly flat planes of the Fermi surface, there are semiclassical closed orbits on the Fermi surface due to slight warping, giving rise to quantized Landau levels. When the field is titled to the next conducting direction, for example, to the b-axis from the a-axis in the case of $(TMTSF)_2 X$, such closed orbits, therefore Landau levels, disappear, giving rise to an anomaly. In terms of this effect they interpret an anomalous kink of magnetoresistance reported for $(DMET)_2 I_3$ when the field is titled from the a- to b-axis by $15°$ [9.130]. They succeeded to find a similar anomaly in $(TMTSF)_2 ClO_4$ [9.129].

Lebed et al. advanced another interpretation to this type of the anomaly [9.131]. For the above-mentioned configuration the Boltzmann equation yields that $\rho_{cc}(H)$ becomes singular when the magnetic field is parallel to the group velocity specific to an inflexion point. For instance, when the single-electron energy assumes $\hbar^2 k_x^2 / 2m - 2t\cos(bk_y)$, $(k_x, \pm\pi/2b, 0)$ is an inflexion point. This field direction is identical to that which gives the kink of the resistivity according to *Osada* et al. Therefore the solution of *Lebed* et al. may be a mathematical expression of the intuitive picture of *Osada* et al.

9.6.3 Magnetic Oscillations in α-$(ET)_2 MHg(SCN)_4$

The isostructural family of organic conductors α-$(ET)_2 MHg(SCN)_4$, with M: K, Rb and Tl, shows very interesting, but complicated magnetic oscillations in resistivity and magnetization [9.132]. This family of materials have a pair of warped planes and a cylinder as its Fermi surface [9.133, 134]. They have a transition to a mysterious Desity Wave (DW), probably SDW, a state at about 8 K [9.135, 136]. Above this temperature they exhibit a series of resistivity peaks due to AMRO (Chap.5) coming from the cylindrical part of the Fermi surface as a function of the tilt angle of the magnetic field from the c-axis which is the least conducting direction. At temperatures below the transition, a series of dips appear instead of peaks [9.137]. After *Kartsovnik* et al. this is considered a commensurability resonance coming from a quasi 1-D band [9.138, 139], as stated in the preceding section. The latter band is not a quasi-1-D plane existing in the normal state since it vanishes after the density-wave formation, but the one which arises from the cylinderical part after reconstruction of the band due to the density wave. With the new DW wave vector $\mathbf{Q}$, the new Fermi surface is formed by overlaping the cylinderical part after displacing it by $n\mathbf{Q}$, with n being intergers; this process is considered to give rise to new quasi 1-D planes and a few additional small cylinders.

In the case of the K salt, for example, the magnetoresistance has a clear kink at H = 23 Tesla and in the magnetic field above this value the Shubnikov-de Haas oscillation due to the normal-state cylinder exhibits qualitative increase of its amplitude [9.140]. This has been considered to be due to the disappearance of the density wave for H > 23 Tesla, probably due to the magnetic breakthrough, which allows a closed orbit to link all parts of the Fermi surface [9.141, 142].

Another interpretation of the dipping pattern for T < 8 K and H < 23 Tesla was proposed by *Yoshioka* [9.143]: The carriers in the cylindrical part remain as in the normal state even in the density state. The dip then appears when the DW periodicity and the magnetic periodicity become commensurate because the above-mentioned carriers increase the conductivity.

Recently, from results of magnetization [9.144, 145] and their own de Haas-van Alphen measurements, *Sasaki* et al. [9.146] argued that the SDW extends up to 23 Tesla all the way from zero to about 11 K, expanding the resitively obtained SDW phase. They observed an anomay in the resistivity at this new boundary as well, but did not find any anomaly of the magnetization at the resistively obtained phase boundary [9.147, 148]. This is reminiscent of the phase diagram of $(TMTSF)_2 ClO_4$ at high fields, which is still controversial.

10. Fullerene Superconductors

A new kind of molecule-based superconductors was initiated by the discovery of superconductivity in alkali-metal-doped fullerene ($A_3 C_{60}$), A representing either K or Rb. In these two cases, the transition temperature T_c is 18 K and 29 K, respectively. The constituent molecule C_{60} has a so-called **soccer-ball shape**, and the crystal is composed of an fcc structure. It is not customary to regard the fullerene molecule which consists solely of carbon atoms, as an organic molecule. The types of molecular structure of the **fullerene molecule** belong to a category different from those of the TTF-type molecule. In the case of fullerene compounds, the crystals are three dimensional. In spite of these differences, the fullerene superconductors exhibit a number of features that are also exhibited by **organic superconductors** treated in this monograph. The similarity between these superconductors is derived from the fact that the principal conducting carriers are π electrons. In both cases, big molecules each with a large number of freedom are assembled, forming relatively narrow bands with a low electron density. Because of this close similarity, we consider doped fullerene superconductors in this chapter. The electronic structure and properties of the Buckminster fullerene C_{60} and its compounds are reviewed in Sects. 10.1 and 2, and the superconducting properties are described in Sect. 10.3. The fullerene relatives of C_{60}, such as C_{24}, C_{70}, etc., are not treated in this book, since superconductivity has only been found in C_{60} compounds.

10.1 Structure and Electronic Properties of C_{60}

The **Buckminster fullerene** C_{60} is a molecule with 60 carbon atoms that occupy the vertices formed by the intersections of 20 hexagonal and 12 pentagonal faces to build a cage (Fig. 10.1). All of the atoms in the molecule are equivalent, and the molecular shape is very close to a sphere. The molecule belongs to the I_h symmetry group, and the high symmetry provides a high degeneracy in the electronic structure. Each carbon atom has three C-C bonds and is surrounded by two hexagons and one pentagon. It possesses a π electron and three σ electrons that form an sp^2 hybrid orbital.

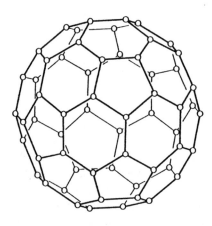

Fig. 10.1. Buckminster fullerene C_{60}

The 180 σ electrons occupy 90 bonding orbitals to form paired spins. The π electrons stay intermediately between the bonding and antibonding states of σ electrons with an energy gap of more than 10 eV. The 60 π electrons occupy 30 bonding π orbitals, and a gap exists between the filled bonding orbitals and the unfilled one with an energy of ≈ 2 eV. The Highest Occupied Molecular Orbital (HOMO) is that of the π states, with 5-fold degeneracy and h_u symmetry, while the Lowest Unoccupied Molecular Orbital (LUMO) has 3-fold degeneracy and t_{1u} symmetry [10.1-3]. The electronic levels are illustrated in Fig. 10.2.

The C-C bonds within a C_{60} molecule are divided into two categories, that of the bonds at the boundaries between pairs of hexagons and that of the bonds at the boundaries between hexagon-pentagon pairs. The former are close to double bonds, with a length of 0.140 nm, while the latter are close to single bonds, with a length of 0.146 nm. These lengths are significantly shorter than the shortest C-C nearest-neighbor distance between the molecules, which is about 0.3 nm. This fact led us to construct a model for a C_{60} crystal consisting of well separated C_{60} molecules. As a result, the electronic structure of the crystal is determined mainly by the bonding interactions within individual molecules, and the band width is determined by electron hopping between the C_{60} molecules. According to angle-resolved photoemission spectra, the valence-band structure of C_{60} exhibits many distinctive features and can well be accounted for by a description which consists of electronic states of the individual molecules with small modifications due to band effects [10.4].

C_{60} molecules aggregate to form a cubic crystal with fcc structure and a lattice constant of 1.42 nm at room temperature. The nearest-neighbor molecular center-to-center distance is 1.002 nm, implying a van der Waals separation of 0.29 nm for a calculated diameter of 0.71 nm. Due to the weak binding, C_{60} molecules reorient rapidly and isotropically, and the

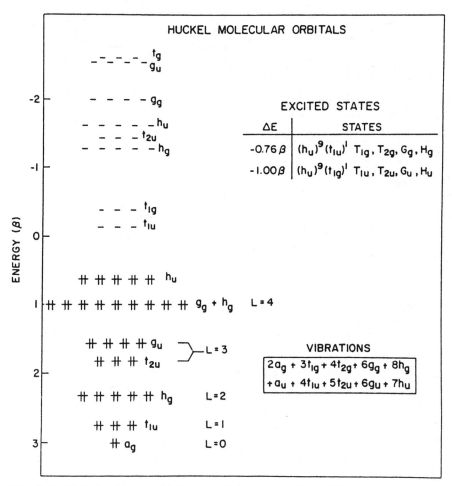

Fig. 10.2. Hückel molecular-orbital energy diagram for C_{60}. From [10.3]

structure is believed to belong to the Fm3m space group above ≈ 260 K. On the low-temperature side, an orientational order develops, and the structure becomes simple cubic, belonging to the space group Pa3 with four C_{60} molecules that occupying the fcc lattice sites [10.5-7].

The band structure of the C_{60} crystal calculated in the Local Density Approximation (LDA) is characterized by an energy gap between the valence band corresponding to HOMO of the molecule and the conduction band corresponding to LUMO [10.8]. Figure 10.3 shows the energy levels of the C_{60} molecules, and the band structure of the fcc C_{60} crystal. The electronic structure expresses the fact that the C_{60} crystal is a semiconductor with a direct gap [10.9]. The calculated energy gap between the h_u and t_{1u} orbitals is ≈ 1.9 eV. If the system had perfect spherical symmetry, the

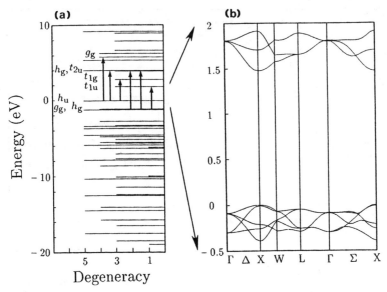

Fig. 10.3. Electronic energy levels of an isolated C_{60} molecule (**a**), and the energy band structure of fcc C_{60} (**b**). The orbitally allowed transitions with excitation energies less than 6 eV are indicated by arrows. From [10.8]

energy levels would be characterized by an angular momentum number l, and the optically-allowed transitions would be from l to $l \pm 1$ states. The energy levels of the C_{60} molecule possess a clear correspondence to the l states. For example, the second- and third-highest occupied states (g_g and h_g) correspond to the $l = 4$ states in the case of spherical symmetry, and h_u corresponds to the $l = 5$ states. The allowed optical transitions are shown by arrows in Fig. 10.3: $h_u \rightarrow t_{1g}$, $h_g \rightarrow t_{1u}$, $h_u \rightarrow h_g$, $g_g \rightarrow t_{2u}$, $h_g \rightarrow t_{2u}$, and $h_u \rightarrow g_g$. Since the energy levels of the C_{60} molecule correspond to the energy bands of the C_{60} crystal, we can expect the optical spectra of the fcc C_{60} crystal to be similar to those of the C_{60} molecule. The excitation energies calculated for the transitions are in qualitative agreement with the observed peaks in photoabsorption measurements [10.11].

Both the valence-band top and the conduction-band bottom are located at the X point in the crystal of the highest $Fm\bar{3}$ symmetry group (Fig. 10.3). The transition between the two, however, is optically forbidden since both the conduction and the valence bands have "ungerade" symmetry under the inversion operation. It should be noted that the rotation of the molecules affects the dispersion of the energy bands since the C_{60} molecule is not perfectly spherical due to the internal bond network. The calculations referred to above are changed due to the molecular rotation at the lattice sites. The position of the conduction minimum changes, and as a result, the energy

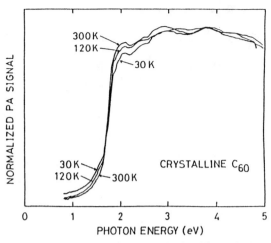

Fig. 10.4. Photoacoustic spectra obtained from single-crystal C_{60}. From [10.13]

gap becomes an indirect one [10.12]. The degeneracies at the special k-points almost disappear.

The effective mass of the conduction band of $Fm\bar{3}$ fcc C_{60} has been calculated to be $1.3m_0$, whereas the valence-band masses are $1.5m_0$ and $3.4m_0$. Using the static dielectric constant $\epsilon = 18 \pm 4$, the ionization for the shallow states are found to be 60 ± 30 meV for the donor states, and 60 ± 30 and 170 ± 70 meV for the acceptor states.

In a single-crystal C_{60}, an optical gap is observed at 1.8 eV in the photoacoustic measurement, as shown in Fig.10.4. This does not correspond to the lowest-allowed transition but it does to the $h_u \rightarrow t_{1u}$ transition [10.14] which is weakly permitted due to the solid-state effect of the C_{60} molecule, and the charge-transfer states accompanied by the t_{1g} states [10.15]. The optical absorption corresponding to the $h_u \rightarrow t_{1g}$ transition in the C_{60} molecule is found at 2.7 eV [10.16]. The optical absorption spectrum obtained for an evaporated C_{60} film is depicted in Fig.10.5. In contrast to the photoacoustic spectrum, a clear absorption edge is not found in this case, although we should notice that the absorbance is presented on a logarithmic scale. In addition to a broad peak appearing near 2.5 eV, the absorption spectrum shows a shoulder near 2.0 eV. This shift is ascribed to the exciton formation and an electron correlation [10.15]. Here we note that the on-site Coulomb energy for C_{60} is estimated to be 1.6 ± 0.2 eV from photoemission and inverse-photoemission spectra [10.18].

With regard to the electrical transport properties, the C_{60} crystal is an insulator with a resistivity of 10^6 $\Omega \cdot$cm or higher at room temperature. The lower resistivities are found in crystals kept away from oxygen gas with the activation energy in the range of $0.15 \div 0.5$ eV, much less than the energy

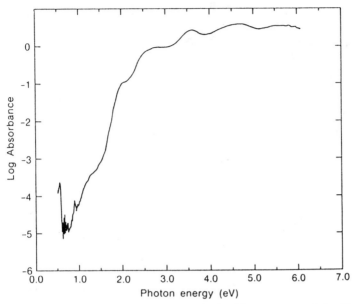

Fig. 10.5. Optical absorption spectrum obtained from evaporated C_{60} film at room temperature. From [10.17]

gap, indicating that C_{60} is not an intrinsic semiconductor. The resistivity is raised significantly in an ambient atmosphere due to carrier trapping caused by the absorption of O_2 [10.19].

10.2 Structure and Electronic Properties of $A_x C_{60}$

In the fcc C_{60} lattice there are two tetrahedral and one octahedral interstitial sites per C_{60}. Tetrahedral and octahedral sites have sufficient room to accommodate spheres of radii of 0.11 nm and 0.21 nm, respectively. When the ionic radii of dopants are smaller than these values, the dopants can be accommodated at the interstitial sites. For a potassium ion, with a radius of 0.13 nm, the phases of the intercalation compounds $K_x C_{60}$ ($x = 1 \div 6$) have been observed. The structures of several compounds have been determined in considerable detail, including fcc $K_3 C_{60}$, body-centered tetragonal $K_4 C_{60}$, and bcc $K_6 C_{60}$ [10.20]. $K_3 C_{60}$ belongs to the space group $Fm\bar{3}m$ with a lattice parameter of 1.424 nm at room temperature. At 300 K, it was demonstrated by a ^{13}C NMR study that C_{60} molecules are dynamically disordered, but the jumping motion is frozen at a low temperature [10.6].

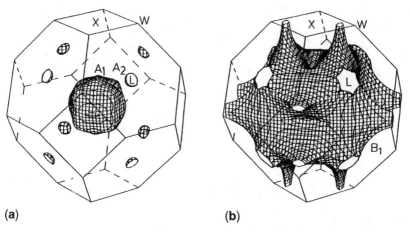

(a)   **(b)**

Fig. 10.6. Fermi surfaces of K_3C_{60}: the surface originating from the lower conduction band (**a**), and the other surface from the upper conduction band (**b**). All surfaces have holes inside. From [10.21]

However, no structural transitions have been observed down to 10 K in X-ray diffraction experiments.

For K_xC_{60}, significant electron transfer from the K atom to the C_{60} molecule takes place, i.e., electrons are transferred from the K 4s orbital to the C_{60} t_{1u} conduction band. The triply degenerate t_{1u} bands can accommodate up to six electrons. Thus, KC_{60}, K_2C_{60} and K_3C_{60} possess unfilled bands, and all are expected to be metallic and ionic, provided that a simple band model is applicable. The electrical conductivity increases with the dopant concentration and shows a maximum for K_3C_{60}.

In the case of K_3C_{60}, three electrons are transferred to the conduction band with t_{1u} symmetry, which is triply degenerate at the Γ point. The Fermi level crosses the lower two of the three bands; about 90 % of the lowest and 60 % of the middle conduction bands being occupied. According to LDA calculations, the lowest band that contributes to conduction yields a Fermi surface which, near the Γ point, contains holes (Fig. 10.6a), while the upper conduction band corresponds a complicated Fermi surface, sketched in Fig. 10.6b. The contributions from these Fermi surfaces to the Fermi-level density of the states are 12 % and 88 %, respectively. As a result, there are both electron-like and hole-like orbits as well as open orbits. It is understood that A_6C_{60} is an insulator, since the three-fold degenerate conduction band is filled.

The electronic structure of pure C_{60} and its compounds have been studied by photoemission and inverse-photoemission measurements of doped C_{60} samples represented by A_xC_{60}, x being the nominal value which ranges from 0 to 6. Isolated photoelectron spectra for K_3C_{60} and K_4C_{60} are displayed in Fig. 10.7 with the corresponding calculated density of states [10.23].

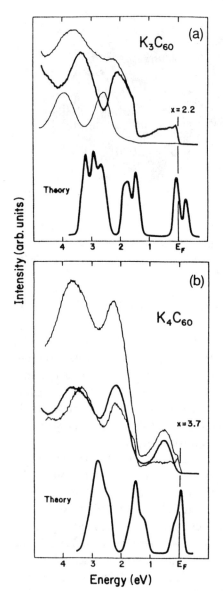

Fig. 10.7. (a) Photoemission spectrum K_3C_{60} (*bold*) deduced from $K_{2.2}C_{60}$ (*upper thin line*) and C_{60} (*lower thin line*) and the calculated density of states of K_3C_{60}. (b) Photoemission spectrum of K_4C_{60} (*bold*) deduced from $K_{3.7}C_{60}$ (*upper thin line*) and K_3C_{60} (*lower thin line*) and the calculated density of states [10.22]

The spectrum of K_3C_{60} shows that K_3C_{60} is a metal with a distinct Fermi-level cut-off, but the LUMO-derived band structure is much broader than predicted by one-electron band theory. Figure 10.7b exhibits that K_4C_{60} is an insulator, although band calculations predict a metallic character. The result is understood by taking into account the effect of electron-electron corrlation: the band is split off by the Hubbard-type interaction with the on-site Coulomb energy of about 1.5 eV. Transport measurements also reveal

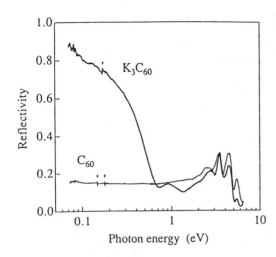

Fig. 10.8. Reflectivity spectra of K_3C_{60} and C_{60}. Vibration modes are marked by arrows. From [10.25]

that K_4C_{60} is an isulator [10.24]. The optical reflectance spectrum of K_3C_{60} is displayed in Fig. 10.8 together with that of C_{60} [10.25]. A clear plasma edge is found in the reflectance spectra of K_3C_{60} at 0.73 eV. The effective mass of the conduction band is obtained as $2.4m_0$, applying a Drude analysis.

We have assumed that complete charge transfer occurs for alkali metals. However, a study of the ESR linewidth that is associated with conduction electrons has revealed that the linewidth varies, depending sensitively on the alkali metals under use. This result is interpreted in terms of incomplete electron transfer from the alkali atoms to C_{60}, with the remaining fraction being greater in the tetrahedral sites than in the octahedral sites of the fcc lattice [10.26]. The idea of incomplete charge transfer has been applied to explain photoemission experiments [10.27].

10.3 Superconductivity of A_3C_{60}

10.3.1 Superconductivity of K_3C_{60} and Rb_3C_{60}

Since the observation of superconductivity in K_3C_{60} [10.28] and Rb_3C_{60} [10.29], supercoductivity has been observed in dozens of other compounds of C_{60}. The transition temperature spans a range of up to 33 K [10.30]. Further, a trace of superconductivity with an onset temperature of 40 K was observed in Cs_3C_{60} under a pressure of 15 kbar [10.31]. However, the C_{60} compounds are unstable under ambient atmosphere. In most cases, weakly linked polycrystalline samples were prepared in sealed ampoules, and their superconducting properties have been characterized by inductive or mag-

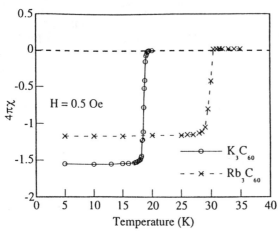

Fig. 10.9. Zero-field-cooled curves for $4\pi\chi$ versus temperature for K_3C_{60} and Rb_3C_{60} single crystals. From [10.32]

netic measurements. In the preparation procedure, the stoichiometry control is not simple, and the proper doping level has been determined by trial and error.

Single-crystal samples, thought to possess good quality, have been prepared by diffusing alkali metal into pristine C_{60} single crystals grown by the sublimation method. Figure 10.9 illustrates the zero-field-cooled **magnetic susceptibility** χ versus temperature for K_3C_{60} and Rb_3C_{60} single crystals. The transition temperatures, defined by the appearance of non-zero χ were found to be 19.3 K and 30 K for K_3C_{60} and Rb_3C_{60}, respectively. The higher than 100 % diamagnetic response of K_3C_{60} is ascribed to uncertainties related to the demagnetization factor that gives values ranging from 0.2 to 0.3. Samples displaying transition widths of less than 1 K ($10 \div 90\%$ shielding) and a high volume fraction are regarded as high-quality superconducting samples. The temperature dependences of the **lower critical field** H_{c1} evaluated by a deviation of the field dependence of the magnetization signal from linearity are displayed in Fig. 10.10, where lines are fits to the standard expression

$$H_{c1}(T) = H_{c1}(0)\left[1 - \left[\frac{T}{T_c}\right]^2\right] . \tag{10.1}$$

The data indicate that $H_{c1}(0)$ is 42 ± 1 Oe for K_3C_{60} and 32 ± 1 Oe for Rb_3C_{60}. These values are smaller than those obtained with powder by a factor of ≈ 3 [10.33]. It should be noted that measurements of $H_{c1}(0)$ on powder samples are susceptible to overestimation as a result of surface defects which can pin flux lines as the field begins to penetrate the sample.

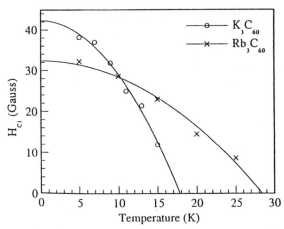

Fig. 10. 10. H_{c1} versus temperature for $K_3 C_{60}$ and $Rb_3 C_{60}$ single crystals. From [10.32]

Single-crystal samples, on the other hand, suffer from underestimation as a result of demagnetization effects. The observation that $H_{c1}(0)$ is larger for $K_3 C_{60}$ is consistent with another report on polycrystalline samples [10.33]. In these cases, we cannot rule out the influence of sample inhomogeneity, in particular for $Rb_3 C_{60}$, due to a lower diffusivity into C_{60} solids. As a result, sample defects seem to play a greater role in $Rb_3 C_{60}$.

The **upper critical fields** have been determined by many groups via magnetic measurements. Among these, an unusual value for $H_{c2}(0)$ of 780 kOe was claimed for $Rb_3 C_{60}$ [10.33]. This exceeds the Pauli weak-coupling limit of 530 kOe. However, later works have yielded the smaller values summarized in Table 10.1 [10.34, 35].

By using the Ginzburg-Landau relation

$$H_{c2}(0) = \frac{\phi_0}{2\pi \xi_{GL}(0)^2} \tag{10.2}$$

obtained in the mean-field approximation, the **coherence length** $\xi_{GL}(0)$ is estimated to be 3.4 nm for $K_3 C_{60}$ and 3.0 nm for $Rb_3 C_{60}$ [10.34]. Here we note that the transition temperatures are rather high, whereas the coherence lengths are on the same order as the lattice spacing. In this situation the effect of thermal fluctuations cannot be dismissed. Based on renormalization theory of the fluctuations that has been developed for high-T_c superconductors [10.36], the value of the coherence length is 2.1 nm for $K_3 C_{60}$ and 1.3 nm for $Rb_3 C_{60}$ [10.37] (Appendix B). These lengths are considerably shorter than those obtained on the basis of a mean-field approximation.

Figure 10.11 plots the resistive transition near T_c for a single crystal of $K_3 C_{60}$ under various magnetic fields. The transition is seen to be rather

Table 10.1. Superconductivity parameters in K_3C_{60}, Rb_3C_{60} and $RbCs_2C_{60}$

Parameter	K_3C_{60}	Rb_3C_{60}	$RbCs_2C_{60}$	Ref.
T_c [K]	19.3	30		10.32
			32.5	10.47
$H_{c1}(0)$ [Oe]	42	32		10.32
$H_{c2}(0)$ [kOe]	280	380		10.34
			500 ± 30	10.47
	175			10.38
$\xi_{GL}(0)$ [nm]	3.4	3.0		10.34
	2.1	1.3		10.37
	4.5			10.38
			2.5 ± 0.1	10.47
λ_L [nm]	800	800		10.42
	480	420		10.44
	600	460		10.46
$2\Delta/k_B T_c$	3.6	3.0		10.42
	5.2	5.3		10.43

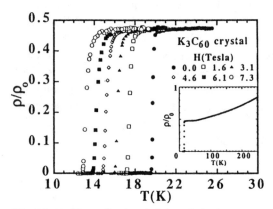

Fig. 10.11. Normalized resistivity of single-crystal K_3C_{60} near T_c for different applied magnetic fields. ρ_0 denotes the resistivity at room temperature. *Inset*: Zero-field temperature dependence of the normalized resistivity. From [10.38]

sharp even in a magnetic field. From the relationship dH_{c2}/dT vs. T, $H_{c2}(0)$ for K_3C_{60} is evaluated to be 175 kOe applying the Werthamer-Helfand-Hohenberg formula [10.39]

$$H_{c2}(0) = 0.69 \left[\frac{dH_{c2}}{dT} \right] T_c \ , \tag{10.3}$$

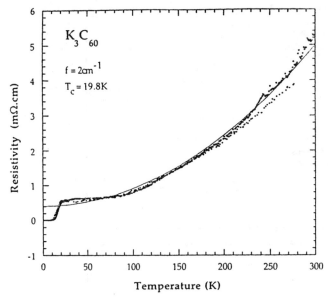

Fig. 10.12. Temperature dependence of the resistivity of K_3C_{60}, measured at microwaves. The *solid line* represents a fit of the data using a functional form of $\rho(T) = a + bT^2$. From [10.41]

and the zero-temperature coherence length $\xi(0) = 4.5$ nm. The value of $\xi(0)$ is significantly larger than values determined by magnetic measurements. It should be noted that, in general, the values of $\xi(0)$ become shorter or $H_{c2}(0)$ increases for granular samples. It is notheworthy, however, that the transition exhibits thermodynamic fluctuations even in single-crystal samples, producing small transient regions of the superconductivity above T_c, i.e., an anomalous increase in the normal-state conductivity known as paraconductivity [10.40].

Electrodynamical responses are useful in investigating polycrystalline conductors. Figure 10.12 exhibits the temperature dependence of the **resistivity** measured at microwave frequency (60 GHz) for K_3C_{60}, giving a room-temperature conductivity of about 200 S/cm [10.41]. The superconductivity transition is found at 19.8 K, but the transition is rather broad. In this case, the normal resistivity exhibits a stronger temperature dependence than that for the single crystal represented in the inset of Fig. 10.11.

The optical conductivity evaluated by the reflectivities both above and below T_c of K_3C_{60} and Rb_3C_{60} is depicted in Fig. 10.13. The optical conductivity below T_c shows a well-defined gap feature which becomes progressively sharper when the temperature is decreased. However, this gap feature is not decreased significantly by raising the temperature from 6 K to T_c, as would be expected according to a simple BCS picture. From optical conductivity, the **superconducting gaps** were obtained as $\Delta = 3.0$ meV

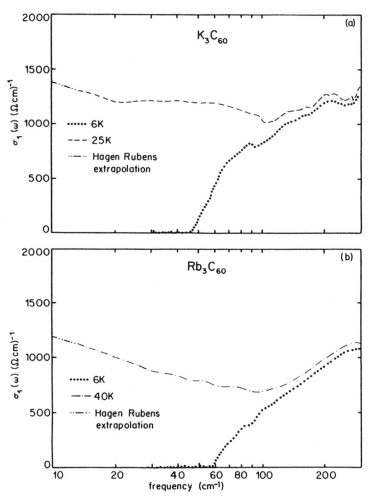

Fig. 10.13. The optical conductivity above and below T_c for K_3C_{60} (**a**) and Rb_3C_{60} (**b**), evaluated from the reflectivity data. From [10.42]

for K_3C_{60} and $\Delta = 3.7$ meV for Rb_3C_{60}. Together with the superconducting transition temperatures $T_c = 19$ and 29 K, this leads to the values 3.6 and 2.98 for the ratio $2\Delta/k_B T_c$ of K_3C_{60} and RbC_{60}, respectively. These values are in good agreement with the weak-coupling BCS result of $2\Delta/k_B T_c = 3.53$. Here, we note that the energy-gap value derived from single-particle tunneling becomes $\Delta = 6.6 \pm 0.4$ meV for Rb_3C_{60}, corresponding to the reduced energy gap of 5.3 [10.43]. The **penetration depth** is derived as $\lambda = 800 \pm 5$ nm for the two compounds. This has been reported to be 480 [10.44] or 600 nm [10.45] for K_3C_{60}, and 420 [10.44] or 460 nm [10.46] for Rb_3C_{60}. Thus, the A_3C_{60} compounds are superconductors of the second kind. The evaluation of the electron mean-free path is affected more strongly by the sample quality with respect to the inhomogeneity and defects. The estimated values are in the range of $1 \div 3$ nm, implying that the system is at the boundary of pure and dirty superconductors.

Studies on the superconducting properties of Cs_2RbC_{60} are rather limited. Some parameters determined by AC susceptibility measurements are listed in Table 10.1 [10.47].

10.3.2 Relationship Between T_c and the Lattice Parameter

The lattice parameter a_0 of A_3C_{60} fullerides can be varied by changing the alkali metals, represented by A. Generally speaking, a_0 increases with an increase of the total volume of the cations A^+. More precisely, the lattice parameters are controlled mainly by the radius of A^+ at a tetrahedral site; that at the octahedral site gives a smaller influence [10.48].

In Fig. 10.14, the T_c values of A_3C_{60} and $A_2A'C_{60}$ are plotted as functions of the lattice parameter. The highest T_c of 33 K is attained with

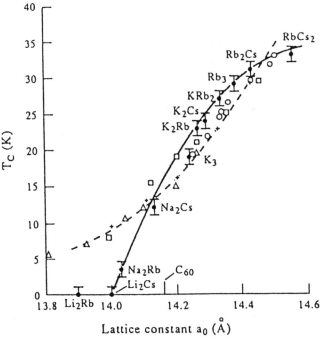

Fig. 10.14. The relationship between T_c and the lattice parameter a_0 for A_3C_{60} (A: Li, Na, K, Rb, Cs and their binary alloys) superconductors. Open triangles and squares are from pressure experiments, and the dotted line represents the T_c-a_0 relationship expected from the simple BCS theory using the density of states due to LDA calculations. The *solid line* is a guide for the eyes. From [10.48]

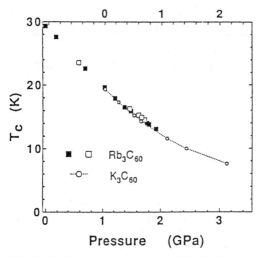

Fig. 10.15. Pressure dependence of T_c for $K_3 C_{60}$ and $Rb_3 C_{60}$. Solid and open symbols correspond to the data taken with increasing and decreasing pressure, respectively. The upper and lower scales are for $K_3 C_{60}$ and $Rb_3 C_{60}$, respectively, where the upper scale is translocated by 1.06 GPa. Both data sets approximately follow the curve $T_c(P) = T_c(0) \exp(-\gamma P)$, where $\gamma = 0.44 \pm 0.03 \, GPa^{-1}$. From [10.33]

$Cs_2 RbC_{60}$ [10.31]. The lattice parameter is also varied by applied pressure. In this case, the pressure affects mostly the inter-ball distance, due to a weak molecular coupling which is susceptible to pressure: the intra-ball bonding is little influenced by a pressure up to 2 GPa. Figure 10.15 shows the pressure dependences of T_c for $K_3 C_{60}$ and $Rb_3 C_{60}$. The curves for these two materials can be made to overlap in the common region of T_c simply by displacing the pressure scale of one with respect to the other about 1 GPa. The simplest hypothesis to account for this coincidence is that T_c is a function only of a_0. The pressure dependence of T_c is superimposed via open triangles ($K_3 C_{60}$) and squares ($Rb_3 C_{60}$) in Fig. 10.14. The dotted line represents the T_c versus a_0 relationship that is expected from simple BCS theory, using the density of states at the Fermi surface calculated by LDA [10.50].

The effects of the lattice parameter and/or the pressure on T_c are very large; recall that T_c increases with an increase of the lattice spacing. It is also asserted that T_c is a function of only the inter-ball distance, and that the identity of the alkali metals is not of critical significance, so that the role of the alkali ions would be reduced to that of an electron donor and an intermolecular spacer. This situation resembles the chemical and/or pressure dependences of the $(TMTSF)_2 X$ and $(ET)_2 X$ superconductors. In accordance with the change of the lattice parameter over the range $a_0 = 1.45$ nm to 1.39 nm, the distance between the carbon atoms of neighboring

C_{60} molecule changes from 0.32 to 0.28 nm, provided that the diameter of the C_{60} molecule remains fixed. It is interesting that superconductivity ceases to exist in the form of Na_3C_{60} with a lattice parameter of $a_0 = 1.419$ nm [10.49]. Detailed structural studies have revealed that the stable phase of Na-containing fulleride at low temperatures is Pa3 simple-cubic, in contrast to the Fm3m fcc phase at high temperatures [10.51]. This should be compared with the situation for K_3C_{60} and Rb_3C_{60} and other $A_2A'C_{60}$ fullerides, in which the appearance of the fcc phase is continued to low temperatures [10.52].

10.3.3 Mechanism of the Superconductivity in A_3C_{60}

As molecule-based superconductors, where π electrons dominate the electronic properties, it is interesting to consider the similarities and differences between A_3C_{60} and $(ET)_2X$ or $(TMTSF)_2X$. There are proposals to explain the mechanism of superconductivity based on phonon-mediated electronic processes. The phonon mechanism can be further subdivided into models that rely upon intramolecular phonons [10.53-55] (Sect. 8.3.6) and those which incorporate intermolecular translational modes [10.56] and librons [10.57] or alkali-C_{60} optic phonons [10.58, 59]. The electronic processes are concerned with an appreciable on-site Coulomb energy and the narrow band [10.60], and somewhat complicated Fermi surfaces (Fig. 10.6) with valley degeneracy [10.61].

In order to clarify the **phonon-mediated process**, it is crucial to identify the isotope-substitution effect: T_c is proportional to $M_i^{-\alpha_i}$ (M_i is the mass of a given atomic constituent and α_i is the isotope-shift exponent). First, we note that changing A in A_3C_{60} from K to Rb causes an increase in T_c in contrast to the atomic-weight increase. This rules out the role of acoustic phonons in the appearance of superconductivity. The effect of a Rb substitution can be understood in terms of the spacing between C_{60} molecules, resulting in a higher density of states at the Fermi level $N(\epsilon_F)$ due to a weak intermolecular electron transfer. With respect to the alkali-C_{60} optical phonons, the rubidium isotope effect was investigated for Rb_3C_{60}. Naturally occuring Rb was replaced with enriched ^{85}Rb and ^{87}Rb atomic species. The results of this investigation showed that changing the mass of the Rb dopant does not change T_c [10.62], or induces a small negative isotope shift of $\alpha_{Rb} = -0.028 \pm 0.036$ [10.63]. These results rule out the possible contribution of alkali-C_{60} optical phonons to the appearance of superconductivity.

On the other hand, the effect of carbon substitution is substantial, although the claimed values of α_c are somewhat scattered. *Ramiretz* et al.

[10.64] reported $\alpha_c = 0.37 \pm 0.05$ for a $(75 \pm 5)\%$ substitution of ^{13}C for ^{12}C in Rb_3C_{60}, while *Ebbesen* et al. [10.65] found the remarkably high value $\alpha_c = 1.4 \pm 0.5$ for Rb_3C_{60} with three different concentrations of ^{13}C. *Zakhidov* et al. [10.66] noted the consistent value $\alpha_c = 1.6 \pm 0.2$. The distribution of α_c values is due partly to the uncertainty in the evaluation of the intrinsic T_c, which is obscured by disorders and inhomogeneity, but also due partly to the difference in the site distribution of ^{13}C within C_{60}, which affects the molecular vibrational mode. A more homogeneous carbon-isotopic distribution seems to yield a smaller carbon-isotope effect, such as $\alpha_c = 0.3 \pm 0.05$ for K_3C_{60} and Rb_3C_{60} [10.67]. In any case, the sizable isotope shifts of T_c suggest the involvement of intra-ball phonons in the pairing mechanism. Neutron spectroscopy has revealed that the radial modes, which allow finite σ-π mixing, play a dominant role in the electron-phonon coupling [10.68].

Based on the evaluation of $N(E_F)$ by the measurement of ^{13}C NMR, the correlation between $N(E_F)$ and T_c has been studied [10.69]. Analyzing the results in terms of the McMillan formula, the electron-phonon coupling constants and a relevant phonon energy larger than 600 K were derived. Accordingly, it has been asserted that the A_3C_{60} superconductor is described within the framework of the conventional phonon-mediated BCS superconductivity.

The increase of $N(E_F)$ for larger intermolecular spacings upon replacement of K by Rb in A_3C_{60} has been revealed by measurements of the spin-lattice relaxation time T_1 of NMR [10.69-71]. $N(E_F)$ of $20 \div 25$ states $eV^{-1}C_{60}^{-1}$ for K_3C_{60} is increased by $\approx 40\%$ for the case of Rb_3C_{60} [10.70]. The relationship between $N(E_F)$ and the lattice parameter of A_3C_{60}, displaying a wide range of T_c from Na_2RbC_{60} ($T_c = 3.5K$, $a_0 = 1.4028nm$) to $RbCs_2C_{60}$ ($T_c = 33K$, $a_0 = 1.4555nm$), is displayed in Fig. 10.16, where $N(E_F)$ has been derived from T_1. This graph is consistent with the LDA calculation [10.72]. The reported results from the static magnetic susceptibility [10.73], thermoelectric power [10.74] and optical reflectance [10.25] are also consistent with the evaluation of $N(E_F)$. On the other hand, the value derived from photoelectron spectroscopy is 1.9 ± 0.1 $eV^{-1}C_{60}^{-1}$ for K_3C_{60} [10.75]. This value is anomalously small, and the reason for this is not yet understood.

NMR can provide crucial tests of superconductivity with regard to the symmetry of the superconductivity wave function. *Tycko* et al. [10.46] reported ^{13}C NMR investigations of K_3C_{60} and Rb_3C_{60} asserting the absence or near-absence of Hebel-Slichter coherence peaks in spin-lattice relaxation. In a later work *Stenger* et al. [10.76] found a Hebel-Slichter peak at a low magnetic field (15kOe), which is suppressed at higher fields (90kOe) in Rb_2CsC_{60}. The presence of a Hebel-Slichter peak makes evident that superconductivity is BCS-like and is associated with an isotropic energy gap at

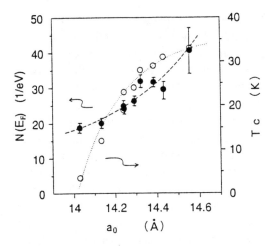

Fig. 10.16. The relationship between the density of states $N(E_F)$ and the lattice parameter a_0. The relation between T_c and a_0 is also shown. From [10.71]

the Fermi surface. The suppression of the peak by magnetic fields higher than half of H_{c2} can be understood in terms of a theoretical description by *Cyrot* [10.77] but 90 kOe is less than 1/5 of H_{c2}. The Hebel-Slichter peak was also found by a muon-spin relaxation experiment in Rb_3C_{60} [10.78]. The spin-exchange scattering of an endohedral muonium ($\mu^+\mu^-$) with thermal excitation provides the coherence peak just below T_c in $1/T_1$. It can be fitted to the conventional Hebel-Slichter theory for the spin relaxation in a superconductor with a broadened BCS density of states.

With regard to the electronic process, **electron correlation** effects within single C_{60} molecules are considered to play a central role in electron pairing [10.60]. The attractive interaction has been evaluated on the basis of the Hubbard model for a single C_{60} molecule and the possibility of singlet superconductivity that depends on the effective intra-ball electron-electron repulsion and the inter-ball hopping amplitude. It was argued also that the isotope substitution effect in a superconductivity transition can be explained based on the phonon-mediated attraction with on-site Coulomb repulsion [10.79].

10.4 Other Fulleride Superconductors

The lattice parameter dependence of T_c for A_3C_{60} implies that cations simply play the role of a spacer, and their identity seems not to be significant. This view has been further supported by the raise of T_c to 29.6 K by ammoniation of Na_2CsC_{60} with T_c of 10.5 K, as $(NH_3)_4Na_2CsC_{60}$, in which the octahedral site is occupied by a tetrahedron of four NH_3 molecules with the Na atom at the center [10.80]. Following this method to expand the lattice, the NH_3 molecule is introduced into NaA_2C_{60} (A = K,

Rb) as $(NH_3)_x NaA_2 C_{60}$ ($x \simeq 0.5 \div 1.0$). The structure of this latter compound is characterized as fcc [10.80]. However, a decrease in T_c was observed with a lattice-parameter expansion in this case. According to the detailed structure analysis, this is not a simple expansion, but the Na-NH$_3$ cluster is accommodated at the octahedral site with Na$^+$ at a position of off-center by $0.05 \div 0.06$ nm.

For dopants other than alkali metal ions, it has been found that alkaline earth metals, such as Ca, Ba and Sr, yield superconductors like simple-cubic $Ca_5 C_{60}$ and orthorhombic $Ba_4 C_{60}$ at $T_c = 8.4$ and 7 K, respectively [10.82, 83]. According to the electronic structure of $A_x C_{60}$, the electron transfer from dopants causes the occupation of bands that correspond to LUMO with relatively little modification of the fullerene orbits. Based on this observation, it is believed that superconductivity in the fullerides with alkali-earth metals is related to the filling of the second LUMO band. By mixing the alkali element with an alkaline earth element, the superconductor $K_3 Ba_3 C_{60}$ with T_c of 5.6 K was formed. It turns out that the salt is a solid solution of $K_6 C_{60}$ and $Ba_6 C_{60}$ with a bcc structure [10.84].

Rare-earth metal dopants, such as Yb and Sm, have been found to yield superconductivity in the form $Yb_{2.75} C_{60}$ with an orthorhombic structure at 6 K and $Sm_x C_{60}$ ($x \leq 3$) at 8 K [10.85, 86]. Similarly to the case with an alkali-metal-doped $A_3 C_{60}$, T_c for the rare-earth metal doped C_{60} increases with increasing lattice parameter, but the salt is not easily studied due to problems related to the stoichiometry and the dopant concentration. Further, various doping trials have been carried out in an attempt to find new superconductors. Combinations of the C_{60} molecule with some alkali metal alloys (RbTl$_{1.5}$ with $T_c = 27.5$ K [10.87], and CsTl$_2$, CsBi or CsHg$_{1.1}$ with T_c close to 30 K [10.88]) exhibit superconductivity. The report of superconductivity with values of $T_c = 42$ K in C_{60} samples doped with Rb-Tl alloy [10.89] is interesting, but these results have not yet been reproduced independently.

11. Design, Synthesis, and Crystal Growth of Organic Metals and Superconductors

This chapter deals with concepts, strategies, methods, and ideas that concern the design of molecular conductors and superconductors from a chemist's point of view. The structural and electronic requirements, and thus the design of organic metals and superconductors, in general, are described in Sects.11.1 and 2. The structure-property relationships of the quasi one-dimensional TMTSF system are treated in Sect.11.3, for the two-dimensional system, mainly the ET system, in Sects.11.4-7, and for the three-dimensional C_{60} system in Sect.11.8. Crystal-growth features are outlined in Sect.11.9 for typical examples. The chemicals dealt with in this chapter, of which the readers of the preceding chapters are probably not much familiar with, are depicted in Fig.11.1 and other appropriate places.

11.1 Birth of TTF, and the Requirements for an Organic Metal

11.1.1 Routes to the TTF Molecule

The discovery of the conductive organic compounds, perylene · halides, by *Akamatu* et al. in 1954 [11.1] was a consequence of studies on radical molecules of the polycyclic aromatic hydrocarbons such as perylene, anthanthrene, violanthrene, etc. in 1930s mainly by *Brass* and *Clar* [11.2], and *Zinke* et al. [11.3]. In those days the perylene · halides were one of the controversial materials; whether they had radical electrons or not [11.3,4]. Observations of the small diamagnetic susceptibility of perylene · halides by *Akamatu* and *Matsunaga* [11.5] and of the high conductivity of the graphite intercalated compounds by *McDonnell* et al. [11.6] in early 1950s were combined and led to the first highly conductive perylene · bromine complexes ($10^0 \div 10^{-3}$ S/cm [11.1]). This research demonstrated the importance of **mobile radical electrons** in solids to realize a high electrical conductivity. These experiments and *Mulliken*'s **Charge Transfer** (CT) **theory** in the 1950s [11.7] had promoted the development of conductive materials. Although a great number of polycyclic aromatic hydrocarbons and their alkyl and alkoxy derivatives were examined by chemical or electrochemical doping methods using halogens, alkali metals, or organic and inorganic electrolytes, the achievements were very limited [11.8-11]. The main reason lies in the chemical instability of the anion or cation radical molecules of conventional aromatic hydrocarbons including graphite. Even though the donor or acceptor strength, and also the stability of a radical molecule increase with increasing size of the molecule among the aromatic hydrocarbons, both the solubility in conventional organic solvents and the ability of single-crystal formation rapidly decrease. Such disadvantages in the stability, solubility, and the donor or acceptor ability have been relaxed by adding a variety of functional groups of different **Hammett** σ to the parent skelton of the aromatic hydrocarbons with keeping the molecular size small, such as *p*-phenylenediamine as an electron donor (D) or TCNQ (TetraCyaNoQuinodimethane) as an electron acceptor (A) molecule.

Structural and physical studies of semiconducting CT complexes of TCNQ from the 1960s to the early 1970s had revealed two main requirements for the conductivity, namely, (1) a **uniform segregated stacking** *of the same kind of component molecules*, and (2) *the fractional CT state* (**uniform partial CT**) *of the molecules* [11.12-17]. The uniform partial CT state ($D^{+\rho} A^{-\rho}$, $0.5 \leq \rho < 1$, ρ: degree of CT, Sect. 11.6) of the component molecules is one of the most important design considerations and is a consequence of the relatively large **on-site Coulomb repulsive energy** U of conventional organic molecules. U is reduced in solids but, still the effective value (U_{eff}) is larger or comparable to the bandwidth W for most of the low-dimensional molecular metals [11.18-21].

Besides these two requirements, the **polarizability** has been one of the key issues for the molecular design in the field of molecular metals and superconductors. *Little'*s theory or the **excitonic superconductivity** demonstrates the importance of polarizability [11.22]. During the course of the study of TCNQ anion radical salts in 1960s; two very interesting works with respect to the molecular design emerged, one is a theory and the other an experiment. *LeBlanc* proposed that U of a TCNQ molecule is decreased by a factor of about $(1-\alpha/r^3)$ by the presence of a polarizable cation, α being the molecular polarizability of the cation, and r is the distance between TCNQ and the cation molecules [11.23]. According to this idea, a variety of cyanine dyes were tested as the countercation of TCNQ anion radical salts [11.24]. The conductivity increases with increasing size of the dye molecules at first, as expected, but it decreases rapidly with a further increase of the size. The rapid decrease of conductivity can be explained by a mismatch in the size; namely, too big a size of the dye molecule destroys the uniformity of the TCNQ segregated column. Therefore, *to increase the molecular polarizability by keeping the molecular size small* became one of the requirements of organic conductors.

From the standpoint of structural organic chemistry, **fulvalene** compounds in which two odd-numbered alternating conjugated rings are connected by a double bond (a in Fig.11.1), have been the key materials for both synthesis and theory with regard to **aromaticity**. The aromatization energy is the resonance stabilization energy of an aromatic hydrocarbon. The stabilization due to aromaticity is conventionally called **Hückel's law**. A conjugated ring having a number of $(4n+2)$ π-electrons is stable, n being an integer [11.25, 26]. Accordingly one of the fulvalenes, heptafulvalene (b in Fig.11.1) was thought to provide a not so stable neutral molecule since each ring contains seven π-electrons, but to give stable mono- and dication molecules since they contain aromatic ring(s) of six $(n=1)$ π-electrons. It was observed that the neutral heptafulvalene molecule, which was synthesized in 1959, has a bond alternation in a solid. However, it gained electron delocalization upon ionization into the cation radical molecule. Nevertheless, the monocation and dication molecules, which have $(6\pi\text{-}7\pi)^+$ and $(6\pi\text{-}6\pi)^{2+}$ electronic structures, respectively, were found to be not so thermodynamically stable.

The delocalization between the carbon $2p_\pi$ electrons and the non-bonding electrons of hetero atoms was extensively studied in the field of heterocyclic aromatic hydrocarbon chemistry in 1960s. It was known that an ethylene group (C=C), which has two $2p_\pi$ electrons, is equal, in the numbers of electrons, to the chalcogen atoms (O, S, Se, Te) or the NH group in terms of chemical substitution (**equi-electron substitution**). Furthermore, the in-

perylene

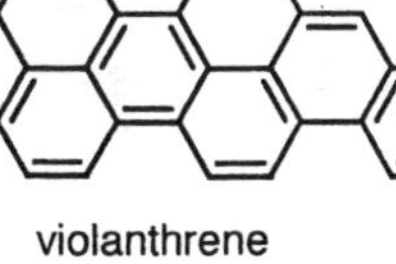

anthanthrene

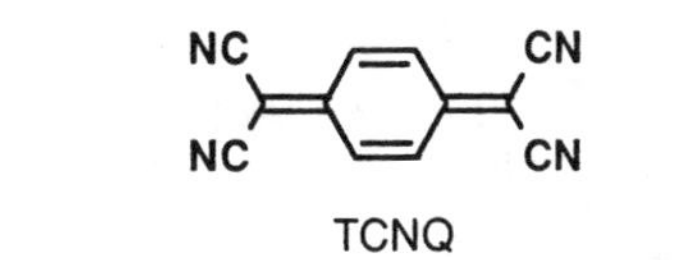

violanthrene

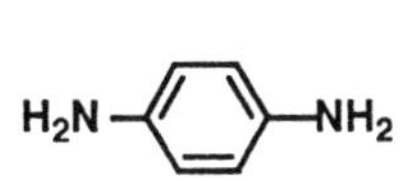

TCNQ

p-phenylenediamine

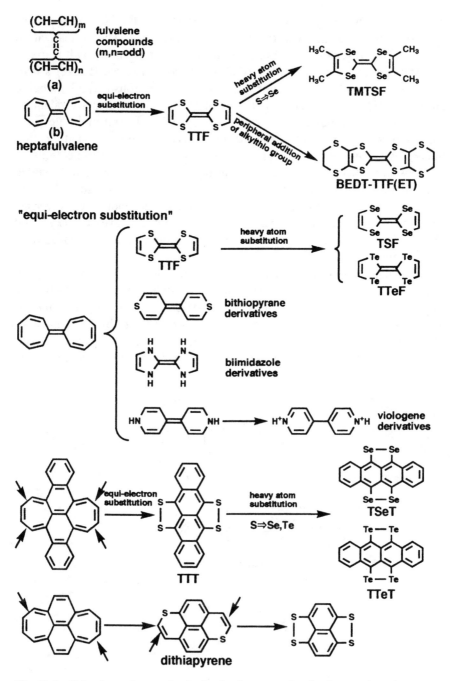

Fig. 11.1. Molecules and strategies in the development of molecular metals and superconductors

troduction of S or Se into a molecule has a big advantage for increasing both the stability and the polarizability by keeping the molecular size small. In 1970, *Coffen* [11.27] and *Wudl* et al. [11.28], synthesized TTF (Tetra-ThiaFulvalene), independently. TTF has a structure equi-electronic with that of heptafulvalene, but it has a superior stability in the neutral and ionized states. In Fig.11.1 other electron-donor molecules which were developed or explained by the concepts of *increased polarizability*, *Hückel's law* and/or *equi-electron substitution* are depicted.

11.1.2 Characteristics of the TTF Molecule

Hereafter, the donor strength, polarizability, molecular U and HOMO of a TTF molecule and the requirements of an organic metal for a TTF CT complex are described. Table 11.1 compares the vertical **ionization potentials** (I^v) of typical donor molecules, of molecular metals and superconductors [11.29-33]. The molecular polarizability (α), and the first **redox potential** $E^1(D)$ [usually measured by the **Cyclic Voltammetry (CV)**], i.e.,

$$E^1(D) = \frac{1}{2}[E^p(D^0 \rightarrow D^+) + E^p(D^+ \rightarrow D^0)]$$
$$= I^{ad}(D) - \Delta E(solv) - C, \tag{11.1}$$

where E^p, I^{ad}, $\Delta E(solv)$ are the redox peak potential, the adiabatic ionization potential, and the difference of solvation energies of neutral and cation molecules, respectively. The constant C is related to the reference electrode and in the case of the Saturated Calomel Electrode (SCE), contains the work function of Hg and the electrode potential of SCE against the hydrogen electrode. Table 11.1 also lists the difference between the first and second redox potentials (ΔE^{12}) and the **CT transition energy** ($h\nu_{CT}$) of a CT complex of sym-TriNitroBenzene (TNB) in chloroform which is given by

$$h\nu_{CT}(D \cdot TNB) = I^v(D) - E^v(TNB) - C', \tag{11.2}$$

where E^v is the vertical **electron affinity**, and C' is mainly the electrostatic Coulomb energy between D^+ and TNB^-.

The I^v, $E^1(D)$ or $h\nu_{CT}(D \cdot TNB)$ values are a measure of the donor strength. Although I^{ad} has been used as a measure of the donor strength, unexpectedly high values of I^{ad} observed for TTF and TSF (TetraSelenaFulvalene) molecules among the tetrachalcogenafulvalene derivatives suggested the inadequency of using I^{ad} until they will be re-examined. The I^v, $E^1(D)$, and $h\nu_{CT}(D \cdot TNB)$ values are usually linearly related to each other. TTF is a

Table 11.1. Characteristics of the component molecules of molecular metals and superconductors

Donor	Ionization potential I [eV][a] molecular polarizability α [10^{-32} m^3][b]	Redox potential[c,d] E^1 [V] ΔE^{12} [V]	hν_{CT} (D·TNB)[e] [10^3 cm^{-1}]
Perylene	I^v = 7.00 (I^{ad} = 6.90) α = 35.2	1.04	22.2
TTF	6.70 (6.26) 16.5	E^1 = 0.37 ΔE^{12} = 0.39	15.4
TMTSF	6.58 (6.27) 30.7	0.45 0.28	15.1
BEDT-TTF (ET)	6.7 (6.30) 37.9	0.53 0.25	16.1
BEDO-TTF (BO)	6.46 (6.12)	0.43 0.25	15.0
TMTTF	6.38 (6.03) 21.0	0.29 0.36	13.9
C$_{60}$	7.6 E^v = 2.65[f]	−0.37[g] 0.42[g]	13.7[h]

[a] I^v, I^{ad}: Vertical and adiabatic ionization potential, respectively [11.29-32]

[b] [11.29] and private commun. with N. Sato (1997)

[c] Versus SCE, Pt electrode, CH$_3$CN, 0.1 M TBA·BF$_4$, 20 ÷ 22°C, 10 ÷ 20 mV/s

[d] For DMET, BEDT-TSF (BETS) and DTEDT, the first redox potentials were reported as 0.52 V (vs. SCE, CH$_3$CN), 0.27 V (vs. Ag/AgNO$_3$, benzonitrile), and 0.37 V (vs. SCE, benzonitrile), respectively

[e] hν_{CT}(D·TNB): Charge transfer absorption of a complex with s-trinitrobenzene (TNB) in CHCl$_3$ at 20 ÷ 22°C

[f] E^v: Vertical electron affinity [11.33]

[g] Versus SCE, Pt electrode, CH$_3$CN/toluene(1/5), 0.1 M TBA·BF$_4$, 20 ÷ 22°C, 10 ÷ 20 mV/s

[h] hν_{CT} of TTF complex in o-dichlorobenzene

donor stronger than perylene by 0.3 eV in the I^v value. The hν_{CT} (D·TNB) and E^1 (D) values are in agreement, along the line of the donor strength.

The electrochemical redox potentials are the least adequate parameters from the point of physics to measure the donor (also acceptor) strength among them due to the complicated and significant contributions of the solvation term to E^1 (D) in (11.1). However, the E^1 (D or A) values are most frequently employed in the literature, since the I^v values are not always available for most of the donor molecules, and the $h\nu_{CT}$ (D·TNB) values are difficult to measure for extremely strong electron donors including alkali metals. Of course, in many cases, the redox process is irreversible [E^p ($D^0 \rightarrow D^+$) − E^p ($D^+ \rightarrow D^0$) > 59mV], and only the E^p ($D^0 \rightarrow D^+$) value can be utilized as a measure of the donor strength. However, strictly speaking, a comparison of the donor strength should be done only with E^1 (D) values. In addition, we should be reminded that the E^1 (D) values [also in E^1 (A) for an acceptor molecule and any polarographic data] are sensitive to the experimental conditions such as the reference electrode, solvent, the kinds and concentration of the supporting electrolyte, temperature, scan speed, etc., but especially to the reference electrode and solvent. So, it is highly recommended to examine the validity of the measured E^1 (D or A) value by a plot of E^1 (D or A) and $h\nu_{CT}$ (D·TNB or pyrene·A) data. Furthermore, it should be remembered that the solvation energy in E^1 (D or A) values, ΔE(solv), varies largely from system to system (such as TTF and p-phenylenediamine systems for a donor, and p-benzoquinone, TCNQ and DCNQI systems for an acceptor) even under the same conditions [11.34].

ΔE^{12} is a measure of both U and the thermodynamic stability of radical molecules in a solution (not in a solid). In order to yield a highly conductive metal, a small ΔE^{12} is favorable, though it means that a dispropotionation reaction, such as $2D^+ \rightarrow D^{2+} + D^0$, easily takes place in solution and to obtain crystals of good quality becomes difficult. The ΔE^{12} values of TTF,

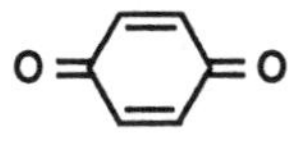

p-benzoquinone

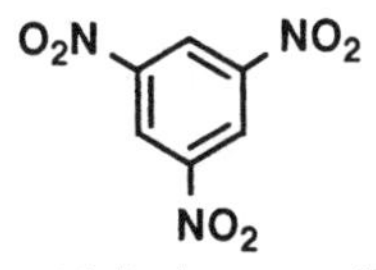

sym-trinitrobenzene (TNB)

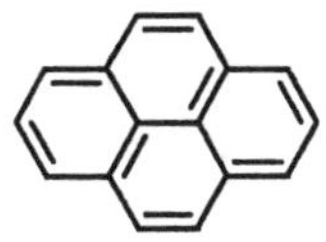

pyrene

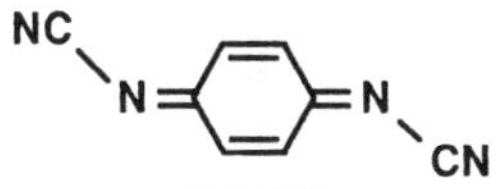

DCNQI

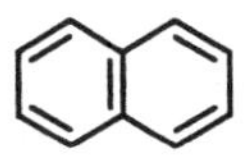

naphthalene

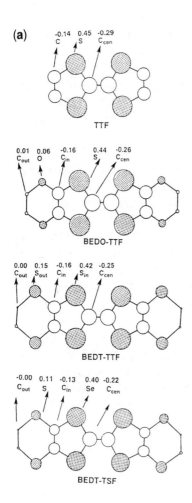

(a)

TTF

BEDO-TTF

BEDT-TTF

BEDT-TSF

TMTSF (TetraMethylTetraSelenaFulvalene) and ET (BEDT-TTF, BisEthyleneDiThio-TTF) suggest that the U values decrease in this order. The molecular polarizability of TTF, which is calculated from the difference of the ionization potentials in the gas and the solid phase of TTF, is nearly half of that of TMTSF. The exact relation between ΔE^{12} or U, and the formation and T_c of a superconductor are still not known.

HOMO of TTF has the b_{1u} symmetry, and the charge density of the atomic π orbitals calculated by the extended Hückel method are exhibited in Fig. 11.2. It is noteworthy that the sulfur atoms produce a charge density high enough to construct a two-dimensional electronic structure, though it has not yet been realized that TTF molecules yield a two-dimensional crystal structure with comparable face-to-face and side-by-side interactions.

In order to have a partial CT state for a DA-type 1:1 complex, *a closely matched redox difference between D and A ($-0.02 \leq \Delta E_{DA} \leq$*

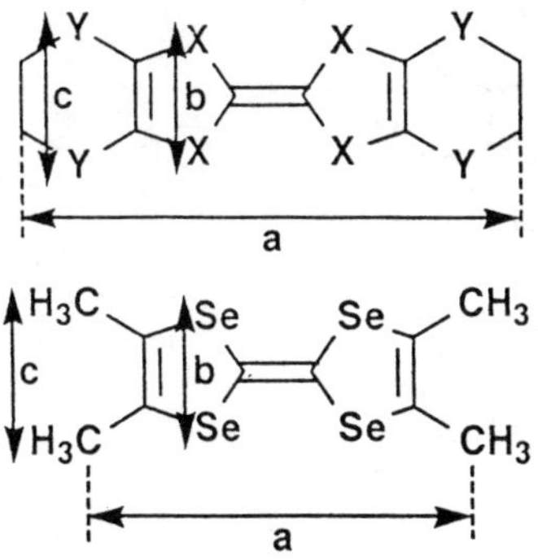

a: molecular length
  between the carbon atoms of
  terminal ethylene(or methyl) groups

b: molecular width
  between the innner chalcogen atoms

c: molecular width
  between the outer chalcogen atoms
  (or carbon atoms of methyl groups)

	a / Å	b / Å	c / Å
X=S Y=O BEDO-TTF(BO)	10.1	2.99	2.90
X=S Y=S BEDT-TTF(ET)	11.8	2.96	3.54
X=Se Y=S BEDT-TSF(BETS)	12.1	3.19	3.52
TMTSF	9.15	3.15	3.12

Fig. 11.2. (a) Coefficients of the Highest Occupied Molecular Orbital (HOMO) by the extended Hückel method including d-orbitals of sulfur and selenium atoms but excluding f-orbitals of selenium atoms, and (b) molecular geometry of electron donor molecules based on the molecular structures in $(BO)_2 I_3 (H_2O)_3$, $(ET)_2 ClO_4 (TCE)_{0.5}$, $(BETS)_2 GaCl_x$, and $(TMTSF)_2 ClO_4$

+0.36V) was found empirically to be important [11.21, 35, 36]. The oxidation potential of TTF (0.37V vs. SCE) indicates that the partial CT state is achieved with an acceptor of the reduction potentials in the range of +0.01 ÷ +0.39V vs. SCE, where TCNQ and p-chloranil reside. For achieving a partial CT state in a cation or anion radical salt, in which ΔE_{DA} is less than −0.02, *a stoichiometry other than 1:1* is the necessary condition except for the C_{60} complex (Sect. 11.8). In this sence, it is not adequate to stress much the importance of the limits of the ionization potential of the donor [11.35, 37] or the electron affinity of the acceptor, since even a very weak donor, naphthalene (I^v = 8.15eV) [11.8], and a very weak acceptor,

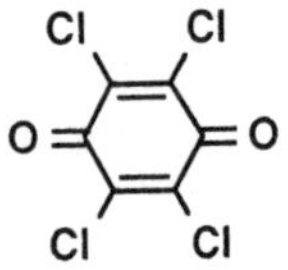

Cl   Cl

p-chloranil

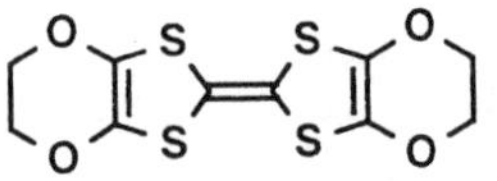

BEDO-TTF

TTP    M(dmit)$_2$

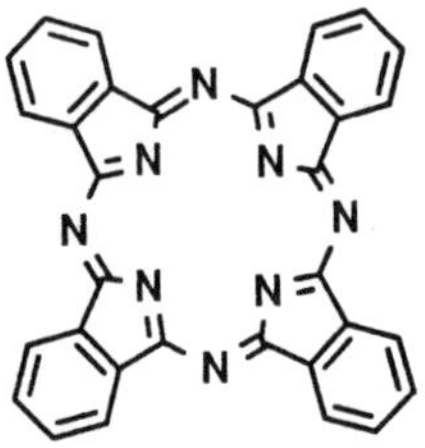

phthalocyanine (PC)

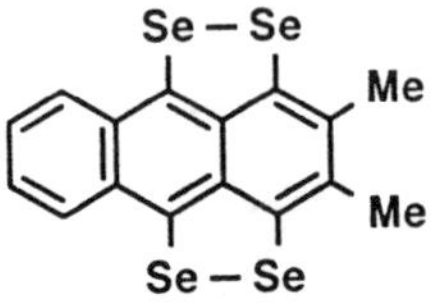

Se — Se
Me
Me
Se — Se

dimethyltetraselenaanthracene

C_{60}, yield conducting radical salts, though the stability of the ion-radical species under ambient conditions decreases considerably.

To achieve a uniformly segregated column for conduction, no generally acceptable concepts or methods have been developed yet, though information on the achievement of segregated stacking or self-assembled stacking has been accumulated.

With increasing the dimensionality and/or π-conjugation these two essential requirements are modified to include the layered structure (ET compounds) or even alternating stacking (C_{60} compounds) and ca. $1/4 \leq \rho < 1$ {ET, BEDO-TTF (BO, BisEthyleneDiOxy-TTF), TTP (TetraThiaPentalene) derivatives, M(dmit)$_2$ and phthalocyanine compounds [11.38-40]} or even ρ being an integer (ET, TTP, dimethyl-tetraselenoanthracene [11.41-44] and C_{60} compounds), as will be discussed in Sect. 11.6.

In order to have more general information concerning the CT complex formation and the molecular design of molecular conductors, the literature should be consulted [11.45-52], in addition to [11.17-21, 26, 36-38].

11.2 Suppression of the Transition to an Insulating State

From a practical point of view, the starting point of molecular superconductors is to find a way to suppress the transition to an insulating state inherent in a low-dimensional metal, by eliminating the nesting of the Fermi surface. A few strategies have been proposed for this purpose. The two major ones are (i) the introduction of **disorder** and (ii) the increase of the **dimensionality** of the Fermi surface. Several novel organic metals of low-dimensionality were known to retain the metallic character down to very low temperatures. The structure-property relationships of these complexes have led to the above-mentioned important strategies for molecular and crystal designs.

11.2.1 Disorder

There are two kinds of disorders from the point of molecular design and crystal preparation. The first one (1a) is tentatively termed **internal disorder** which is provided by one constituent component constructing the low-dimensional Fermi surface. The second one (1b) is the **external disorder** which is created by the counterpart that does not contribute to the electron transport directly. Since the bandwidth of organic metals is rather narrow, the effect of disorder is expected to be large.

Examples of a donor · acceptor type are TCNQ complexes of trans-Di-EthylDiMethyl-TSF (DEDMTSF) [11.53], dithiapyrene (Fig. 11.1), and some mixed crystals [11.53]. A good contrast between a sharp metal-insulator transition at 57 K in TMTSF · TCNQ and a smeared out one at 28 K in DEDMTSF · TCNQ demonstrates the importance of *low-symmetric component molecules* [11.53]. The highly-conductive metallic nature of a TCNQ complex of dithiapyrene down to 4 K strongly supported the importance. In this complex, the low-symmetric donor molecules induce positional disorder of less than 10 % in the crystal [11.54]. In addition, in the mixed crystal of TMTSF · (dimethyl-TCNQ)$_{0.75}$ (methyl-TCNQ)$_{0.25}$, the metal-insulator transition is smeared out markedly by the static disorder of the acceptor part [11.53]. In the case of DEDMTSF · TCNQ, the positive Seebeck coefficient in the metallic region indicates that the donor part which creates the positional disorder dominates the transport (1a). The dithiapyrene complex has negative, and the above-mentioned mixed crystal has positive Seebeck coefficients. Thus, the low-dimensional Fermi surface derived from the TCNQ or TMTSF column plays an essential role in the transport, and is thought to be modulated by the disorder of the counterpart so as to avoid perfect nesting (2b).

Me

NC ... CN

NC ... CN

Me

dimethyl-TCNQ

Et—Se ... Se—Me

Me—Se ... Se—Et

trans-diethyldimethyl-TSF

Se—Se

F

Se—Se

2-fluoro-tetraselenatetracene

Me

NC ... CN

NC ... CN

methyl-TCNQ

Et—N$^+$ ... N$^+$—Et

DEPE

Et—N$^+$ ... N$^+$—Et

DEPA

Among the radical salts, there are many examples. Interesting cases are (2-fluoro-TetraSelenoTetracene (2-fluoro-TSeT))$_2$ Br [11.55], metal and metal-free phthalocyanine iodide [11.40, 56-59] and (DEPE and DEPA)· (TCNQ)$_{4.5}$(H$_2$O)$_x$ [11.60-63]. (2-fluoro-TSeT)$_2$ Br exhibits good metallic behavior down to 2 K with high conductivity, though both the interplanar distance and the intermolecular Se···Se contacts reveal a strong one-dimensional feature in this salt. A positional disorder of the donor part was detected by a structural analysis. It is most likely that the conduction path made of the donor part has no periodicity (1a) which prevents the perfect nesting of the one-dimensional Fermi surface. The PhthaloCyanine (PC) cation radical salts have been studied extensively from several viewpoints: a high conductivity with a low oxidation state (PC$^{+1/3}$), one-dimensionality, d-π electron interaction, a model of the Kondo effect, polymer composites, etc. [11.64, 65]. In common, these PC compounds have an essentially one-dimensional conduction column composed of metal or metal-free PC molecules piling up with an alternation of ca. 39° to the neighboring PC molecules in the column. In the iodide salt, iodide ions exist as triiodide in the channel parallel to the conduction column (Fig. 11.3). The I$_3$ channel has superstructures incommensurate to the PC column. It is thought that the metallic nature retained down to low temperatures (Fig. 11.3) is ascribable to the non-periodic modulation by the iodine lattice (1b). The TCNQ radical salts discovered by *Ashwell* [11.61] are some of the most interesting materials from the point of crystal design, regarding the disorder. The transport property of (DEPE)(TCNQ)$_{4.5}$(H$_2$O)$_x$ significantly depends on the content of water. When x=0 it is semiconducting with $\sigma_{RT} = 2 \cdot 10^{-3}$ S/cm. With

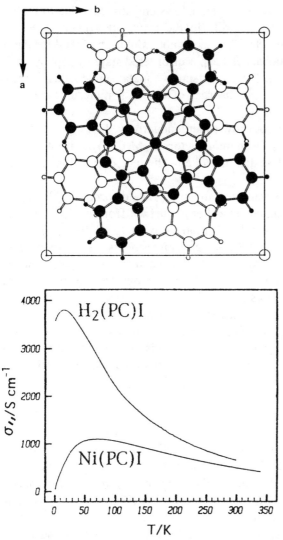

Fig. 11.3. Temperature dependence of the conductivity and the crystal structure viewed parallel to the stacking direction of one-dimensional organic metal [nickel phthalocyanine iodide, Ni(PC)I]. The temperature dependence of the conductivity of metal-free PC iodide, H_2(PC)I, is compared. Redrawn after [11.57, 58]

increasing water content, the compound becomes a good semiconductor with $\sigma_{RT} = 40 \div 60$ S/cm for x = 0.88. Surprisingly, it is metallic down to the extremely low temperature of 30 mK for 0.88 > x > 0.2. The water in the crystal comes from the not-dried solvent (CH_3CN) and occupies non-periodic positions in the crystal. It is most likely that the disorder of included solvents modulates the low-dimensional Fermi surface in an appropriate way (1b). Even in a two-dimensional system, disorder due to the included solvent plays an essential role in the transport property. $(ET)_2 \cdot (ClO_4)(1, 1, 2\text{-trichloroethane(TCE)})_{0.5}$ does not exhibit any metal-insulator transition when the TCE molecules keep a disordered state by rapidly cooling ($-5K/min$) down to 0.8 K [11.66]. It shows a metal-semiconductor transition below $40 \div 50$ K when the disorder of the TCE molecules had vanished by slow cooling ($-5K/h$).

All these examples did not show superconductivity. However, it should be remembered that all the ET superconductors of $M(CF_3)_4$ (solvent) (Table 1.1) may contain more or less disorder due to the included asymmetric solvent (2b).

These examples demonstrate that an appropriate disorder is able to prevent a system with a low-dimensional Fermi surface from perfect nesting. In connection with this, we should mention γ-ray or X-ray irradiation as a source of disorder (1a) [11.67, 68]. A light dose smears out or suppresses the Peierls transition by $110K/(1\%$ defect concentration) in the case of TMTSF·dimethyl-TCNQ, but a heavy dose destroys the metallic nature since the irradiation deteriorates crystals.

A difficult issue in the first strategy is *to know the quality and quantity of the disorder enough to suppress the metal-insulator transition but not so strong to kill the superconductivity*. This strategy is important to develop one-dimensional superconductors other than the TMTSF family.

11.2.2 Dimensionality Increase

The two main strategies from the chemical point of view are (a) a *heavy-atom substitution* proposed by *Engler* and *Patel* [11.69], *Cowan* et al. [11.70], and others (Fig. 11.1), and (b) the *peripheral addition of an alkyl-chalcogeno group* due to *Saito* et al. [11.71, 72] (Fig. 11.1). A physicist would consider the application of pressure [11.73]. These methods were found to be effective to suppress the metal-insulator transition and to induce the superconducting state.

The use of the Se or Te analogues of TTT (TetraThioTetracene)[1] or TTF was thought to increase the interstack interaction because of the larger van der Waals (vdW) radius of Se ($1.90\,\text{Å}$) or Te ($2.06\,\text{Å}$) than that of S ($1.80\,\text{Å}$) [11.74]. The high atomic polarizability of Se or Te is a favorable factor to reduce the U value in a CT complex. According to this **heavy-atom substitution method**, tetraseleno- and tetratellurotetracene (TSeT, TTeT, Fig. 11.1) cation radical salts were synthesized [11.75]. However, it turned out that these salts were of rather low-dimensional nature due to the

---

[1] Also called TetraThioNaphthacene (TTN)

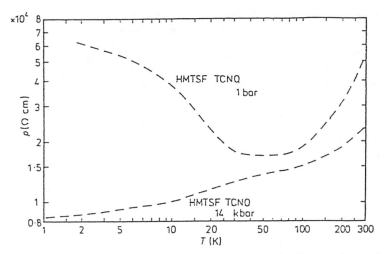

Fig. 11.4. Logρ vs. logT for HMTSF·TCNQ at 1 bar and 14 kbar. Redrawn after [11.80]

specific stacking of the radical molecules. Some of them showed extremely high conductivity under pressure ($>10^6$ S/cm at 4K, 8kbar for (TSeT)$_2$X (X: Cl, Br)), but no superconductivity was detected [11.76].

As for the fulvalene derivatives, a TSF molecule gave a TCNQ complex exhibiting a metal-insulator transition at a lower temperature than that of TTF·TCNQ [11.69]. However, in many TSF complexes the donor has not given an appropriate donor packing so as to increase the dimensionality. On the other hand, a few CT complexes of TSF derivatives were found to retain the metallic property down to low temperatures [e.g., HMTSF· TCNQ (Fig.11.4), HMTSF·TNAP, TMTSF·TCNQ] [11.77-81] mainly due to the increased dimensionality as a consequence of the short intermolecular Se···N atomic contacts (Fig. 11.5) [11.82]. It is presumed that the methyl or hexamethylene groups of TMTSF or HMTSF, respectively, contribute to an arrangement of the donor molecules in an appropriate donor packing. The Peierls transition in HMTSF·TCNQ (Fig.11.4) or TMTSF·dimethyl-TCNQ [11.83] could completely be suppressed by pressure. A precise study of the newly appearing metallic phase in the latter complex, to be briefly described in the next section, has opened a way to investigate TMTSF cation radical salts and the first superconducting family [11.84, 85].

The iodine atom (vdW radius: 1.98Å) has been employed to give rise to increased intermolecular interactions. The high polarizability, with a not so strong electronegativity and a large atomic orbital of the iodine atom is an advantageous feature for the formation of conductive materials. Pristine iodine molecules become metallic and superconductive under a high pressure [11.86]. Furthermore, an insulating tetraiodo-p-benzoquinone (p-iodanil) becomes metallic at ca. 35 GPa and very conductive at ca. 52 GPa

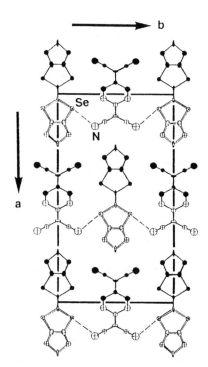

Fig. 11.5. A view normal to the ab-plane of the crystal packing in HMTSF·TCNQ. Short atomic contacts (Se···N, C···N) are indicated by dotted lines. Redrawn after [11.82]. The HMTSF molecules are centered around the I-symmetry element at 0,0,0 and the TCNQ molecules are centered around I at 0, ½, ½

[11.87-89]. A significant improvement in conductivity has been noticed on replacing hydrogen with bromine and iodine on the TCNQ part in TTT·TCNQ and TTF·TCNQ systems [11.35]. These examples are iodine-substituted acceptor molecules of quinone or TCNQ type. On the other hand, several iodine-substituted TTF derivatives and their metallic complexes, such as iodo-EDT-TTF·[Pd(dmit)₂] and (diiodo-EDT-TTF)₄M·(CN)₄ (M: Ni, Pd, Pt), have been prepared [11.90, 91]. However, no convincing evidence about the suppression of the Peierls transition has yet been presented, since Pd(dmit)₂ might be responsible for the metallic nature down to 4.2 K in the iodo-EDT-TTF complex and a metal-insulator transition takes place at 80 ÷ 100 K in the M(CN)₄ complexes. Since the HOMO coefficients of iodine atom(s) of these donors are very small [11.90, 91], it is difficult to gain enough intermolecular transfer interactions to suppress the Peierls transition even with short atomic contacts which, though, favor to control the molecular packing in solids. The HOMO and LUMO coefficients of the iodine atom(s) in the donor and acceptor molecules, respectively, as well as the position of iodine atom(s) in a molecule are critical parameters in utilizing iodo-substituted molecules. So far, it is more efficient to implant iodine in the acceptor molecules than in the donor molecules to have significant intermolecular interactions.

NC

CN

NC

CN

TNAP

Se   Se

Se   Se

HMTSF

iodo-EDT-TTF

tetraiodo-p-benzoquinone

diiodo-EDT-TTF

BETS

TOF

The **peripheral addition of alkylchalcogeno groups** to the TTF skeleton has proved to be a successful method to control the electronic and structural dimensionality of a crystal through the molecular geometry and the HOMO coefficients. By choosing various kinds of inner and outer chalcogen atoms and alkyl groups, a rich variety of donor molecules and their complexes with different structural, chemical and physical properties have been developed [11.37, 48-52]. The electronic dimensionality of a radical salt which was derived from the Se or Te analogues of ET is roughly estimated on the basis of the molecular viewpoint: the HOMO coefficient and molecular geometry (Fig. 11.2). The ratio b/c of ET analogues in Fig. 11.2b is a geometrical parameter to be considered for the design of the dimensionality. The ratio of ET (0.83) is considerably smaller than that of BO (1.03) or BETS (0.90). The HOMO coefficients of **inner chalcogen atoms**, which are in the fulvalene ring, are more than twice larger than those of the **outer chalcogen atoms** in the outer six-membered rings of the ET analogues, so the short atomic contact with inner chalcogen atoms is effective by, at most, about four times that of the outer chalcogen atoms. When outer chalcogen atoms are larger in size than the inner ones, such efficient short contacts by inner chalcogen atoms will be weakened along the **side-by-side** direction. Accordingly weak **face-to-face** interactions are needed to meet the two-dimensional electronic structure. However, the tuning of the magnitude of particular transfer interactions is not an easy task. So, it is more

409

likely to yield a one-dimensional metal along the face-to-face direction or an insulator of a Mott type rather than a two-dimensional metal.

On the contrary, when the outer chalcogen atoms are smaller in size than the inner ones such as in BO and BETS, large side-by-side intermolecular interactions are expected with the aid of inner chalcogen atomic contacts. In BO salts approximately equal transfer interactions along two different oblique directions are created due to short face-to-face CH$\cdots$O contacts, as is described in Sect. 11.7.3, which results in a two-dimensional electronic structure. In contrast, in BETS compounds the side-by-side intermolecular interactions are expected to surpass the face-to-face ones and tend to yield much one-dimensional nature along the side-by-side direction. The two-dimensional metals of BETS (Sect. 5.5.1) have a bandwidth wider than ET metals with the same crystal phase; many of the κ-salts of BETS are good metals [11.92-95] while the λ-type ones exhibit a temperature dependence of the resistivity [11.96] similar to that of κ-(ET)$_2$Cu(NCS)$_2$.

A simple consideration in the framework of BCS theory tells us the disadvantage of using heavy molecules or atoms to raise T_c if T_c is represented as $T_c \approx M^{-1/2}$ with M being the molecular weight (M = 447.9 for TMT SF). According to this argument a molecule of small molecular weight is desirable and oxygen analogues of TTF (TOF M = 140.1) are promising compared to TSF (M = 391.9) and TTeF (Fig. 11.1, M = 586.5) derivatives, but they have not been successfully synthesized [11.97]. For the peripheral substitution of TTF with the alkylchalcogeno groups, BO (M = 320.4) is expected to produce superconductors with T_c higher than ET (M = 384.7). However, a strong self-aggregation ability of the BO molecules (Sect. 11.7.3) gives rise to a two-dimensional metal with a wide bandwidth and a low density of states at the Fermi level, and hence results in low T_c superconductors (Table 1.1) [11.98, 99]. So the availability of high-T_c BO superconductors depends on how one can weaken the self-aggregation ability of the BO molecules and soften the crystals of BO complexes. The following fact, however, contradicts the above consideration. The superconducting transition temperature observed in λ-(BETS)$_2$GaCl$_4$ ($\approx 8K$) [11.96] is not much different from those of the κ-phase of ET salts and much higher than those of the TMTSF salts, in spite of the exceeding molecular weight of BETS (M = 572.3). This suggests the inadequacy of $T_c \approx M^{-1/2}$. However, a general tendency is noticed that the highest T_c value of superconductors so far known is higher with a smaller M of the basic molecule among the TTF based superconductors.

11.3 The Development of TMTSF Salts

TMTSF was synthesized by *Bechgaard* et al. using highly toxic CSe_2 and H_2Se [11.100]. An improved synthetic route, which does not use both CSe_2 and H_2Se, was developed later [11.101]. The coefficient of HOMO calculated by the extended Hückel method exhibits that the charge density for Se is larger than that for the carbon atom of the central C=C. Therefore, the intermolecular interactions by means of short Se···Se atomic contacts are efficient to increase the dimensionality. However, the molecular geometry, especially the lengths of b and c of TMTSF in Fig. 11.2b, suggests that the intermolecular side-by-side Se···Se contacts are rather difficult to attain owing to the bulkiness of the methyl groups of the TMTSF molecules. TMTSF can be a donor stronger than TTF, but weaker than TMTTF, judging from I and $h\nu_{CT}$(D·TNB) (Table 11.1). The ΔE^{12} value of TMTSF is smaller than those of TTF and TMTTF (Table 11.1), which implies considerable weakening of the electron-electron correlation in a TMTSF molecule. The molecular polarizability of TMTSF is almost twice that of TTF and 1.5 times that of TMTTF (Table 11.1). The oxidation potential of TMTSF requires an appropriate acceptor for the partial CT state in the range of $+0.09 \div +0.47V$ vs. SCE. Near the lower boundary of the partial CT region, TCNQ is allocated, and it gives **monotropic complex isomers**: One is a 1:1 red insulator with $\rho = 0.21$ and the other is a 1:1 black metal with $\rho = 0.57$ [11.102, 103]. In the partial CT region closer to the neutral side, TMTSF·dimethyl-TCNQ resides with $\rho = 0.5$. This complex exhibits an extremely high conductivity under pressure at low temperatures ($>10^5$ S/cm) [11.83]. Based on the positive thermoelectric power of this complex, the TMTSF cation radical salts of $(TMTSF)_2X$ were extensively studied and the first superconductor of $X = PF_6$ was reported by *Jérome* et al. in 1980 [11.84].

The Fermi surface of $(TMTSF)_2X$ was found to be not closed, but open due to the lack of enough side-by-side transfer interactions with fair warping. The degree of warping depends on the size of X, which affects the length of an intermolecular Se···Se atomic contact along the b-axis (Fig. 11.6). *Williams* and co-workers observed fairly good linear relations between the anion volume, unit cell volume and the average intercolumn Se···Se distance $[(2d_7 + d_9)/3]$ [11. 104, 105]. They tried to correlate the transition temperature of $(TMTSF)_2X$ (SDW, anion order-disorder, superconducting) with the anion volume (Fig. 11.7). While *Kistenmacher* plotted the transition temperature against the van der Waals ion size of the anion (Fig. 11.8) [11.106]. In both cases the metal-insulator transition temperature due to an order-disorder of the tetrahedral anion is almost linearly related to the anion volume (size). The SDW transition temperature is insensitive to

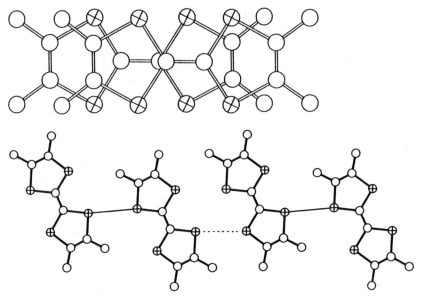

Fig. 11.6. Projection views of a donor molecule dimer and side-by-side Se· · ·Se atomic contacts in (TMTSF)$_2$ ClO$_4$. The *solid line* (d$_9$) is the distance shorter than the sum of van der Waals radii (<3.80 Å). d$_7$ is indicated by the *dotted line*

the anion volume (size). It was noted that these salts need some pressure (P$_c$ in Table 1.1) to suppress the metal-insulator transition since they do not have enough two-dimensionality due to the large anion volume (size). Interestingly enough, P$_c$ is linearly related to the anion size (Fig. 11.8). However, the ambient-pressure superconductor (the ClO$_4$ salt) deviates considerably from the present line of thinking.

Since the TTF and TTT derivatives of heavy atoms such as Se and Te have both small U and big intracolumn interactions, they usually give good

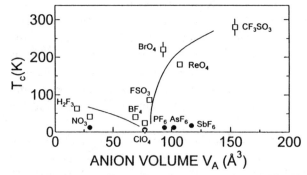

Fig. 11.7. Electronic transition temperature of (TMTSF)$_2$ X vs. anion volume V$_A$. Lines are guides for the eye. Open squares, closed circles and open triangle indicate transitions due to anion-ordering, SDW and superconductivity, respectively. Redrawn after [11.104]

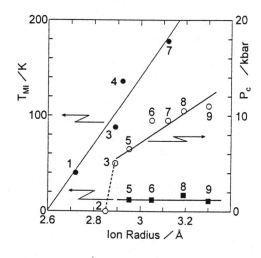

Fig. 11.8. Metal-insulator phase transition (T_{M-I}: closed symbols) and critical pressure (P_c: open circles) as a function of the anion radius. Redrawn after [11.50, 106]. (1) BF_4, (2) ClO_4, (3) FSO_3, (4) F_2PO_2, (5) PF_6, (6) AsF_6, (7) ReO_4, (8) SbF_6, (9) TaF_6

conductors due to the increase of the bandwidth, the decrease of electron-phonon interaction, and the decrease of electron-electron correlation. These characteristics are advantageous for yielding superior metals. Actually a number of molecular metals have been developed on the basis of the TSF, TTeF, TSeT, TTeT and their derivatives [11.48, 70, 73, 75-85, 107-111]. However, these characteristics are not favorable factors for the realization of high-T_c superconductors because of the decrease of both the density of states and the electron-phonon coupling. No further drastic increase of T_c is expected in the TMTSF family.

11.4 The Development of ET Salts

An ET molecule was first synthesized by *Mizuno* et al. starting from an electrochemical reduction of CS_2 [11.112] and later from 1,3,4,6- tetrathia-pentalene-2,5-dione [11.71]. A most convenient method for a large-scale preparation was developed by *Steimecke* et al. starting from the reduction of CS_2 with Na in N,N-dimethylformamide (DMF) [11.113-115] that yields dmit. The structural isomer of dmit, which is obtained as a by-product, has to be carefully eliminated from the reaction product. It was thought that the contamination with the structural isomer of ET derived from the isomer of dmit may affect both conductivity and superconductivity. In order to avoid the contamination by the structural isomer of dmit, which, in fact, can be eliminated by a careful column chromatography in a few reaction steps to ET, a different route starting from 1,2-ethanedithiol was developed [11.

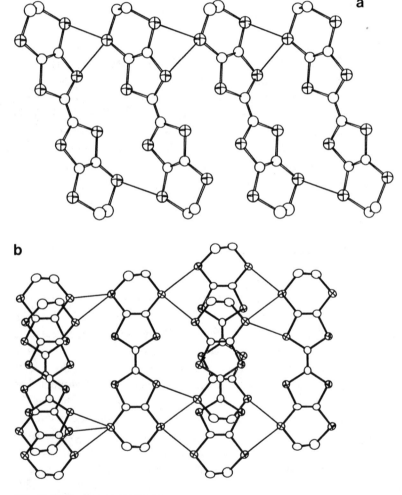

1,3,4,6-tetrathiapantalene-2,5-dione   1,2-ethanedithiol

116]. However, the deuterated ET is rather difficult to synthesize by this scheme.

The donor ability of ET is weakened by the addition of the ethylenedithio groups and is weaker than TMTSF but comparable or weaker than TTF, as judged on the basis of the I^v and $h\nu_{CT}$ values in Table 11.1. This is consistent with the complex formation of TCNQ. Namely, TTF forms a metal having a partial CT state, while ET forms 1:1 isomeric complexes;

Fig. 11.9. Caption on next page

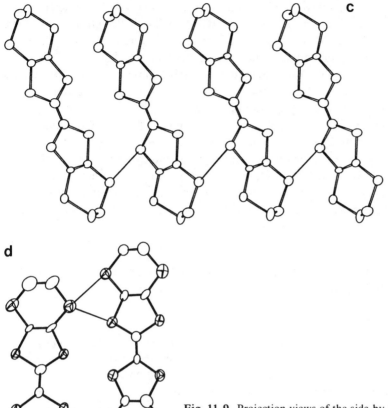

c

d

Fig. 11.9. Projection views of the side-by-side inter-molecular $S \cdots S$ stomic contacts in (**a**) α-$(ET)_2 \cdot NH_4 Hg(SCN)_4$; (**b**) β-$(ET)_2 IBr_2$; (**c**) θ-$(ET)_2 I_3$; and (**d**) κ-$(ET)_2 Cu(NCS)_2$. *Solid lines* indicate $S \cdots S$ distances shorter than the sum of van der Waals radii ($< 3.60 \text{Å}$)

one is in the neutral state and the other one is in a partial CT state with TCNQ [11.117], indicating that ET is a donor weaker than TTF. The oxidation potential of ET indicates that the partial CT state is achieved in a 1:1 complex of the TTF·TCNQ type with an acceptor of the reduction potential in the range of $+0.17 \div +0.55$V vs. SCE, where TCNQ resides near the lower boundary and so isomeric complexes are obtained (the **complex isomerization, neutral-ionic phase transition** and the requirement of a partial CT state in the organic CT complex share the common basic concept [11.36, 46, 118-124]). The ΔE^{12} value of ET is two-thirds of that of TTF or TMTTF, which indicates that the electron-electron correlation of ET is significantly reduced. The coefficient of HOMO (b_{1u} symmetry, Fig. 11.2) exhibits that the carbon atoms of the central C=C bond have a charge density smaller than those of the inner sulfur atoms (S_{in}). Furthermore, the outer sulfur atoms (S_{out}) have a significant charge density which implies that the delocalization of the π-electrons of HOMO reduces the electron-electron correlation. It should be emphasized that not only the numbers of intermolecular S···S atomic contacts but also the kinds of such contacts (S_{in}···S_{in}, S_{in}···S_{out}, S_{out}···S_{out}) influence the degree of dimensionality, since both S_{in}···S_{in}, which is usually not attainable, and S_{in}···S_{out} are much more efficient than an S_{out}···S_{out} contact to produce transverse transfer interactions (Fig. 11.9). The molecular polarizability (α) of ET is about twice that of TTF and even higher than that of TMTSF.

The above-mentioned electronic features indicate that ET should be a superior donor to yield highly conductive CT complexes among the donors in Table 11.1. However, it has turned out that the argument is not straightforward due to the structural peculiarities inherent in an ET molecule.

11.5 Origin of the Two-Dimensionality and the Polymorphism of ET Compounds

In contrast to the TTF and TMTSF molecules, the neutral ET molecule is non-planar, and the central C_6S_8 plane is deformed in the shape of a boat (Fig. 11.10a) [11.125]. On the formation of the complex it becomes almost flat, except for terminal ethylene groups which are not on the molecular plane and thermally disordered at higher temperatures (Fig. 11.10b). The main conformations of one of the terminal ethylene groups of an ET molecule are illustrated by A and C in Fig. 11.11 (A and B are equivalent with

a

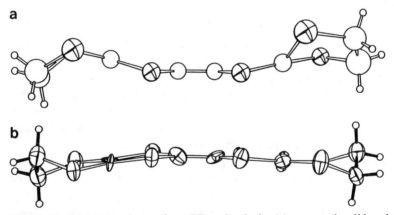

b

Fig. 11.10. Molecular shape of an ET molecule in (**a**) a neutral solid and in (**b**) κ-$(ET)_2$ Cu(CN)[N(CN)$_2$] (*staggered form*) to demonstrate the molecular shape change on complex formation. Courtesy of H. Yamochi

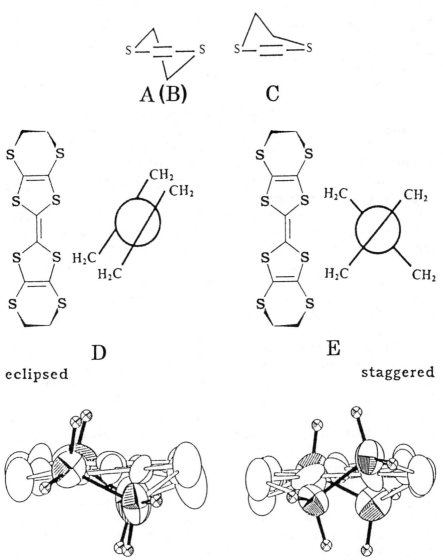

Fig. 11.11. The main conformations of a terminal ethylene group of an ET molecule (A - C) and the relation of the two ethylene groups (D: eclipsed, E: staggered). Projection views of an ET molecule along the long molecular axis for the eclipsed and staggered ones in κ-(ET)$_2$ Cu(CN)[N(CN)$_2$]. S–=–S in A(B) and C, and a big circle in D and E represent the C$_6$S$_8$ plane viewed along the long molecular axis of an ET molecule. Courtesy of H. Yamochi

opposite conformations of the ethylene group) [11.126]. The relation between two terminal ethylene groups is either eclipsed (D in Fig. 11.11) or staggered (E in Fig. 11.11) in the A or B conformation. It was pointed out by the Argonne group that the ethylene conformation is one of the key pa-

S S S S
S S S S

C$_6$S$_8$ plane

rameters in determining the physical and structural properties that include the superconductivity of ET compounds (Table 11.4) [11.126].

The ET molecules in a complex tend to pile up one after the other with a sliding overlap so as to minimize the steric hindrance caused by the ethylene group(s). They leave behind cavities along the direction of the long molecular axis, where counter anions and sometimes solvent molecules reside (Fig.5.2). This tendency weakens the proximate face-to-face π-intermolecular interactions leading to a comparatively small transfer integral along the stacking (or pseudo-stacking) axis, $t_{\parallel}$. Therefore, the conductivity of an ET complex is not as high as that of TTF or TMTSF complexes.

On the other hand, the ET molecule shows a strong tendency to form a variety of proximate intermolecular S$\cdots$S atomic contacts along the side-by-side direction (short molecular axis, Fig.11.9) leading to an increment of $t_{\perp}$. This feature provides the increment of the electronic dimensionality in the ET complex. In Sects.4.3.3 and 5.1.3 the arrangement of ET molecules were noticed to play an important role in increasing the dimensionality and the warping of the band; for example, in the case of β-(ET)$_2$X the effective transverse transfer energy is much larger in relation to the longitudinal one, leading to a closed Fermi surface, than that in (TMTSF)$_2$X. This is because the transverse transfer energies in the former take identical signs while those in the latter do not do so. Significant donor-anion interactions are frequently recognized by short atomic contacts between the ethylene hydrogens of ET and anion atoms around the anion openings, or hollow space (Sect.11.7.1) in the anion layer. However, the donor-donor interactions through the anion opening are small due to the fact that the electron densities on the outer ethylene groups are negligible (Fig.11.2) and the distance between donor molecules along this direction is large. Therefore, the transfer integral along this direction ($t'_{\perp}$) is the smallest ($t_{\parallel}, t_{\perp} > t'_{\perp}$).

These competing differences of donor-donor and donor-anion intermolecular interactions and the large conformational freedom of the ethylene groups together with the rather flexible molecular framework provide a variety of ET complexes. They show one- or two-dimensional character, polymorphism, clathrate of the solvent, a variety of molecular compositions, or complex isomerism. Even with a particular anion, an ET molecule forms a variety of single crystals of CT complexes. They have the same composition but different crystal structures (**polymorphism**) or with different compositions (sometimes inclusion of a solvent occurs) depending on the crystal-growth conditions. Furthermore, these different single crystals sometimes grow together and should be separated from the morphology observed under a microscope or by inspecting the physical properties such as the ESR signal and the electrical conductivity.

11.6 Stoichiometry, Valence and Band Filling

Most of the one-dimensional metals of TMTSF have a particular charge (+1/2). A similar feature is observed in the quasi one-dimensional metal of DMET. On the other hand, two-dimensional metals of ET and BO have a variety of stoichiometries and donor charges. In addition, a wide variation in stoichiometry and charge is observed in the three-dimensional C_{60} metal. In this section, the stoichiometry of the complex and the valence state of the component molecules, which are closely connected with the position of the Fermi level, hence with the Density Of States at the Fermi level, DOS(ϵ_F), are described by taking ET complexes as a typical example.

11.6.1 Stoichiometry

The common **stoichiometry** of an ET complex (Table 11.2) with the mono-anion X is DX, D_3X_2, D_3X_2(solvent)$_n$ (n=2), D_2X and D_2X(solvent)$_n$ (n = ½, 1, 2, 3). As may be inferred from the Tables 11.1,2, the most common stoichiometry is D_2X among the donor-based organic superconductors. A few exceptions are accounted: γ-(ET)$_3$(I$_3$)$_{2.5}$ [11.41], (ET)$_3$Cl$_2 \cdot$2H$_2$O [11.127], (BO)$_3$[Cu$_2$(NCS)$_3$] [11.98] and (DTEDT)$_3$[Au(CN)$_2$] [11.128]. The D_3X stoichiometry is very rare in ET salts. With dianion Y, ET gives D_3Y_2, D_2Y, D_4Y and D_4Y(solvent)$_n$ (n=1). Only four ET superconductors have been developed with D_4Y or D_4Y(solvent) [11.129-133]. The D_3Y stoichiometry is not common among ET salts. No superconducting salts with dianions have been developed with other donor molecules. For the tri-anion Z, not so many examples are known, but the superconducting ET salt of Fe(ox)$_3$ belongs to this class with the stoichometry of D_4Z(solvent) [11.134].

In general, multiple stoichiometries are accessible with increasing both the kinds of intermolecular interactions and the dimensionality of the system. Especially with increasing the **self-assembling ability** of a molecule like phthalocyanine, BO, TTPs, and M(dmit)$_2$ [11.38-40, 135, 136], complicated stoichiometries often appear. The simple stoichiometries (1:1, 3:2, 2:1) observed for TMTSF or ET complexes with an inorganic anion or an

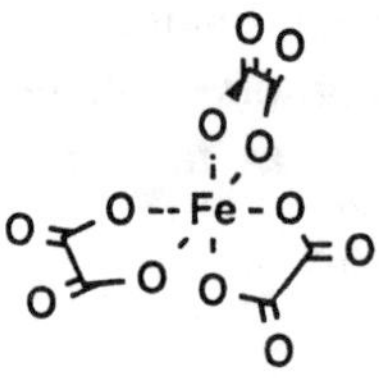

Fe(ox)₃

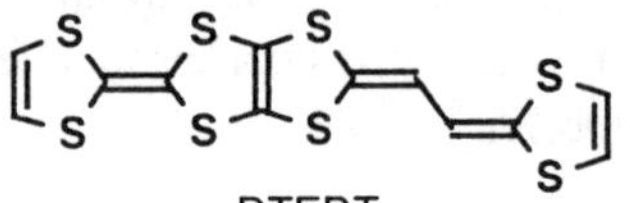

DTEDT

organic acceptor is in clear contrast with the frequently observed complicated stoichiometries (1.75:1, 1.8:1, 2.25:1, 2.4:1) for BO complexes [11.38]. Such an incommensurability of stoichiometry is a consequence of the strong self-assembling ability and sometimes results in the formation of a **misfit lattice** [11.137], which might not be advantageous for having a high T_c. Even though simple 1:1 and 1:2 stoichiometries are found in dmit superconductors (Table 1.3), a number of incommensurate stoichiometries have been noted for the salts with alkali and organic cations [11.135, 136].

Another factor which gives rise to multiple stoichiometry is termed the **multi-redox** (or **multi-stage redox**) [11.138]. It means the easiness of multiple redox processes under an ambient condition. In conventional donor or acceptor molecules the multiplicity of a redox process is single (perylene, symtrinitrobenzene) or double (TTF, ET, TCNQ, M(dmit)₂). A high multiplicity of the redox process has been accomplished by several methods; (1) to extend the π-conjugation (TTPs, polycyclic aromatic hydrocarbons), (2)

Table 11.2. Common stoichiometry of ET cation radical salts and estimated average charge of ET (ρ) (solv: solvent, ⊖: most common, ⊙: common, ○: fairly common, Δ : rare, ×: no sample)

Anion	1 : 1 or 1 : 1 : solv	3 : 2 or 3 : 2 : solv	2 : 1 or 2 : 1 : solv	3 : 1	4 : 1 or 4 : 1 : solv
X^{-1}	○	⊙	⊖	Δ	Δ
	$\rho = 1$	2/3	1/2	1/3	1/4
Y^{-2}	×	○	⊙	Δ	⊖
	2	4/3	1	2/3	1/2
Z^{-3}	×	×	×	×	⊖
	3	2	3/2	1	3/4

cobalt tetraphenylporphyrin

biferrocene

radialene

triptycene

cyclophane

to combine a number of π-units or simple redox units (polyamines, bi-ferrocenes, cyclophanes, triptycenes, radialenes, cytochrome C_3 [11.139]), (3) to have degenerate HOMO or LUMO (C_{60}), (4) to decrease the HOMO-LUMO gap by combining donor and acceptor moieties by a σ or π bond (donor-σ (or π)-acceptor molecule [11.140]), and (5) to include transition metals (metallophthalocyanines, cobalt tetraphenylporphyrin). These materials may have small U values and are expected to be good candidates for organic metals with a high conductivity. Especially the high multiplicity of the HOMO or LUMO degeneracy is advantageous for having a high DOS(ϵ_F) value.

It was pointed out by the Argonne group that there is an obvious relation between the anion volume and the stoichiometry of ET salts for the tetrahedral anion [11.51]. They defined the relative volume (V_{rel}) of tetrahedral anion as

$$V_{rel} = (R_i + 2R_0)^3 , \qquad (11.3)$$

where R_i and R_0 are the ionic radii (in Å) of the inner and outer atoms, respectively. They observed a nearly linear correlation between the relative tetrahedral anion volume and the unit-cell volume of the 3:2 salts. They noticed that small anions such as FSO_3 ($V_{rel} = 14.7 Å^3$), HSO_4 ($15.3 Å^3$), BF_4 ($16.4 Å^3$) and ClO_4 ($18.4 Å^3$) lead to either 3:2 or 2:1:0.5 stoichiometries. Large anions [$GaCl_4$ ($61.6 Å^3$) to InI_4 ($116.2 Å^3$)] lead to a 2:1 stoichiometry exclusively, while the larger anion TlI_4 ($125.8 Å^3$) leads to a 2:1:1, $(ET)_2 \cdot$

$(TlI_4)(I_3)$. Anions with intermediate volumes, such as BrO_4 (22.2Å^3), tend to form all three different stoichiometries, i.e., 2:1, 2:1:0.5 and 3:2. In addition, they pointed out that the number of donor molecules per anion tends to be larger for the larger anions.

A study of the complex formation which is based on the anion size, is significantly important in order to obtain a particular stoichiometry and a preferred phase. From this point of view, the tetrahedral anions which have a size between IO_4 (26.5Å^3) and $GaCl_4$ (61.6Å^3), are necessary to understand the property-structure-stoichiometry relationship in detail.

For other combinations of donors and anions, no extensive study has been performed that is concerned with the controlling factors which favor a particular stoichiometry.

11.6.2 Valence State

The dication ET^{2+} exists in DX_2 (X: ClO_4, BF_4) [11.141, 142] and D_3Y_2 (Y: $ZnCl_4$, $MnCl_4$) [11.143, 144], in which ET^{1+} coexists. The monocation state is also observed in several DX or D_2Y complexes. Except for the high conductivity of 0.4 S/cm of $(ET)_3(ZnCl_4)_2$ [11.143] and the metallic nature of δ-$(ET)I_3(TCE)_{1/3}$; $\sigma_{RT} = 10 \div 20$ S/cm down to 130K [11.41], all these materials are insulators. The common charge of an ET molecule in a metallic salt is +2/3 and +1/2. Rarely a charge of +1 (*vide ante*), +5/6 [γ-$(ET)_2(I_3)_{2.5}$], +3/4 {$(ET)_4[Fe(ox)_3H_2O \cdot benzonitrile(BN)]$} or +1/4 {$(ET)_4[Re_6X_5Cl_9(X: S, Se)]$} [11.145] have been reported. Two of them are superconductors and the last two are metallic salts that consist of the lowest charged ET molecules so far known. (However, the possibility that γ-$(ET)_2(I_3)_{2.5}$ is not a unique phase has not been excluded. In addition, there is the possibility of an anion $[Fe(ox)_3H_3O \cdot BN]^{2-}$ instead of $[Fe(ox)_3H_2O \cdot BN]^{3-}$ and $[Re_6X_5Cl_9Cl]^{2-}$ instead of $[Re_6X_5Cl_9]^{1-}$, which lead to the most common +1/2 charge of an ET molecule).

These results indicate that a metallic band can be constructed with $ET^{+\rho}$ molecules having a ρ value, at least, in the range of $1/4 \le \rho \le 1$. Since no substantiating reports on the exact stoichiometry of δ-$(ET)I_3(TCE)_{1/3}$ or the physical properties sensitive to ρ such as the thermoelectric power have been published so far, it is more reliable to take the maximum ρ value as 5/6. Then, the band filling corresponds to the 7/12 ($\rho = 5/6$) to 7/8 ($\rho = 1/4$) full band, provided both a uniformly segregated donor stacking (no strong dimerization) and $U_{eff} < W$. Figure 11.12 shows DOS of typical α, β, κ and θ-phases of ET salts, which have been calculated with the extended Hückel method including the d-orbitals of sulfur atoms on the basis of crystal structures at room temperature. The Fermi levels of salt composed of $ET^{+\rho}$ ($\rho = 1/4$, 1/2 and 5/6) for each phase are indicated in the figure.

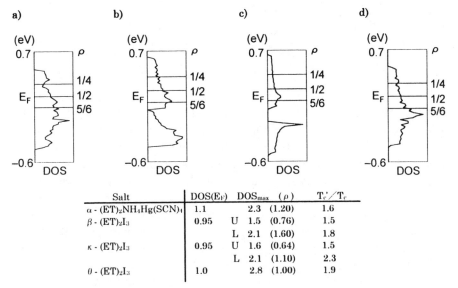

Fig. 11.12. Calculated Density Of States (DOS) of ET superconductors: (**a**) α-$(ET)_2NH_4Hg(SCN)_4$, (**b**) β-$(ET)_2I_3$, (**c**) κ-$(ET)_2I_3$ and (**d**) θ-$(ET)_2I_3$ based on crystal structures at room temperature by the extended Hückel method which includes d-orbitals of sulfur atoms. The Fermi levels for the degree of CT (ρ) of ET molecules of 1/4, 1/2 and 5/6 are depicted by solid lines. The table below the figure presents the calculated DOS for $\rho = 1/2$, DOS(ϵ_F), the maximum of DOS (DOS$_{max}$) with its ρ value in parentheses and T_c'/T_c $\{= \exp[1/DOS(\epsilon_F) - 1/DOS_{max}]/V$, assuming V to be 1 eV$\}$. U and L for DOS$_{max}$ mean the upper and lower bands, respectively. Courtesy of H. Sasaki

Salt	DOS(E_F)	DOS$_{max}$	(ρ)	T_c'/T_c
α - $(ET)_2NH_4Hg(SCN)_4$	1.1	2.3	(1.20)	1.6
β - $(ET)_2I_3$	0.95	U 1.5	(0.76)	1.5
		L 2.1	(1.60)	1.8
κ - $(ET)_2I_3$	0.95	U 1.6	(0.64)	1.5
		L 2.1	(1.10)	2.3
θ - $(ET)_2I_3$	1.0	2.8	(1.00)	1.9

Here a quantitative discussion of the magnitude of DOS that have been obtained by the extended Hückel method may not be meaningful. However, a qualitative comparison leads to the followings. It is seen that the calculated DOS(ϵ_F) value for the 7/12 ($\rho = 5/6$) full band is higher than that for the 7/8 ($\rho = 1/4$), regardless of the phases. It should be noticed that the shape of DOS of the HOMO band as a function of energy is a simply decreasing function when the charge of ET is less than $+1/2$. So, a state of ET less oxidized than $+1/2$ is not beneficial with respect to having a high T_c. Even though the maximum of the DOS value varies from salt to salt; roughly speaking, it appears in the lower part of the HOMO band for the α- and θ-phases and in the lower part of the upper HOMO band for the β- and κ-phases (Fig. 11.12). This indicates that a highly oxidized state of ET with more than $+1/2$ is preferable to raise T_c with respect to DOS(ϵ_F).

The M(dmit)$_2$ molecule is known to change its valence from 0 to -2. The most common valence of metallic and superconducting dmit complexes is $-1/2$, though the exact valence states of TTF and EDT-TTF complexes of the M(dmit)$_2$ type has not been settled yet. Charges of $-1/3$ and $-2/5$ have been deduced for metallic compounds {TBA (TetraButylAmmon-

EDT-TTF

ium)$\}_{0.33}$[Pd(dmit)$_2$] [11.135, 146] and K$_{0.4}$[Ni(dmit)$_2$] [11.147]. So the degree of CT in metallic dmit compounds is $1/3 \leq \rho \leq 1/2$, which corresponds to the 1/6 ($\rho = 1/3$) to 1/4 ($\rho = 1/2$) filled band on grounds of the one-band assumption. DOS calculations, which have not yet been reported, will reveal whether the less or highly reduced state of M(dmit)$_2$ is the preferred one.

As mentioned, the most common ρ value for superconductors, except for the C$_{60}$ family, is 1/2, and this number is not a favorable one to have a high DOS(ϵ_F) value. This particular ρ value is also unfavorable with respect to the electron correlation. The ρ value of 1/2 corresponds to a 1/4 or 3/4 filled band of an acceptor- or donor-type conductor, respectively, if U$_{eff}$ < W, and the 1/4- and 3/4-filled electronic structures are known to be sensitive to U$_{eff}$. In the dimerized system, to which many organic superconductors of the donor type belong, the valence band derived from HOMO is splitted into upper and lower HOMO bands due to the dimerization (see β and κ-salts of ET as examples, Fig. 11.12). As a result the upper HOMO band is 1/2 filled if U$_{eff}$ < W. The half-filled band is known to be one of the electronic structures that is most sensitive to U$_{eff}$ [11.148]. So, many ET salts become Mott insulators since W$_U$, which is the bandwidth of the upper HOMO band, is not so wide compared with those of the salts that contain selenium donors (Sect. 11.7.5).

If one could tune the charge of the counterpart widely without changing much the crystal structure, T$_c$ of ET materials is expected to increase at most 1.9 times in the range of $\rho \leq 1$ according to both the calculated maximum DOS and simple BCS theory, as indicated in Fig. 11.12 by the number of $T'_c/T_c = \exp\{[1/DOS(\epsilon_F) - 1/DOS_{max}]/V\}$ assuming V to be 1 eV. Even though a band structure with an extremely high DOS(ϵ_F), such as the one calculated for κ-(ET)$_2$I$_3$ with $\rho = 1.10$, is sensitive to nesting leading to CDW, SDW or other modulations, it seems not to be very difficult to raise T$_c$ more than 1.5 times by tuning the band filling according to Fig. 11.12.

11.6.3 Band Filling

The principal objects of **band filling** are (a) to tune the Fermi level in a metal, and (b) to convert a semiconductor into a metal. Remarkable success was accomplished in high-T$_c$ oxide superconductors which were derived from Mott insulators. Their T$_c$ values were tuned by this method through changing both the oxygen content and the charge of the constituent elements, to some extent, while keeping the crystal structure unchanged. On the other hand, such a method has not been successfully applied to organic CT crystals. Even in systems of aromatic hydrocarbons-halogens or -alkali metals, no continuous band filling has been realized. Only a few discrete stoichiometries or a narrow range of a mixed region are stable, and an excess doping usually destroys the crystals.

A few examples are known for controlling the continuous band filling though not so widely in the range as follows:

a1) Replacement of some portion of the constituent by a guest molecule which has different charge, ionization potential or electron affinity with a small modification of the molecular structure: (N-methylphenazinium)$_{1-x}$(phenazine)$_x$TCNQ [11.149], (TTF)$_{1-x}$(TSF)$_x$TCNQ [11.150], (ET) (TCNQ)$_{1-x}$(F$_2$TCNQ)$_x$ [11.151], (ET)$_y$[(MnCl$_4$)$_{1-x}$(FeCl$_4$)$_x$] [11.152], (TMTSF)$_{2-x}$(TMTTF)$_x$ClO$_4$ [11.153], κ-(ET)$_{2-x}$(BEDT-STF)$_x$Cu[N(CN)$_2$] Br [11.154]. In the last two examples, the superconductivity diminishes rapidly with increasing x, maybe due to the increase of the static disorder. In the last case, where the guest donor has two inner selenium atoms in one of the five-membered rings, the mixed crystals exhibit a less pronounced semiconductive anomaly above 90 K with increasing x (Fig. 11.13). Above x = 0.18 no anomaly in the resistivity was detected. These results are indicative that an increment of x induces not only disorder but also an increment of bandwidth since BEDT-STF molecules may have intermolecular interactions larger than ET molecules have. The increment of bandwidth reduces the electron-electron correlation, that is thought to be the reason for the resistivity anomaly above 90 K, which results in the reduced density of states and T_c.

a2) Usage of a mixed counterpart of different size. (This is not an actual band filling control. However, it is effective to modify the crystal structure subtlely. In this sense the present method shares the common concept

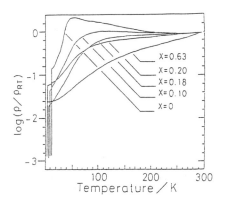

Fig. 11.13. Temperature dependence of the electrical resistivity of κ-(ET)$_{2-x}$(BEDT-STF)$_x$Cu[N(CN)$_2$]Br. Redrawn after [11.154]

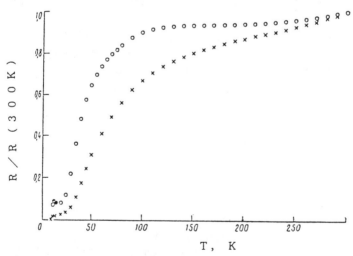

BEDT-STF

F₂TCNQ

N-methylphenazinium

phenazine

of **lattice pressure** or **chemical pressure**, and has often been applied to organic superconductors): β-$(ET)_2(I_3)_{1-x}(IBr_2)_x$ [11.155], κ-$(ET)_2$Cu[N·$(CN)_2$]Cl$_{1-x}$Br$_x$ [11.156, 157], λ-$(BETS)_2$GaX$_{4-x}$X$'_x$ [11.158], $A_{3-x}A'_x C_{60}$ (Sect. 11.8). The first example shows a rapid deterioration of the superconductivity by doping IBr$_2$, that was ascribed to the static disorder in the anion side. The content of bromine in the κ-$(ET)_2$Cu[N(CN)$_2$]Cl$_{1-x}$Br$_x$ was controllable by changing the crystal-growth conditions (such as the kinds of electrolyte and applied current) to yield salts of x = 0.1 ÷ 0.2, 0.5 and 0.8 [11.156, 157]. *Kushch* et al. claimed that the x = 0.5 salt is not a mixed crystal (alloy) but a unique phase based on X-ray studies which indicated no disorder in the anion layer. The temperature dependence of the resistivity is different from pure κ-$(ET)_2$Cu[N(CN)$_2$]Br (Fig. 11.14). The x = 0.5 salt

Fig. 11.14. Temperature dependence of the resistivity of κ-$(ET)_2$Cu[N(CN)$_2$]Cl$_{0.5}$Br$_{0.5}$; measuring the current in the ac-plane (X) and perpendicular to the ac-plane (O). Redrawn after [11.156]

exhibits no drastic change of resistivity down to 100 K and gives T_c of 11.3 K. However, the x = 0.8 salt, which is a mixed crystal (alloy) of x = 0.5 and x = 1, reveals both superconductivity at a temperature lower than does the x = 0.5 salt, and the static disorder on the anion side. Only C_{60} superconductors $A_{3-x}A'_xC_{60}$ show a remarkable success in raising T_c by modifying the lattice constants (Sect. 11.8). It is very curious why the static disorder inherent in a mixed cation of $A_{3-x}A'_xC_{60}$ does not destroy superconductivity.

b) As for the conversion from a Mott insulator into a metal by band filling, κ'-$(ET)_2Cu_2(CN)_3$ is the only one which reveals superconductivity by changing several hundred ppm of Cu^{1+} of κ-$(ET)_2Cu_2(CN)_3$ (a Mott insulator) to Cu^{2+}, while keeping the crystal structure unchanged [11.159]. The Mott insulator κ-$(ET)_2Cu_2(CN)_3$ (Cu = +1), was prepared by the Argonne group. They were able to convert it into a superconductor by applying pressure (T_c = 2.8K, 1.5kbar) [11.160, 161]. A subtle change of the valence state of Cu in the anion $Cu_2(CN)_3$ was found by *Komatsu* et al. to be effective to convert the Mott insulator into a superconductor [11.159]. Figure 11.15a presents the change of T_c at ambient pressure with the content of Cu^{2+}, which was measured via the intensity of the ESR signal of Cu^{2+}. Thereafter T_c increased from the initial stage to 5.1 K at $400 \div 430$ ppm of Cu^{2+}, then decreased with increasing Cu^{2+}. The magnetic susceptibility (Fig. 11.15b) decreased in the initial stage, too, indicating a considerable decrease of U_{eff} by a slight shift of the Fermi level from the exact 3/4 (or 1/2) full band. A further increase of the paramagnetic species (Cu^{2+}) seemed to

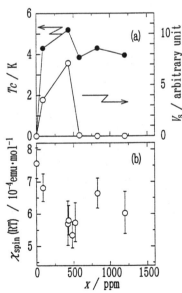

Fig. 11.15. The dependence on the content of Cu^{2+} (x) of (a) the superconducting transition temperature (T_c) defined as an onset of magnetically modulated microwave absorption and the relative volume fraction (V_S) and (b) the spin susceptibility at room temperature of ET molecules of κ'-$(ET)_2Cu_2(CN)_3$. Courtesy of T. Komatsu

destroy superconductivity since the volume fraction decreased from 7.2% at around $Cu^{2+} = 400$ ppm to almost 0 for $500 \div 600$ ppm (Fig. 11.15a).

Since the replacement of $+2$ species at the Cu^{1+} site corresponds to a decrease of the ρ-value of an ET molecule, this is not an advantageous direction for band filling with respect to $DOS(\epsilon_F)$. A study of the substitution of an anion MX by a less cationic metal M' and/or a more anionic ligand X' with least structural modification, such as $(ET)_y[(MnCl_4)_x \cdot (FeCl_4)_{1-x}]$ [11.152], where 5000 ppm of $FeCl_4$ converted the semiconducting $(ET)_2(MnCl_4)$ into a metallic one, should be investigated to pursuit a high DOS.

It is interesting to see the difference between the metallic nature of the metals derived from the same Mott insulator; one by pressure and the other one by doping (Fig. 11.16). The figure compares the pressure dependence of the resistivity behavior of $(ET)_2Cu_2(CN)_3$ salts containing Cu^{2+} of 0 ppm (κ-salt, Fig. 11.16a) and 400 ppm (κ'-salt, Fig. 11.16b). The κ-salt exhibits a pronounced hump of the resistivity followed by a rapid resistivity drop of the order of magnitude of more than five at around 13 K, which may be a transition from a Mott insulator to a metal. With increasing pressure, the hump appears at a higher temperature and the rapid decrease of the resistivity is smeared out. On the other hand, the κ'-salt does not show both the hump and drop of the resistivity down to the superconducting transition. Since the resistivity hump observed at higher temperatures in some κ-type ET salts and the λ-salt of BETS is most likely due to strong electron correlation, the absence of such a hump in κ'-$(ET)_2Cu_2(CN)_3$ would be an indication of the decrement of U_{eff} by the shift of the Fermi level of the upper HOMO band from 1/2.

11.7 Packing Pattern, and the Other Factors Correlating to T_c

The energy dispersion and Density Of States (DOS) of a band depend on the intermolecular interactions in the crystal. Utilizing energy dispersion and DOS, the Fermi surface and $DOS(\epsilon_F)$ can be predicted when the valence state of the component molecules is known. Since the intermolecular interactions are governed by how the component molecules pack in a crystal, it is very important to know the controlling factors which favor particular molecular packing in order to design a metal or have a high T_c of the superconductor. A few other factors relating to T_c, such as the effective volume, the softness, the effective mass and the electron correlation, are described from the point of view of the molecular and crystal designs.

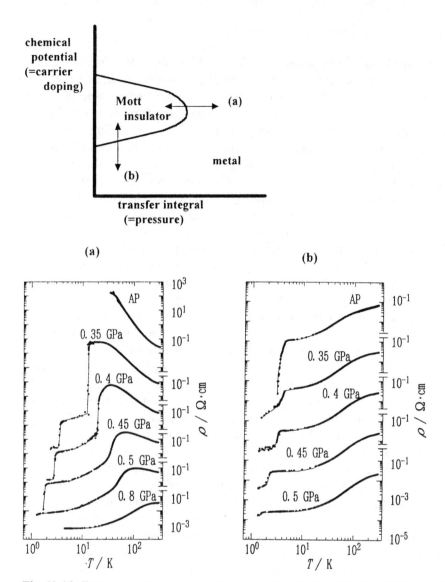

Fig. 11.16. Temperature dependence of the resistivity of (**a**) κ-$(ET)_2 Cu_2 (CN)_3$ and (**b**) κ'-$(ET)_2 Cu_2 (CN)_3$ (Cu^{2+} = 430ppm) under various pressures. The indicated pressures have been measured at room temperature. The solid lines are guides to the eye. Upper figure shows a schematic phase diagram of a Mott insulator. The abscissa represents intermolcular transfer interaction and the ordinate the chemical potential. A Mott insulator can be converted into a metal either by applying pressure (a) or carrier doping (b). Courtesy of T. Komatsu

11.7.1 Packing Pattern of Donors

The molecular **packing pattern** of ET complexes depends on many factors, i.e., the intermolecular interactions among ET molecules, among acceptor (or anion) molecules, between ET and acceptor (or anion) molecules, the shape and size of acceptor (or anion) molecules, the conformation of ethylene groups, and the Madelung energy and transfer energy gain.

Emge et al. studied the intermolecular interactions of β-(ET)$_2$X (X: I$_3$, AuI$_2$, ICl$_2$, etc.) with regard to the ethylene groups from the structural point of view [11.162, 163]. With the aid of neutron-diffraction measurements, they observed short **H-bonding** interactions of the type CH$\cdots$halide and concluded that these interactions play a major role in determining the crystal cohesive forces between donors and anions, thereby controlling the compound formation. Although their conclusion seems too much exaggerated with regard to the role played by the CH$\cdots$halide contacts on the cohesive energy, such as the local interactions around the ethylene groups are important factors in controlling the packing pattern of donors.

Whangbo et al. examined the energies of the chalcogen$\cdots$chalcogen contact interactions, the CH$\cdots$donor interactions, and the CH$\cdots$anion contact interactions based on ab-initio calculations according to a simple model systems [11.164, 165]. On the basis of their calculations, the attractive CH$\cdots$donor and CH$\cdots$anion interactions provide the energy required for the π-framework bending and the six-membered ring conformational change. They give rise to short-range chalcogen$\cdots$chalcogen contacts which are repulsive. This demonstrates the energetical importance of CH$\cdots$donor (donor-donor) and CH$\cdots$anion interactions in the compound formation, although the calculated result according to which the chalcogen$\cdots$chalcogen contacts are repulsive, is difficult to be accepted.

A structural comparison with the results of BO complexes by *Horiuchi* et al. [11.38] brought to light that a donor-donor interaction among the ET molecules is not a predominant factor for the molecular packing as follows. For BO materials there are mainly three kinds of intermolecular interactions; the conventional π-π interactions, and the $S_{in}\cdots S_{in}$ and CH$\cdots$O contacts. The presence of oxygen at the position of S_{out} of an ET molecule provides two advantages for **self-aggregation**; i.e., (i) the CH$\cdots$O contacts make BO molecules to form face-to-face stacks which slightly move along the short molecular axis only (Fig. 11.17); that is commonly observed in many BO complexes. This overlapping mode is not a favorable one, owing to the phase problem, to gain large transfer interactions along the stacking direction. Instead large transfer interactions are produced along two different oblique directions to give rise to a two-dimensional Fermi surface. The sliding overlap pattern is similar to those of the α- and θ-phases of ET salts, but it is quite different from those of typical TMTSF

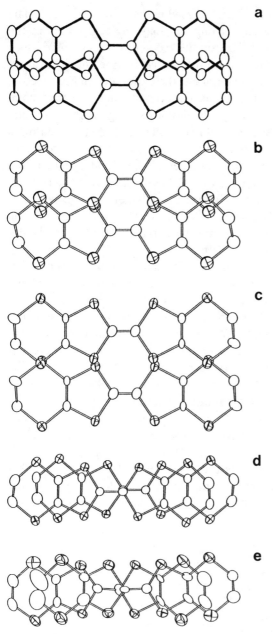

Fig. 11.17. Projection views of donor molecule dimer of (a) $(BO)_2 I_3 (H_2O)_3$, (b) α-$(ET)_2 NH_4 Hg(SCN)_4$, (c) θ-$(ET)_2 I_3$, (d) β-$(ET)_2 IBr_2$ and (e) κ-$(ET)_2 Cu(NCS)_2$. Courtesy of H. Yamochi

(Fig. 11.6), β- and κ-phases of ET salts, as illustrated in Fig. 11.17. The donor molecules overlap by a shift along the long molecular axis in the latter three cases. (ii) The small size of the oxygen atom compared to a sulfur atom at the S_{out} site generates a preferential feature for S_{in} to form strong side-by-side $S_{in} \cdots S_{in}$ contacts since the ratio b/c in Fig. 11.2b is large. As a consequence, donor stacking patterns are not so varied compared to those of ET materials; the typical donor packing patterns are only three, and many BO complexes are good two-dimensional metals regardless of size, shape and charge of the counteranion or acceptor molecules [11.38, 164, 165]. Even with a severe disorder of the anion side or in the disordered matrices, the metallic nature of BO compounds is preserved in, for example, Langmuir-Blodgett (LB) films, polymer composites, etc. [11.166-168]. The lighter effective mass deduced from the Shubnikov-de Haas oscillations in $(BO)_2 ReO_4 H_2 O$ may be a reflection of a bandwidth wider than that for ET materials [11.169].

In ET compounds such a self-aggregation ability is weak: this is because $CH \cdots S_{out}$ is weak and the $S_{in} \cdots S_{in}$ formation is prevented by the large size of S_{out} compared to that of BO materials. So, both the face-to-face and side-by-side transfer interactions are small and two-dimensionality is easily broken in ET compounds. Whereas in BETS compounds, the self-aggregation ability is expected to be stronger than in ET since the bulky selenium atoms extend in the transverse direction, and they resides at the inner chalcogen sites (Fig. 11.2). The outer sulfur atoms of BETS afford a relatively smaller contribution than that in ET materials.

The donor-anion atomic contacts have been regarded to play a crucial role in governing the packing pattern and the conformation of the donor molecule in the ET compound [11.162-165]. However, a detailed analysis of the crystal structures of ambient-pressure ET superconductors revealed that significantly shorter $CH \cdots$ anion contacts could not solely determine the donor packing picture, since such significantly shorter contacts are ob-

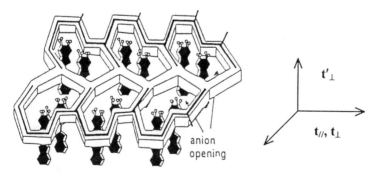

Fig. 11.18. Schematic view of κ-$(ET)_2 Cu(NCS)_2$ indicating anion openings and transfer interactions ($t_{||}$, $t_\perp$, $t'_\perp$)

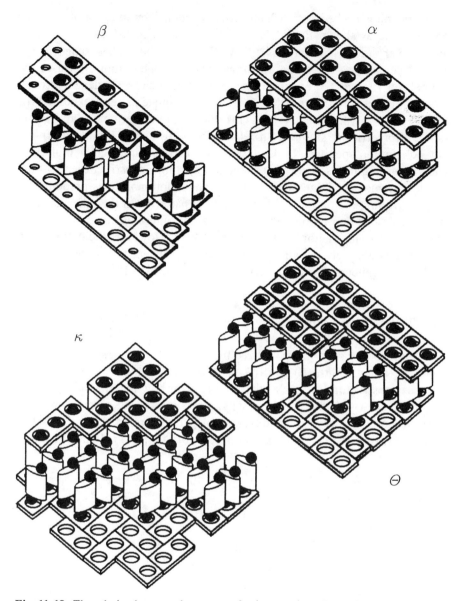

Fig. 11.19. The relation between the pattern of anion openings (key hole) and donor (key) packing pattern of α-, β-, θ-, and κ-types of ET complexes. Courtesy of H. Yamochi

served in κ-$(ET)_2 X$, X: $Cu(NCS)_2$, $Ag(CN)_2 H_2 O$, $Cu(CN)[N(CN)_2]$; but not in those of κ-$(ET)_2 X$, X: I_3, $Cu_2 (CN)_3$, $Cu[N(CN)_2]Br$; and the pattern of contacts is not common even among the κ-type salts [11.170].

The packing pattern of donors was found correlated deeply to the core pattern of the **anion opening** according to *Yamochi* et al. [11.170]. The anion opening is the space which is not occupied by anion atoms in the anion layer, as shown in Fig.11.18 schematically. In the case of a thick anion layer such as $MHg(SCN)_4^-$ (M: NH_4, K), the hollow space in the anion layer acts as the anion openings. The hydrogen atom closest to the anion layer fits into the opening. The position of such an ethylene hydrogen atom projected onto the anion layer produces a unique pattern, exhibited for α, β, θ, and κ types (Fig.11.19). This donor-anion intermolecular interaction is geometrically similar to a relation between key (ET molecules) and key-hole (anion opening). In the case of polymerized anions, ET molecules pack according to a scheme that the anion openings construct thereby, and so the intermolecular interactions among the anion molecules predominate as a controlling factor for the donor packing pattern. On the other hand, in the case of discrete anions, the anion$\cdots$anion intermolecular interactions are not the driving factor for the donor packing. However, the donor$\cdots$anion contacts do solely not allow to interpret the existence of so many phases of the ET compounds with the very same anion, such as I_3^-.

11.7.2 Effective Volume

The β-type donor stacking is realized by using the symmetric linear counter anion I_3^- (anion length: $10.1 \div 10.2 \text{Å}$), AuI_2^- (9.4Å) and IBr_2^- ($9.3 \div 9.4 \text{Å}$); they are isostructural to each other. To explain how the size of the linear anion affects T_c of these β-phase salts, a few interesting correlations have been proposed between T_c and structural parameters, such as the *lattice pressure* (Fig.5.39) [11.171], the *unit-cell volume* [11.172] and the *anion*

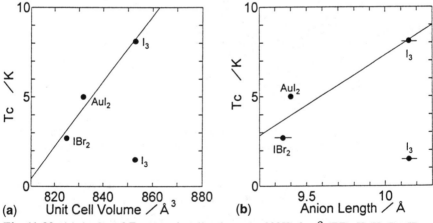

Fig. 11.20. (a) A plot of T_c vs. unit-cell volume (at 298K) for β-$(ET)_2$X (X: I_3, IBr_2, AuI_2). Redrawn after [11.172]. (b) A plot of T_c vs. anion length (vdW) for β-$(ET)_2$X (X: I_3, AuI_2, IBr_2). Redrawn after [11.173]

length (Fig. 11.20). The lattice pressure, or sometimes termed the **chemical pressure**, is a hypothetical pressure to represent the lattice compression, and is used as one of the parameters to measure the effect of a chemical substitution on the physical properties among isomorphous crystals.

A linear anion longer than I_3^- might yield a β-phase salt having a unit-cell volume larger than that of β_H-$(ET)_2 I_3$ and give a more negative lattice pressure resulting in a β-phase salt with T_c higher than 8 K (Figs. 5.39 and 11.20). However, tri-iodide is the longest one among the symmetric linear polyhalides (however, a linear I_5^- is known to be seldom found in a crystal). Consequently, a search for longer anions among a variety of metal halides and metal pseudo-halides commenced. In these studies Cu, Ag, Hg, Ni, Pd, etc. were the metals, and CN, OCN, SCN, SeCN, etc. were the pseudo-halides.

The first superconductor with T_c above 10 K was prepared with the anion of $Cu(NCS)_2^-$ in 1988 by *Urayama* et al. [T_c = 10.4K for the salt of h_8-ET (H-salt, Fig. 5.60) and 11.2K for that of d_8-ET (D-salt)] [11.174]. However, the salt has no β- but only a κ-phase; furthermore, the anion of $Cu(NCS)_2$ is neither symmetric nor linear, but polymerized (Fig. 5.7).

Table 11.3. Comparison of the structural features of $(ET)_2 X$ salts. (V_{cell}: Unit-cell volume, V_{anion}: Calculated anion volume, V_{eff}: Effective volume obtained with (11.4), and V_{mes2}: The volume of the space for the $C_6 S_8$ skeletons of the ET molecules plus the intra-donor-layer overlap space between them per two carriers

	T_c [K]	V_{cell}/Z [Å^3]	V_{anion} [Å^3]	V_{eff} [Å^3]	V_{mes2} [Å^3]
α-$(ET)_2 KHg(SCN)_4$	0.3	998.5	--	--	955
α-$(ET)_2 NH_4 Hg(SCN)_4$	0.8 ÷ 1.8	1004	--	--	959
$(ET)_2 ReO_4$	2 (4kbar)	794.7	106	689	--
β_L-$(ET)_2 I_3$	1.5, 2	853	148.3	705	952
β-$(ET)_2 IBr_2$	2.7	825	133	692	949
κ-$(ET)_2 I_3$	3.6	843.8	148.3	696	969
θ-$(ET)_2 I_3$	3.6	846.5	148.3	698	945
κ-$(ET)_2 Cu_2 (CN)_3$	4.1	847.0	152	695	1019
β-$(ET)_2 AuI_2$	4.9	831.8	124.5	707	952
κ-$(ET)_2 Ag(CN)_2 \cdot H_2 O$	5	828.4	--	--	983
β_H-$(ET)_2 I_3$	8.1	853	148.3	705	952
κ-$(ET)_2 Cu(NCS)_2$	10.4	844.0	130	714	997
κ-$(ET)_2 Cu(CN)[N(CN)_2]$	11.2	832.0	120	712	1003
κ-$(ET)_2 Cu[N(CN)_2]Br$	11.8	829.3	109	720	1013
κ-$(ET)_2 Cu[N(CN)_2]Cl$	12.8 (0.3kbar)	824.8	107	718	--

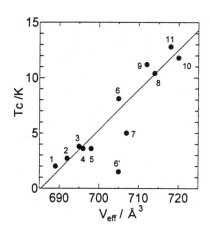

Fig. 11.21. A plot of T_c vs. V_{eff} of 1: $(ET)_2 ReO_4$, 2: β-$(ET)_2 IBr_2$, 3: κ'- $(ET)_2 \cdot Cu_2 (CN)_3$, 4: κ-$(ET)_2 I_3$, 5: θ-$(ET)_2 I_3$, 6: β_H-$(ET)_2 I_3$, 6': β_L- $(ET)_2 I_3$, 7: β-$(ET)_2 AuI_2$, 8: κ-$(ET)_2 \cdot Cu(NCS)_2$, 9: κ-$(ET)_2 Cu(CN) \cdot [N(CN)_2]$, 10: κ-$(ET)_2 \cdot Cu[N(CN)_2]Br$, 11: κ-$(ET)_2 Cu[N(CN)_2]Cl$

As ET superconductors have both a variety of donor packing patterns and anions, a new structural parameter which correlates with T_c has been examined. The unit-cell volume, the anion length and lattice pressure are not adequate since, for instance, a variety of superconductors with both different phases and T_c's ($1.5 \div 8.1 K$) were obtained on the basis of the same anion I_3^- and no clear relation was observed between T_c and the unit-cell volume in Table 11.3.

The **effective volume** V_{eff} was proposed by *Saito* et al. [11.175] as the space which ET molecules fill in a unit cell for one conduction electron

$$V_{eff} = \frac{V_{cell} - V_{anion}}{N}, \qquad (11.4)$$

where V_{cell}, V_{anion} and N are the unit-cell volume, the approximated anion volume, and the number of carriers per formula unit of a CT complex at room temperature, respectively (Table 11.3). This concept evolved from the idea that a low density of ET molecules in a crystal induces a narrow bandwidth and thereby a high $DOS(\epsilon_F)$. Although the calculated V_{eff} contains an uncertainty due to the inaccuracy of V_{anion}, a linear relation exists between T_c and V_{eff} among different phases and anions (Fig. 11.21). According to (11.4) and Fig. 11.21, metallic salts of α-$(ET)_2 MHg(SCN)_4$ (M: K, NH$_4$) which have a very large anion, were prepared with the expectation of a high T_c [11.176]. The use of large anions such as MHg(SCN)$_4^-$ increases V_{cell} considerably and is hoped to result in a large V_{eff}. However, the thick and bulky anions increase V_{anion} sufficiently to suppress V_{eff} and give a low T_c [11.177]. As a consequence, *a big anion which forms a thin anion layer* has been proposed as an appropriate anion-design strategy [11.178]. The substantial reduction of T_c in these salts (Table 11.3) seems also to suggest the importance of the interlayer interactions ($t'_{\perp}$) through an anion opening.

T_c of ET complexes were improved by the discovery of new κ-type salts by the Argonne group in 1990. They used large, yet thin anions $Cu[N(CN)_2]X$ (X: Cl [11.179], Br [11.180]), where dicyanamide ($N\equiv C$-N-$C\equiv N)^-$ bridges Cu^{1+} to form a polymer, and X attaches to Cu^{1+} as a pendant. For X: Cl, the salt becomes superconducting with $T_c = 12.8$ K under a pressure of 0.03 GPa (13.1K for the D-salt). For X: Br, the salt is an ambient-pressure superconductor with $T_c = 11.8$ K ($T_c = 11.3$K for D-salt). Another superconductor with T_c above 10 K was prepared by *Komatsu* et al. [11.181] in an attempt to replace the halogen X of $Cu[N(CN)_2]X$ by CN, but the anion was found to be $Cu(CN)[N(CN)_2]^-$ instead of $Cu[N(CN)_2](CN)^-$. Cu^{1+} has been connected to a ligand CN instead of $N(CN)_2$ which now plays the role of a pendant. T_c is 11.2 K (12.3 K for the D-salt). These T_c values cited here are the maximum values of T_c defined as the mid-point of the resistivity jump. The reported T_c values sometimes differ from those cited above, probably due to the different definitions of T_c, different measuring conditions or the impure materials used. {T_c data defined by magnetization measurements using fluctuation renormalization theory are 10.9 ± 0.3 and 10.6 ± 0.3 for the H- and D-salts of κ-$(ET)_2[N(CN)_2]Br$ and 8.7 ± 0.2 and 9.0 ± 0.2K for the H- and D-salts of κ-$(ET)_2Cu(NCS)_2$, respectively (Table 1.1) [11.182]}.

The large **thermal contraction of organic crystals** is one of the factors reducing T_c. The unit-cell volume of κ-$(ET)_2Cu(NCS)_2$ contracts by 63.6 Å^3 (or 31.8Å^3 per ET dimer) from room temperature to 20 K. This corresponds roughly to a 10 K reduction in T_c according to the calculation based on Fig. 11.21. In order to keep the thermal contraction small, the use of *a structurally two- to three-dimensional anion layer* is effective [11.52, 175, 178]. Comparison of the temperature dependences of the lattice parameters for κ-$(ET)_2Cu(NCS)_2$ and κ-$(ET)_2Cu[N(CN)_2]Br$ revealed that the magnitude of the thermal contraction along a particular direction depends on whether or not there are tight intermolecular atomic contacts. Short atomic contacts between ethylene groups and anion molecules prevent thermal contraction along the long axis of ET molecules. The anion···anion interactions play an important role in the thermal contraction. For example, the b-axis of κ-$(ET)_2Cu(NCS)_2$, the $Cu(NCS)_2$ anions being bonded to one another, exhibited very small thermal contraction. On the other hand, since anion polymers are isolated from each other along the c-axis, the contraction of this axis is large. Therefore, in order to keep the thermal contraction small, it is desirable to use a structurally two-dimensional anion layer which is able to provide short anion···donor contacts. However, this hardens the crystal and decreases the electron-phonon coupling (*vide infra*).

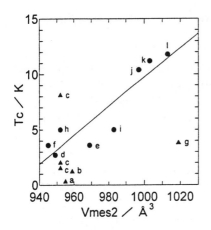

Fig. 11.22. A plot of V_{mes2} for $(ET)_2 X$. (a) α-$(ET)_2 KHg(SCN)_4$; (b) α-$(ET)_2 NH_4 \cdot Hg(SCN)_4$; (c) β-$(ET)_2 I_3$; (d) β-$(ET)_2 IBr_2$; (e) κ-$(ET)_2 I_3$; (f) θ-$(ET)_2 I_3$; (g) κ'-$(ET)_2 Cu_2 (CN)_3$; (h) β-$(ET)_2 AuI_2$; (i) κ-$(ET)_2 Ag(CN)_2 H_2 O$; (j) κ-$(ET)_2 Cu(NCS)_2$; (k) κ-$(ET)_2 Cu(CN)[N(CN)_2]$; (l) κ-$(ET)_2 \cdot Cu[N(CN)_2]Br$. The line indicates the result of least-squares fit of the plot omitting the points a-c and g, given by (11.5). Courtesy of H. Yamochi

The difficult point of (11.4) lies in an estimation of V_{anion}. Especially, V_{anion} has a large ambiguity in its approximation for complicated polymerized anions. To solve the problem, a structural parameter derived solely from crystallographic parameters is necessary. A V_{mes2} value has thus been proposed by *Yamochi* et al. [11.170], which corresponds to the sum of the volume of the **space for $C_6 S_8$ skeletons** of the donor molecules and that of the intra-donor-layer overlap space between them per two carriers (Table 11.3). In other words, this is the space which contributes to the carrier distribution most effectively. Figure 11.22 includes only the ambient-pressure superconductors of various phases. Though α-$(ET)_2 MHg(SCN)_4$ (M: K,a; NH_4,b), β-$(ET)_2 I_3$ (c) and κ'-$(ET)_2 Cu_2 (CN)_3$ (g) exhibited considerable deviation due to an SDW formation, a thick anion layer, disorder of the ethylene conformation and disorder of the anion layer, respectively, others are represented by the linear relationship

$$T_c \text{ [K]} = 0.131 V_{mes2} \text{ [Å}^3\text{]} - 121 . \tag{11.5}$$

Summarizing Sects. 11.7.1 and 2, it can roughly be said that the area of the basal plane of an anion opening determines the magnitude of T_c when the layer is thin and the pattern of anion openings governs the packing pattern of donors. It is desirable to find suitably large anions which construct large anion openings that correspond to loose packing of donor molecules. The concept of *lattice pressure, unit-cell volume, anion length* and *effective volume* are related to not only the density of states but also the softness of the crystal. *Whangbo* et al. have tried to evaluate the softness of a crystal as follows.

11.7.3 Softness

Whangbo and co-workers pointed out the importance of the softness of a crystal in connection with T_c of organic superconductors [11.51, 164, 165, 183-185]. The softness of a lattice provides a large electron-phonon coupling constant λ. They estimated the λ values by employing the observed T_c values of β-$(ET)_2X$ (X: I_3, IBr_2, AuI_2) with $\mu^* = 0.1$, $\langle \omega \rangle / \omega_0 = 0.3$ and $\theta_D = 200$ K via the **McMillan's formula**

$$
T_c = \frac{\theta_D}{1.45} \exp\left[-1.04 \frac{1 + \lambda}{\lambda - \mu^*(1 + \lambda\langle\omega\rangle/\omega_0)}\right]. \tag{11.6}
$$

where $\langle\omega\rangle$ and ω_0 are the average and maximum frequencies of the phonon band, respectively. The λ values increase with T_c for all β-$(ET)_2X$ (Table 11.4). The distances of both the CH$\cdots$anion contacts and CH$\cdots$CH (between neighboring ET molecules) contacts increase in the order

$$
\beta\text{-}(ET)_2IBr_2 \; < \; \beta\text{-}(ET)_2AuI_2 \; < \; \beta_H\text{-}(ET)_2I_3 \;. \tag{11.7}
$$

By taking the lengths of these contacts as a measure of the lattice softness, they observed good agreement with the tendency of softening of the lattice and the increase of the λ values. On the other hand, the DOS(ϵ_F) values calculated on the basis of the low-temperature crystal structures did not show good correlation with the T_c values (Table 11.4).

For κ-type salts, they also advanced the importance of lattice softness in having a high T_c, although a simple geometrical parameter could not be extracted from the results of the various κ-type salts. The T_c values of κ-$(ET)_2I_3$ and κ-$(MDT\text{-}TTF)_2AuI_2$ higher than that of κ-$(DMET)_2AuBr_2$ have been explained by contacts of CH$\cdots$I softer than those of CH$\cdots$Br. T_c of κ-$(ET)_2I_3$ lower than that of κ-$(MDT\text{-}TTF)_2AuI_2$ was thought to be related to the presence of stiff CH$\cdots$donor [CH$\cdots$S, CH$\cdots$C(sp^2)] contacts in the former. In spite of the harder contacts of CH$\cdots$C, CH$\cdots$N, CH$\cdots$S and/or CH$\cdots$Br than CH$\cdots$I, and also the calculated DOS(ϵ_F)'s for them are not much different from κ-$(ET)_2Cu(NCS)_2$, κ-$(ET)_2Cu\cdot$[N(CN)$_2$]Br and κ-$(ET)_2I_3$, the T_c values of the former two are higher than the last one. This result being opposite from the standpoint of lattice softness is difficult to understand.

MDT-TTF

Table 11.4. Calculated electron-phonon coupling constant (λ), density of states at Fermi level DOS(ϵ_F), and ethylene conformation (E: eclipsed, S: staggered)

	λ^a	DOS(ϵ_F)^a	DOS(ϵ_F)^b	Ethylene conformation	
		$-120\,K$	RT	RT^c	120 K
β_L-(ET)$_2$I$_3$	0.37	3.66	0.95	no	S&E
β-(ET)$_2$IBr$_2$	0.43	3.69	0.87	E	E
β-(ET)$_2$AuI$_2$	0.52	3.48	0.96	E	E
β_H-(ET)$_2$I$_3$	0.62	3.53	0.95	E	E
θ-(ET)$_2$I$_3$			1.0	E	
κ-(ET)$_2$I$_3$		7.09^d	0.90	E	E
κ-(ET)$_2$Cu$_2$(CN)$_3$			1.1	no(S)	S
κ'-(ET)$_2$Cu$_2$(CN)$_3$			1.1	S	S
κ-(ET)$_2$Cu(NCS)$_2$		7.39	0.89	no(S)	S
κ-(ET)$_2$Ag(CN)$_2\cdot$H$_2$O			1.0	S&E	
κ-(ET)$_2$Cu(CN)[N(CN)$_2$]			1.0	S&E	S&E
κ-(ET)$_2$Cu[N(CN)$_2$]Br		7.27	0.98	no(E)	E
κ-(ET)$_2$Cu[N(CN)$_2$]Cl		7.26	0.91	no	E
α-(ET)$_2$NH$_4$Hg(SCN)$_4$			1.1	E	

[a] From [11.185] DOS(ϵ_F) is in e/eV (unit cell) calculated by the extended Hückel tight-binding method which excludes d-orbitals of sulfur atoms based on the crystal structures at low temperatures and ambient pressure [β_L-(ET)$_2$I$_3$, β-(ET)$_2$IBr$_2$, β-(ET)$_2$AuI$_2$] or at 1.5 kbar [β_H-(ET)$_2$I$_3$]

[b] In states·eV^{-1}·molecule^{-1}·spin^{-1} calculated by the extended Hückel tight-binding method, which includes d-orbitals of the sulfur atoms, based on the crystal structures at room temperature and under ambient pressure

[c] No(S or E) means the ethylene groups are disordered but are close to the S or E conformation, respectively

[d] From [11.184]

The dependence of the inverse isotope effect on T_c was similarly interpreted by the lattice softness as following [11.183, 184]: The C-D bond effectively longer than a C-H bond may provide a softer enviroment for ethylene$\cdots$anion and ethylene$\cdots$ donor interactions, which thus results in a larger λ value and a higher T_c. A similar but different term, the **geometrical chemical isotope effect**, was proposed by *Toyota* et al. to interpret the isotope effect; namely, the chemical substitution of H atoms by D atoms exerts in these salts a minute change of the local structure around the terminal ethylene groups. This results in a chemical pressure change between the H- and D-salts [11.186]. These proposals will be validated after a precise crystal-structure analysis of the H- and D-salts of several ET superconductors, most of which except κ-(ET)$_2$ Cu[N(CN)$_2$]Br show the inverse isotope effect in the superconducting phase.

A quantitative measure of the lattice softness has not been developed yet. The lattice softness is closely related to *Pearson's* Hard and Soft Acid and Base (HSAB) [11.187]. HSAB was utilized to interpret the coordination pattern in an anion of MHg(SCN)$_4$ [11.176]. Further adoption of the concept of HSAB to design organic conductors and superconductors, and to raise T_c is definitely important.

11.7.4 Effective Mass

Since DOS(ϵ_F) is proportional to the effective mass of electrons for the two-dimensional system, the T_c value can be a function of the effective mass. *Caulfield* et al. observed a sharp decrease of the effective mass with increasing pressure up to $3 \div 4$ kbar, and a gradual decrease above 5 kbar by the Shubnnikov-de Haas measurements under pressure on κ-(ET)$_2$ Cu(NCS)$_2$ [11.188]. T_c of this salt decreases almost linearly with pressure (Fig. 5.45). An increase of pressure is expected to broaden the bandwidth that causes both the decreases of the effective mass due to the increase of transfer interactions and of quasi-particle interactions which is known to be larger, in general, for a narrower bandwidth. They postulated that the initial decrease of the effective mass with pressure represents the suppression of a component of the quasiparticle interactions [11.188].

According to their observation and postulation, it is desirable to have a large effective mass at the Fermi level and large quasiparticle interactions to enhance T_c. In addition, it is important to have such an effect not by pressure but by chemical procedures from the standpoint of molecular design. It is said that quasiparticle interactions become large in the system of a narrow bandwidth composed of molecules with large on-site Coulomb repulsion.

11.7.5 Electron Correlation vs. Superconductivity

Although the four 10K-class ET superconductors with polymerized anions $\{(\kappa$-(ET)$_2$ A, A: Cu(NCS)$_2$, Cu(CN)[N(CN)$_2$], Cu[N(CN)$_2$]X (X: Cl, Br)$\}$ have similar structural aspects, their transport properties differ apparently (Fig. 11.23). κ-(ET)$_2$ Cu(CN)[N(CN)$_2$] shows a monotonical decrease of resistivity down to low temperatures. On the other hand, κ-(ET)$_2$ Cu(NCS)$_2$ and κ-(ET)$_2$ Cu[N(CN)$_2$]Br have a semiconductor-like region above 70 K.

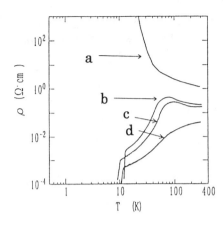

Fig. 11.23. Temperature dependence of the resistivity of (**a**) κ-(ET)$_2$ Cu[N(CN)$_2$]Cl, (**b**) κ-(ET)$_2$ Cu[N(CN)$_2$]Br, (**c**) κ-(ET)$_2$ Cu·(NCS)$_2$ and (**d**) κ-(ET)$_2$ Cu(CN)[N(CN)$_2$]

κ-(ET)$_2$ Cu[N(CN)$_2$]Cl exhibits a similar semiconducting character above ca. 42 K.

The resistivity hump above 70 K that is observed in κ-(ET)$_2$ Cu(NCS)$_2$ and κ-(ET)$_2$ Cu[N(CN)$_2$]Br has been ascribed to one of the characteristic features of κ-type donor packing since all the following κ-type salts reveal such a queer hump: (ET)$_2$ X (X: Cu(NCS)$_2$, Cu[N(CN)$_2$]Br, Cu[N(CN)$_2$]· Br$_{0.5}$Cl$_{0.5}$, Ag(CN)$_2$H$_2$O, Hg$_{2.89}$Br$_8$ at ambient pressure and X: Cu· [N(CN)$_2$]Cl, Cu$_2$(CN)$_3$ under pressure), (DMET)$_2$ AuBr$_2$ and (MDT-TTF)$_2$ AuI$_2$. However, not all κ-type salts show the hump; namely, κ-(ET)$_2$· X (X: I$_3$, Cu(CN)[N(CN)$_2$]) and κ'-(ET)$_2$Cu$_2$(CN)$_3$. Furthermore, κ-(BETS)$_2$X (X: SbF$_6$, TaF$_6$, Cu[N(CN)$_2$]Br, GaCl$_4$, FeCl$_4$, CF$_3$SO$_3$) does not exhibit the hump, but λ-(BETS)$_2$X (X: GaCl$_4$, GaBr$_4$, FeCl$_4$, FeBr$_4$, InCl$_4$) does so [11.96]. Therefore, the resistivity hump is not ascribable to the particular κ-type donor packing.

The origin of this queer hump of the resistivity is controversial [11. 189-193]. Strong electron correlation [11.189], mixed valence of Cu ions [11.190], freezing of the ethylene disorder (Table 11.4) [11.191], structural effect including the abnormal lattice dilation change [11.192], a contribution of the thermally excited carriers from flat portions of the occupied bands to the unoccupied bands [11.193], and an imperfection of the crystals have been proposed, but still this behavior is not fully understood.

Magnetic susceptibility measurements reveal that the magnitude of the spin susceptibility χ_{spin}, at room temperature, of these four compounds (Table 11.5) is much smaller than those of the Mott insulators of ET compounds. χ_{spin} is more than $8 \cdot 10^{-4}$ emu/mol for Mott insulators such as α'-(ET)$_2$X [X: AuBr$_2$, CuCl$_2$, Ag(CN)$_2$, IBr$_2$, Au(CN)$_2$, $(8.5 \div 10) \cdot 10^{-4}$], β'-(ET)$_2$X [X: ICl$_2$, AuCl$_2$, $9.6 \cdot 10^{-4}$] and (ET)$_2$GaCl$_4$ $(8 \cdot 10^{-4})$, but less than $6 \cdot 10^{-4}$ emu/mol for the 10K superconductors $(4.5 \div 5.5 \cdot 10^{-4})$. This is much close to those of metallic ET compounds $(<7 \cdot 10^{-4}$ emu/mol) as, for

Table 11.5. Spin susceptibility at room temperature and calculated band parameters of Mott Insulators (MI), Metals and Superconductors (MS) of ET compounds

		$10^4 \chi_{spin}$ [emu/mol]	Band parameter [eV][a]				
			W	W_L	W_U	ΔE	$W_U/\Delta E$
MI	β'-(ET)$_2$AuCl$_2$	9.6	0.83	0.42	0.25	0.53	0.48
MI	β'-(ET)$_2$BrICl		0.83	0.38	0.26	0.54	0.49
MI	β'-(ET)$_2$ICl$_2$	9.6	0.85	0.38	0.27	0.55	0.49
MI	ET·TCNQ(triclinic)	–	0.93	0.53	0.41	0.46	0.89
MI	κ-(ET)$_2$Cu(CN)$_3$	7.1÷7.6	1.00	0.33	0.50	0.45	1.11
MS	κ'-(ET)$_2$Cu$_2$(CN)$_3$	5.3÷6.8	1.00	0.33	0.49	0.46	1.08
MS	κ-(ET)$_2$Cu[N(CN)$_2$]Cl	4.5	1.10	0.32	0.56	0.51	1.10
MS	κ-(ET)$_2$Cu[N(CN)$_2$]Br	4.5÷5.5	1.08	0.32	0.55	0.49	1.13
MS	κ-(ET)$_2$Cu(CN)[N(CN)$_2$]	4.6	1.03	0.32	0.52	0.45	1.15
MS	κ-(ET)$_2$Ag(CN)$_2$·H$_2$O	4.0	1.06	0.32	0.54	0.47	1.16
MS	κ-(ET)$_2$Cu(NCS)$_2$	4.5÷4.6	1.08	0.36	0.57	0.46	1.24
MS	β-(ET)$_2$AuI$_2$	3.4	1.07	0.45	0.57	0.50	1.14
MS	β-(ET)$_2$I$_3$	4.6	1.03	0.41	0.59	0.48	1.22
MS	β-(ET)$_2$IBr$_2$		1.07	0.45	0.60	0.49	1.22
MS	κ-(ET)$_2$I$_3$		1.15	0.35	0.61	0.49	1.24

[a] Calculated by the extended Hückel tight-binding method, which includes d-orbitals of sulfur atoms, based on the crystal structures at room temperature and ambient pressure

example, α-(ET)$_2$I$_3$ ($4.2 \cdot 10^{-4}$), β-(ET)$_2$X [X: I$_3$, AuI$_2$, ($3.4 \div 4.6) \cdot 10^{-4}$] and (ET)$_2ClO_4(TCE)_{0.5}$ ($4.2 \cdot 10^{-4}$). Moreover, the temperature dependence of χ_{spin} is less temperature sensitive than those of Mott insulators of the ET compounds. Therefore, as far as the magnetic susceptibility concerns, the electronic states of these four 10K superconductors may not be those of typical Mott insulating state above 70 K.

Due to the dimerization of ET molecules of the β-, β'-, or κ-type packing, the calculated HOMO band is split into upper- and lower HOMO bands (Sect. 5.14, and Figs. 5.31 and 11.12). The upper HOMO band is half filled provided that U_{eff} is negligible compared to the bandwidth of the upper HOMO band (W_U). The energy splitting of the upper and the lower HOMO bands (ΔE) corresponds to the dimerization energy of a dimerized ET pair. The on-site Coulomb repulsion of a dimer (U_{dimer}) can be expressed as [11.194],

$$U_{dimer} = \Delta E + \frac{1}{2}[U - (U^2 + 4\Delta E^2)^{1/2}] . \tag{11.8}$$

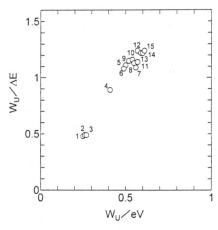

Fig. 11.24. A relation between dimerization energy (ΔE) and bandwidth of the upper band (W_U) of ET compounds having dimerized ET molecules. (1) β'-$(ET)_2 AuCl_2$; (2) β'-$(ET)_2 BrICl$; (3) β'-$(ET)_2 ICl_2$; (4) $ET \cdot TCNQ$ (triclinic); (5) κ-$(ET)_2 Cu_2 (CN)_3$; (6) κ'-$(ET)_2 Cu_2 (CN)_3$; (7) κ-$(ET)_2 Cu[N(CN)_2]Cl$; (8) κ-$(ET)_2 Cu[N(CN)_2]Br$; (9) κ-$(ET)_2 Cu(CN)[N(CN)_2]$; (10) κ-$(ET)_2 Ag(CN)_2 \cdot H_2 O$; (11) β-$(ET)_2 Cu(NCS)_2$; (12) β-$(ET)_2 AuI_2$, (13) β-$(ET)_2 I_3$, (14) β-$(ET)_2 IBr_2$; (15) κ-$(ET)_2 I_3$. Courtesy of H. Sasaki

Since U of an ET molecule is much larger than the ΔE values of ET molecules, U_{dimer} can be approximated by ΔE. So, the **Mott criterion** is modified as *when the ΔE value is sufficiently larger than the bandwidth of the upper HOMO band ($\Delta E > W_U$), the electronic system turns into a Mott-insulating state.*

The calculated band parameters based on the extended Hückel method including the d-orbitals of sulfur atoms of ET are summarized for the above four 10 K superconductors together with those of ET compounds of Mott insulators and the superconductors [κ-$(ET)_2 I_3$ and β-$(ET)_2 X$; $X = AuI_2$, I_3, IBr_2] in Table 11.5 [11.159, 195]. The ratio of the magnitudes of ΔE and W_U can distinguish whether the system is a typical Mott insulator or not [11.195]. A plot of $W_U/\Delta E$ vs. W_U (Fig. 11.24) demonstrates the following aspects: (1) A nearly linear relation exists between the W_U/Δ and W_U values among the ET compounds with a dimerized donor pair. (2) Typical Mott insulators reside at the lower-left side ($1 \div 4$ in Fig. 11.24), and good metals ($12 \div 15$) are located at the upper-right side. (3) The 10 K superconductors ($7 \div 9$, 11) reside between the Mott insulators and good metals with $W_U/\Delta E$ around $1.1 \div 1.2$. (4) A Mott insulator having $W_U/\Delta E$ close to unity is able to be converted to a metal and a superconductor by a light band-filling change [κ-$(ET)_2 Cu_2 (CN)_3$ (5) $\rightarrow$ κ'-$(ET)_2 Cu_2 (CN)_3$ (6)]. (5) Figure 11.24 strongly suggests that materials showing a resistivity hump have rather small $W_U/\Delta E$ values which diplay a strong electron correlation.

11.8 Development of C_{60} Salts

The mass production of a new type of carbon of spherical shape, namely C_{60} (I_h symmetry, radius: ≈ 7Å), by *Krätchmer* et al. [11.196] has allowed the physical properties of C_{60} and its CT complexes in the solid state to be studied. Soot containing C_{60} was generated by a DC arc discharge between graphite electrodes under a reduced pressure of purified He (≈ 100 torr). Soxhlet extraction with toluene gave an about 10% (w/w) yield of the extract, from which an about 50% (w/w) yield of pure C_{60} was separated by column chromatography on neutral alumina with hexane/toluene or on activated charcoal with pure toluene. Recently, nonchromatographic purification methods for C_{60} by selective complexation with calixarenes [11.197, 198] or by reversible addition to silica-supported dienes [11.199] were reported. Since these methods require no chromatographic operations, they would be applicable to a factory scale process to reduce the commercial price of purified C_{60}. The C_{60} powder thus obtained usually contains solvents which can be eliminated by sublimation. Single crystals of C_{60} were prepared by gradient sublimation (550 to 600°C, 10^{-6} torr) in a sealed quartz tubing.

HOMO of a C_{60} molecule consists of a five-fold degenerate h_u state, and LUMO is a triply degenerate t_{1u} state (Sect. 10.1). The HOMO-LUMO gap is 1.8 eV. The vertical ionization potential (I^v) of a C_{60} molecule was estimated to be 7.6 eV (compare TMTSF: 6.58eV, ET: 6.7eV, perylene: 7.00eV) which indicates that the C_{60} molecule is a very weak electron donor (Table 11.1) [11.32]. The vertical electron affinity E^V of a C_{60} molecule deduced from UPS or threshold photodetachment of cold C_{60}^{1-} ($E^V = 2.650 \pm 0.050$eV) suggests that a C_{60} molecule is a fairly strong acceptor [11.200]. The adiabatic electron affinity E^{ad} estimated from both the CT absorption band and the redox potential is, however, rather low ($E^{ad} = 2.10 \div 2.21$ eV), and the acceptor strength of C_{60} is concluded to be comparable with that of the weak electron acceptor tetracyanobenzene ($E^{ad} = 2.15$eV) [11.201]. The reduction potential measured in CH_3CN/toluene is -0.38V vs. SCE, which corresponds to -0.44V vs. SCE in CH_3CN. Hence, in order to ionize the C_{60} molecule completely, an electron donor with $E^1(D) \leq$ ca.-0.5V vs. SCE in CH_3CN is necessary. Such strong electron donors with largely negative $E^1(D)$ values are rare for organic materials, e.g., $Fe^I(C_5H_5)(C_6Me_6)$ ($E^1(D) = -1.55$V vs. SCE in DMF) [11.202], cobaltocene [$E^1(D) = -0.95$V vs. SCE in CH_3CN] [11.203], Tetrakis (DimethylAmino)Ethylene (TDAE, -0.75V vs. SCE in CH_3CN) [11.204].

Metallic or superconducting CT complexes of C_{60}, which have so far been obtained, exhibit features deviating from those of the TMTSF or ET superconductors in that (i) they have enough transfer interactions even

NC, CN

NC CN

tetracyanobenzene

Me Fe Me
Me — —— Me
Me Me

$Fe^I(C_5H_5)(C_6Me_6)$

Co

cobaltocene

Me   Me
Me - N   N - Me

Me - N   N - Me
Me   Me

TDAE

S   S

S   S

OMTTF

along the alternately stacking direction of C_{60} and the alkali ion, and (ii) the degree of CT is integral. The former characteristics may be caused by the three-dimensional nature of the C_{60} complexes. These C_{60} CT complexes are the radical anion salts, where the countercations are alkali and alkaline earth ions, which are small in size compared with the C_{60} molecule. This large size difference between the components results in a rather common alternating and three-dimensional features among the alkali-metal C_{60} complexes. The latter characteristics may be due to the instability of the phase with incommensurate stoichiometry that is caused by a weak self-assembling ability of C_{60} molecules and indicate that U_{eff} is smaller than W in the C_{60} complexes. However, this point is not clearly understood yet, as noted by *Ramirez* [11.205]. The accuracy of the stoichiometry determination is a subject of intense debate, and the precise stoichiometry of the superconducting phase remains an important open question [11.205]. The difficulty lies in the fact that the metallic and superconducting C_{60} compounds are very sensitive to ambient conditions and easily decompose in air.

AC_{60} (fcc for A: K, Rb, Cs) is a semiconductor and turns into the one-dimensional polymer of C_{60} (orthorhombic) reversibly by slow cooling below 400 K (Fig. 11.25). The polymer of KC_{60} is metallic down to ca. 50

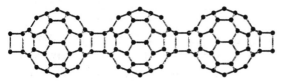

Fig. 11.25. One-dimensional polymer of C_{60} in AC_{60} (orthorhombic)

K while that of RbC_{60} is a semiconductor under ambient pressure. Both become metallic down to 4.2 K under pressure; however, no superconductivity has been detected so far [11.206]. A_2C_{60} (fcc only A: Na), A_4C_{60} (bct for M: K, Rb, Cs) and A_6C_{60} (fcc for A: Na, bcc for A: K, Rb, Cs) are insulators. A_5C_{60} has not been produced. A solid solution has been obtained only with Na_xC_{60} in the range of $1 < x < 3$.

Tanigaki et al. [11.207] summarized features of the phase stability of A_3C_{60}: (i) The stability of the fcc A_3C_{60} phase is increased by the octahedral(O)-site occupation with larger alkali-metals as follows: $A(T)_2Cs(O)C_{60}$ > $A(T)_2Rb(O)C_{60}$ > $A(T)_2K(O)C_{60}$, where $A(T)$ means an alkali-metal occupying the tetrahedral(T)-site. (ii) The T-site occupation with Cs^+ (r = 1.70 Å) makes the fcc A_3C_{60} phases unstable. Actually no fcc stable phases exist for $Cs(T)_2Cs(O)C_{60}$ and $Na(T)Cs(T)Cs(O)C_{60}$, and $K(T)Cs(T)Cs(O)\cdot C_{60}$ is unstable. The only exception has been found for $Rb(T)Cs(T)Cs(O)\cdot C_{60}$, where Rb^+ accommodated in one of the T-sites expands the other T-site so that it can be filled by a Cs^+ atom without too much loss of stability. (iii) The Li^+ and Na^+ ions preferentially occupy T-sites. The occupation of the O-sites with either Li^+ or Na^+ having a small ionic radius leads to an instability of the fcc A_3C_{60} phase. So, $Li_2(T)A(O)C_{60}$ and $Na(T)_2A(O)C_{60}$ are the predominant stable phases.

T_c of an fcc A_3C_{60} superconductor was found to increase with an increase of the lattice parameter ℓ of A_3C_{60} (Fig. 10.14) by *Fleming* et al. [11.208] and *Tanigaki* et al. [11.209]. A plot of T_c vs. the effective volume of C_{60} ($V_{eff} = V_{cell} - V_{cation}$) [11.210] or volume per one C_{60} molecule [11.211] also shows almost the same result as that of Fig. 10.14, since the V_{cation} values are so small compared to V_{cell} of A_3C_{60}.

The relation in Fig. 10.14 is interpreted in the same way as that for β-$(ET)_2X$ salts (Sect. 11.7.2); namely, the increase of the lattice parameter corresponds to an increase of the distance among the C_{60} molecules in A_3C_{60}. Therefore, the transfer integral and the bandwidth decrease results in an increase of $DOS(\epsilon_F)$. In addition, a softening of the lattice is expected by the increase of the lattice parameter. The relation between T_c and the lattice parameter in Fig. 10.14 holds as long as the superconducting phase is fcc and the charge of C_{60} is -3.

Though the ammoniation of $CsNa_2C_{60}$ gives rise to T_c that is expected from the empirical relation in Fig. 10.14, that of NaK_2C_{60} or $NaRb_2C_{60}$ yields $(NH_3)_xNaK_2C_{60}$ or $(NH_3)_xNaRb_2C_{60}$ ($x \simeq 0.8$) having fcc with a drastically lower T_c value than expected [11.211]. A structural analysis of $(NH_3)_xNaRb_2C_{60}$ revealed the presence of off-centered alkali ions in the octahedral site. It is expected that the disorder due to the off-centered alkali ions removes the degeneracy of the t_{1u} orbital and gives rise to a low T_c [11.211].

Several C_{60} superconductors not belonging to the fcc and sc A_3C_{60} have been prepared such as Ba_4C_{60}, Ca_5C_{60}, $A_3Ba_3C_{60}$ (A: K,Rb) and Sr_6C_{60} (Table 1.4). The superconductivity of these complexes is based on the second LUMO (t_{1g}). Furthermore, a mild doping with alkali metals to neutral CT complexes of C_{60} and various donor molecules, which have a two-dimensional layer of C_{60} molecules, gave superconductors: $K_x \cdot$ OMTTF$\cdot C_{60} \cdot$benzene ($T_c = 17 \div 18.8K$), $Rb_x \cdot$OMTTF$\cdot C_{60} \cdot$benzene ($T_c = 23 \div 26K$) [11.212]. It is important to know whether the degeneracy of LUMO (t_{1s} or t_{1u}) of C_{60} in these complexes is broken or not on the complex formation for further development of C_{60} superconductors.

11.9 Crystal Growth

In this section, a general description of the crystal-growth method, especially electrocrystallization, is presented at first, and then some specific examples are described in more detail.

11.9.1 Method of Crystal Growth of a Molecular Complex

Even though the quality of single crystals depends on the cleanliness of the apparatus and the purity of the starting materials (supporting electrolytes, halides, alkali metals, solvents, gas molecules, etc.), readers may obtain information on the purification method for conventional chemicals from [11.213], and we do not refer to them further. As for the donor and acceptor molecules, gradient sublimation on Teflon [11.214] is the most convenient scheme for the final step of purification of materials which can be sublimed, such as TTF, TMTTF and TMTSF. Others listed in Table 11.1, except for C_{60}, are purified by a combination of recrystallization and chromatographic methods.

CT complexes are prepared mainly by the following three redox reactions: (i) electrocrystallization (galvanostatic and potentiostatic), (ii) a direct reaction of donors (D) and acceptors (A) in the gaseous, liquid or solid phase, and (iii) metathesis usually in a solution ($D \cdot X + M \cdot A \rightarrow D \cdot A$). In the latter two cases, single crystals are produced by the diffusion, concentration, slow cooling, or slow cosublimation methods [11.215].

Electrocrystallization is performed with a variety of glass cells, as shown in Fig.11.26, see also [11.51]. Strictly speaking, the potentiostatic method is the proper way, in which a three-compartments cell is employed and one of the compartments contains the reference electrode, such as SCE

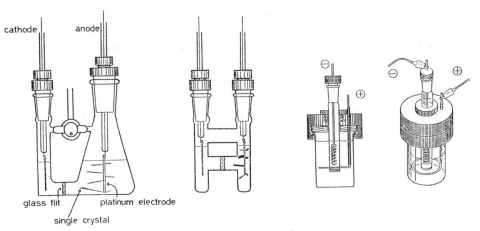

cathode anode

glass flit platinum electrode

single crystal

Fig. 11.26. Cells for electrocrystallization [11.215, 216]

or Ag/AgCl. However, this method is troublesome when a large number of crystal-growth runs are performed for a long period of time due to the following: (1) the contamination through the use of a reference electrode cell, and (2) the limited space for the experiment. The galvanostatic method is much more convenient than the potentiostatic one from these points of view. An H cell (20 ml or 50 ml capacity) and an Erlenmeyer-type cell (100 ml) with a fine-porocity glass-frit equipped with two platinum wire electrodes (1 and 2 mm in diameter) have been used. The main procedures are as follows: (i) Put a small magnetic stirring bar into each compartment and add donor molecules (10 to 20 mg for the ET case in a 20 ml cell) to the anode compartment and a sufficient amount of supporting electrolyte (0.1 ÷ 0.2 M) to the cathode compartment. When the electrolyte is soluble it can be added to both compartments. (ii) Evacuate the cell, fill it with inert gas (N_2 or Ar), and repeat this procedure several times. (iii) Add the solvent which had been distilled under inert gas prior to its use, employ syringe, and dissolve the materials by stirring for a while (half a day to overnight) under inert gas at room temperature. When the materials are thermally stable, heating (or reflux) saves time. Then take the magnetic stirring bars off, and the mixture is kept until the undissolved materials subside. When the electrolytes are light- or air-sensitive, it is better to add them after dissolving the donor portion. (iv) Insert platinum electrodes to each compartment. The anode and cathode are connected to a constant-current power supply and electrolyzed under dark condition which is often critical.

There are so many factors and tricks to grow single crystals of good quality. The important factors besides both the purity and the concentration are the kinds of solvent, the surface of the electrode, the current density, and temperature. TetraHydroFurane (THF), CH_2Cl_2, TCE, chlorobenzene,

CH$_3$CN and benzonitrile are commonly utilized solvents. The addition of 1÷10 v/v% ethanol occasionally accelerates the crystal growth. It is recommended to elute the halogenated solvent through basic activated alumina just prior to the use in order to eliminate a trace of acid that is easily produced especially in TCE. It is worth looking at the report of *Anzai* et al., which treats the solvent effects on the crystal growth for several particular combinations of donors and electrolytes [11.216, 217]. Regarding the surfaces of electrodes each research group has a special treatment such as burning (but not melting) or polishing with very fine powder. The electrode surfaces can be treated by applying a current to switch the polarity in a 1 M H$_2$SO$_4$ solution [11.51]. Low-current density is desirable (0.1÷2μA for a 2-mm$^\phi$ electrode; 0.05÷1 μA/cm^2). In general, good crystals grow at the temperature above 5° and below 25°C. However, the appropriate current density and temperature vary from salt to salt; e.g., better crystals of κ-(ET)$_2$Cu[N(CN)$_2$]Br are needed at lower current densities and temperatures compared to those of κ-(ET)$_2$Cu$_2$(CN)$_3$.

The advantage and disadvantage of the electrocrystallization method are:

- This is a clean redox reaction without taking advantage of oxidizing or reducing chemical reagents. However, it needs a supporting electrolyte and the reaction of a supporting electrolyte is difficult to predict, especially in the case of a galvanostatic reaction at a high current density.
- The rate of crystal growth is, to some extent, controllable by changing the temperature and current density. As a result, CT crystals composed of radical species which are unstable in solution, can be grown by applying a high current at low temperatures; e.g., salts of fluoranthene (−30°C, 2mA, Ni electrode), naphthalene, azulene [11.8, 218].
- Compared to the diffusion method, the growing period is rather short in the electrocrystallization method (2÷3 months vs. a few tenth of a second to a few weeks). Thus, the optimum condition of crystal growth for a particular salt are more or less easy to be attained.
- In some cases the partial CT state is artificially attained [11.219, 220].
- With this method it is difficult to obtain a D·A-type complex compared with the direct reaction method since a soluble electrolyte of D·X or M·A is needed in a large quantity.

azulene

fluoranthene

Superconducting single crystals of good quality were prepared mainly by the electrochemical redox process of TTF analogues or a dmit system. In some cases, direct chemical oxidation with iodine in gas or with $TBA \cdot I_3$ or $TBA \cdot IBr_2$ in solution yielded superconducting crystals with a better superconducting quality [11.221, 222]. Better-quality single crystals of $(TTF) \cdot [Ni(dmit)_2]_2$ were obtained by the diffusion method of the methathesis reaction rather than electrocrystallization [11.223].

C_{60} superconductors were prepared by chemical doping in the gas phase [11.224], a direct reaction with metal or metal azide in solution [11.225, 226], or a disproportionation reaction between C_{60} and $A_6 C_{60}$ [11.227]. The best C_{60} superconductors so far produced, were obtained by doping with single crystals of C_{60} molecules [11.228, 229]. Information on the structure and stoichiometry of the resultant doped C_{60} material has not been reported.

11.9.2 Typical Examples of Single-Crystal Growth

$(TMTSF)_2 X$

Black shiny single crystals with needle shape ($30 \times 0.7 \times 0.1$ or $12 \times 2 \times 1$ mm^3) of $(TMTSF)_2 ClO_4$ were prepared by electro-oxidation of TMTSF ($\approx 10^{-3}$ M) in the presence of an electrolyte: $TBA \cdot ClO_4$ (≈ 0.1M) from THF, $CH_2 Cl_2$, or TCE at $1 \div 2$ μA (Fig. 11.27a). Other $(TMTSF)_2 X$ complexes were prepared by means of $TBA \cdot X$. The crystals grew mainly on the electrode but also on the surface of the cell [11.230].

I_3 Salts of ET

β-phase crystals were prepared, usually together with α-$(ET)_2 I_3$, by electro-oxidation of ET at a current[2] of ≈ 20 μA/cm^2 with $TBA \cdot I_3$ in TCE or benzonitrile [11.231]. Under a high current density (64μA/cm^2) in TCE the γ-phase was obtained as the major product with the δ-phase as minor one [11.231].

A direct chemical oxidation by I_2 in benzonitrile provided α- and β-phases from benzonitrile when equimolar I_2 was utilized [11.231]. By using 2 molar excess I_2 or equimolar TetraEthylAmmonium (TEA) $\cdot CII_4$ in benzonitrile, the γ- phase showed up. When benzonitrile solutions of ET and I_2 in a molar ratio of 1:2 was mixed at 80°C followed by a slow cooling (1.4°C/h) to room temperature, the ϵ-phase came up, whereas the molar ratios of 1:5 and 1:10 produced the ζ-phase [11.232]. Both the ϵ- and

---

[2] A high current density was often utilized in the early days, but good-quality crystals have been obtained at lower current density

Fig. 11.27. Single crystals of (**a**) (TMTSF)$_2$ ClO$_4$ and (**b**) κ-(ET)$_2$ Cu(NCS)$_2$

ζ-phases converted to a β_t-phase superconductor after tempering (Sect. 5.2.8) [11.232-234]. The oxidation of ET by excess TBA·I$_3$ in TCE or by excess TBA·IBr$_2$ in benzonitrile yielded the α- and β- phases of (ET)$_2$I$_3$ or (ET)$_2$IBr$_2$, respectively.

The θ- and κ-phases were prepared by electro-oxidation of ET with TBA·I$_3$ in the presence of a small amount of TBA·AuI$_2$ (molar ratio: 1:0.05) in THF [11.235]. The main product was the α-phase, and the θ- and κ-phases are the minor ones. Later, the θ- and κ-phases were prepared with TBA·I$_3$ as an electrolyte. The θ-phase was made with mixed electrolytes (TBA·I+CdI$_2$) in THF where the α-phase was co-produced. The use of TCE as a solvent gave the semiconducting (ET)$_4$Cd$_2$I$_6$ [11.236]. The θ-phase is usually composed of twinned domains with approximately orthorhombic symmetry.

κ-(ET)$_2$Cu(NCS)$_2$

Single crystals are hexagonal thin plates with the dimensions of 2÷5 mm along the b-axis, 1÷2 mm along the c-axis, and 0.05÷0.1 mm along the a*-axis. They were prepared by electro-oxidation of ET in the presence of CuSCN, KSCN and 18-crown-6 ether from TCE, TCE + 10 v/v% ethanol or benzonitrile [11.174, 237, 238].

For example, 120 mg of KSCN, 70 mg of CuSCN, 210 mg of 18-crown-6 ether and 30 mg of ET were placed in the anode compartment in a 100-ml Elrenmeyer cell which had been filled with 100 ml of TCE or a mixture of TCE plus 1÷10 v/v% ethanol. The electrocrystallization was carried out according to the procedure described above. After one month, single crystals of typical size (5×1.5×0.1 mm^3) were harvested (Fig. 11.27). The mixture of TBA·SCN and CuSCN or a white solid Cu(NCS)$_2$ · K(18-crown-6 ether) was used as an electrolyte, too. Two other phases have been known. One is α-(ET)$_2$Cu(NCS)$_2$ which was grown at a low temperature (10°C) and has a metal-insulator transition below 200 K. The other is the semiconducting (ET)Cu$_2$(SCN)$_3$ where the cation radical dimer (ET$^+$)$_2$ is encapsulated in the necklace of polymerized anion atoms. The last one was obtained by employing a mixture of TBA·SCN and CuSCN as an electrolyte together with κ-(ET)$_2$Cu(NCS)$_2$. κ-(ET)$_2$Cu(NCS)$_2$ is the most easily prepared salt among the 10 K class superconductors.

κ-(ET)$_2$Cu[N(CN)$_2$]Br

Five different electrolytes were developed for crystal growth: (1) CuBr + Na[N(CN)$_2$] + 18-crown-6 ether, (2) CuBr + TetraPhenylPhosphonium (TPP) [N(CN)$_2$], (3) TPP$_2$Cu$_2$[N(CN)$_2$]$_3$Br, (4) Cu[N(CN)$_2$] + TBA·Br, and (5) TPP$_4$Cu[N(CN)$_2$]$_2$ + TBA·Br [11.239]. The second one is the most convenient one. The crystal is a thick rhombus of 1×1×0.3 mm^3. Usually the crystal growth of D-salt is easier than that of H-salt for an unknown reason. However, single crystals of the D-salt having reproducibly constant physical properties are rather difficult to obtain: Maybe this is due to its location in the proximaty of the boundary between a Mott insulator and a metal [11.240, 241].

κ-(ET)$_2$Cu[N(CN)$_2$]Cl

The following six kinds of electrolytes were examined: (1) Na·N(CN)$_2$ + CuCl + 18-crown-6 ether, (2) Cu[N(CN)$_2$] + KCl + 18-crown-6 ether, (3) TPP N(CN)$_2$ + CuCl, (4) TPP·N(CN)$_2$ + CuCl + KCl, (5) TPP·N(CN)$_2$ + CuCl + tetrahexylammonium·Cl, and (6) TPP·N(CN)$_2$ + CuCl + TPP·Cl, in TCE + 5÷20 v/v% ethanol [11.238]. There are at least three kinds of salts; (rhombic, superconductor), α-(ET)$_2$CuCl$_2$ (rectangular) and (ET)Cu·

[N(CN)$_2$]$_2$ (needle) together with polycrystals not well characterized. The superconductor was prepared in the electrolytes (2 or 3) and the electrolyte 3 is superior in producing the crystals. The most critical factor of the success in the crystal growth of κ-(ET)$_2$Cu[N(CN)$_2$]Cl was the electrode in use. The growing of this salt is strongly electrode dependent. The best way of crystal growth of this salt is rather trivial; to find out and use repeatedly a particular electrode which constantly affords good crystals.

κ-(ET)$_2$Cu(CN)[N(CN)$_2$] and κ-(ET)$_2$Cu$_2$(CN)$_3$

The following five kinds of electrolytes were examined in benzonitrile + ca. 10 v/v% ethanol or water: (1) CuCN+KCN+18-crown-6 ether, (2) CuN(CN)$_2$+KCN+18-crown-6 ether, (3) CuCN+TPP·N(CN)$_2$+KCN, (4) CuCN+Na[N(CN)$_2$]+18-crown-6 ether, and (5) CuCN+TPP·N(CN)$_2$ [11.159]. The first one yielded κ-(ET)$_2$Cu$_2$(CN)$_3$. In the second case, the use of a small amount of KCN gave only the semiconductor θ-(ET)$_2$Cu$_2$(CN)[N(CN)$_2$]$_2$, while κ'-(ET)$_2$Cu$_2$(CN)$_3$ was always obtained when the amount of KCN was $0.5 \div 2$ times that of Cu[N(CN)$_2$]. The third one gave the θ-phase salt as a major product, and κ-(ET)$_2$Cu$_2$(CN)$_3$ as a minor one. The fourth and fifth ones produced the superconductors κ-(ET)$_2$Cu(CN)[N(CN)$_2$] and κ'- (ET)$_2$Cu$_2$(CN)$_3$ together with the θ-phase salt, which is the major product. No optimum condition for obtaining the κ-(ET)$_2$Cu(CN)[N(CN)$_2$] salt has yet been established.

M(dmit)$_2$ Salt

Single crystals (needles, $1\times0.1\times0.02$ mm^3) of TTF[Ni(dmit)$_2$]$_2$ were prepared by the diffusion method between (TTF)$_3$(BF$_4$)$_2$ and TBA·[Ni(dmit)$_2$] in CH$_3$CN at 40°C for 15 days [11.223]. The amounts of both component molecules were adjusted to have a saturated CH$_3$CN solution. The electrocrystallization of TTF and TBA·Ni(dmit)$_2$ also produced single crystals of this salt but with poor quality.

Single crystals of TetraMethylAmmonium (TMA) [Ni(dmit)$_2$]$_2$ (black platelets) were obtained by electro-oxidation of TMA·[Ni(dmit)$_2$] in the presence of TMA·ClO$_4$ in CH$_3$CN at 0.6 μA [11.243].

Appendix

A. Angle-Dependent Magnetoresistance Oscillation

This section of the appendix provides a theoretical description of the Angle-dependent MagnetoResistance Oscillation (AMRO) in a two-dimensional metal with a nearly cylindrical Fermi surface. For simplicity, we assume the following band-energy relation

$$\epsilon_{\mathbf{k}} = \frac{\hbar^2 (k_x{}^2 + k_y{}^2)}{2m} - 2t\cos(ck_z) , \tag{A.1}$$

where k_x and k_y are the conducting-plane components of the crystal wave vector $\mathbf{k}$, and c and k_z are, respectively, the spacing between neighboring conducting planes and the component of the wave vector perpendicular to the plane; m is the effective electron mass in the conducting plane. The transfer energy t is assumed to be much smaller than the Fermi energy $\epsilon_F = \hbar^2 k_F{}^2/2m$, k_F being the Fermi wave number in the case we can neglect t. An intersection of the Fermi surface in the extended Brillouin zone with the $k_x k_z$-plane is depicted in Fig. A.1.

When the applied magnetic field H is tilted by the angle φ from the k_z-direction to the k_x-direction, the trajectories of semiclassically closed

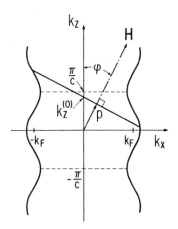

Fig. A.1. Vertical thick curves denote the intersection of the Fermi surface with the $k_x k_z$-plane. Horizontal dashed lines shown the Brillouin zone boundaries. The oblique thick line perpendicular to the magnetic field H is the plane on which the semiclassical electron orbit lies in the field. The plane is specified by $k_z^{(0)}$ in the field. From [A.1]

orbits are given by the intersection in the **k**-space of the Fermi surface with the plane normal to the field, defined by equation:

$$k_x \sin\varphi + k_z \cos\varphi = p \equiv k_z^{(0)} \cos\varphi \qquad (A.2)$$

where $k_z^{(0)}$ denotes the point of intersection at the k_z-axis with the orbital plane. By employing the polar coordinates k and θ in the $k_x k_y$-plane, the area S_k of this orbit in the **k**-space can be calculated with a precision of up to the first order of ϵ_F in the following way:

$$S_k \cos\varphi = \int_0^\pi \{k_F^2 + (4mt/\hbar^2)\cos[c(k_z^{(0)} - k_F \tan\varphi \cos\theta)]\}d\theta$$

$$= \pi k_F^2 + (4\pi mt/\hbar^2)\cos(ck_z^{(0)}) J_0(ck_F \tan\varphi) , \qquad (A.3)$$

where J_0 is the Bessel function. Since $J_0(z) \approx (2/\pi z)^{1/2}\cos(z-\pi/4)$ for $z \geq 1$ the $k_z^{(0)}$-dependent term in (A.3) vanishes periodically for values of φ that satisfy

$$ck_F \tan\varphi = \pi(n - 1/4) , \qquad (A.4)$$

n being an integer. This is remarkable because for these values of φ all the orbits on the Fermi surface have an identical value of the orbital area S_k. Therefore, the eigenenergies of electrons near the Fermi energy are completely discretized into Landau levels or, in other words, the Landau-band-width vanishes. In this situation, since the levels formed in the energy range near ϵ_F loses the dependence on the **k**-component parallel to the field, the group velocity in the field direction at the Landau states vanishes. In other words, the effective mass becomes infinite in the field direction. Then, the component of the magnetoresistance in the field direction should become very large. Magnetoresistance is considered to be very much enhanced around these angles at low temperatures. Therefore, the peak angles of the magnetoresistance should be given by (A.4).

Yagi et al. carried out a semiclassical calculation of the conductivity tensor $\sigma_{\alpha\beta}$ according to the same model [A.2]. The Boltzmann equation for the distribution function of electrons which are subject to both applied electric and magnetic fields, can be solved. This leads to

$$\sigma_{\alpha\beta} = \frac{e^2}{4\pi^3 \hbar^2} \int dk_H \int_0^{2\pi} d\phi \int_0^{+\infty} d\phi' m^* v_\alpha(\phi, k_H)$$

$$\times v_\beta(\phi-\phi', k_H) \frac{m^*}{\omega} e^{-\phi'/\omega\tau} . \tag{A.5}$$

This is called the **Shockley tube integral** [A.3]. Here, integration over ϕ' is carried out along the semiclassical trajectory of the electron. v_α is the α component of the group velocity of an electron on the Fermi surface, defined as

$$v_\alpha = \frac{1}{\hbar} \frac{\partial\epsilon}{\partial k_\alpha}\bigg|_{\epsilon=\epsilon_F} , \tag{A.6}$$

and m* is the cyclotron mass

$$m^* = \frac{\hbar}{2\pi} \oint \frac{dk_\parallel}{v_\perp} , \tag{A.7}$$

where the integral is taken over the closed orbit. $k_\parallel$ is the wave vector parallel to the magnetic field, and $v_\perp$ is the component of the electron group velocity in the plane normal to the field. The phase variable ϕ is defined as

$$\phi = \frac{\hbar}{m^*} \int^{\mathbf{k}} \frac{dk_\parallel}{v_\perp} , \tag{A.8}$$

where integration is performed along the electron trajectory, and τ is the scattering time. Note $\omega = \omega_0 \cos\varphi$ with $\omega_0 = eH/mc$, φ being the angle between the field and the c-axis. The calculated result converted into ρ_{zz} reproduces AMRO as in Fig. A.2. In the present model, oscillations do not appear in ρ_{xx} and ρ_{yy}. The weak oscillations observed in those components take their origin in the lower symmetry of the material.

For the more general case of conducting planes where each plane has a convex, closed form of the Fermi surface, and the planes are coupled by weak interplaner transfer energies, *Peschansky* et al. demonstrated that the relation (A.4) holds with different coefficients depending on the angle of

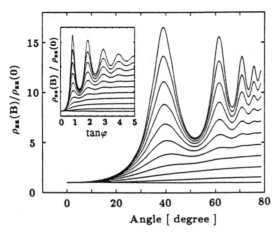

Fig. A.2. Depedence of the angle φ on the zz component of the resistivity tensor for different magnetic fields. From top to bottom the value of $\omega_0 \tau$ varies from 4 to 0 with 0.4 step. The inset is the replot against $\tan\varphi$. Parameter values are $\epsilon_F/t = 100$ and $mta^2/\hbar^2 = 0.045$. Temperature is zero. From [A.2]

the plane in which the magnetic field is rotated from the c-axis to the ab-plane [A.4]. According to them, k_F in (A.4) should be replaced by the half $k_H^{(max)}$ of the maximum size of the two-dimensional Fermi surface that is measured in the direction of the in-plane magnetic-field component on the $k_x k_y$-plane [Ref.A.5, Fig.1]. With this knowledge, one can construct the two-dimensional Fermi surface from the values of $k_H^{(max)}$ for a sufficient number of directions: if one draws a line in each direction of the field from the coordinate origin and adds the normal to the line at the point being away from the origin by $k_H^{(max)}$, the internal envelope of the normals gives the Fermi surface. *Kartsovnik* et al. first succeeded in getting the Fermi surface of β-(ET)$_2$IBr$_2$ by this method [A.5]. They assumed that the interlayer energy (A.1) is given by $-2t\cos(ck_z + u_x k_x + u_y k_y)$ and that in this case (A.4) is modified to

$$ck_H^{(max)}\tan\varphi = \pi[n - 1/4 + (\mathbf{k}_\parallel^{max} \cdot \mathbf{u})] , \qquad (A.9)$$

where $\mathbf{k}_\parallel^{(max)}$ is the $\mathbf{k}$-vector on the 2D Fermi surface in the $k_x k_y$-plane at which its magnitude becomes maximimum. This new ingredient provides the information on $\mathbf{u}$.

According to a fully quantum-mechanical theory of AMRO [A.6], the angular oscillation appears in the matrix element of the z-component of the current operator between the Landau states of the adjacent layers.

B. Superconductivity Transition in a Magnetic Field

Organic superconductors known to date are type-II superconductors, into which quantized magnetic vortices penetrate in the field region above the lower critical field H_{c1}. For three-dimensional superconductors, the mean-field critical fields H_{c1} and H_{c2} can be determined by the magnetization and the electrical resistivity measurements [B.1]. In the case of low-dimensional superconductors, the mean-field transition that corresponds to H_{c2} is obscured due to thermal fluctuation effects [B.2]: the resistive transition becomes too broad to define the mean-field transition point, and the magnetization versus temperature curve is significantly rounded in the H_{c2} transition region. Furthermore, the properties of superconductors in a magnetic field near H_{c2} are influenced by statistical mechanics and the dynamics of vortices [B.3]. In this section, the influence of thermal fluctuations is described, since it concerns directly to usual procedure of superconductivity characterization of organic salts, through which the upper critical field H_{c2}, the Ginzburg-Landau (GL) coherence lengthes $\xi_{\parallel}$ (in-plane) and $\xi_{\perp}$ (interplane) are evaluated.

For two-dimensional superconductors, for which $\xi_{\perp}$ is much shorter than the interlayer spacing s, thermal fluctuation hinders to establish superconductivity of long-range order. The presence of a high magnetic field enhances this tendency, because the magnetic field itself restricts the electron motion in directions perpendicular to the magnetic lines of force. As a consequence, H_{c2} which is introduced in the mean-field theory, cannot be determined from experimental data, since the measured curves become so smooth even in the vicinity of the transition. Meanwhile, a boundary between a moving vortex state and a pinned vortex state has been found at a lower temperature, which discriminated the change in the slope of the temperature dependence of resistivity and/or the onset of irreversible magnetization.

According to *Lawrence* and *Doniach*, layered superconductors are viewed as two-dimensional superconductors in combination with Josephson tunneling between adjacent layers [B.4]. Then, the Ginzburg-Landau (GL) free energy is given by

$$F = \sum_i \left[\alpha|\psi_i|^2 + \frac{\beta}{2}|\psi_i|^4 + \frac{1}{2m^*}\left| \left(-i\hbar\nabla - \frac{2e\mathbf{A}}{c}\right)\psi_i \right|^2 \right.$$
$$\left. + \eta|\psi_{i+1}-\psi_i|^2 \right] + \frac{B^2}{8\pi} \tag{B.1}$$

where the index i runs over the superconducting layers. The first three terms represent the superconductivity of each layer with the GL order parameter ψ_i, the effective mass m^*, the vector potential **A**, a positive constant β and

$$\alpha = \frac{\hbar^2}{4m^*\xi_\parallel^2}\epsilon \; , \tag{B.2}$$

$$\epsilon = \frac{T - T_{c0}}{T_{c0}} \; , \tag{B.3}$$

where $\xi_\parallel$ ($\xi_\perp$) denotes the intraplane (interplane) coherence length at 0 K, and T_{c0} denotes T_c in the absence of a magnetic field. The fourth term accounts for the Josephson coupling between the superconducting layers. The fifth term represents the magnetic energy accompanying the magnetic-flux density **B**.

In a magnetic field, the GL equation with respect to the order parameter ψ is reduced to the form of the Schrödinger equation for a changed particle, provided that the $|\psi|^4$ term is neglected [B.1]. In order to deal with the superconducting state with a finite value of $|\psi|$, *Ikeda* et al. have renormalized the GL mass parameters related to the Landau levels, which take into account the $|\psi|^4$ term. Thus, the transport and thermodynamic behaviors of the superconductor in the transition region in a magnetic field can be described by the fluctuation renormalization theory based on the GL functional [B.5].

As a consequence, the temperature dependence on the magnetization is given by

$$\begin{aligned}
\frac{\partial M}{\partial T} = \frac{H_{c2}(0)}{8\pi\kappa^2 T_c} \frac{g_3\sqrt{\lambda}}{2h} \Bigg\{ & E_0\left[-1 + h\frac{\partial\mu_{0R}}{\partial h}(\mu_{0R} + \tfrac{1}{2}\lambda)E_0^2\right] \\
& + 2E_2\left[1 + 2h(\mu_{0R} + 4h + \tfrac{1}{2}\lambda)E_2^2\right] \\
& - E_1\left[1 - h\frac{\partial\mu_{1R}}{\partial h}(\mu_{1R} + \tfrac{1}{2}\lambda)E_1^2\right] \\
& + \sum_{n=2}(n+1)\left[-E_n + E_{n+1} + 2h[\mu_{0R} + 2(n+1)h + \tfrac{1}{2}\lambda]E_{n+1}^3\right] \\
& + h\frac{\partial\mu_{0R}}{\partial h}\sum_{n=2}(\mu_{0R} + 2nh + \tfrac{1}{2}\lambda)E_n^3 \Bigg\} ,
\end{aligned} \tag{B.4}$$

where

$$H_{c2}(0) = \frac{\phi_0}{2\pi \xi_\parallel^2} ,$$

$$h = \frac{B}{H_{c2}(0)} , \quad \lambda = \left[\frac{2\xi_\perp}{s}\right]^2 ,$$

$$g_3 = \frac{k_B B}{\Delta C \phi_0 \xi_\perp} ,$$

$$\kappa = \frac{H_{c2}(0)}{\sqrt{8\pi T_{c0} \Delta C}} ,$$

$$E_n \simeq \frac{1}{\sqrt{\mu_{nR}(\mu_{nR} + \lambda)}} ,$$

and μ_{nR} is the renormalized electron mass at the n-th Landau level, which is a function of T, T_{c0}, h, g_3 and λ, calculated by the method described in [B.5], ΔC represents the gap of the specific heat at the transition, ϕ_0 is the fluxoid, and s is the interlayer spacing. The intraplane and interplane conductivities in a magnetic field applied normal to the two-dimensional plane are represented by

$$\sigma_\perp = \sigma_{0\perp} + \Delta \sigma_\perp \tag{B.5}$$

$$\sigma_{0\perp} = \frac{e^2}{2\hbar\xi_\perp} h^2 \sum_{n=0} \frac{n+1}{(\mu_{n+1R} - \mu_{nR})^2} (f_n + f_{n+1} - 2f_{n+\frac{1}{2}}) , \tag{B.6}$$

$$\sigma_\parallel = \frac{e^2}{32\hbar\xi_\perp} \left(\frac{\xi_\perp}{\xi_\parallel}\right)^2 h \sum_{n=0} f_n^3 , \tag{B.7}$$

with

$$f_n = \left[\mu_{nR}(1 + \mu_{nR}/\lambda)\right]^{-1/2} , \quad f_{n+\frac{1}{2}} = \left[\bar{\mu}_{nR}(1 + \bar{\mu}_{nR}/\lambda)\right]^{-1/2} ,$$

$$\bar{\mu}_{nR} = (\mu_{nR} + \mu_{(n+1)R})/2 .$$

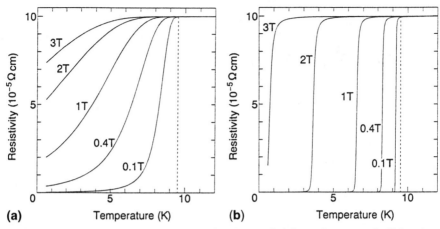

Fig. B.1. Temperature dependence of the in-plane resistivity under magnetic filds calculated with (**a**) $\xi_\perp = 0.3$ nm, $\xi_\parallel = 2.9$ nm, $s = 1.5$ nm representing a two-dimensional superconductor and with (**b**) $\xi_\perp = \xi_\parallel = 10$ nm, $s = 1.5$ nm representing a three-dimensional superconductor. For calculation, $\rho_n = 10^{-4}\,\Omega\cdot$ cm and $\Delta C = 520$ mJ/K $\cdot$ mol are used

$\Delta\sigma_\perp$ is the non-Gaussian term [B.5], but the relative contribution of $\Delta\sigma_\perp$ to $\sigma_\perp$ tends to vanish with decreasing temperature.

The effect of thermal fluctuations found in a two-dimensional superconductor with $\xi_\perp < s < \xi_\parallel$ is illustrated in Fig. B.1. Taking into account the normal-state resistivity ρ_n, the measured in-plane resistivity ρ can be written as

$$\rho^{-1} = \sigma_\parallel + \rho_n^{-1} \,. \tag{B.8}$$

For a two-dimensional superconductor with $\xi_\perp = 0.3$ nm, $\xi_\parallel = 2.9$ nm and $s = 1.5$ nm corresponding to κ-(ET)$_2$Cu(NCS)$_2$, the resistive transition characteristics in a magnetic field are displayed in Fig. B.1a. Note that the theory is applicable in the vicinity of the transition, and zero resistivity can be achieved by the contribution of vortex pinning. This contrasts to the three-dimensional case with $\xi_\parallel = \xi_\perp = 10$ nm, as shown in Fig. B.1b. The calculated results are obtained by means of (B.8) with $T_c = 9.0$ K. $\Delta C = 520$ mJ/K. ρ_n^{-1} is assumed to be 10^4 S/cm, independent of temperature. The measured magnetization and resistivity in the transition region in a magnetic field have been analyzed on the basis of these relations for κ-(ET)$_2$Cu[N(CN)$_2$]Br and κ-(ET)$_2$Cu(NCS)$_2$ [B.6, 7].

On the other hand, the effect of fluctuations on the transport coefficients in a magnetic field near the mean-field transition temperature T_c(H) has been treated on grounds of the time-dependent GL theory in the self-consistent Hartree approximation [B.8]. In this case, interactions between the fluctuations are considered and thereby the divergence due to Gaussian fluctuations is removed in agreement with experiments. In a magnetic field H sufficiently strong so that the paired quasi-particles are efficiently limited to their lowest Landau level, the temperature and field dependence of physical quantities shows a scaling behavior in the variable

$$t_G = \frac{T - T_c(H)}{(TH)^n} \tag{B.9}$$

with $n = 2/3$ and $1/2$ for three-dimensional case and two-dimensional case, respectively. Then the scaling function of various thermodynamic and transport quantities for the three-dimensional case is given by

$$\Xi_i = \left(\frac{T^2}{H}\right)^{\epsilon_i} F_i\left[A \frac{T - T_c(H)}{(TH)^{2/3}}\right]. \tag{B.10}$$

Ξ_i is a shorthand for the measured quantities: $\Xi_1 = M/H$, $\Xi_2 = \sigma$, $\Xi_3 = U_\phi/H$ (U_ϕ/ϕ_0: Ettingshausen coefficient), $\Xi_4 = C/T$, $\epsilon_1 = \epsilon_2 = \epsilon_3 = 1/3$, and $\epsilon_4 = 0$. F_i denotes the scaling functions. The coefficient A is given by

$$A = \left[\frac{\phi_0 \xi_c H_{c2}^2}{8\pi k_B \kappa^2 T_c^{3/2}}\right]^{3/2} \tag{B.10}$$

where ξ_c and H_{c2} are the GL coherence length and the upper critical field, both at 0 K and for fields applied perpendicular to the two-dimensional plane. κ is the GL parameter. If $\Xi_i(H/T^2)^{\epsilon_i}$ is plotted versus t_G, then convergence into a single curve, independent of H, is expected [B.9]. For the two-dimensional case, similar universal curves for $M/(TH)^{1/2}$ vs $[T - T_c(H)]/(TH)^{1/2}$ and $\sigma/(T/H)^{1/2}$ vs $[T - T_c(H)]/(TH)^{1/2}$ are expected [B.7, 10, 11].

References

Chapter 1

1.1   H. Akamatu, H. Inokuchi, Y. Matsunaga: Nature **173**, 168-169 (1954)
1.2   R.G. Kepler, P.E. Bierstedt, R.E. Merrifield: Phys. Rev. Lett. **5**, 503-504 (1960)
1.3   L.B. Coleman, M.J. Cohen, D.J. Sandman, F.G. Yamagishi, A.F. Garito, A.J. Heeger: Solid State Commun. **12**, 1125 (1973)
1.4   H.J. Keller (ed.): *Low-Dimensional Cooperative Phenomena* (Plenum, New York 1974)
       J.T. Devreese, R.P. Evrard, V.E. Van Doren (eds): *Highly Conducting One-Dimensional Solids* (Plenum, New York 1979)
       L. Alcacer (ed.): *The Physics and Chemistry of Low-Dimensional Solids* (Reidel, Dordrecht 1979)
       S. Kagoshima, H. Nagasawa, T. Sambongi: *One-Dimensional Conductors*, Springer Ser. Solid-State Sci., Vol. 72 (Springer, Berlin, Heidelberg 1988)
1.5   W.A. Little: Phys. Rev. **134**, A1416-1424 (1964)
1.6   Conference proceedings: Mol. Cryst. Liq. Cryst. **78**, No. 1-4 (1982) (Boulder 1981); J. Physique **44**, C3-1983 (1983) (Les Arcs 1982); Mol. Cryst. Liq. Cryst. **117-121** (1985) (Abano Terme 1984); Physica B **143** (1986) (Yamada 1986); Synth. Met. **17-18** (1987) (Kyoto 1986); Synth. Met. **27** (1989) (Sante Fe 1988); Synth Metals **41-43** (1991) (Tübingen 1990); Synth. Met. **55-57** (1993) (Göteborg 1992); Synth. Met. **69-71** (1995) (Seoul 1994); Synth. Met. **84-86** (1997) (Snowbird 1996)
1.7   D. Jérome, A. Mazaud, M. Ribault, K. Bechgaard: J. Physique Lett. **41**, L95-98 (1980)
1.8   R.L. Greene, E.M. Engler: Phys. Rev. Lett. **45**, 1587 (1980)
1.9   S.S.P. Parkin, E.M. Engler, R.R. Schumaker, R. Lagier, V.Y. Lee, J.C. Scott, R.L. Greene: Phys. Rev. Lett. **50**, 270-273 (1983)
1.10   E.B. Yagubskii, I.F. Shchegolev, V.N. Laukhin, P.A. Kononovich, M.V. Kartsovnic, A.V. Zvarykina, L.I. Bubarov: JETP Lett. **39**, 12-15 (1984)
1.11   K. Murata, M. Tokumoto, H. Anzai, H. Bando, G. Saito, K. Kajimura, T. Ishiguro: J. Phys. Soc. Jpn. **54**, 1236-1239 (1985)
1.12   V.N. Laukhin, E.E. Kostyuchenko, Yu.V. Sushko, I.F. Shchegolev, E.B. Yagubskii: JETP Lett. **41**, 81 (1985)
1.13   H. Urayama, H. Yamochi, G. Saito, K. Nozawa, T. Sugano, M. Kinoshita, S. Sato, K. Oshima, A. Kawamoto, J. Tanaka: Chem. Lett. **1988**, 55 (1988)
1.14   A.M. Kini, U. Geiser, H.H. Wang, K.D. Carson, J.M. Williams, W.K. Kwok, K.G. Vandervoort, J.E. Thompson, D.L. Stupka, M.H. Whangbo: Inorg. Chem. **29**, 2555 (1990)

1.15   U. Geiser, A.J. Schultz, H.H. Wang, D.M. Watkins, D.L. Stuyka, J.M. Williams, J.E. Schirber, D.L. Overmer, D. Jung, J.J. Nova, M.H. Whangbo: Physica C **174**, 475 (1991)

1.16   H. Ito, M. Watanabe, Y. Nogami, T. Ishiguro, T. Komatsu, G. Saito, N. Hosoito: J. Phys. Soc. Jpn. **60**, 3230-3233 (1991)
H. Ito, Y. Nogami, T. Ishiguro, T. Komatsu, G. Saito, N. Hosoito: Jpn. J. Appl. Phys. **7**, 419-425 (1992)

1.17   H.H. Wang, K.D. Carlson, U. Geiser, W.W. Kwok, M.D. Vashon, J.E. Thompson, N.F. Larsen, G.D. McCabe, R.S. Husler, J.M. Williams: Physica C **166**, 57 (1990)

1.18   H. Ito, M.V. Kartsovnik, H. Ishimoto, K. Kono, H. Mori, N.D. Kushch, G. Saito, T. Ishiguro, S. Tanaka: Synth. Met. **70**, 899-902 (1995)

1.19   T. Sasaki, N. Toyota: Solid State Commun. **82**, 447 (1992)

1.20   D. Jérome, H.J. Schultz: Adv. Phys. **31**, 299 (1982)

1.21   H. Kobayashi, T. Udagawa, H. Tomita, K. Bun, T. Naito, A. Kobayashi: Chem. Lett. **1993**, 1559 (1993)
A. Kobayashi, T. Udagawa, H. Tomita, T. Naito, H. Kobayashi: Chem. Lett. **1993**, 2179 (1993)

1.22   F. Goze, V.N. Laukhin, L. Brossard, A. Audouard, J.P. Ulmet, S. Askenazy, T. Naito, H. Kobayashi, A. Kobayashi, M. Tokumoto, D. Cassouix: Europhys. Lett. **28**, 427-431 (1994)

1.23   K. Kikuchi, M. Kikuchi, T. Namiki, K. Saito, I. Ikemoto, K. Murata, T. Ishiguro, K. Kobayashi: Chem. Lett. **1987**, 931-932 (1987)

1.24   G.C. Papavassiliou, G.A. Mousdis, J.S. Zambounis, A. Terzis, A. Hountas, B. Hilti, C.W. Mayer, J. Pfeiffer: Synth. Met. B **27**, 379-383 (1988)

1.25   Y. Misaki, N. Higuchi, H. Fujiwara, T. Yamabe, T. Mori, H. Mori, S. Tanaka: Angew. Chem. Int'l Edn. **34**, 1222 (1995)

1.26   L. Brossard, M. Ribault, M. Bousseau, L. Valade, P. Cassoux: Physica B **143**, 378-380 (1986)

1.27   A.F. Hebard, M.J. Rosseinsky, R.C. Haddon, D.W. Murphy, S.H. Glarum, T.T.M. Palstra, A.P. Ramirez, A.R. Kortan: Nature **350**, 600 (1991)

1.28   M.J. Rosseinsky, A.P. Ramirez, S.H. Glarum, D.W. Murphy, R.C. Haddon, A.F. Hebard, T.T.M. Palstra, A.R. Kortan, S.M. Zahurak, A.V. Makhija: Phys. Rev. Lett. **66**, 2830 (1991)

1.29   R.L. Greene, G.B. Street, L.J. Suter: Phys. Rev. Lett. **34**, 577 (1975)

1.30   J. Tsukamoto: Adv. Phys. **41**, 509 (1992)
T. Ishiguro, H. Kanaeko, J.P. Pouget, J. Tsukamoto: Synth. Met. **69**, 37 (1995)

1.31   J.M. Williams, H.H. Wang, T.J. Emge, U. Geiser, M.A. Beno, P.C.W. Leung, K.D. Carlson, R.J. Thorn, A.J. Schultz, M.-H. Whangbo: Proc. Inorg. Chem. **34**, 51-218 (1987)
J.R. Ferraro, J.M. Williams: *Introduction to Synthetic Electrical Conductors* (Academic, New York 1987)
J.M. Williams, J.R. Ferraro, R.J. Thorn, K.D. Carlson, U. Geiser, H.H. Wang, A.M. Kini, M.-H. Whangbo: *Organic Superconductors (Including Fullerenes)* (Prentice Hall, Englewood Cliffs, NJ 1992)

1.32   B.K. Vainshtein, V.M. Fridkin, V.L. Indenbom: *Structure of Crystals*, 2nd edn., Modern Crystallography 2 (Springer, Berlin, Heidelberg 1995)

1.33   L.N. Bulaevskii: Adv. Phys. **37**, 443 (1988)
1.34   J. Wosnitza: Int'l Mod. Phys. B **7**, 2707-2741 (1992)
        J. Wosnitza: *Fermi Surfaces of Low-Dimensional Organic Metals and Superconductors* (Springer, Berlin, Heidelberg 1996) p.169

Chapter 2

2.1   R.S. Mulliken: J. Am. Chem. Soc. **74**, 811-824 (1952)
2.2   P.Y. Yu, M. Cardona: *Fundamentals of Semiconductors* (Springer, Berlin, Heidelberg 1966)
2.3   See, for example, J.A. Pople, D.L. Beveridge: *Approximate Molecular Orbital Theory* (McGraw-Hill, New York 1970)
        T. Yonezawa, C. Nagata, H. Kato, A. Imamura, K. Morokuma: *Introduction to Quantum Chemistry* (Kagakudojin, Kyoto 1963) [in Japanese]
2.4   C.A. Coulson: Proc. Roy. Soc. (London) **164**, 383-396 (1938)
        H.C. Longuet-Higgins, L. Salem: Proc. Roy. Soc. (London) A **251**, 172-185 (1959)
2.5   A.J. Berlinsky, J.F. Carolan, L. Weiler: Solid State Commun. **15**, 795-801 (1974)
2.6   M.J. Rice: In *Low Dimensional Cooperative Phenomena*, ed. by H.J. Keller (Plenum, New York 1974) pp.23-34
2.7   R.E. Peierls: *Quantum Theory of Solids* (Oxford Univ. Press, London 1955) p.108
2.8   S. Kagoshima, H. Nagasawa, T. Sambongi: *One-Dimensional Conductors*, Springer Ser. Solid-State Sci., Vol.72 (Springer, Berlin, Heidelberg 1988)
2.9   D. Allender, J.W. Bray, J. Bardeen: Phys. Rev. B **9**, 119-129 (1974)
2.10   H. Fröhlich: Proc. Roy. Soc. (London) A **223**, 296-305 (1954)
2.11   H. Fukutome, M. Sasai: Prog. Theor. Phys. **67**, 41-67 (1982)
2.12   P.A. Lee, T.M. Rice, P.W. Anderson: Phys. Rev. Lett. **31**, 462-465 (1973); also Solid State Commun. **14**, 703-709 (1974)
2.13   T. Ishiguro, H. Sumi, S. Kagoshima, K. Kajimura, H. Anzai: J. Phys. Soc. Jpn. **48**, 456 (1980)
2.14   A. Andrieux, H.J. Schultz, D. Jérome, K. Bechgaard: Phys. Rev. Lett. **43**, 227-230 (1979)
2.15   S. Megtert, R. Comès, C. Vettier, R. Pynn, A.F. Garito: Solid State Commun. **37**, 875-877 (1981)
2.16   D.B. Tanner, C.S. Jacobsen, A.F. Garito, A.J. Heeger: Phys. Rev. B **13**, 3381-3404 (1976)
2.17   C. Kittel: *Quantum Theory of Solids* (Wiley, New York 1963) Chap.12
2.18   P.M. Chaikin, T. Holstein, M.Ya. Azbel: Phil. Mag. B **48**, 457-473 (1983)
2.19   P.M. Chaikin, R.L. Greene: Phys. Today **39**, 24-32 (May 1986)
2.20   A.J. Heeger, A.F. Garito: In *Chemistry and Physics of One-Dimensional Metals*, ed. by H.J. Keller (Plenum, New York 1977) pp.87-136
        E. Ehrenfreund, A.J. Heeger: Phys. Rev. B **16**, 3830-3833 (1977)
2.21   G. Soda, D. Jérome, M. Weger, J. Alizon, J. Gallice, H. Robert, J.M. Febre, L. Giral: J. Physique **38**, 931-948 (1977)
2.22   J.C. Bonner, M.E. Fisher: Phys. Rev. **135**, A640 (1964)
2.23   H. Shiba: Phys. Rev. B **6**, 930 (1972)

2.24 J.B. Torrance, Y. Tomkiewicz, B.D. Silverman: Phys. Rev. B **15**, 4738-4749 (1977)

2.25 S. Kagoshima, T. Ishiguro, H. Anzai: J. Phys. Soc. Jpn. **40**, 2061-2071 (1976)
K.S. Khanna, J.P. Pouget, R. Comes, A.F. Garito, A.J. Heeger: Phys. Rev. B **16**, 1468-1479 (1977)

2.26 J. Kondo, K. Yamaji: J. Phys. Soc. Jpn. **43**, 424-436 (1978)

2.27 J.B. Torrance, B.A. Scott, F.B. Kaufman: Solid State Commun. **17**, 1369-1373 (1975)
J.B. Torrance: In *Chemistry and Physics of One-Dimensional Metals*, ed. by H.J. Keller (Plenum, New York 1977) pp. 137-166

2.28 J. Solyom: Adv. Phys. **28**, 201-303 (1979)

2.29 S. Tomonga: Prog. Theor. Phys. **5**, 349 (1950)

2.30 J.M. Luttinger: J. Math. Phys. **4**, 1154 (1963)

2.31 T. Chakraborty, P. Pietiläinen: *The Quantum Hall Effects, Fractional and Integral*, 2nd edn., Springer Ser. Solid-State Sci., Vol. 85 (Springer, Berlin, Heidelberg 1995)

2.32 P. Fulde: *Electron Correleations in Molecules and Solids*, 3rd edn., Springer Ser. Solid-State Sci., Vol. 100 (Springer, Berlin, Heidelberg 1995)

2.33 J. Hubbard: Proc. Roy. Soc. (London) A **281**, 401 (1964)

Chapter 3

3.1 K. Bechgaard, C.S. Jacobsen, K. Mortensen, J.H. Pederson, N. Thorup: Solid State Commun. **33**, 1119-1125 (1980)

3.2 D. Jerome, A. Mazaud, M. Ribault, K. Bechgaard: J. Physique Lett. **41**, L95-98 (1980)

3.3 N. Thorup, G. Rindorf, H. Soling, K. Bechgaard: Acta Cryst. B **37**, 1236-1240 (1981)
G. Rindorf, H. Soling, N. Thorup: Acta. Cryst. B **38**, 2805-2808 (1982)

3.4 F. Wudl, E. Aharon-Shalom, D. Nalewajek, J.V. Waszeczak, W.M. Walsh Jr., L.W. Rupp Jr., P.M. Chaikin, R. Lacoe, M. Burns, T.O. Pehler, J.M. Williams, M.A. Beno: J. Chem. Phys. **76**, 5497 (1982)

3.5 R.C. Lacoe, S.A. Wolf, P.M. Chaikin, F. Wudl, E. Aharon-Shalom: Phys. Rev. **27**, 1947-1950 (1983)

3.6 S. Cox, R.M. Boysel, D. Moses, F. Wudl, J. Chen, S. Ochsenbein, A.J. Heeger, W.M. Walsh Jr., L.W. Rupp Jr.: Solid State Commun. **49**, 259-263 (1984)

3.7 K. Mortensen, C.S. Jacobsen, A. Lindegaard-Andersen, K. Bechgaard: J. Physique **44**, C3-963-968 (1983)

3.8 P. Batail, L. Ouahab, J.B. Torrance, M.L. Pylman, S.S.P. Parkin: Solid State Commun. **55**, 597-600 (1985)

3.9 L. Pauling: *The Nature of the Chemical Bond* (Cornell Univ. Press, Ithaca, NY 1960)

3.10 A. Bondi: J.Phys. Chem. **68**, 441 (1964)

3.11 C.S. Jacobsen, D.B. Tanner, K. Bechgaard: Phys. Rev. B **28**, 7019-7032 (1983)

3.12 P.M. Grant: J. Physique **44**, C3-847-857 (1983)

3.13 N. Kinoshita, M. Tokumoto, H. Anzai, T. Ishiguro, G. Saito, T. Yamabe, H. Teramae: J. Phys. Soc. Jpn. **53**, 1504-1512 (1984)

3.14   P.C. Stein, P. Bernier, C. Lenoir: Physica B **143**, 491-493 (1986)

3.15   H.J. Pedersen, J.C. Scott, K. Bechgaard: Phys. Rev. B **24**, 5014-5025 (1981)
        S. Frandrois, C. Coulon, P. Delhaes, D. Chasseau, C. Hauw, J. Gaultier, J.M.
        Fabre, L. Giral: Mol. Cryst. Liq. Cryst. **79**, 307 (1982)

3.16   C.S. Jacobsen, D.B. Tanner, K. Bechgaard: Phys. Rev. Lett. **46**, 1142-1145 (1981)

3.17   D. Jérome: In *Physics and Chemistry of Electrons and Ions in Condensed Matter*, ed.
        by J.A. Acrivos, N.F. Mott, A.D. Yoffe (Reidel, Dordrecht 1984) pp.595-624

3.18   T. Ishiguro, K. Kajimura, H. Bando, K. Murata, H. Anzai: Mol. Cryst. Liq.
        Cryst. **119**, 19-26 (1985)

3.19   C.S. Jacobsen, K. Mortensen, M. Weger, K. Bechgaard: Solid State Commun. **38**,
        423-428 (1981)

3.20   K. Murata, H. Anzai, G. Saito, K. Kajimura, T. Ishiguro: J. Phys. Soc. Jpn. **50**,
        3529-3530 (1981)

3.21   R.L. Greene, P. Haen, S.Z. Huang, E.M. Engler, M.-Y. Choi, P.M. Chaikin:
        Mol. Cryst. Liq. Cryst. **79**, 183 (1982)

3.22   K. Mortensen: Solid State Commun. **44**, 643-647 (1982)

3.23   W. Kang, S.T. Hannahs, P.M. Chaikin: Phys. Rev. Lett. **69**, 2827-2838 (1992)

3.24   K. Behnia, M. Ribault, C. Lenoir: Europhys. Lett. **25**, 285 (1994)

3.25   T. Osada, A. Kawasumi, S. Kagoshima, N. Miura, G. Saito: Phys. Rev. Lett. **66**,
        1525-1528 (1991)
        T. Osada, S. Kagoshima, N. Miura: Phys. Rev. B **46**, 1812 (1992)

3.26   N.J. Naughton, O.H. Chung, M. Chaparala, X. Bu, P. Coppens: Phys. Rev. Lett.
        67, 3712-3715 (1991)

3.27   A.G. Lebed: JETP Lett. **43**, 174-177 (1986)

3.28   A.G. Lebed: Synth Metals **70**, 993-996 (1995)

3.29   K. Maki: Phys. Rev. B **45**, 5111-5113 (1992)
        A. Bjelis, K. Maki: Phys. Rev. B **44**, 6791 (1991)

3.30   Y. Sun, K. Maki: Synth. Met. **70**, 991-992 (1995)

3.31   P.M. Chaikin: Phys. Rev. Lett. **69**, 2831-2834 (1992)

3.32   A.T. Zheleznyak, V.M. Yakovenko: Synth. Met. **70**, 1005-1008 (1995)

3.33   S.P. Strong, D.G. Clarke, P.W. Anderson: Phys. Rev. Lett. **73**, 1007-1010 (1994)

3.34   G.M. Danner, W. Kang, P.M. Chaikin: Phys. Rev. Lett. **72**, 3714-3717 (1994)

3.35   K. Mortensen, Y. Tomkiewicz, K. Bechgaard: Phys. Rev. **25**, 3319-3325 (1982)

3.36   W.M. Walsh, Jr., F. Wudl, G.A. Thomas, D. Nalewajek, J.J. Hauser, P.A. Lee,
        T. Poehler: Phys. Rev. Lett. **45**, 829-832 (1980)

3.37   A. Andrieux, D. Jerome, K. Bechgaard: J. Physique Lett. **42**, L87-91 (1982)

3.38   J.C. Scott, H.J. Pedersen, K. Bechgaard: Phys. Rev. B **24**, 475-477 (1981)

3.39   J.B. Torrance, H.J. Pedersen, K. Bechgaard: Phys. Rev. Lett. **49**, 881-884 (1982)

3.40   J.M. Delrieu, M. Roger, Z. Toffano, E. Wope Mbougue, R. Saint James, K.
        Bechgaard: Physica B **143**, 412-416 (1986)

3.41   T. Takahashi, Y. Maniwa, H. Kawamura, G. Saito: Physica B **143**, 417-421 (1986)

3.42   D. Jérome: Mol. Cryst. Liq. Cryst. **79**, 155-182 (1982)

3.43   R. Brusetti, M. Ribault, D. Jérome, K. Bechgaard: J. Physique **43**, 801-808 (1982)

3.44   J.P. Pouget, G. Shirane, K. Bechgaard, J.M. Fabre: Phys. Rev. B **27**, 5203-5206
        (1983)

3.45   S. Kagoshima, T. Yasunaga, T. Ishiguro, H. Anzai, G. Saito: Solid State Commun.
        46, 867-870 (1983)

3.46   K. Bechgaard, K. Carneiro, M. Olsen, F.B. Rasmussen, C.S. Jacobsen: Phys. Rev. Lett. **46**, 852-855 (1981)

3.47   H. Bando, K. Kajimura, H. Anzai, T. Ishiguro, G. Saito: Mol. Cryst. Liq. Cryst. **119**, 41-44 (1985)

3.48   H. Schwenk, K. Andres, F. Wudl: Phys. Rev. B **29**, 500-502 (1984)

3.49   T. Takahashi, D. Jérome, K. Bechgaard: J. Physique Lett. **43**, L565-573 (1982)

3.50   K. Mortensen, C.S. Jacobsen, A. Lindergaard-Andersen, K. Bechgaard: J. Physique **44**, C3-963-968 (1983)

3.51   R. Moret, J.P. Pouget, R. Comès, K. Bechgaard: J. Physique **44**, C3-957 (1983)

3.52   S.S.P. Parkin, E.M. Engler, R.R. Schumaker, R. Lagier, V.Y. Lee, J.C. Scott, R.L. Greene: Phys. Rev. Lett. **50**, 270-273 (1983)

3.53   R. Moret, S. Ravy, J.P. Pouget, R. Comès, K. Bechgaard: Phys. Rev. Lett. **57**, 1915-1918 (1986)

3.54   F. Wudl: In *Chemistry and Physics of One Dimensional Metals*, ed by H.J. Keller (Plenum, New York 1977) p.233

3.55   B. Liautard, S. Peytaven, G. Brun, M. Maurin: J. Physique **44**, C3-951-956 (1983)

3.56   K. Mortensen, E.M. Conwell, J.M. Fabre: Phys. Rev. B **28**, 5856-5862 (1983)
       K. Mortensen, E.M. Engler: Mol. Cryst. Liq. Cryst. **119**, 293-296 (1985)

3.57   H. Fukuyama: Synth. Met. **19**, 63-68 (1987)

3.58   J. Kondo: Physica **123B**, 169 (1984)

3.59   A. Maaroufi, S. Flandrois, G. Fillion, J.P. Morand: Mol. Cryst. Liq. Cryst. **119**, 311-315 (1985)

3.60   C. Coulon, J.C. Scott, R. Laversanne: Mol. Cryst. Liq. Cryst. **119**, 307-310 (1985)

3.61   F. Creuzet, D. Jérome, A. Moradpour: Mol. Cryst. Liq. Cryst. **119**, 297-302 (1985)

3.62   L. Balicas, K. Behnia, W. Kang, E. Canadell, P. Auban-Senzier, D. Jérome, M. Ribault, J.M. Fabre: J. Phys. I (France) **4**, 1539-1544 (1994)

3.63   D. Jerome: Science **252**, 1059 (1991)

3.64   T. Ishiguro, K. Murata, K. Kajimura, N. Kinoshita, H. Tokumoto, M. Tokumoto, T. Ukachi, H. Anzai, G. Saito: J. Physique **44**, C3-831-838 (1983)

3.65   P. Garoche, R. Brusetti, D. Jérome, K. Bechgaard: J. Physique Lett. **43**, L 147-152 (1982)

3.66   T. Ishiguro, T. Ukachi, K. Kato, K. Murata, M. Tokumoto, H. Tokumoto, H. Anzai, G. Saito: J. Phys. Soc. Jpn. **52**, 1585-1592 (1983)

3.67   H. Schwenk, K. Andres, F. Wudl, E. Aharon-Shalom: J. Physique **44**, C3 1041-1045 (1983)

3.68   W.M. Walsh Jr., F. Wudl, E. Aharon-Shalom, L.W. Rupp Jr., J.M. Vandenberg, K. Andres, J.B. Torrance: Phys. Rev. Lett. **49**, 885-888 (1982)

3.69   R.L. Greene, E.M. Engler: Phys. Rev. Lett. **45**, 1587-1590 (1980)

3.70   M.-H. Whangbo, J.M. Williams, A.J. Schultz, T.J. Emge, M.A. Beno: J. Am. Chem. Soc. **109**, 90-94 (1987)

3.71   F. Gross, H. Schwenk, K. Andres, F. Wudl, S.D. Cox, J. Brennan: Phys. Rev. B **30**, 1282-1284 (1984)

3.72   M. Takigawa, H. Yasuoka, G. Saito: J. Phys. Soc. Jpn. **56**, 873-876 (1987)

3.73   L.C. Hebel, P.C. Slichter: Phys. Rev. **113**, 1504-1519 (1959)

3.74   Y. Hasegawa, H. Fukuyama: J. Phys. Soc. Jpn. **56**, 877-880 (1987)

3.75 C. Bourbonnais, F. Creuzet, D. Jérome, K. Bechgaard, A. Moradpour: J. Phys. Lett. **45**, L755-765 (1984)

3.76 K. Andres, F. Wudl, D.B. McWhan, G.A. Thomas, D. Nalewajek, A.L. Stevens: Phys. Rev. Lett. **45**, 1449-1452 (1980)

3.77 D. Mailly, M. Ribault, K. Bechgaard: J. Physique **44**, C3-1037-1040 (1983)

3.78 H. Schwenk, K. Andres, F. Wudl: Solid State Commun. **49**, 723-726 (1984)

3.79 K. Maki: Physica **1**, 21, 127 (1964)

3.80 K. Murata, M. Tokumoto, H. Anzai, K. Kajimura, T. Ishiguro: Jpn. J. Appl. Phys. **26**, Suppl. 26-3, 1367-1368 (1987)

3.81 A.M. Clogston: Phys. Rev. Lett. **9**, 266 (1962)

3.82 M. Tinkham: *Introduction to Superconductivity* (McGraw-Hill, New York 1975) Chap.5

3.83 H. Schwenk, K. Andres, F. Wudl, E. Aharon-Shalom: Solid State Commun. **45**, 767-769 (1983)

3.84 C. More, R. Roger, J.P. Sorbier, D. Jérome, M. Ribault, K. Bechgaard: J. Phys. Lett. **42**, L313-317 (1981)

3.85 D. Jérome, H.J. Schulz: Adv. Phys. **31**, 299-490 (1982)

3.86 A. Fournel, C. More, G. Roger, J.P. Sorbier, C. Blanc: J. Physique **44**, C3-879-884 (1983)

3.87 H.K. Ng, T. Timusk, J.M. Delrieu, D. Jérome, K. Bechgaard, J.M. Fabre: J. Physique Lett. **43**, L513-519 (1983)

3.88 D. Djurek, D. Jérome, K. Bechgaard: J. Phys. **17**, 4179-4192 (1984)

3.89 J.E. Eldridge, C.C. Homes, F.E. Bates, G.S. Bates: Phys. Rev. B **32**, 5156-5162 (1985)

3.90 M-Y. Choi, P.M. Chaikin, R.L. Greene: Phys. Rev. B **34**, 7727-7732 (1986)

3.91 M.-Y. Choi, P.M. Chaikin, S.Z. Huang, P. Haen, E.M. Engler, R.L. Greene: Phys. Rev. B **25**, 6208-6217 (1982)

3.92 S. Tomic, D. Jérome, D. Mailly, M. Ribault, K. Bechgaard: J. Phys. **44**, C3-1075 (1983)

3.93 C. Coulon, P. Delhaès, J. Amiell, J.P. Manceau, J.M. Fabre, L. Giral: J. Physique **43**, 1721 (1982)

3.94 Y. Hasegawa, H. Fukuyama: J. Phys. Soc. Jpn. **55**, 3717 (1986)

Chapter 4

4.1 J.C. Slater: Phys. Rev. **82**, 538-541 (1951)

4.2 A.W. Overhauser: Phys. Rev. **128**, 1437-1452 (1962)

4.3 J.P. Pouget: Chemica Scripta **17**, 85-91 (1981)
    J.P. Pouget, R. Moret, R. Comès, K. Bechgaard, J.M. Fabre, L. Giral: Mol. Cryst. Liq. Cryst. **79**, 129-143 (1982)

4.4 A. Andrieux, D. Jérome, K. Bechgaard: J. Physique Lett. **42**, L87-91 (1982)

4.5 J.C. Scott, H.J. Pedersen, K. Bechgaard: Phys. Rev. B **24**, 475-477 (1981)

4.6 K. Mortensen, Y. Tomkiewicz, K. Bechgaard: Phys. Rev. B **25**, 3319-3325 (1982)

4.7 J.B. Torrance, H.J. Pedersen, K. Bechgaard: Phys. Rev. Lett. **49**, 881-884 (1982)

4.8 W.M. Walsh Jr., F. Wudl, E. Aharon-Shalom, L.W. Rupp Jr., J.M. Vandenberg, K. Andres, J.B. Torrance: Phys. Rev. Lett. **49**, 885-888 (1982)

4.9   J.M. Delrieu, M. Roger, Z. Toffano, E. Wope Mbougue, R. Saint James, K. Bechgaard: Physica B **143**, 412-416 (1986)

4.10   T. Takahashi, Y. Maniwa, H. Kawamura, G. Saito: Physica B **143** 417-421 (1986)

4.11   J.M. Delrieu, M. Roger, Z. Toffano, A. Moradpour, K. Bechgaard: J. Physique **47**, 839-861 (1986)

4.12   T. Takahashi, Y. Maniwa, H. Kawamura, G. Saito: J. Phys. Soc. Jpn. **53**, 1364-1373 (1986)

4.13   Yu.A. Bychkov, L.P. Gor'kov, I.E. Dzyaloshinskii: Sov. Phys. JETP **23**, 489-501 (1966)

4.14   J. Solyom: Adv. Phys. **28**, 201-303 (1979)

4.15   B. Horovitz, H. Gutfreund, M. Weger: Solid State Commun. **39**, 541-545 (1981)

4.16   H. Gutfreund, B. Horovitz, M. Weger: J. Physique **44**, C3-983-989 (1983)

4.17   V.J. Emery, R. Bruinsma, S. Barisic: Phys. Rev. Lett. **48**, 1039-1043 (1982)

4.18   V.J. Emery: J. Phys. **44** C3-977-982 (1983)

4.19   R. Moret, S. Ravy, J.P. Pouget, R. Comès, K. Bechgaard: Phys. Rev. Lett. **57**, 1915-1918 (1986)

4.20   D. Jérome, A. Mazaud, M. Ribault, K. Bechgaard: J. Physique Lett. **41**, 95-98 (1980)

4.21   N. Thorup, G. Rindorf, H. Soling, K. Bechgaard: Acta. Cryst. B **37**, 1236-1240 (1981)

4.22   K. Bechgaard, C.S. Jacobsen, K. Mortensen, H.J. Pedersen, N. Thorup: Solid State Commun. **33**, 1119-1125 (1980)

4.23   K. Yamaji: J. Phys. Soc. Jpn. **51**, 2787-2797 (1982)

4.24   C.S. Jacobsen, D.B. Tanner, K. Bechgaard: Phys. Rev. Lett. **46**, 1142-1145 (1981)

4.25   K. Yamaji: J. Phys. Soc. Jpn. **52**, 1361-1372 (1983)

4.26   J.F. Kwak: Phys. Rev. B **26**, 4789-4792 (1982)

4.27   C.S. Jacobsen, D.B. Tanner, K. Bechgaard: Phys. Rev. B **28**, 7019-7032 (1983)

4.28   K. Kikuchi, I. Ikemoto, K. Yakushi, H. Kuroda: Solid State Commun. **42**, 433-435 (1982)

4.29   K. Yamaji: J. Chem. Soc. Jpn. 406-409 (1986) (in Japanese)

4.30   D. Jérome: In *The Physics and Chemistry of Low Dimensional Solids*, ed. by L. Alcacer (Reidel, Dordrecht 1980) pp. 123-142

4.31   T. Ishiguro, H. Sumi, S. Kagoshima, K. Kajimura, H. Anzai: J. Phys. Soc. Jpn. **48**, 456-463 (1980)

4.32   K. Murata, M. Tokumoto, H. Anzai, K. Kajimura, T. Ishiguro: Jpn. J. Appl. Phys. **26**, Suppl. 26-3, 1367-1368 (1987)

4.33   R.L. Greene, P. Haen, S.Z. Huang, E.M. Engler, M.Y. Choi, P.M. Chaikin: Mol. Cryst. Liq. Cryst. **79**, 183 (1982)

4.34   L.P. Gor'kov, D. Jérome: J. Physique Lett. **46**, 643-646 (1985)

4.35   X. Huang, K. Maki: Phys. Rev. B **39**, 6459-6464 (1989)

4.36   A.I. Buzdin, L.N. Bulaevskii: Sov. Phys. Usp. **27**, 830-844 (1984)

4.37   P.M. Grant: J. Physique **44**, C3-847-857 (1983)

4.38   M.-H. Whangbo, W.M. Walsh Jr., R.C. Haddon, F. Wudl: Solid State Commun. **43**, 637-639 (1982)

4.39   T. Mori, A. Kobayashi, Y. Sasaki, H. Kobayashi: Chem. Lett. **1982**, 1923-1926 (1982)

4.40   L. Ducasse, M. Abderrabba, J. Hoarau, M. Pesquer, B. Gallois, J. Gaultier: J. Phys. C **19**, 3805-3820 (1986)

4.41   A.J. Berlinsky, J.F. Carolan, L. Weiler: Solid State Commun. **15**, 795-801 (1974)

4.42   B. Horovitz, H. Gutfreund, M. Weger: Phys. Rev. B **12**, 3174-3185 (1975)

4.43   J.B. Torrance, B.A. Scott, B. Welber, F.B. Kaufman, R.E. Seiden: Phys. Rev. B **19**, 730-741 (1975)

4.44   L. Chen, K. Maki, A. Virosztek: Physica B **143**, 444-446 (1986)

4.45   K. Yamaji: J. Phys. Soc. Jpn. **55**, 860-864 (1986)

4.46   B. Gallois, A. Meresse, J. Gaultier, R. Moret: Mol. Cryst. Liq. Cryst. **131**, 147-161 (1985)

        B. Gallois, J. Gaultier, C. Hauw, T.-D. Lamcharfi, A. Filhol: Acta Cryst. B **42**, 564-575 (1986)

        B. Gallois, J. Gaultier, A. Filhol, C. Vettier: Mol. Cryst. Liq. Cryst. **148**, 279-293 (1987)

4.47   J.E. Eldridge, G.S. Bates: Mol. Cryst. Liq. Cryst. **119**, 183-190 (1985)

4.48   T. Timusk: In *Low-Dimensional Conductors and Superconductors*, ed. by D. Jérome, L.G. Caron (Plenum, New York 1987) pp.275-284

4.49   K. Kornelsen, J.E. Eldridge, G.S. Bates: Phys. Rev. B **35**, 9162-9167 (1987)

4.50   L. Degiorgi, M. Dressel, A. Schwartz, B. Alavi, G. Grüner: Phys. Rev. Lett. **76**, 3838-3841 (1996)

4.51   E.W. Fenton, G.C. Aers: In *Low-Dimensional Conductors and Superconductors*, ed. by D. Jérome, L.G. Caron (Plenum, New York 1987) pp.285-294

4.52   P.M. Chaikin, P. Haen, E.M. Engler, R.L. Greene: Phys. Rev. B **24**, 7155-7161 (1981)

4.53   M. Roger, J.M. Delrieu, E. Wope Mbougue: Phys. Rev. B **34**, 4952-4955 (1986)

4.54   T. Takahashi, H. Kawamura, T. Ohama, Y. Maniwa, K. Murata, G. Saito: J. Phys. Soc. Jpn. **58**, 703-709 (1989)

4.55   J. Bardeen, L.N. Cooper, J.R. Schrieffer: Phys. Rev. **108**, 1175-1204 (1957)

4.56   T. Moriya: *Spin Fluctuation in Itinerant Elecron Magnetism*, Springer Ser. Solid-State Sci., Vol.56 (Springer, Berlin, Heidelberg 1985)

4.57   R. Moret, J.P. Pouget, R. Comès, K. Bechgaard: Phys. Rev. Lett. **49**, 1008-1012 (1982)

4.58   D.R. Guy, G.S. Boebinger, E.M. Marseglia, R.H. Friend, K. Bechgaard: J. Phys. C **16**, 691-698 (1983)

4.59   L. Ducasse, A. Fritsch: Synth. Met. **27**, B1-B5 (1988)

4.60   S. Jafarey: Phys. Rev. B **16**, 2584-2592 (1977)

4.61   K. Yamaji: In *Superconductivity in Magnetic and Exotic Materials*, ed. by T. Matsu-bara, A. Kotani, Springer Ser. Solid-State Sci., Vol.52 (Springer, Berlin, Heidelberg, 1984) pp.149-166; J. Phys. Soc. Jpn. **53**, 2189-2192 (1984)

4.62   K. Yamaji: J. Phys. Soc. Jpn. **56**, 1841-1854 (1987). The − sign in front of s in Eq.(2.8) and in the caption of Fig. 2 should be changed to a + sign

4.63   Y. Hasegawa, H. Fukuyama: Physica B **143**, 447-449 (1986)

4.64   J.F. Kwak, J.E. Schirber, P.M. Chaikin, J.M. Williams, H.-H. Wang, L.Y. Chiang: Phys. Rev. Lett. **56**, 972-975 (1986)

4.65   L.J. Azevedo, J.E. Schirber, J.M. Williams, M.A. Beno, D.R. Stephens: Phys. Rev. B **30**, 1570-1572 (1984)

        J.F. Kwak, L.J. Azevedo: Comment Solid State Phys. **11**, 245-255 (1985)

4.66   G. Montambaux: Phys. Rev. B **38**, 4788-4795 (1988)

4.67   J.R. Cooper, M. Miljak, G. Delplanque, D. Jérome, M. Weger: J. Physique **38**, 1097-1103 (1977)

4.68   M. Ribault, D. Jérome, J. Tuchendler, C. Weyl, K. Bechgaard: J. Physique Lett. **44**, 953-961 (1983)

4.69   K. Oshima, M. Suzuki, K. Kikuchi, H. Kuroda, I. Ikemoto, K. Kobayashi: J. Phys. Soc. Jpn. **53**, 3295-3298 (1984)

4.70   P.M. Chaikin, M.-Y. Choi, J.F. Kwak, J.S. Brooks, K.P. Martin, M.J. Naughton, E.M. Engler, R.L. Greene: Phys. Rev. Lett. **51**, 2333-2336 (1983)

4.71   R. Brusetti, M. Ribault, D. Jérome, K. Bechgaard: J. Physique **43**, 801-808 (1982)

4.72   D. Jérome: In *Physics and Chemistry of Electrons and Ions in Condensed Matter*, ed. by J.A. Acrivos (Reidel, Dordrecht 1984) pp. 595-624

4.73   R.L. Greene, E.M. Engler: Phys. Rev. Lett. **45**, 1587-1590 (1980)

4.74   K. Machida: J. Phys. Soc. Jpn. **50**, 2195-2202 (1981)
       K. Machida, T. Matsubara: J. Phys. Soc. Jpn. **50**, 3231-3239 (1981)
       K. Machida: Appl. Phys. A **35**, 193-217 (1984)

4.75   J. Keller: J. Magn. Magn. Mat. **28**, 193-198 (1982)

4.76   T. Takahashi, T. Harada, Y. Kobayashi, K. Kanoda, K. Suzuki, K. Murata: Synth. Met. **41-43**, 3985-3988 (1991)

4.77   K. Nomura, Y. Hosokawa, N. Matsunaga, M. Nagasawa, T. Sambongi, H. Anzai: Synth. Met. **70**, 1295-1296 (1994)

4.78   J.C. Lasjaunias, K. Biljakovic, P. Monceau, K. Bechgaard: Solid State Commun. **84**, 297-300 (1992)

4.79   J.C. Lasjaunias, K. Biljakovic, F. Nad, P. Monceau. K. Bechgaard: Phys. Rev. Lett. **72**, 1283-1286 (1994)

4.80   S. Takada: J. Phys. Soc. Jpn. **53**, 2193-2196 (1984)

4.81   I. Tüttö, A. Zawadovski: Phys. Rev. Lett. **60**, 1442-1445 (1988)

4.82   Y. Suzumura, T. Saso, H. Fukuyama: Jpn. J. Appl. Phys. **26**, Suppl. 26-3, 589-590 (1987)

4.83   T. Osada, N. Miura, I. Ogura, G. Saito: Phys. Rev. Lett. **58**, 1563-1566 (1987)

4.84   S. Tomic, J.R. Cooper, D. Jérome, K. Bechgaard: Phys. Rev. Lett. **62**, 462-465 (1989)

4.85   T. Sambongi, K. Nomura, T. Shimizu, K. Ichimura, N. Kinoshita, M. Tokumoto, H. Anzai: Solid State Commun. **72**, 817-819 (1989)

4.86   W. Kang, S. Tomic, J.R. Cooper, D. Jérome: Phys. Rev. B **41**, 4862-4865 (1990)

4.87   N. Hino, T. Sambongi, K. Nomura, M. Nagasawa, M. Tokumoto, H. Anzai, N. Kinoshita, G. Saito: Symth. Met. **40**, 275-281 (1991)

4.88   M. Nagasawa, T. Sambongi, K. Nomura, H. Anzai: Solid State Commun. **93**, 33-39 (1995)

4.89   K. Nomura, T. Shimizu, K. Ichimura, T. Sambongi, M. Tokumoto, H. Anzai, N. Kinoshita: Solid State Commun. **72**, 1123-1126 (1989)

4.90   J.M. Delrieu, N. Kinoshita: C.R. Acad. Sci. (Paris) Ser. II **310**, 891-896 (1990)

4.91   G. Grüner: Rev. Mod. Phys. **66**, 1-24 (1994)

4.92   T. Takahashi, H. Kawamura, T. Ohyama, Y. Maniwa, K. Murata, G. Saito: J. Phys. Soc. Jpn. **58**, 703-709 (1989)

4.93   M. Nagasawa, T. Sambongi, K. Nomura, H. Anzai: J. Phys. Soc. Jpn. **62**, 3974-3978 (1993)

4.94   G.M. Danner, P.M. Chaikin, S.T. Hannahs: Phys. Rev. B **53**, 2727-2731 (1996)

4.95   W.G. Clark, M.E. Hanson, W.H. Wong, B. Alavi: Physica B **194-196**, 285-286 (1994)

4.96   S. Tomic, J.R. Cooper, W. Kang, D. Jérome, K. Maki: J. Phys. I (France) **1**, 1603- 1625 (1991)

4.97   T. Takahashi, F. Creuzet, D. Jérome, J.M. Fabre: J Physique **44**, C3-1095-1097 (1983)

4.98   L. Balicas, K. Behnia, W. Kang, E. Canadell, P. Auban-Senzier, D. Jérome, M. Ribault, J.M. Fabre: J. Phys. I (France) **4**, 1539-1549 (1994)

4.99   T. Nakamura, T. Nobutoki, Y. Kobayashi, T. Takahashi, G. Saito: Synth. Met. **70**, 1293-1294 (1995)

4.100 T.J. Emge, H.H. Wang, P.C.W. Leung, P.R. Rust, J.D. Cook, P.L. Lackson, K.D. Carlson, J.M. Williams, M.-H. Whangbo, E.L. Venturini, J.E. Schirber, L.J. Azevedo, J.R. Ferraro: J. Am. Chem. Soc. **108**, 695-702 (1986)

4.101 K. Kanoda, T. Takahashi, T. Tokiwa, K. Kikuchi, K. Sato, I. Ikemoto, K. Kobayashi: Phys. Rev. B **38**, 39-43 (1988)

4.102 K. Kanoda, S. Okui, T. Takahashi, K. Kikuchi, K. Sato, I. Ikemoto, K. Kobayashi: In *Physics and Chemistry of Organic Superconductors*, ed. by G. Saito, S. Kagoshima, Springer Proc. Phys., Vol.51 (Springer, Berlin, Heidelberg 1990) pp.242-246

4.103 K. Kanoda, Y. Kobayashi, T. Takahashi, T. Inukai, G. Saito: Phys. Rev. B **42**, 8678-8681 (1990)

4.104 J.S. Brooks, X. Chen, S.J. Klepper, S. Valfells, G.J. Athas, Y. Tanaka, T. Kinoshita, N. Kinoshita, M. Tokumoto, H. Anzai, C.C. Agosta: Phys. Rev. B **52**, 14457-14478 (1995)

4.105 T. Sasaki, H. Sato, N. Toyota: Synth. Met. **41-43**, 2211-2214 (1991)

4.106 F.L. Pratt, T. Sasaki, N. Toyota, K. Nagamine: Phys. Rev. Lett. **74**, 3892-3895 (1995)

4.107 A.A. House, N. Harrison, S.J. Blundell, J. Deckers, J. Singleton, F. Herlach, W. Hayes, J.A.A.J. Perenboom, M. Kurmoo, P. Day: Phys. Rev. B **53**, 9127-9136 (1996)

Chapter 5

5.1   G. Saito, T. Enoki, K. Toriumi, H. Inokuchi: Solid State Commun. **42**, 557-560 (1982)

5.2   P.C.W. Leung, T.J. Emge, M.A. Beno, H.H. Wang, J.M. Williams, V. Patrick, P. Coppens: J. Am. Chem. Soc. **107**, 6184-6191 (1985)

5.3   H. Kobayashi, A. Kobayashi, Y. Sasaki, G. Saito, T. Enoki, H. Inokuchi: J. Am. Chem. Soc. **105**, 297-298 (1983)

5.4   H. Kobayashi, T. Mori, R. Kato, A. Kobayashi, Y. Sasaki, G. Saito, H. Inokuchi: Chem. Lett **1983**, 581-584 (1983)

5.5   S.S.P. Parkin, E.M. Engler, R.R. Schumaker, R. Lagier, V.Y. Lee, J. Voiron, K. Carneiro, J.C. Scott, R.L. Greene: J. Physique **44**, C3-791-797 (1983)

5.6   I. Hennig, K. Bender, D. Schweitzer, K. Diez, H. Endres, H.J. Keller, A. Gleiz, H.W. Helberg: Mol. Cryst. Liq. Cryst. **119**, 337-441 (1985)

5.7 K. Bender, I. Henning, D. Schweitzer, K. Dietz, H. Endres, H.J. Keller: Mol. Cryst. Liq. Cryst. **108**, 359 (1984)

5.8 H. Kobayashi, R. Kato, A. Kobayashi, Y. Nishino, K. Kajita, W. Sasaki: Chem. Lett. **1986**, 2017-2020 (1986)

5.9 A. Kobayashi, R. Kato, H. Kobayashi, S. Moriyama, Y. Nishino, K. Kajita, W. Sasaki: Chem. Lett. **1987**, 459-462 (1987)

5.10 E.B. Yagubskii, I.F. Shchegolev, V.N. Laukhin, P.A. Kononovich, M.V. Kartsovnic, A.V. Zvarykina, L.I. Bubarov: JETP Lett. **39**, 12-15 (1984)

5.11 R.P. Shibaeva, V.F. Kaminskii, E.B. Yagubskii: Mol. Cryst. Liq. Cryst. **119**, 361-373 (1985)

5.12 H. Kobayashi, R. Kato, A. Kobayashi, M. Tokumoto, H. Anzai, T. Ishiguro, G. Saito: Chem. Lett. **1985**, 1924 (1985)

5.13 H. Kuroda, K. Yakushi, H. Tajima, A. Ugawa, Y. Okawa, A. Kobayashi, R. Kato, H. Kobayashi, G. Saito, Synth. Met. **27**, A491-498 (1988)

5.14 H. Urayama, H. Yamochi, G. Saito, S. Sato, A. Kawamoto, J. Tanaka, T. Mori, Y. Maruyama, H. Inokuchi: Chem. Lett. **1988**, 463-466 (1988)
G. Saito, H. Urayama, H. Yamochi, K. Oshima: Synth. Met. **27**, A331-340 (1988)

5.15 H. Yamochi, T. Nakamura, T. Komatsu, N. Matsukawa, T. Inoue, G. Saito, T. Mori, M. Kusunoki, K. Sakaguchi: Solid State Commun. **82**, 101-105 (1992)

5.16 U. Geiser, A. J. Schultz, H. H. Wang, D. M. Watkins, D. L. Stupka, J. M. Williams, J. E. Schirber, D. L. Overmyer, D. Jung, J. J. Nova, M. H. Whangbo: Physica C **174**, 475-486 (1991)

5.17 L. I. Buravov, N. D. Kushch, V. N. Laukhin, A. G. Khomenko, E. B. Yagubskii, M. V. Kartsovnik, L. P. Rozenberg, R. P. Shibaeva, M. A. Tanatar, V. S. Yefanov, V. V. Dyakin, V. A. Bondarenko: J. Physique **4**, 441-451 (1994)

5.18 H. Mori, S. Tanaka, M. Oshima, G. Saito, T. Mori, Y. Maruyama, H. Inokuchi: Bull. Chem. Soc. Jpn. **63**, 2183-2190 (1990)

5.19 H. Mori, S. Tanaka, K. Oshima, M. Oshima, G. Saito, T. Mori, Y. Maruyama, H. Inokuchi: Solid State Commun. **74**, 1261-1264 (1990)

5.20 T. Mori, A. Kobayashi, Y. Sasaki, H. Kobayashi, G. Saito, H. Inokuchi: Bull. Chem. Soc. Jpn. **57**, 627-633 (1984)

5.21 T. Mori, A. Kobayashi, Y. Sasaki, H. Kobayashi, G. Saito, H. Inokuchi: Chem. Lett. **1984**, 957-960 (1984)

5.22 T. Mori, A. Kobayashi, Y. Sasaki, R. Kato, H. Kobayashi: Solid State Commun. **53**, 627-631 (1985)

5.23 K. Kajita, Y. Nishio, S. Moriyama, W. Sasaki, R. Kato, H. Kobayashi, A. Kobayashi: Solid State Commun. **64**, 1279-1284 (1987)

5.24 K. Oshima, T. Mori, H. Inokuchi, H. Urayama, H. Yamochi, G. Saito: Phys. Rev. B **38** 938-941 (1988); Synth. Met. **27**, A165-170 (1988)

5.25 M.-H. Whangbo, M.A. Beno, P.C.W. Leung, T.J. Emge, H.H. Wang, J. Williams: Solid State Commun. **59**, 813-818 (1986)

5.26 D. Schweitzer, E. Balthes, S. Kahlich, H. J. Keller, W. Strunz, W. Biberacher, A.G.M. Jansen, E. Steep: Synth. Met. **70**, 857-860 (1995)

5.27 W. Kang, G. Montambaux, J.R. Cooper, D. Jérome, P. Batail, C. Lenoir: Phys. Rev. Lett. **62**, 2559-2562 (1989)

5.28 M.V. Kartsovnik, P.A. Kononovich, V.N. Laukhin, I.F. Shchegolev: JETP Lett. **48**, 541-544 (1988)

5.29 J. Caulfield, W. Lubczynski, F.L. Pratt, J. Singleton, D.Y.K. Ko, W. Hayers, M. Kurmoo, P. Day: J. Phys. Condens. Matter **6**, 2911-2924 (1994)

5.30 M.V. Kartsovnik, G.Yu. Logvenov, H. Ito, T. Ishiguro, G. Saito: Phys. Rev. B **52**, R15715-15718 (1995)

5.31 Y. Yamauchi, M.V. Kartsovnik, T. Ishiguro, M. Kubota, G. Saito: J. Phys. Soc. Jpn. **65**, 354-357 (1996)

5.32 E. Ohmichi, H. Ito, T. Ishiguro, T. Komatsu, G. Saito: J. Phys. Soc. Jpn. **66**, 310 (1997)

5.33 K. Oshima, K. Arai, H. Yamazaki, K. Kato, Y. Maruyama, K. Yakushi, T. Mori, H. Inokuchi, H. Mori, S. Tanaka: Physica C **185-189**, 2689 (1991)

5.34 T. Terashima, S. Uji, H. Aoki, M. Tamura, M. Kinoshita, M. Tokumoto: Solid State Commun. **91**, 595-598 (1994)
      M. Tamura, H. Kuroda, S. Uji, H. Aoki, M. Tokumoto, A.G. Swanson, J. S. Books, C.C. Agosta, S.T. Hannas: J. Phys. Soc. Jpn. **63**, 615-622 (1994)

5.35 T. Osada, A. Kawasumi, R. Yagi, S. Kagoshima, N. Miura, M. Oshima, H. Mori, T. Nakamura, G. Saito: Solid State Commun. **75**, 901-905 (1990)

5.36 T. Osada, R. Yagi, A. Kawasumi, S. Kagoshima, N. Miura, M. Oshima, G. Saito: Phys. Rev. B **41**, 5428 (1990)

5.37 J.S. Brooks, S.J. Klepper, C.C. Agosta, M. Tokumoto, N. Kinoshita, Y. Tanaka, S. Uji, H. Aoki, A.S. Perel, G.J. Athas, X. Chen, D.A. Howe, H. Anzai: Physica B **184**, 489-493 (1993)

5.38 S. Uji, H. Aoki, M. Tokumoto, T. Kinoshita, N. Kinoshita, Y. Tanaka, H. Anzai: Phys. Rev. B **49**, 732-735 (1994)

5.39 V.N. Laukhin, A. Audouard, H. Rakoto, J.M. Broto, F. Goze, G. Coffe, L. Brossard, J.P. Redoules, M.V. Kartsovnik, N.D. Kushch, L.I. Buravov, A.G. Khomenko, E.B. Yagubskii, S. Askenazy: Physica B **211**, 282-285 (1995)

5.40 J. Wosnitza: *Fermi Surfaces of Low-Dimensional Organic Metals and Superconductors*, Springer Tracts Mod. Phys., Vol.134 (Springer, Berlin, Heidelberg 1996)

5.41 M.V. Kartsovnik, V.N. Laukhin, S.I. Pesotskii, I.F. Schegolev, V.M. Yakovenko: J. Physique I **2**, 89-99 (1990)

5.42 K. Kajita, Y. Nishino, T. Takahashi, R. Kato, H. Kobayashi, W. Sasaki, A. Kobayashi, Y. Iye: Solid State Commun. **70**, 1189-1193 (1989)

5.43 K. Yamaji: J. Phys. Soc. Jpn. **58**, 1520-1523 (1989)

5.44 R. Yagi, Y. Iye, T. Osada, S. Kagoshima: J. Phys. Soc. Jpn. **59**, 3069 (1990)

5.45 T. Sasaki, H. Sato, N. Toyota: Synth. Met. **41-43**, 2211 (1991)

5.46 Y. Iye, R. Yagi, N. Hanasaki, S. Kagoshima, H. Mori, H. Fujimoto, G. Saito: J. Phys. Soc. Jpn. **63**, 674-684 (1994)

5.47 M.V. Kartsovnik, A.E. Kovalev, N.D. Kushch: J. Physique **3**, 1187 (1993)

5.48 J. Singleton, F.L. Pratt, M. Doporto, J.M. Caulfield, S.O. Hill, T.J.B.M. Janssen, I. Deckers, G. Pitsi, F. Herlach, W. Hayes, J.A.A.J. Perenboom, M. Kurmoo, P. Day: Physica B **184**, 470-480 (1993)

5.49 N. Toyota, E.W. Fenton, T. Sasaki, M. Tachiki: Solid State Commun. **72**, 859 (1989)

5.50 V.A. Merzhanov, E.E. Kostyuchenko, O.E. Faber, I.F. Shchegolev, E.B. Yagubskii: Sov. Phys. - JETP **62**, 165-167 (1985)
      K. Mortensen, J.M. Williams, H.H. Wang: Solid State Commun. **56**, 105-110 (1985)

5.51　T. Mori, H. Inokuchi: J. Phys. Soc. Jpn. **57**, 3674-3677 (1988)

5.52　R. Liu, H. Ding, J.C. Campuzano, H.H. Wang, J.M. Williams, K.D. Carlson: Phys. Rev. B **51**, 13000-13004 (1995)

5.53　Y.C. Jean, Y. Lou, H.L. Yen, K.M. O'Broen, R.N. West, H.H. Wang, K.D. Carlson, J.M. Williams: Physica C **221**, 399 (1994)

5.54　H. Kino, H. Fukuyama: J. Phys. Soc. Jpn. **64**, 2726-2729 (1995)

5.55　H. Kobayashi, T. Mori, R. Kato, A. Kobayashi, Y. Sasaki, G. Saito, H. Inokuchi: Chem. Lett. **1983**, 581-584 (1983)

　　　B.K. Vainshtein: *Fundamental of Crystals*, 2nd edn., Modern Crystallography 1 (Springer, Berlin, Heidelberg 1994)

5.56　R. Lavaersanne, J. Amiell, P. Delhaes, D. Chasseau, C. Haw: Solid State Commun. **52**, 177-181 (1984)

5.57　K. Kajita, T. Ojiro, H. Fujii, Y. Nishio, H. Kobayashi, A. Kobayashi, R. Kato: J. Phys. Soc. Jpn. **61**, 23-26 (1992)

5.58　Y. Hasegawa, H. Fukuyama: Physica B **184**, 498-502 (1993)

5.59　H. Mori, S. Tanaka, K. Oshima, G. Saito, T. Mori, Y. Maruyama, H. Inokuchi: Synth. Met. **41-43**, 2013-2018 (1991)

5.60　H.H. Wang, K.D. Carson, U. Geiser, W.K. Kwok, M.D. Vashon, J.E. Thompson, N.F. Larsen, G.D. Mccabe, R.S. Hulscher, J.M. Williams: Physica C **166**, 57 (1990)

5.61　H. Ito, M.V. Kartsovnik, H. Ishimoto, K. Kono, H. Mori, N.D. Kushch, G. Saito, T. Ishiguro, S. Tanaka: Synth. Met. **70**, 899 (1995)

5.62　T. Sasaki, N. Toyota, M. Tokumoto, N. Kinoshita, H. Anzai: Solid State Commun. **75**, 93-96 (1990)

5.63　V.N. Laukhin, A. Audouard, H. Rakoto, J.M. Broto, F. Goze, G. Coffe, L. Brossard, J.P. Redoules, M.V. Kartsovnik, N.D. Kushch, L.I. Bravov, A.G. Khomenko, E.B. Yagubskii, S. Askenazy, P. Pari: Physica B **211**, 282-285 (1995)

5.64　L.N. Bulaevskii: Adv. Phys. **37**, 433-470 (1988)

5.65　W. Kang, D. Jerome, C. Lenodr, P. Batail: J. Phys. Cond. Matter **2**, 1665-1668 (1990)

5.66　H. Kusuhara, Y. Sakata, Y. Ueba, K. Toda, M. Kaji, T. Ishiguro: Solid State Commun. **74**, 251-255 (1990)

5.67　Y. Watanabe, T. Sasaki, H. Sato, N. Toyota: J. Phys. Soc. Jpn. **60**, 2118-2121 (1991)

5.68　T. Doi, K. Oshima, H. Yamazaki, H. Maruyama, H. Maeda, A. Koizumi, H. Kimura, M. Fujita, Y. Yunoki, H. Mori, S. Tanaka, H. Yamochi, G. Saito: J. Phys. Soc. Jpn. **60**, 1441-1444 (1991)

5.69　N. Toyota, T. Sasaki, H. Sato, Y. Watanabe: Physica C **178**, 339-344 (1991)

5.70　A.V. Skripov, A.P. Stepanov: Physica C **197**, 89-94 (1992)

5.71　N. Toyota, T. Sasaki: Solid State Commun. **74**, 361 (1990)

5.72　L.I. Buravov, A.V. Zvarykina, N.D. Kushch, V.N. Laukhin, V.A. Merzanov, A.G. Komenko, E.B. Yagubskii: Sov. Phys. JETP **68**, 182-185 (1989)

5.73　V. Kataev, G. Winkel, D. Khomskii, D. Wohlleben, W. Crump, K.F. Tebbe, J. Haen: Solid State Commun. **83**, 435 (1992)

5.74　H. Mayaffre, P. Wzietek, S. Charfi-Kaddour, C. Lenir, D. Jérome, P. Batail: Physica B **206**, **207**, 767-770 (1995)

5.75   A. Kawamoto, K. Miyagawa, Y. Nakazawa, K. Kanoda: Phys. Rev. Lett. **74**, 3455-3458 (1995)

5.76   H. Ito, T. Ishiguro, M. Kubota, G. Saito: J. Phys. Soc. Jpn. **65**, 2987-2993 (1996)

5.77   H.H. Wang, K.D. Karlson, U. Geiser, A.M. Kini, A.J. Schultz, J.M. Williams, L.K. Montgomery, W.K. Kwok, U. Welp. K.G. Vandervoort, S.J. Boryshuk, A.V. Strieby Crouch, J.M. Kommers, D.M. Watkins, J.E. Schirber, D.L. Overmyer, D. Jung, J.J. Novoa, M.H. Whangbo: Synth. Met. **41-43**, 1983-1990 (1991)

5.78   K. Miyagawa, A. Kawamoto, Y. Nakazawa, K. Kanoda: Phys. Rev. Lett. **75**, 1174-1177 (1995)

5.79   U. Welp, S. Fleshler, W.K. Kwok, G.W. Crabtree, K.D. Carlson, H.H. Wang, U. Geiser, J.M. Williams, V.M. Hitsman: Phys. Rev. Lett. **69**, 840-843 (1992)

5.80   H. Posselt, H. Müller, K. Andres, G. Saito: Phys. Rev. B **49**, 15849-15852 (1994)

5.81   M.A. Tanatar, T. Ishiguro, H. Ito, M. Kubota, G. Saito: Phys. Rev. B **55**, 12529 (1997)

5.82   M. Tokumoto, H. Anzai, T. Ishiguro, G. Saito, H. Kobayashi, R. Kato, A. Kobayashi: Synth. Met. **19**, 215-220 (1987)

5.83   H. Kino, H. Fukuyama: J. Phys. Soc. Jpn. **64**, 4523-4526 (1995)

5.84   M. Héritier, S. Charfi-Kaddour, P. Ordon: Synth. Met. **70**, 1025-1026 (1995)

5.85   B. Koch, H.P. Geserich, W. Ruppel, D. Schweitzer, K.H. Dietz, H.J. Keller: Mol. Cryst. Liq. Cryst. **119**, 343-346 (1985)

5.86   C.S. Jacobsen, J.M. Williams, H.H. Wang: Solid State Commun. **54**, 937-941 (1985)

5.87   H. Kuroda, K. Yakushi, H. Tajima, A. Ugawa, Y. Okawa, A. Kobayashi, R. Kato, H. Kobayashi, G. Saito: Synth. Met. **27**, A491-498 (1988)

5.88   M.J. Rice, N.O. Lipari: Phys. Rev. Lett. **38**, 437-439 (1977)

5.89   H. Tajima, K. Yakushi, H. Kuroda, G. Saito: Solid State Commun. **56**, 159-163 (1985)
       H. Tajima, H. Kanbara, K. Yakushi, H. Kuroda, G. Saito: Solid State Commun. **57**, 911-914 (1986)

5.90   K. Kornelsen, J.E. Eldridge, H.H. Wang, J.M. Williams: Solid State Commun. **74**, 501 (1990); ibid **76**, 1009 (1990)

5.91   J.E. Eldridge, K. Kornelsen, H.H. Wang, J.M. Williams, A.V. Strieby Crouch, D.M. Watkins: Solid State Commun. **79**, 583-589 (1991)

5.92   K. Kornelsen, J.E. Eldridge, H.H. Wang, H.A. Charlier, J.M. Williams: Solid State Commun. **81**, 343-349 (1992)

5.93   M. Dressel, J.E. Eldridge, H.H. Wang, U. Geisers, J.M. Williams: Synth. Met. **52**, 201-211 (1992); ibid. **55-57**, 2923-2926 (1993)

5.94   K. Yakushi, H. Kanabara, H. Tajima, H. Kuroda, G. Saito, T. Mori: Chem. Lett. **1987**, 4251 (1987)

5.95   R.M. Vlasova, S.Ya. Priev, V.N. Semkin, R.N. Lyubovskaya, E.I. Zhilyaeva, E.B. Yagubskii, V.M. Yartsev: Synth Metals **48**, 129-142 (1992)

5.96   J.E. Eldridge, C.C. Homes, J.M. Williams, A.M. Kini, H.H. Wang: Spectrochim. Acta A **51**, 947-960 (1995)

5.97   M.E. Kozlov, K.I. Pokhodnia, A.A. Yurchenko: Spectrochim. Acta A **43**, 323 (1987); ibid A **45**, 437 (1989)

5.98   J.M. Williams, H.H. Wang, T.J. Emge, U. Geiser, M.A. Beno, P.C.W. Leung, K.D. Carlson, R.J. Thorn, A.J. Schultz, M.-H. Whangbo: Proc. Inorg. Chem.

35, 51-218 (1987)

J.R. Ferraro, J.M. Williams: *Introduction to Synthetic Electrical Conductors* (Academic, New York 1987)

5.99   H. Schwenk, F. Gross, C.-P. Heidmann, K. Andres, D. Schweitzer, H. Keller: Mol. Cryst. Liq. Cryst. **119**, 329-335 (1985)

H. Schwenk, C.P. Heidmann, F. Gross, E. Hess, K. Andres, D. Schweitzer, H.J. Keller: Phys. Rev. B **31**, 3138-3140 (1985)

5.100   G.W. Crabtree, K.D. Carlson, L.N. Hall, P.T. Copps, H.H. Wang, T.J. Emge, M.A. Beno, J.M. Williams: Phys. Rev. B **30**, 2958-2960 (1984)

5.101   M. Tokumoto, H. Bando, H. Anzai, G. Saito, K. Murata, K. Kajimura, T. Ishiguro: J. Phys. Soc. Jpn. **54**, 869-872 (1985)

5.102   A.M. Clogston: Phys. Rev. Lett. **9**, 266 (1962)

5.103   G.R. Stewart, J. O'Rourke, G.W. Crabtree, K.D. Carlson, H.H. Wang, J.M. Williams, F. Gross, K. Andres: Phys. Rev. B **33**, 2046-2048 (1986)

5.104   V.N. Laukhin, E.E. Kostyuchenko, Yu.V. Sushko, I.F. Shchegolev, E.B. Yagubskii: JETP Lett. **41**, 81 (1985)

5.105   K. Murata, M. Tokumoto, H. Anzai, H. Bando, G. Saito, K. Kajimura, T. Ishiguro: J. Phys. Soc. Jpn. **54**, 2084 (1985)

5.106   L.J. Azevedo, E.L. Venturini, J.E. Schirber, J.M. Williams, H.H. Wang, T.J. Emge: Mol. Cryst. Liq. Cryst. **119**, 389-392 (1985)

5.107   R.L. Greene, E.M. Engler: Phys. Rev. Lett. **45**, 1587-1590 (1980)

5.108   M. Tokumoto, K. Murata, H. Bando, H. Anzai, G. Saito, K. Kajimura, T. Ishiguro: Solid State Commun. **54**, 1031-1034 (1984)

5.109   T.J. Emge, P.C.W. Leung, M.A. Beno, A.J. Schultz, H.H. Wang, L.M. Sowa, J.M. Williams: Phys. Rev. B **30**, 6780-6782 (1984)

5.110   Y. Nogami, S. Kagoshima, T. Sugano, G. Saito: Synth. Met. **16**, 367-377 (1986)

5.111   K. Yamaji: Solid State Commun. **61**, 413-417 (1987)

5.112   A.J. Schultz, M.A. Beno, H.H. Wang, J.M. Williams: Phys. Rev. B **33**, 7823-7826 (1986)

5.113   V.N. Molchanov, R.P. Shibaeva, V.F. Kaminskii, E.B. Yagubskii, V.I. Simonov, B.K. Vainshtein: Dokl. Akad. Nauk. USSR **286**, 637 (1986)

A.J. Schulz, H.H. Wang, J.M. Williams: J. Am. Chem. Soc. **108**, 7853-7855 (1986)

5.114   B. Hazmic, G. Creuzet, C. Lenoir: Europhys. Lett. **3**, 373-378 (1978)

5.115   M. Tokumoto, H. Anzai, K. Murata, K. Kajimura, T. Ishiguro: Japan J. Appl. Phys. **26**, Suppl. 26-3, 1977-1982 (1987)

5.116   A.V. Gudenko, V.V. Ginodman, P.A. Kononovich, V.N. Laukhin: Sov. Phys. JETP **71**, 350-353 (1990)

5.117   B. Hamzic, G. Creuzet, C. Lenoir: J. Phys. F **17**, 2267 (1987)

5.118   K. Murata, M. Ishibashi, Y. Honda, M. Tokumoto, N. Kinoshita, H. Anzai: J. Phys. Soc. Jpn. **58**, 3469-3472 (1989)

5.119   T.J. Kistenmacher: Solid State Commun. **63**, 977 (1987)

5.120   B. Korin-Hamzic, L. Ferro, J. R. Cooper: Phys. Rev. B **41**, 11646-11648 (1990)

5.121   C.-P. Heidmann, K. Andres, D. Schweitzer: Physica B **143**, 357-359 (1986)

K. Andres, H. Schwenk, H. Veith: Physica B **143**, 334-337 (1986)

5.122   F. Creuzet, G. Creuzet, D. Jerome, D. Schweitzer, H.J. Keller: J. Physique Lett. **46**, L1079-1085 (1985)

5.123 V.B. Ginodman, A.V. Gudenko, P.A. Kononovich, V.N. Laukhin, I.F. Shchegolev: JETP Lett. **44**, 673-676 (1986)

5.124 W. Kang, D. Jérome, C. Lenoir, P. Batail: Synth. Met. **27**, A353-359 (1988)

5.125 S. Kagoshima, Y. Nogami, M. Hasumi, H. Anzai, M. Tokumoto, G. Saito, N.
Mori: Solid State Commun. **62**, 1177-1180 (1989)

5.126 J.E. Schirber, L.J. Azevedo, J.F. Kwak, E.L. Venturini, P.C.W. Leung, M.A.
Beno, H.H. Wang, J.M. Williams: Phys. Rev. **33**, 1987-1989 (1986)

5.127 K. Murata, N. Toyota, M. Tomumoto, H. Anzai, G. Saito, K. Kajimura, S.
Morita, Y. Muto, I. Ishiguro: Physica B **143**, 366-368 (1986)

5.128 K. Murata, M. Tokumoto, H. Bando, H. Tanino, H. Anzai, N. Kinoshita, K. Kajimura, G. Saito, T. Ishiguro: Physica B **135**, 515-519 (1985)

5.129 M. Tokumoto, H. Anzai, H. Bando, G. Saito, N. Kinoshita, K. Kajimura, T. Ishiguro: J. Phys. Soc. Jpn. **54**, 1669-1672 (1985)

5.130 K.D. Carlson, G.W. Crabtree, L. Nunez, H.H. Wang, M.A. Beno, U. Geiser,
M.A. Firestone, K.S. Webb, J.M. Williams: Solid State Commun. **57**, 89-92
(1986)

5.131 V.N. Laukhin, S.I. Pesozkii, E.B. Yagubskii: JETP Lett. **45**, 392 (1987)

5.132 H. Shimahara: Phys. Rev. B **50**, 12760 (1994)

5.133 B. Rothamel, L. Forro, J.R. Copper, J.S. Schilling, M. Weger, P. Bele, H.
Brunner, D. Schweitzer, H.J. Keller: Phys. Rev. B **34**, 704-712 (1986)

5.134 F. Creuzet, C. Bourbonnais, G. Creuzet, D. Jerome, D. Schweitzer, H.J. Keller:
Physica B **143**, 363-365 (1986)

5.135 J.M. Williams, T.J. Emge, H.H. Wang, M.A. Beno, P.T. Copps, L.N. Hall,
K.D. Carlson, G.W. Crabtree: Inorg. Chem. **23**, 3839 (1984)

5.136 H.H. Wang, M.A. Beno, U. Geiser, M.A. Firestone, K.S. Webb, L. Nunez,
G.W. Crabtree, K.D. Carlson, J.M. Williams, L.J. Azevedo, J.F. Kwak, J.E.
Schirber: Inorg. Chem. **24**, 2465-2466 (1985)

5.137 M. Tokumoto, H. Bando, K. Murata, H. Anzai, N. Kinoshita, K. Kajimura, T.
Ishiguro: Synth. Met. **13**, 9-20 (1986); Physica B **143**, 338-342 (1986)

5.138 H. Tanino, K. Kato, M. Tokumoto, H. Anzai, G. Saito: J. Phys. Soc. Jpn. **54**,
2390 (1985)

5.139 H. Kobayashi, R. Kato, A. Kobayashi, M. Tokumoto, H. Anzai, T. Ishiguro, G.
Saito: Chem. Lett. **1985**, 1294 (1985)

5.140 T.J. Emge, H.H. Wang, M.A. Beno, P.C.W. Leung, M.A. Firestone, H.C. Jen-
kins, J.D. Cook, K.D. Carlson, J.M. Williams, E.L. Venturini, L.J. Azevedo,
J.E. Schirber: Inorg. Chem. **24**, 1736 (1985)

5.141 H. Schwenk, S.S.P. Parkin, V.Y. Lee, R.L. Greene: Phys. Rev. B **34**, 3156-3160
(1986)

5.142 C.-P. Heidmann, H. Veith, K. Andres, H. Fuchs, K. Polborn, E. Amberger: Solid
State Commun. **57**, 161-163 (1986)

5.143 G.R. Stewart, J.M. Williams, H.H. Wang, L.N. Hall, M.T. Perozzo, K.D. Carlson: Phys. Rev. B **34**, 6509-6510 (1986)

5.144 K. Andres, H. Schwenk, H. Veith: Physica B **143**, 334 (1986)

5.145 A. Nowack, U. Poppe, M. Weger, D. Schweitzer, H. Schwenk: Z. Phys. B **68**,
41-47 (1987)

5.146 M.E. Hawley, K.E. Gray, B.D. Terris, H.H. Wang, K.D. Carlson, J.M. Williams: Phys. Rev. Lett. **57**, 629-632 (1986)

5.147 H. Anzai, M. Tokumoto, K. Takahashi, T. Ishiguro: J. Cryst. Growth **91**, 225-228 (1988)

5.148 P.W. Anderson: J. Phys. Chem. Solids **11**, 26 (1959)

5.149 Y. Hasegawa, H. Fukuyama: J. Phys. Soc. Jpn. **55**, 3717 (1986)

5.150 L. Zuppiroli: In *Low-dimensional Conductors and Superconductors* ed. by D. Jérome, L.G. Caron (Plenum, New York 1987) p. 307

5.151 M. Tokumoto, I. Nishiyama, K. Murata, H. Anzai, T. Ishiguro, G. Saito: Physica B **143**, 372-374 (1986)

5.152 M.-Y. Choi, P.M. Chaikin, S.Z. Huang, P. Haen, E.M. Engler, R.L. Greene: Phys. Rev. B **25**, 6208-6217 (1982)

5.153 A.V. Zvarykina, P.A. Kononovich, V.N. Laukhin, V.N. Molchanov, S.I. Pesotskii, V.I. Simonov, R.P. Shibaeva, I.F. Shchegolev, E.B. Yagubskii: JETP Lett. **43**, 329-332 (1986)

5.154 G.O. Baram, L.L. Burabov, L.C. Degtariev, M.E. Kozlov, V.N. Laukhin, E.E. Laukhina, V.G. Orischenko, K.I. Pokhodnia, N.M.K. Scheinkmann, R.P. Shibaeva, E.B. Yagubskii: JETP Lett. **44**, 293 (1986)

5.155 D. Schweitzer, P. Bele, H. Brunner, E. Gogu, U. Haeberlen, I. Henning, I. Klutz, R. Sweitlik, H.J. Keller: Z. Phys. B **67**, 489-495 (1987)

5.156 N.V. Avramenko, A.V. Zvarykina, V.N. Laukhina, E.E. Laukhin, R.B. Lyubovskii, R.P. Shibaeva: JETP Lett. **48**, 472-476 (1988)

5.157 H. Urayama, H. Yamochi, G. Saito, K. Nozawa, T. Sugano, M. Kinoshita, S. Sato, K. Oshima, A. Kawamoto, J. Tanaka: Chem. Lett. **1988**, 55 (1988)

5.158 K. Murata, Y. Honda, H. Anzai, M. Tokumoto, K. Takahashi, N. Kinoshita, T. Ishiguro: Synth Metals **27**, A263-270 (1989)

5.159 J.E. Schirber, E.L. Venturini, A.M. Kini, H.H. Wang, J.R. Whiworth, J.M. Williams: Physica C **152**, 157 (1988)

5.160 H. Ito, T. Ishiguro, T. Komatsu, N. Matsukawa, G. Saito, H. Anzai: J. Superconductivity **7**, 667-669 (1994)

5.161 D.E. Fernel, C.J. Allen, R.C. Haddon, S.V. Chicherster: Phys. Rev. B **42**, 8694-8697 (1990)

5.162 H. Ito, M. Watanabe, Y. Nogami, T. Ishiguro, T. Komatsu, N. Hosoito, G. Saito: J. Phys. Soc. Jpn. **60**, 3230-3233 (1991)
H. Ito, Y. Nogami, T. Ishiguro, T. Komatsu, G. Saito, N. Hosoito: Jpn. J. Appl. Phys. **7**, 419 (1992)

5.163 W.K. Kwok, U. Welp, K.D. Carlson, G.W. Graebner, K.G. Vandervoot, H.H. Wang, A.M. Kini, J.M. Williams, D.L. Stupka, L.K. Montgomery, J.E. Thompson: Phys. Rev. B **42**, 8686 (1990)

5.164 M. Tinkham: Physica C **235-240**, 3-8 (1994)
R. Ikeda: Int'l J. Mod. Phys. B **10**, 601-634 (1996)

5.165 S. Ravy, J.P. Pouget, C. Lenoir, P. Batail: Solid State Commun. **73**, 37 (1990)

5.166 K. Oshima, R.C. Yu, P.M. Chaikin, H. Urayama, H. Yamochi, G. Saito: In *The Physics and Chemistry of Organic Superconductors*, ed. by G. Saito, S. Kagoshima, Springer Proc. Phys., Vol. 51 (Springer, Berlin, Heidelberg 1990) 276-279
N. Toyota, T. Sasaki, H. Sato, Y. Watanabe: Physica C **178**, 339 (1991)

5.167 F.L. Pratt, J. Caulfield, L. Cowey, J. Singleton, M. Doporto, W. Hayes, J.A.A. J. Petenboom, M. Kurmoo, P. Day: Synth. Met. **55-57**, 2289 (1993)

5.168 M. Tokumoto, H. Anzai, K. Takahashi, K. Murata, N. Kinoshita, T. Ishiguro: Synth. Met. **27**, A305-310 (1988)

5.169 H. Mayaffre, P. Wzietek, D. Jérome, C. Lenoir, D. Batail: Phys. Rev. Lett. **75**, 4122 (1995)

5.170 K. Oshima, H. Urayama, H. Yamochi, G. Saito: J. Phys. Soc. Jpn. **57**, 730-733 (1988)
K. Murata, Y. Honda, H. Anzai, M. Tokumoto, K. Takahashi, N. Kinoshita, T. Ishiguro, N. Toyota, T. Sakai, Y. Muto: Synth. Met. **27**, A341-346 (1988)

5.171 B. Andraka, J.S. Kim, G.R. Stewart, K.D. Carlson, H.H. Wang, J.M. Williams: Phys. Rev. B **40**, 11345-11347 (1989)

5.172 J.E. Graebner, R.C. Haddon, S.V. Chichester, S.H. Glarum: Phys. Rev. B **41**, 4808-4810 (1990)

5.173 T. Sugano, K. Nozawa, H. Hayashi, K. Kinoshita, K. Terui, T. Fukasawa, H. Takenouchi, S. Mino, H. Urayama, G. Saito, N. Kinoshita: Synth. Met. A **27**, 325-330 (1988)

5.174 J. Wosnitiza, X. Liu, D. Schweitzer, H.J. Keller: Phys. Rev. B **50**, 12747-12751 (1994)

5.175 Y. Maruyama, T. Inabe, H. Urayama, H. Yamochi, G. Saito: Sollid State Commun. **67**, 35 (1988)

5.176 M.E. Hawley, K.E. Gray, B.D. Terris, H.H. Wang, K.D. Carlson, J.M. Williams: Phys. Rev. Lett. **57**, 629 (1986)

5.177 K.E. Kornelsen, J.E. Eldridge, H.H. Wang, J.M. Williams: Solid State Commun. **76**, 1009-1013 (1990)

5.178 R. Zamboni, D. Schweitzer, H.J. Keller: Solid State Commun. **73**, 41-44 (1990)

5.179 M. Dressel, O. Klein, G. Grüner, K.D. Carlson, H.H. Wang, J.M. Williams: Phys. Rev. B **50**, 13603-13615 (1995)

5.180 D. Achkir, M. Poirier, C. Bourbonnais, G. Quirion, C. Lenoir, P. Batail, D. Jérome: Phys. Rev. B **47**, 11595 (1993)

5.181 D.R. Harshman, R.N. Kleinman, R.C. Haddon, S.V. Chichester-Hicks, M.L. Kaplan, L.W. Rupp, T. Pfis, D.L. Williams, D.B. Mitzi: Phys. Rev. Lett. **64**, 655 (1991)

5.182 L.P. Le, G.M. Luke, B.J. Sternlief, W.D. Wu, Y.J. Uemura, J.H. Brewer, T.M. Riseman, C.E. Stronach, G. Saito, H. Yamochi, H.H. Wang, A.M. Kini, K.D. Carlson, J.M. Williams: Phys. Rev. Lett. **68**, 1923 (1992)

5.183 T. Takahashi, K. Kanoda, G. Saito: Jpn. J. Appl. Phys. **7**, 414 (1992)
K. Kanoda, K. Akiba, K. Suzuki, T. Takahashi: Phys. Rev. Lett. **65**, 1271 (1990)

5.184 M. Lang, N. Toyota, T. Sasaki, H. Sato: Phys. Rev. Lett. **69**, 1443 (1992); Phys. Rev. B **46**, 5822 (1992)

5.185 A.F. Hebbard, T.A. Fiory, M.P. Siegal, J.M. Phillips, R.C. Haddon: Phys. Rev. B **44**, 9753 (1991)

5.186 T. Takahashi, T. Tokiwa, K. Kanoda, H. Urayama, H. Yamochi, G. Saito: Physica C **153-155**, 487-488 (1988)

5.187 H. Mayaffre, P. Wzietek, D. Jerome, S. Brazovskii: Phys. Rev. Lett. **76**, 4951 (1996)

5.188 V.D. Kuznetsov, V.V. Metlushko, L.A. Epanechnikov, E.F. Makarov, E.B. Yagubskii, N.D. Kushch: JETP Lett. **52**, 293-296 (1990)

5.189 A.C. Mota, G. Juri, P. Visani, A. Pollini, T. Teruzzi, K. Aupke, B. Hilti: Physica C **185-189**, 343-348 (1991)

5.190 Y. Kopelevich, A. Gupta, P. Esquinazi, C.P. Heidmann, H. Müller: Physica C **183**, 345-354 (1991)

5.191 T. Sato, T. Ishiguro, Y. Ueba: Synth. Met. **55-57**, 2839-2844 (1993)

5.192 M. Yoshizawa, Y. Nakamura, T. Sasaki, N. Toyota: Solid State Commun. **89**, 701-704 (1994)

5.193 H. Haneda, T. Ishiguro: Physica C **274**, 81 (1997)

5.194 W.E. Lawrence, S. Doniach: In *Proc. 12th Int'l Conf. on Low Temperature Physics*, Kyoto 1970, ed. by E. Kanda (Keigaku, Tokyo 1971) p.361

5.195 P.A. Mansky, P.M. Chaikin, R.C. Haddon: Phys. Rev. Lett. **70**, 1323-1326 (1993); Phys. Rev. B **50**, 15929-15944 (1994)

5.196 P. Müller: Physica C **235-240**, 289-292 (1994)

5.197 T. Ishiguro, H. Ito, Y. Nogami, Y. Ueba, H. Kusuhara: In *Organic Superconductivity*, ed. by V. Z. Kresin, W. A. Little (Plenum, New York 1990) pp.123-131
H. Kusuhara, Y. Sakata, Y. Ueba, K. Tada, K. Kaji, T. Ishiguro: Solid State Commun. **74**, 251 (1990)

5.198 C.E. Campos, J.S. Brooks, P.J.M. van Bentum, S.J. Klepper, P. Sandhu, M. Tokumoto, T. Kinoshita, N. Kinoshita, Y. Tanaka, H. Anzai: Physica B **211**, 293-296 (1995)

5.199 M. Kund, J. Lehrke, W. Biberacher, A. Lerf, K. Andres: Synth. Met. **70**, 949-950 (1995)

5.200 M. Lang, R. Modler, F. Steglich, N. Toyota, T. Sasaki: Physica B **194-196**, 2005 (1994)

5.201 S. Ullah, A.T. Dorsey: Phys. Rev. B **44**, 262 (1991)

5.202 Y. Nogami, J.P. Pouget, H. Ito, T. Ishiguro, G. Saito: Solid State Commun. **89**, 113 (1994

5.203 S.M. DeSoto, C.P. Slichter, A.M. Kini, H.H. Wang, U. Geizer, J.M. Williams: Phys. Rev. B **52**, 10364 (1995)

5.204 K. Kanoda, K. Miyagawa, A. Kawamoto, Y. Nakazawa: Phys. Rev. B **54**, 74 (1996)

5.205 A.J. Legett: Rev. Mod. Phys. **47**, 331 (1975)

5.206 H. Mayaffre, P. Wzietek, D. Jérome, C. Lenoir, P. Batail: Europhys. Lett. **28**, 205 (1994)

5.207 A. Kawamoto, K. Miyagawa, Y. Nakazawa, K. Kanoda: Phys. Rev. Lett. **74**, 3455 (1995)

5.208 V. Kataev, G. Winkel, D. Khomskii, D. Wohlleben, W. Crump, K.F. Tebe, J. Hahn: Solid State Commun. **83**, 435 (1992)

5.209 M. Dressel, G. Grüner: Mol. Cryst. Liq. Cryst. **284**, 107 (1996)

5.210 H. Ito: J. Phys. Soc. Jpn. **64**, 3018 (1995)

5.211 M.W. Coffey, J.R. Clem: Phys. Rev. B **45**, 9872 (1992)

5.212 G. Blatter, M.V. Feigel'man, V.B. Geshkenbein, A.I. Larkin, V.M. Vinokur: Rev. Mod. Phys. **66**, 1124 (1994)

5.213 Yu.V. Sushko, H. Ito, T. Ishiguro, G. Saito: Solid State Commun. **87**, 997 (1993)

5.214 H. Posselt, K. Andres, G. Saito: Physica B **204**, 159 (1995) and references therein

5.215 Yu.V. Sushko, H. Ito, T. Ishiguro, G. Saito: J. Phys. Jpn. **62**, 3372 (1993)

5.216 J.E. Schirber, D.L. Overmyer, K.D. Carlson, J.M. Wiliams, A.M. Kini, H.H.

Wang, H.A. Charlier, B.J. Love, D.M. Watkins, G.A. Yaconi: Phys. Rev. B **44**, 4666 (1991)

5.217   U. Geiser, A.J. Schultz, H.H. Wang, D.M. Watkins, D.L. Stupka, J.M. Williams, J.E. Schirber, D.L. Overmyer, D. Jung, J.J. Novoa, M.H. Whangbo: Physica C **174**, 475 (1991)

5.218   V.A. Bondarenko, R.A. Petrasiov, M.A. Tanatar, V.S. Yefanov, N.D. Kushch: Physica C **235-240**, 2467 (1994)

5.219   V.A. Bondarenko, Yu.V. Sushko, V.I. Barchuk, V.S. Yafenov, V.V. Dyakin, M.A. Tanatar, N.D. Kushch, E.B. Yagubskii: Synth. Met. **55-57**, 2386 (1993)

5.220   N.D. Kushch, L.I. Buravov, A.G. Kohmenko, E.B. Yagubskii, L.P. Rosenberg, R.P. Shibaeva: Synth. Met. **53**, 155 (1993)

5.221   N.D. Kushch, L.I. Buravov, A.G. Khomenko, S.I. Pesotskii, V.N. Laukhin, E.B. Yagubskii, R.P. Shibaeva, V.E. Zavodnik, L.P. Rosenberg: Synth. Met. **72**, 181 (1995)

5.222   S.S.P. Parkin, E.M. Engler, R.R. Schumaker, R. Lagier, V.Y. Lee, J.C. Scott, R.L. Greene: Phys. Rev. Lett. **50**, 270-273 (1983)

5.223   K. Carneiro, J.C. Scott, E.M. Engler: Solid State Commun. **50**, 477-481 (1984)

5.224   S. Ravy, R. Moret, J.P. Pouget, R. Comes, S.S.P. Parkin: Phys. Rev. B **33**, 2048-2051 (1986)

5.225   R.N. Lyubovskaya, R.B. Lubovskii, R.P. Shibaeva, M.Z. Aidoshina, L.M. Gol′denberg, L.P. Rozenberg, M.L. Khidekel, Yu.F. Shul′pyakov: JETP Lett. **42**, 468-472 (1985)

5.226   R.N. Lyubovskaya, E.I. Zhilyaeva, S.I. Pesotskii, R.B. Lyubuskii, L.O. Atovmyan, O.A. D′yachenko, T.G. Takhirov: JETP Lett. **46**, 188 (1987)

5.227   R.N. Lyubovskaya, R.B. Lyubuskii, M.K. Makoba, S.I. Pesotskii: JETP Lett. **51**, 361 (1990)

5.228   J.E. Schirber, D.L. Overmyer, E.I. Venturini, H.H. Wang, K.D. Carlson, W.K. Kwok, S. Kleinjan, J.M. Williams: Physica C **161**, 412 (1989)

5.229   J.A. Schlueter, U. Geiser, H.H. Wang, M.E. Kelly, J.D. Dudek, J.M. Williams, D. Naumann, T. Roy: Mol. Cryst. Liq. Cryst. **284**, 195-202 (1996)

5.230   J.A. Schlueter, K.D. Carlson, J.M. Williams, U. Geiser, H.H. Wang, U. Welp, W.-K. Kwok, J.A. Fendrich, J.D. Dudek, C.A. Achenbach, P.M. Keane, A.S. Komosa, D. Naumann, T. Roy, J.E. Schivber, W.R. Bayless: Physica C **230**, 378-384 (1994)

5.231   U. Geiser, J.A. Schlueter, K.D. Carlson, J.M. Williams, H.H. Wang, W.-K. Kwok, U. Welp, J.A. Fendrich, J.D. Dudek, C.D. Dudek, C.A. Achenbach, A.S. Komsa, P.M. Keane, D. Naumann, T. Roy, J.E. Schirber, W.R. Bayless, J. Ren, M.-H. Whangbo: Synth. Met. **70**, 1105 (1995)

5.232   H. Taniguchi, H. Sato, Y. Nakazawa, K. Kanoda: Phys. Rev. B **53**, R8879 (1996)

5.233   Y. Nakazawa, A. Kawamoto, K. Kanoda: Phys. Rev. B **52**, 12890 (1995)

5.234   C.E. Campos, J.S. Brooks, P.J.M. van Bentum, J.A.A.J. Perenboom, S.J. Klepper, P.S. Sandhu, S. Valfells, Y. Tanaka, T. Kinoshita, N. Kinoshita, M. Tokumoto, H. Anzai: Phys. Rev. B **52**, R7014 (1995)

5.235   T. Hirayama, K. Manabe, H. Akimoto, H. Ishimoto, T. Ishiguro, H. Mori and G. Saito: Czech. J. Phys. **46**, Suppl.2, 813 (1996)

5.236   H. Sato, H. Taniguchi Y. Nakazawa, A. Kawamoto, K. Kato, K. Kanoda: Synth. Met. **70**, 915 (1996)

5.237 T. Mori, H. Inokuchi: Chem. Lett. **1987**, 1657-1660 (1987); Solid State Commun. **64**, 335-337 (1987)

5.238 H. Mori, I. Hirabayashi, S. Tanaka, T. Mori, Y. Maruyama, H. Inokuchi: Solid State Commun. **80**, 411 (1991)

5.239 T. Mori, K. Kato, Y. Maruyama, H. Inokuchi, H. Mori, I. Hirabayashi, S. Tanaka: Solid State Commun. **82**, 177 (1992)

5.240 A. Kobayashi, R. Kato, T. Naito, H. Kobayashi: Synth. Met. **56**, 2078 (1993)

5.241 L.K. Montogomery, T. Burgin, C. Husting, L. Tilley, J.C. Huffmann, K.D. Carlson, J.D. Dudek, G. A. Yaconi, U. Geiser, J.M. Williams: Mol. Cryst. Liq. Cryst. **211**, 283 (1992)

5.242 H. Kobayashi, T. Udagawa, H. Tomita, K. Bun, T. Naito, A. Kobayashi: Chem. Lett. **1993**, 1559 (1993)

5.243 A. Kobayashi, T. Udagawa, H. Tomita, T. Naito, H. Kobayashi: Chem. Lett. **1993**, 2179 (1993)

5.244 F. Goze, V.N. Laukhin, L. Brossard, A. Audouard, J.P. Ulmet, S. Askenazy, T. Naito, H. Kobayashi, A. Kobayashi, M. Tokumoto, P. Cassoux: Europhys. Lett. **28**, 427-431 (1994)

5.245 H. Kobayashi, H. Tomita, T. Naito, A. Kobayashi: Synth. Met. **70**, 867 (1995)

5.246 H. Yamochi, S. Horiuchi, G. Saito, M. Kusunoki, K. Sakaguchi, T. Kikuchi, S. Sato: Synth. Met. **55-57**, 2096-2101 (1993)
       S. Horiuchi, H. Yamochi, G. Saito, K. Sakaguchi, M. Kusunoki: J. Am. Chem. Soc. **118**, 8604-8622 (1996)

5.247 M.A. Beno, H.H. Wang, A.M. Kini, K.D. Carlson, U. Geiser, W.K. Kwok, J.E. Thompson, J.M. Williams, J. Ren, M.-H. Whangbo: Iorg. Chem. **29**, 1599 (1990)

5.248 S. Kalich, D. Schweitzer, I. Heinen, En Lan Song, B. Nuber, H.J. Keller, K. Winzer, H.W. Helberg: Solid State Commun. **80**, 191 (1991)

5.249 L.I. Buravov, A.K. Kohmenko, N.D.Kushch, V.N. Laukhin, A.I. Schegolev, E.B. Yagubskii, L.P. Rozenberg, R.P. Shibaeva: J. Physique I 2, 529 (1992)

Chapter 6

6.1   K. Kikuchi, I. Ikemoto, K. Kobayashi: Synth. Met. **19**, 551 (1983)

6.2   K. Murata, K. Kikuchi, T. Takahashi, K. Kobayashi, Y. Honda, K. Saito, K. Kanoda, T. Tokiwa, H. Anzai, T. Ishiguro: J. Mol. Elect. **4**, 173-179 (1988)

6.3   M.Z. Aldoshina, L.O. Atovmyam, L.M. Goldenberg, O.N. Krasochka, R.N. Lyubovskaya, M.L. Khidekel: Dokl. Nauk SSR **289**, 860 (1986). (The authors used DMEDT-DSDTF for DMET)

6.4   K. Kanoda, T. Takahashi, K. Kikuchi, K. Saito, I. Ikemoto, K. Kobayashi: Synth. Met. B **27**, 385-390 (1988)

6.5   K. Kikuchi, K. Murata, Y. Honda, T. Namiki, K. Saito, H. Anzai, K. Kobayashi, T. Ishiguro, I. Ikemoto: J. Phys. Soc. Jpn. **55**, 4241-4244 (1987)

6.6   K. Kikuchi, K. Murata, Y. Honda, T. Namiki, K. Saito, T. Ishiguro, K. Kobayashi, I. Ikemoto: J. Phys. Soc. Jpn. **55**, 3435-3439 (1987)

6.7   K. Kikuchi, Y. Honda, Y. Ishikawa, K. Saito, I. Ikemoto, K. Murata, H. Anzai, T. Ishiguro: Solid State Commun. **66**, 405-408 (1988)

6.8 K. Kanoda, K. Kato, A. Kawamoto, K. Oshima, T. Takahashi, K. Kikuchi, K. Saito, I. Ikemoto: Synth. Met. **55-57**, 2309 (1993)

6.9 K. Yamaji: Solid State Commun. **61**, 413-417 (1987) and Synth. Met. A **27**, 115-119 (1988)

6.10 K. Kikuchi, K. Murata, Y. Honda, T. Namiki, K. Saito, K. Kobayashi, T. Ishiguro, I. Ikemoto: J. Phys. Soc. Jpn. **55**, 2527-2528 (1987)

6.11 G.C. Papavassiliou, G.A. Mousdis, J.S. Zambounis, A. Terzis, A. Hountas, B. Hilti, C.W. Mayer, J. Pfeiffer: Synth. Met. B **27**, 379-383 (1988)

6.12 A.M. Kini, M.A. Beno, D. Son, H.H. Wang, K.D. Carlson, L.C. Potter, U. Welp, B.A. Vogt, J.M. Williams: Solid State Commun. **69**, 503-507 (1989)

6.13 P. Delhaes, J. Amiell, S. Flandrois, L. Ducasse, A. Fritsch, B. Hilti, C.W. Mayer, J. Zambounis, G.C. Papavassiliou: J. Physique **51**, 1179 (1990)

6.14 K. Kanoda, K. Kato, Y. Kobayashi, M. Kato, T. Takahashi, K. Oshima, B. Hilti, J. Zambounis: Synth. Met. **55-57**, 2871 (1993)

6.15 K. Kanoda, Y. Tsubokura, K. Ikeda, T. Takahashi, N. Matsukawa, G. Saito, H. Mori, T. Mori, B. Hilti, J.S. Zambounis: Synth. Met. **55-57**, 2865 (1993)

6.16 J.S. Zambounis, C.W. Mayer, K. Hauenstein, B. Hilti, W. Hofherr, J. Pfeiffer, M. Burkle, G. Rihs: Adv. Mater. **4**, 33 (1992)

6.17 T. Ichimiya, K. Yamada: J. Phys. Soc. Jpn. **65**, 1764 (1996)

6.18 Y. Kobayashi, T. Nakamura, T. Takahashi, K. Kanoda, B. Hilti, J.S. Zambounis: Synth. Met. **70**, 871 (1995)

6.19 Y. Misaki, N. Higuchi, H. Fujiwara, T. Yamabe, T. Mori, H. Mori, S. Tanaka: Angew. Chem. Int'l Edn. **34**, 1222 (1995)

Chapter 7

7.1 L. Brossard, M. Ribault, L. Valade, P. Cassoux: Physica B **143**, 378-380 (1986)

7.2 L. Brossard, M. Ribault, L. Valade, P. Cassoux: Phys. Rev. B **42**, 3935-3943 (1990)

7.3 M. Bousseau, L. Valade, J.P. Legros, P. Cassoux, M. Garbauskas, L.V. Interrante: J. Am. Chem. Soc. **108**, 1908-1916 (1986)

7.4 A. Kobayashi, H. Kim, Y. Sasaki, R. Kato, H. Kobayashi: Solid State Commun. **62**, 57-64 (1987)

7.5 E. Canadell, I.E.-I. Rachidi, S. Eavy, J. P. Pouget, L. Brossard, J. P. Legros: J. Physique **50**, 2967-2981 (1989)

7.6 S. Raby, J.P. Pouget, L. Valade, J.P. Legros: Europhys. Lett. **9**, 391-396 (1989)

7.7 L. Brossard, H. Hurdequint, M. Ribault, L. Valade, J.P. Legros, P. Cassoux: Synth. Met. B **27**, 157-162 (1988)

7.8 L. Brossard, M. Ribault, L. Valade, P. Cassoux: J. Physique **50**, 1521-1534 (1984)

7.9 L. Brossard, E. Canadell, L. Valade, P. Cassoux: Synth. Met. **70**, 1045 (1995)

7.10 L. Brossard, E. Canadell, L. Valade, P. Cassoux: Phys. Rev. B **47**, 1647-1650 (1993)

7.11 C. Bourbonnais, P. Wzietek, D. Jérome, F. Creuzet, L. Valade, P. Cassoux: Europhys. Lett. **6**, 177 (1988)

7.12 A. Vainrub, E. Canadell, D. Jérome, P. Bernier, T. Nunes, M.F. Bruniquel, P. Cassoux: J. Physique **51**, 2465-2476 (1990)

7.13 K. Yamaji, S. Abe: J. Phys. Soc. Jpn. **5**, 4237-4240 (1987)

7.14 R. Ramakumar, K. Yamaji: Phys. Rev. B **56**, 795 (1997

7.15 H.L. Liu, D.B. Tanner, A.E. Pullen, K.A. Abboud, J.R. Reynolds: Phys. Rev. B **53**, 10557-10568 (1996)

7.16 H. Kim, A. Kobayashi, Y. Sasaki, R. Kato, H. Kobayashi: Chem. Lett. **1987**, 1799-1802 (1987)
A. Kobayashi, H. Kim, Y. Sasaki, R. Kato, H. Kobayashi, S. Moriyama, Y. Nishio, K. Kajita, W. Sasaki: Chem. Lett. **1987**, 1819-1822 (1987)

7.17 K. Kajita, Y. Nishio, S. Moriyama, R. Kato, H. Kobayashi, W. Sasaki: Solid State Commun. **65**, 361-363 (1988)

7.18 E. Canadell, S. Raby, J.P. Pouget, L. Brossard: Solid State Commun. **75**, 633-638 (1990)

7.19 A. Kobayashi, H. Kobayashi, A. Miyamoto, R. Kato, R.A. Clark, A.E. Underhill: Chem. Lett. **1991**, 2163 (1991)

7.20 H. Kobayashi, K. Bun, T. Naito, R. Kato, A. Kobayashi: Chem. Lett. **1992**, 1909 (1992)

7.21 A. Kobayashi, A. Sato, K. Kawano, T. Naito, H. Kobayashi, T. Watanabe: J. Mater. Chem. **5**, 1671 (1995)

7.22 M. Inokuchi, H. Tajima, T. Ohta, H. Kuroda, A. Kobayashi, A. Sato, T. Naito, H. Kobayashi: J. Phys. Soc. Jpn. **65**, 538-544 (1996)

7.23 H. Tajima, S. Ikeda, A. Kobayashi, H. Kuroda, R. Kato, H. Kobayashi: Solid State Commun. **86**, 7 (1993)

7.24 H. Tajima, S. Ikeda, M. Inokuchi, T. Ohta, A. Kobayashi, T. Sasaki, N. Toyota, R. Kato, H. Kobayashi, H. Kuroda: Synth. Met. **70**, 1051-1052 (1995)

Chapter 8

8.1 J. Bardeen, L.N. Cooper, J.R. Schrieffer: Phys. Rev. **108**, 1175-1204 (1957)

8.2 H. Fröhlich: Phys. Rev. **79**, 845-856 (1950)

8.3 E. Maxwell: Phys. Rev. **78**, 477 (1950)

8.4 C.A. Reynolds, B. Serin, W.H. Wright, L.B. Nesbitt: Phys. Rev. **78**, 487 (1950)

8.5 C. Kittel: *Quantum Theory of Solids* (Wiley, New York 1963) Chap. 7

8.6 O. Madelung: *Introduction to Solid-State Theory*, Springer Ser. Solid-State Sci., Vol. 2 (Springer, Berlin, Heidelberg 1981) p. 182

8.7 J. Bardeen, L.N. Cooper, J.R. Schrieffer: Phys. Rev. **106**, 162-164 (1957)

8.8 B.D. Josephson: Phys. Lett **1**, 251-253 (1962); in *Superconductivity* ed. by R.D. Parks (Dekker, New York 1969) Chap. 9

8.9 G. Rickaysen: in *Superconductivity* ed by. R.D. Parks (Dekker, New York 1969) Chap. 2

8.10 J.R. Schrieffer: *The Theory of Superconductivity* (Benjamin, Reading, MA 1964) Chap. 8

8.11 M. Tinkham: *Introduction to Superconductivity* (McGraw-Hill, New York 1996) Chap. 3

8.12 P. Morel, P.W. Anderson: Phys. Rev. **129**, 1263-1271 (1962)

8.13 L.N. Cooper: Phys. Rev. **104**, 1189-1190 (1956)

8.14 A.B. Pippard: Proc. Roy. Soc. (London) A **216**, 547-568 (1953)

8.15   J.C. Wheatley: Rev. Mod. Phys. **47**, 415-470 (1975)

8.16   E. Dagotto: Rev. Mod. Phys. **66**, 763-840 (1994)

8.17   F. Steglich: In *Theory of Heavy Fermions and Valence Fluctuations* ed. by T. Kasuya, T. Saso, Springer Ser. Solid-State Sci., Vol.62 (Springer, Berlin, Heidelberg 1985) pp.23-44

8.18   H.R. Ott: Progr. Low Temp. Phys., Vol.XI, ed. by D.F. Brewer (North-Holland, Amsterdam 1987) Chap.5

8.19   A.M. Clogston: Phys. Rev. Lett. **9**, 266-267 (1962)

8.20   B.S. Chandrasekhar: Appl. Phys. Lett. **1**, 7-8 (1962)

8.21   W.A. Little: Phys. Rev. **134**, A1416-1424 (1964)

8.22   F. London: *Superfluids*, Vol.1 (Wiley, New York 1950)

8.23   W. Little: J. Physique **44**, C3-819-825 (1983)

8.24   V. Ginzburg: Sov. Phys. JETP **20**, 1549-1550 (1965); Contemp. Phys. **9**, 355-374 (1968)

8.25   R.A. Ferrell: Phys. Rev. Lett. **13**, 330-332 (1964)

8.26   T.M. Rice: Phys. Rev. **140**, A1889-1891 (1965)

8.27   V.J. Emery: Synth. Met. **13**, 21-27 (1986)

8.28   D.J. Scalapino, E. Loh, Jr., J.E. Hirsch: Phys. Rev. B **35**, 6694-6698 (1987)

8.29   M.T. Béal-Monod, C. Bourbonnais, V.J. Emery: Phys. Rev. B **34**, 7716-7720 (1986)

8.30   W. Kohn, J.M. Luttinger: Phys. Rev. Lett. **15**, 524-526 (1965)

8.31   J. Friedel: Adv. Phys. **3**, 446-507 (1954)

8.32   K. Miyake, S. Schmidt-Rink, C.M. Varma: Phys. Rev. B **34**, 6554-6556 (1986)

8.33   C. Bourbonnais, L.G. Caron: Europhys. Lett. **5**, 209-215 (1988)

8.34   H. Shimahara, S. Takada: J. Phys. Soc. Jpn. **57**, 1044-1055 (1988)

8.35   H. Shimahara: J. Phys. Soc. Jpn. **58**, 1735-1747 (1989)

8.36   J.E. Hirsch: Phys. Rev. Lett. **54**, 1317-1320 (1985)

8.37   J.E. Hirsch, E. Loh, D.J. Scalapino, S. Tang: Physica C **153**, 549-554 (1988)

8.38   S.R. White, D.J. Scalapino, R.L. Sugar, N. Bickers, R. Scalettar: Phys. Rev. B **39**, 839-842 (1989)

8.39   D.J. Scalapino: In *High Temperature Superconductivity - the Los Alamos Symposium* (1989), ed. by K.S. Bedell, D. Coffey, D.E. Deltzer, D. Pines, J.R. Schrieffer (Addison-Wesley, Redwood City 1990) pp.314-372

8.40   E. Dagotto, J. Riera: Phys. Rev. Lett. **70**, 682-685 (1993)

8.41   Y. Ohta, T. Shimozato, R. Eder, S. Maekawa: Phys. Rev. Lett. **73**, 324-327 (1994)

8.42   C. Gros, R. Joynt, T.M. Rice: Phys. Rev. B **36**, 381-393 (1987)

8.43   Q.P. Li, B.E.C. Koltenbah, R. Joynt: Phys. Rev. B **48**, 437-455 (1993)

8.44   T. Husslein, I. Morgenstern, D.M. Newns, P.C. Pattnaik, J.M. Singer, H.G. Matuttis: Phys. Rev. B **54**, 16179-16182 (1996)

8.45   K. Kuroki, H. Aoki, T. Hotta, Y. Takada: Phys. Rev. B **55**, 2764-2767 (1997)

8.46   T. Giamarchi, C. Lhuillier: Phys. Rev. B **43**, 12943-12951 (1991)

8.47   T. Nakanishi, K. Yamaji, T. Yanagisawa: J. Phys. Soc. Jpn. **66**, 294-297 (1997)

8.48   M. Takigawa, H. Yasuoka, G. Saito: J. Phys. Soc. Jpn. **56**, 873-876 (1987)

8.49   Y. Hasegawa, H. Fukuyama: J. Phys. Soc. Jpn. **56**, 877-880 (1987)

8.50   T. Takahashi, T. Tokiwa, K. Kanoda, H. Urayama, H. Yamochi, G. Saito: Physica C **153-155**, 487-488 (1988)

8.51 N. Mayaffre, P. Wzietek, D. Jérome, L. Lenoir, P. Batail: Phys. Rev. Lett. **75**, 4122-4125 (1995)

8.52 S.M. DeSoto, C.P. Slichter, A.M. Kini, H.H. Wang, U. Geiser, J.M. Williams: Phys. Rev. B **52**, 10364-10368 (1995)

8.53 K. Kanoda, K. Miyagawa, A. Kawamoto, Y. Nakazawa: Phys. Rev. B **54**, 76-79 (1996)

8.54 T. Takahashi, Y. Kobayashi, T. Nakamura, K. Kanoda, B. Hilti, J.S. Zambounis: Physica C **235-240**, 2461-2462 (1994)

8.55 D.R. Harshman, R.N. Kleiman, R.C. Haddon, S.V. Chichester-Hicks, M.L. Kaplan, L.W. Lupp, Jr., T. Pfiz, D.L. Williams, D.B. Mitzi: Phys. Rev. Lett. **64**, 1293-1296 (1990)

8.56 K. Holczer, D. Quinlivan, O. Klein, G. Grüner, F. Wudl: Solid State Commun. **76**, 499-501 (1990)

8.57 K. Kanoda, K. Akiba, K. Suzuki, T. Takahashi, G. Saito: Phys. Rev. Lett. **65**, 1271-1274 (1990)

8.58 L.P. Le, G.M. Luke, B.J. Sternlib, W.D. Wu, Y.J. Uemura, J.H. Brewer, T.M. Riseman, C.E. Stronach, G. Saito, H. Yamochi, H.H. Wang, A.M. Kini, K.D. Carlson, J.M. Williams: Phys. Rev. Lett. **68**, 1923-1926 (1992)

8.59 J.P. Pouget: Chemica Scripta **17**, 85-91 (1981)

8.60 J.P. Pouget, R. Moret, R. Comès, K. Bechgaard, J.M. Fabre, L. Giral: Mol. Cryst. Liq. Cryst. **79**, 129-143 (1982)

8.61 M. Weger, M. Caveh, H. Gutfreund: Solid State Commun. **37**, 421-423 (1981)

8.62 A.J. Berlinsky, J.F. Carolan, L. Weiler: Solid State Commun. **15**, 795-801 (1974)

8.63 R. Comès, G. Shirane: in *Highly Conducting One-Dimensional Solids*, ed. by J.T. Devreese, R.P. Evrard, V.E. van Doren (Plenum, New York 1979) pp. 17-67

8.64 N.O. Lipari, M.J. Rice, C.B. Duke, R. Bozio, A. Girlando, C. Pecile: Int'l J. Quant. Chem. Symp. **11**, 583-594 (1977)

8.65 M.J. Rice, N.O. Lipari: Phys. Rev. Lett. **38**, 437-439 (1977)

8.66 R. Bozio, A. Girlando, C. Pezile: Chem. Phys. Lett. **52**, 503-508 (1977)

8.67 J.E. Hirsch: Phys. Rev. Lett. **51**, 296-299 (1983)

8.68 D. Baeriswyl, K. Maki: Phys. Rev. B **31**, 6633-6642 (1985)

8.69 S. Mazumdar, S.D. Dixit: Phys. Rev. Lett. **51**, 292-295 (1983)

8.70 T. Izuyama, M. Saitoh: Solid State Commun. **16**, 549-552 (1975)

8.71 K. Yamaji: Solid State Commun. **61**, 413-417 (1987)

8.72 A. Girlando, F. Marzola, C. Pecile, J.B. Torrance: J. Chem. Phys. **79**, 1075-1085 (1983)

8.73 A. Painelli, A. Girlando, C. Pecile: Solid State Commun. **52**, 801-806 (1984)

8.74 T. Mori, A. Kobayashi, Y. Sasaki, H. Kobayashi, G. Saito, H. Inokuchi: Chem. Lett. **1984**, 957-960 (1984)

8.75 T. Mori, A. Kobayashi, Y. Sasaki, H. Kobayashi: Chem. Lett. **1982**, 1923-1926 (1982)

8.76 K. Kikuchi: Private commun. (1988)

8.77 M.-H. Whangbo: Private commun. (1989)

8.78 L. Ducasse: Private commun. (1990)

8.79 L. Pietronero, S. Strässler: Europhys. Lett. **18**, 627-633 (1992)

8.80 M. Meneghetti, R. Bozio, C. Pecile: J. Physique **49**, 1377-1387 (1986)

8.81   T. Takahashi, D. Jérome, F. Masin, J.F. Fabre, L. Giral: J. Phys. C **17**, 3777-3792 (1984)

8.82   T. Mori, A. Kobayashi, Y. Sasaki, H. Kobayashi, G. Saito, H. Inokuchi: Chem. Lett. **1984**, 957-960 (1984)

8.83   C.S. Jacobsen, D.B. Tanner, J.M. Williams, H.H. Wang: Synth. Met. **19**, 125-130 (1987)

8.84   H. Tajima, H. Kanbara, K. Yakushi, H. Kuroda: Solid State Commun. **57**, 911-914 (1986)

8.85   M.J. Rice: Phys. Rev. Lett. **37**, 36-39 (1976)

8.86   M.J. Rice, L. Pietronero, P. Brüesch: Solid State Commun. **21**, 757-760 (1977)

8.87   N. Sato, G. Saito, H. Inokuchi: Chem. Phys. **76**, 79-88 (1983)

8.88   S.S. Shaik, M.-H. Whangbo: Inorg. Chem. **25**, 1201-1209 (1983)

8.89   C.B. Duke, N.O. Lipari, L. Pietronero: Chem. Phys. Lett. **30**, 415-420 (1975)

8.90   M.E. Kozlov, K.I. Pokhodnia, A.A. Yurchenko: Spectrochimica Acta A **45**, 437-444 (1989)

8.91   J.C.R. Faulhaber, D.Y.K. Ko, P.R. Briddon: Synth. Met. **60**, 227-232 (1993); g_i in this reference is larger by $\sqrt{2}$ than in the present book by definition

8.92   J. Shumway, S. Chattopadhyay, S. Satpathy: Phys. Rev. B **53**, 6677-6681 (1996); the values of the coupling constants are estimated using the observed values of the normal mode frequencies

8.93   S. Sugai, H. Mori, H. Yamochi, G. Saito: Phys. Rev. B **47**, 14374-14379 (1993)

8.94   G.V. Kamarchuk, A.V. Khotkevich, M.E. Kozlov, K.I. Pokhodnia: Synth. Met. **55-57**, 2933-2938 (1993)

8.95   V.M. Yartsev: Private commun. (1996)

8.96   J. Kübler, M. Weger, C.B. Sommers: Solid State Commun. **62**, 801-805 (1987)

8.97   M.V. Kartsovnik, P.A. Kononovich, V.N. Lauhin, I.F. Shchegolev: JETP Lett. **48**, 541-544 (1988)

8.98   W. Kang, G. Montambaux, J.R. Cooper, D. Jérome, P. Batail, C. Lenoir: Phys. Rev. Lett. **62**, 2559-2562 (1989)

8.99   J. Kübler, C.B. Sommers: In *The Physics and Chemistry of Organic Superconductors*, ed. by G. Saito, S. Kagoshima, Springer Proc. Phys., Vol. 51 (Springer, Berlin, Heidelberg 1990) pp. 208-211

8.100   K. Murata, N. Toyota, Y. Honda, T. Sasaki, M. Tokumoto, H. Bando. H. Anzai, Y. Muto, T. Ishiguro: J. Phys. Soc. Jpn. **57**, 1540-1543 (1988)

8.101   N. Toyota, T. Sasaki, K. Murata, Y. Honda, M. Tokumoto, H. Bando, H. Anzai, T. Ishiguro, Y. Muto: J. Phys. Soc. Jpn. **57**, 2616-2619 (1988)

8.102   I.D. Parker, D.D. Pigram, R.H. Friend, M. Kurmoo, P. Day: Synth. Met. **27**, A387-392 (1988)

8.103   K. Yamaji: J. Phys. Soc. Jpn. **52**, 1361-1372 (1983)

8.104   A. Ugawa, G. Ojima, K. Yakushi, H. Kuroda: Phys. Rev. B **38**, 5122-5125 (1988)

8.105   K. Nozawa, T. Sugano, H. Urayama, H. Yamochi, G. Saito, M. Kinoshita: Chem. Lett. **1988**, 617-620 (1988)

8.106   T. Mori, H. Inokuchi: Solid State Commun. **64**, 335-337 (1987)

8.107   K. Oshima, T. Mori, H. Inokuchi, H. Urayama, H. Yamochi, G. Saito: Phys. Rev. B **38**, 938-941 (1988)

8.108   M. Yoshimura, H. Shigekawa, H. Nejoh, G. Saito, Y. Saito, A. Kawazu: Phys. Rev. B **43**, 13590 - 13503 (1991)

8.109 D.R. Talham, M. Kurmoo, P. Day, D.S. Obertelli, I.D. Parker, R.H. Friend: J. Phys. C **19**, L383-L388 (1986)

8.110 V.A. Merzhanov, E.E. Kostyuchenko, O.E. Faber, I.F. Shchegolev, E.B. Yagubskii: Sov. Phys. JETP **62**, 165-167 (1985)

8.111 B. Rothaemel, L. Forro, J.R. Cooper, J.S. Schilling, M. Weger, P. Bele, H. Brunner, D. Schweitzer, H.J. Keller: Phys. Rev. B **34**, 704-712 (1986)

8.112 Y.-N. Xu, W.Y. Ching, Y.C. Jean, Y. Lou: Phys. Rev. B **52**, 12946-12950 (1995)

8.113 J. Caulfield, W. Lubczynski, F. L. Pratt, J. Singleton, D. Y. K. Ko, W. Hayes, M. Kurmoo, P. Day: J. Phys. Condensed Matter **6**, 2911-2924 (1994)

8.114 B. Andraka, G. B. Stewart, K. D. Karlson, H. H. Wang, M. D. Vashon, J. M. Williams: Phys. Rev. B **42**, 9963-9966 (1990)

8.115 Y. Nakazawa, K. Kanoda: Phys. Rev. B **53**, R8875-R8878 (1996)

8.116 M. Tokumoto, K. Murata, H. Bando, H. Anzai, K. Kajimura, T. Ishiguro: Physica B **143**, 338-342 (1986)

8.117 H. Kobayashi, R. Kato, A. Kobayashi, T. Mori, H. Inokuchi: Solid State Commun. **60**, 473-480 (1986). According to private communication, the scale for angles must be corrected by multiplying by the factor of 10^{-1} .

8.118 G. Saito: Chemistry and Industry (in Japanese) **41**, 649-654 (1988)

8.119 G. Saito, H. Urayama, H. Yamochi, K. Oshima: Synth. Met. **27**, A331-340 (1988)

8.120 A.M. Kini, U. Geiser, H.H. Wang, K.D. Carlson, J.M. Williams, W.K. Kwok, K.G. Vandervoort, J.E. Thompson, D.L. Stupka, D. Jung, M.-H. Whangbo: Inorg. Chem. **29**, 2555-2557 (1990)

8.121 J.M. Williams, A.M. Kini, H.H. Wang, K.D. Carlson, U. Geiser, L.K. Montgomery, G.J. Pyrka, D.M. Watkins, J.M. Kommrs, S.J. Boryschuk, A.V. Strieby-Crouch, W.K. Kwok, J.E. Schirber, D.L. Overmyer, D. Jung, M.-H. Whangbo: Inorg. Chem. **29**, 3272-3274 (1990)

8.122 U. Welp, S. Fleshler, W.K. Kwok, G.W. Crabtree, K.D. Carlson, H.H. Wang, U. Geiser, J.M. Williams, V.M. Hitsman: Phys. Rev. Lett. **69**, 840-843 (1992)

8.123 Yu. V. Sushko, V.A. Bondarenko, R.A. Petrosov, N.D. Kushch, E.B. Yagubskii: Physic C **185-189**, 2683-2684 (1991)

8.124 K. Kanoda: Kotai-Butsuri (Solid Sate Physics, in Japanese) **30**, 240-254 (1995)

8.125 T. Takahashi: In *Research Reports on New Superconducting Materials and High Temperature Oxide Superconductors (1984-1987)*, ed. by S. Nakajima, H. Fukuyama (1988) pp. 175-180

8.126 P. Auban-Senzier, C. Bourbonnais, D. J rome, C. Lenior, P. Batail, E. Canadell, J.P. Buisson, S. Lefrant: J. Phys. I (France) **3**, 871-885 (1993)

8.127 A.M. Kini, J.D. Dudek, K.D. Carlson, U. Geiser, R.A. Klemm, J.M. Williams, K.R. Lykke, J.A. Schlueter, H.H. Wang, P. Wurz, J.R. Ferraro, G.A. Yaconi: P hysica C **204**, 399-405 (1993)

8.128 K.D. Carlson, J.M. Williams, U. Geiser, A. Kini, H.H. Wang, R.A. Klemm, S.K. Kumar, J.A. Schlueter, J.R. Ferraro, K.R. Lykke, P. Wurz, D.H. Parker, J.D.M. Sutin: Inorg. Chem. **114**, 10069-10071 (1992)

8.129 K.D. Carlson, A.M. Kini, J.A. Schlueter, U. Geiser, R.A. Klemm, J.M. Williams, J.D. Durek, M.A. Calera, K.R. Lykke, H.H. Wang, J.R. Ferraro: Physica C **215**, 195-204 (1993)

8.130 A.M. Kini, K.D. Carlson, H.H. Wang, J.A. Schlueter, J.D. Durek, S.A. Sirchio, U. Geiser, K.R. Lykke, J.M. Williams: Physica C **264**, 81-94 (1996)

8.131 R.T. Scalettar, N.E. Bickers, D.J. Scalapino: Phys. Rev. B **40**, 197-200 (1989)
       F. Marsiglio: Phys. Rev. B **42**, 2416 - 2425 (1990)
8.132 See references in [8.126]
8.133 D. Pedron, R. Bozio, M. Meneghetti, C. Pecile: Proc. Europ. Conf. on Molecular
       Electronics, Padova, Italy (1992)
8.134 M. Meneghetti, R. Bozio, I. Zanon, C. Pecile, C. Ricotta, M. Zanetti: J. Chem.
       Phys. **80**, 6210-6224 (1984)
8.135 A.M. Kini, M.A. Beno, D. Son, H.H. Wang, K.D. Carlson, L.C. Porter, U.
       Welp, B.A. Vogt, J.M. Williams, D. Jung, M. Evain, M.-H. Whangbo, D.L.
       Overmyer, J.E. Schirber: Solid State Commun. **69**, 503-507 (1989)
8.136 K. Kanoda, T. Takahashi, K. Kikuchi, K. Saito, I. Ikemoto, K. Kobayashi: Phys.
       Rev. B **39**, 3996-4003 (1989)
8.137 K. Kikuchi, Y. Honda, Y. Ishikawa, K. Saito, I. Ikemoto, K. Murata, H. Anzai,
       T. Ishiguro, K. Kobayashi: Solid Sate Commun. **66**, 405-408 (1988)
8.138 H. Urayama, H. Yamochi, G. Saito, K. Nozawa, T. Sugano, M. Kinoshita, S. Sato,
       K. Oshima, A. Kawamoto, J. Tanaka: Chem. Lett. **1988**, 55-58 (1988)
8.149 A. Kobayashi, H. Kim. Y. Sasaki, R. Kato, H. Kobayashi: Solid State Commun.
       62, 57-64 (1987)
8.140 C.M. Varma, J. Zaanen and K. Raghavachari: Science **254**, 989-992 (1991)
8.141 M. Schlüter, M. Lanno, M. Needels, G.A. Baratoff, D. Tomanek: Phys. Rev.
       Lett. **68**, 526-529 (1992)
8.142 Y. Asai, Y. Kawaguchi: Phys. Rev. B **46**, 1265-1268 (1992)
8.143 W.E. Pickett: In *Solid State Physics* **48**, 226-346 (Academic, San Diego 1994)
8.144 S.J. Duclos, R.C. Haddon, S.H. Glarum, A.F. Hebard, K.B. Lyons: Science **254**,
       1625-1627 (1991)
8.145 K. Prassides, C. Christides, M.J. Rosseinsky, J. Tomkinson, D.W. Murphy,
       R.C. Haddon: Europhys. Lett. **19**, 629-635 (1992)
8.146 A.P. Ramirez: Condensed Matter News **3**, 9-22 (1994)
8.147 I.K. Yanson: Sov. Phys. JETP **39**, 506-513 (1974)
8.148 A.G.M. Jansen, A.P. van Gelder, P. Wyder: J. Phys. C **13**, 6073-6118 (1980)
8.149 A. Nowack, M. Weger, D. Schweitzer, H.J. Keller: Solid State Commun. **60**,
       199-202 (1986)
8.150 A. Nowack, U. Poppe, M. Weger, D. Schweitzer H. Schwenk: Z. Physik B **68**,
       41-47 (1987)
8.151 G. Bergmann, D. Rainer: Z. Physik **263**, 59-68 (1973)
8.152 U. Schmelzer, E.L. Bokhenov, B. Dorner, J. Kalus, G.A. McKenzie, I. Nathan-
       iec, G.S. Pawley, E.F. Sheka: J. Phys. C **14**, 1025-1041 (1981)
8.153 L. Pintschovius, H. Ritschel, T. Sasaki, H. Mori, S. Tanaka, N. Toyota, M. Lang,
       S. Steglich: Europhys. Lett. **37**, 627-632 (1997)
8.154 W.L. McMillan: Phys. Rev. **167**, 331-344 (1968)
8.155 G.M. Eliashberg: Sov. Phys. JETP **11**, 696-702 (1960); ibid. **12**, 1000-1002 (1961)
8.156 Y. Nambu: Phys. Rev. **117**, 648-663 (1960)
8.157 P.B. Allen, R.C. Dynes: Phys. Rev. B **12**, 905-922 (1975)
8.158 S. Barisic, S. Brazovskii: in *Recent Developments in Condensed Matter Physics*,
       Vol. 1, ed. by J.T. Devreese (Plenum, New York 1981) pp. 327-342
8.159 S. Barisic: J. Physique **44**, C3-991-996 (1983)
8.160 B. Horovitz, H. Gutfreund, M. Weger: Solid State Commun. **39**, 541-545 (1981)

8.161 H. Gutfreund, B. Horovitz, M. Weger: J. Physque **44**, C3-983-989 (1983)

8.162 S. Nakajima: J. Phys. Soc. Jpn. **56**, 871-872 (1987)

8.163 S. Mazumdar: Solid State Commun. **66**, 427-430 (1988)

8.164 S. Mazumdar, S.N. Dixit, A.N. Bloch: Phys. Rev. B **30**, 4842-4845 (1984)

8.165 K. Nasu: J. Phys. Soc. Jpn. **54**, 1933-1943 (1985); Physica B **143**, 229-233 (1986); Phys. Rev. B **35**, 1748-1763 (1987)

8.166 T. Mori: Unpublished (1984)

8.167 H. Suhl, B.T. Matthias, L.R. Walker: Phys. Rev. Lett. **3**, 552-554 (1958)

8.168 J. Kondo: Progr. Theor. Phys. **29**, 1-9 (1963)

8.169 K. Yamaji, S. Abe: J. Phys. Soc. Jpn. **56**, 4237-4240 (1987)

8.170 K. Yamaji: Solid State Commun. **64**, 1157-1160 (1987)

8.171 E. Canadell, I.E.-I. Rachidi, S. Ravy, J.P. Pouget, L. Brossard, J.P. Legros: J. Physique **50**, 2967-2981 (1989)

8.172 C.A. Hayward, D. Poilblanc, R.M. Noack, D.J. Scalapino, W. Hanke: Phys. Rev. Lett. **75**, 926-929 (1995)

8.173 K. Yamaji, Y. Shimoi: Physica C **222**, 349-360 (1994)

8.174 R.M. Noack, S.R. White, D.J. Scalapino: Europhys. Lett. **30**, 163-168 (1995)

8.175 R.M. Noack, S.R. White, D.J. Scalapino: Physica C **270**, 281-296 (1996)

8.176 N. Nagaosa: Solid State Commun. **94**, 495-498 (1995)

8.177 R. Ramakumar, Y. Tanaka, K. Yamaji: Phys. Rev. B **56**, 795-801 (1997)

8.178 R. Kato: Private commun. (1996)

8.179 L.P. Gor'kov: Sov. Phys. JETP **9**, 1364-1367 (1959); ibid. **10**, 998-1004 (1960)

8.180 L.P. Gor'kov, D. Jérome: J. Phys. Lett. **46**, 643-646 (1985)

8.181 X. Huang, K. Maki: Phys. Rev. B **32**, 6459-6464 (1989)

8.182 A.I. Buzdin, L.N. Bulaevskii: Sov. Phys. Usp. **27**, 830-844 (1984)

8.183 K. Murata, M. Tokumoto, A. Anzai, K. Kajimura, T. Ishiguro: J. Appl. Phys. **26**, Suppl.26-3, 1367-1368 (1987)

8.184 K. Murata, N. Toyota, M. Tokumoto, H. Anzai, G. Saito, K. Kajimura, S. Morita, Y. Muto, T. Ishiguro: Physica **143**B, 366-368 (1986)

8.185 V.N. Laukhin, S.I. Pesotskii, E.B. Yagubskii: JETP Lett. **45**, 501-504 (1987)

8.186 A.G. Lebed: JETP Lett. **44**, 114-117 (1986)

8.187 L.I. Burlachkov, L.P. Gor'kov, A.G. Lebed': Europhys. Lett. **4**, 941-946 (1987)

8.188 R. Brusetti, M. Ribault, D. Jérome, K. Bechgaard: J. Physique **43**, 801-808 (1982)

8.189 N. Dupuis, G. Montambaux, C.A.R. Sá de Melo: Phys. Rev. Lett. **70**, 2613-2616 (1993)

8.190 N. Dupuis: Phys. Rev. B **51**, 9074-9083 (1995)

8.191 N. Dupuis: J. Phys. I (France) **5**, 1577-1613 (1995)

8.192 H. Shimahara: Phys. Rev. B **50**, 12760-12765 (1994)

8.193 Y. Hasegawa, M. Miyazaki: J. Phys. Soc. Jpn. **65**, 1028-1033 (1996)

8.194 A.G. Lebed, K. Yamaji: Physica C **282-287**, 1859-1860 (1997)

8.195 I.J. Lee, A.P. Hope, M.J. Leone, M. J. Naughton: Synth. Met. **70**, 747-750 (1995)

8.196 M.J. Naughton, I.J. Lee, P.M. Chaikin, G.M. Danner: Synth. Met. **85**, 1481-1485 (1997)

8.197 I.J. Lee, M.J. Naughton, G.M. Danner, P.M. Chaikin: Phys. Rev. Lett. **78**, 3555-3558 (1997)

8.198 S. Kivelson, O.L. Chapman: Phys. Rev. B **28**, 7236-7243 (1983)

8.199 M. Kertesz, R. Hoffmann: Solid State Commun. **47**, 97-102 (1983)

8.200 S. Aono, K. Nishikawa, M. Kimura, H. Kawabe: Synth. Met. **17**, 167-172 (1987)
8.201 T. Yamabe, K. Tanaka, K. Ohzeki: Solid State Commun. **44**, 823-825 (1982)
8.202 M.S. Dresselhaus, G. Dresselhaus, K. Sugihara, I.L. Spain, H.A. Goldberg: *Graphite Fibers and Filaments*, Springer Ser. Mat. Sci., Vol.5 (Springer, Berlin, Heidelberg 1988)
8.203 A. Tachibana, T. Inoue, T. Yamabe, K. Hori: Int'l J. Quantum Chem. **30**, 575-579 (1986)
8.204 A. Tachibana: Synth. Met. **19**, 105-110 (1987)

Chapter 9

9.1   J.F. Kwak, J.E. Schirber, R.L. Greene, E.M. Engler: Phys. Rev. Lett. **46**, 1296-1299 (1981)
9.2   C. Kittel: *Quantum Theory of Solids* (Wiley, New York 1963) Chap.11
9.3   J.F. Kwak: Mol. Cryst. Liq. Cryst. **79**, 111-122 (1982)
9.4   H. Bando, K. Oshima, M. Suzuki, H. Kobayashi, G. Saito: J. Phys. Soc. Jpn. **51**, 2711-2712 (1982)
9.5   K. Kajimura, H. Tokumoto, M. Tokumoto, K. Murata, T. Ukachi, H. Anzai, T. Ishiguro, G. Saito: J. Physique **44**, C3-1059-1062 (1983)
9.6   B. Brusetti, P. Garoche, D. Jérome, K. Bechgaard: J. Physique Lett. **43**, 147-152 (1982)
9.7   T. Takahashi, D. Jérome, K. Bechgaard: J. Physique **45**, 945-952 (1984)
9.8   L.J. Azevedo, J.M. Williams, S.J. Compton: Phys. Rev. B **28**, 6600-6602 (1983)
9.9   J.F. Kwak, J.E. Schirber, P.M. Chaikin, J.M. Williams, H.-H. Wang, L.Y. Chiang: Phys. Rev. Lett. **56**, 972-975 (1986)
9.10 H. Schwenk, S.S.P. Parkin, R. Schumaker, R.L. Greene, D. Schweitzer: Phys. Rev. Lett. **56**, 667-670 (1986)
9.11 T. Chakraborty, P. Pietiläinen: *The Fractional Quantum Hall Effect*, 2nd edn., Springer Ser. Solid-State Sci., Vol.85 (Springer, Berlin, Heidelberg 1995)
9.12 M. Ribault, D. Jérome, J. Tuchendler, C. Weyl, K. Bechgaard: J. Physique Lett. **44**, 953-961 (1983)
9.13 K. Oshima, M. Suzuki, K. Kikuchi, H. Kuroda, I. Ikemoto, K. Kobayashi: J. Phys. Soc. Jpn. **53**, 3295-3298 (1984)
9.14 P.M. Chaikin, M.-Y. Choi, J.F. Kwak, J.S. Brooks, K.P. Martin, M.J. Naughton, E.M. Engler, R.L. Greene: Phys. Rev. Lett. **51**, 2333-2336 (1983)
9.15 M. Ribault: Mol. Cryst. Liq. Cryst. **119**, 91-95 (1985)
9.16 M.J. Naughton, J.S. Brooks, L.Y. Chiang, R.V. Chamberlin, P.M. Chaikin: Phys. Rev. Lett. **55**, 969-972 (1985)
9.17 F. Pesty, P. Garoche, K. Bechgaard: Phys. Rev. Lett. **55**, 2495-2498 (1985)
9.18 P. Garoche, F. Pesty: J. Magn. Magn. Mat. **54-57**, 1418-1422 (1986)
9.19 M.-Y. Choi, P.M. Chaikin, R.L. Greene: J. Physique **44**, C3-1067-1070 (1983)
9.20 J.P. Ulmet, P. Auban, A. Khmou, S. Askenazy, A. Moradpour: J. Physique Lett. **46**, 535-542 (1985)
9.21 J.P. Ulmet, A. Khmou, P. Auban, L. Bachere: Solid State Commun. **58**, 753-758 (1986)
9.22 T. Osada, N. Miura, G. Saito: Physica B **143**, 403-405 (1986)

9.23 T. Osada, N. Miura, I. Ogura, G. Saito: Solid State Commun. **64**, 133-136 (1987)

9.24 L.P. Gor'kov, A.G. Lebed': J. Physique Lett. **45**, 433-440 (1984)

9.25 R. Peierls: Z. Physik **80**, 763-791 (1933)

9.26 A.A. Abrikosov, L.P. Gor'kov, I.E. Dzyaloshinskii: *Method of Quantum Field Theory in Statistical Physics* (Prentice-Hall, Englewood Cliffs, NJ 1963) Chap.3

9.27 P.M. Chaikin: Phys. Rev. B **31**, 4770-4472 (1985)

9.28 M. Héritier, G. Montambaux, P. Lederer: J. Physique Lett. **45**, 943-952 (1984)

9.29 G. Montambaux, M. Héritier, P. Lederer: Phys. Rev. Lett. **55**, 2078-2081 (1985)

9.30 G. Montambaux: Doctorat d'Etat ès Sciences Physiques Thesis, Université de Paris-Sud (1985) Chap.IV

9.31 G. Montambaux: In *Low-Dimensional Conductors and Superconductors*, ed. by D. Jérome, L.G. Caron (Plenum, New York 1987) pp.233-242

9.32 P. Delhaès, E. Dupart, J.P. Manceau, C. Coulon, D. Chasseau, J. Gaultier, J.M. Fabre, L. Giral: Mol. Cryst. Liq. Cryst. **119**, 269-276 (1985)

9.33 J.P. Ulmet, L. Bachere, S. Askenazy: Solid State Commun. **67**, 145-149 (1988)

9.34 J.P. Ulmet, L. Bachere, S. Askenazy, J.C. Ousset: Phys. Rev. B **38**, 7782-7788 (1988)

9.35 K. Mortensen, Y. Tomkiewicz, K. Bechgaard: Phys. Rev. **25**, 3319-3325 (1982)

9.36 S.K. Lyo: Phys. Rev. B **29**, 2685-2688 (1984)

9.37 K. Yamaji: Mol. Cryst. Liq. Cryst. **119**, 105-112 (1985)

9.38 K. Yamaji: J. Phys. Soc. Jpn. **54**, 1034-1040 (1985)

9.39 K. Yamaji: Synth. Met. **13**, 29-43 (1986)

9.40 K. Yamaji: J. Phys. Soc. Jpn. **56**, 1841-1854 (1987). The sign of τ_{sin} in Sects.2 and 3 of this article is defined in the opposite way to that in (4.49) in the present monograph. The superscripts attached to N_{QL} values in Fig.12 of the article should be taken to be opposite; e.g., -1^+ and -4^- should read -1^- and -4^+, respectively. The sign in front of s in (2.8) and in the caption of Fig.2 should be corrected to a + sign

9.41 K. Yamaji: J. Phys. Soc. Jpn. **53**, 2189-2192 (1984)

9.42 K. Yamaji: J. Phys. Soc. Jpn. **51**, 2787-2797 (1982)

9.43 R.V. Chamberlin, M.J. Naughton, X. Yan, L.Y. Chiang, S.-Y. Hsu, P.M. Chaikin: Phys. Rev. Lett. **60**, 1189-1192 (1988)

9.44 J.R. Cooper, W. Kang, P. Auban, G. Montambaux, D. Jérome, K. Bechgaard: Phys. Rev. Lett. **63**, 1984-1987 (1989)

9.45 S.T. Hannahs, J.S. Brooks, W. Kang, L.Y. Chiang, P.M. Chaikin: Phys. Rev. Lett. **63**, 1988-1991 (1989)

9.46 J.F. Kwak, J.E. Schirber, P.M. Chaikin, J.M. Williams, H.-H. Wang: Mol. Cryst. Liq. Cryst. **125**, 375-383 (1985)

9.47 L. Brossard, B. Piveteau, D. Jérome, A. Moradpour, M. Ribault: Physica B **143**, 406-408 (1986)

9.48 L.P. Gor'kov, A.G. Lebed: Mol. Cryst. Liq. Cryst. **119**, 73-77 (1985)

9.49 M. Héritier, G. Montambaux, P. Lederer: J. Physique Lett. **46**, 831-836 (1985)

9.50 K. Maki: Phys. Rev. B **33**, 4826-4829 (1986)

9.51 A. Virosztek, L. Chen, K. Maki: Phys. Rev. B **34**, 3371-3376 (1986)

9.52 L. Chen, K. Maki: Phys. Rev. B **35**, 8462-8488 (1987)

9.53 D. Poilblanc, M. Héritier, G. Montambaux, P. Lederer: J. Phys. C **19**, L321-329 (1986)

9.54 D. Poilblanc: Théorie des phases onde de densité de spin induites par le champ magnétiques dans des conducteurs très anisotropes. DSc. Thesis, Université de Paris-Sud (1988) Chap. B

9.55 G. Montambaux, D. Poilblanc: Phys. Rev. B **37**, 1913-1924 (1988)

9.56 G. Montambaux: J. Phys. C **20**, L327-L333 (1987)

9.57 P. Lederer, D. Poilblanc, G. Montambaux: Europhys. Lett. **5**, 151-156 (1988)

9.58 P.M. Chaikin: Workshop on Spin Density Wave at University of California, Los Angeles, CA (1988)

9.59 W. Kang, S.T. Hannahs, P.M. Chaikin: Phys. Rev. Lett. **70**, 3091-3094 (1993)

9.60 S.K. McKerman, S.T. Hannahs, U.M. Scheven, G.M. Danner, P.M. Chaikin: Phys. Rev. Lett. **75**, 1630-1633 (1995)

9.61 M.J. Naughton, R.V. Chamberlin, X. Yan, S.-Y. Hsu, L.Y. Chiang, M. Ya. Azbel, P.M. Chaikin: Phys. Rev. Lett. **61**, 621-624 (1988)

9.62 R.C. Yu, L. Chiang, R. Upasani, P.M. Chaikin: Phys. Rev. Lett. **65**, 2458-2461 (1990)

9.63 A.G. Lebed, P. Bak: Phys. Rev. B **40**, 11433-11436 (1989)

9.64 L.P. Gor'kov, A.G. Lebed: Phys. Rev. B **51**, 3285-3288 (1995)

9.65 T. Osada, S. Kagoshima, N. Miura: Phys. Rev. Lett. **69**, 1117-1120 (1992)

9.66 T. Osada: Kotai-Butsuri (in Japanese) **30**, 843-858 (1995)

9.67 W. Kang, S.T. Hannahs, L.Y. Chiang, R. Upasani, P.M. Chaikin: Phys. Rev. Lett. **65**, 2812-2815 (1990)

9.68 T. Osada, H. Shinagawa, S. Kagoshima, N. Miura: Synth. Met. **56**, 1795-1802 (1993)

9.69 F. Pesty, P. Garoche, M. Héritier: *Physics and Chemistry of Organic Superconductors*, ed. by G. Saito, S. Kagoshima, Springer Proc. Phys. **51**, 87-90 (Springer, Berlin, Heidelberg 1990)

9.70 A.G. Lebed: JETP Lett. **51**, 663-666 (1990); Physica B **169**, 368-371 (1991)

9.71 K. Machida, M. Nakano: J. Phys. Soc. Jpn. **59**, 4223-4226 (1990)

9.72 K. Machida, Y. Hori, M. Nakano: J. Phys. Jpn. **60**, 1730-1742 (1991)

9.73 Y. Hori, K. Machida: J. Phys. Jpn. **60**, 1246-1256 (1992)

9.74 K. Machida, Y. Hasegawa, M. Kohmoto, V.M. Yakovenko, Y. Hori, K. Kishigi: Phys. Rev. B **50**, 921-931 (1994)

9.75 D. Zanchi, G. Montambaux: Phys. Rev. Lett. **77**, 366-369 (1996)

9.76 K. Oshima, H. Okuno, K. Kato, R. Maruyama, R. Kato, A. Kobayashi, H. Kobayashi: Synth. Met. **70**, 861-862 (1995)

9.77 V.M. Yakovenko: Sov. Phys. JETP **66**, 355-365 (1988)

9.78 V.M. Yakovenko: Phys. Rev. Lett. **61**, 2276 (1988)

9.79 J.S. Brooks, N.A. Fortune, M.J. Graf, P.M. Chaikin, L.Y. Chiang, S. Hsu: Synth. Met. **27**, B29-B33 (1988)

9.80 W. Kang, S.T. Hannahs, L.Y. Chiang, R. Upasani, P.M. Chaikin: Phys. Rev. B **45**, 13566-13571 (1992)

9.81 X.D. Shi, W. Kang, P.M. Chaikin: Phys. Rev. B **50**, 1984-1987 (1994)

9.82 T. Osada, N. Miura, I. Oguro, G. Saito: Phys. Rev. Lett. **58**, 1563-1566 (1987)

9.83 I. Tüttö, A. Zawadovski: Phys. Rev. Lett. **60**, 1442-1445 (1988)

9.84 J. Suzumura, T. Saso, H. Fukuyama: Jpn. J. Appl. Phys. **26**, Suppl. 3, 589-590 (1987)

9.85 S.-r. Chang, K. Maki: J. Low Temp. Phys. **66**, 357-365 (1987)

9.86 F. Tsobnang, F. Pestry, P. Garoche: Phys. Rev. B 49, 15110-15121 (1994)

9.87 A.G. Lebed': JETP Lett. **43**, 174-177 (1986)

9.88 L. Chen, K. Maki: Synth. Met. **29**, F493-F498 (1989)

9.89 K. Behnia, L. Belicas, W. Kang, D. Jérome, P. Carretta, Y. Fagot-Revurat, C. Berthier, M. Horvatic, S. Ségransan, L. Hubert, C. Bourbonnais: Phys. Rev. Lett. **74**, 5272-5275 (1995)

9.90 D. Jérome: J. Phys. Soc. Jpn **49**, Suppl. A, 845-856 (1980)

9.91 D. Jérome: Chem. Scripta **17**, 13-17 (1981)

9.92 D. Jérome, H.J. Schulz: Adv. Phys. **31**, 299-490 (1982)

9.93 A. Audouard, J.P. Ulmet, J.M. Fabre: Synth. Met. **70**, 751-752 (1995)

9.94 J.P. Ulmet, A. Narjis, M.J. Naughton, J.M. Fabre: Phys. Rev. B **55**, 3024-3027 (1997)

9.95 A. Audouard, F. Goze, S. Dubois, J.P. Ulmet, L. Brossarad, S. Askenazy, S. Tomic, J.M. Fabre: Europhys. Lett. **25**, 363-368 (1994)

9.96 N.A. Fortune, J.S. Brooks, M.J. Graf, G. Montambaux, L.Y. Chiang, J.A.A.J. Perenboom, D. Althof: Phys. Rev. Lett. **64**, 2054-2057 (1990)

9.97 S. Uji, T. Terashima, H. Aoki, J.S. Brooks, M. Tokumoto, S. Takasaki, J. Yamada, H. Anzai: Phys. Rev. B **53**, 14399-14405 (1996)

9.98 H. Shinagawa, S. Kagoshima, T. Osada, N. Miura: Physica B **201**, 490-492 (1994)

9.99 H. Shinagawa, S. Kagoshima, T. Osada, N. Miura: Presented at the Meeting of Phys. Soc. Jpn. (March 1996)

9.100 H. Shinagawa: Magmetic quantum oscillations in the quasi one-dimensional conductor $(TMTSF)_2 X$, Dissertation, University of Tokyo (1997)

9.101 D.A. Howe, C.C. Agosta, C.H. Mielke, S.A. Ivanov, T.J. Coffey: Bull. Am. Phys. Soc. **41**, B12-9 (1996)

9.102 M.J. Naughton, G. Montambaux: Synth. Met. **41-43**, 3995-3998 (1991)

9.103 C.C. Agosta, D.A. Howe, M.A. Antia, S.A. Ivanov, C.H. Mielke, F.M. Morgan: In *High Magnetic Fields in the Physics of Semiconductors*, ed. D. Heiman (World Scientic, Singapore 1995) pp. 538-541

9.104 K. Yamaji: J. Phys. Soc. Jpn. **55**, 1424-1427 (1986). The resistivity, given by Eq. (10) of this paper, due to the electron-optical phonon interaction takes no account of the Umklapp process. *Osada* et al pointed out in [9.105] that when one considers the latter, the oscillatory behavior disaapears but that the resistivity due to the interaction with acoustic phonons keeps the oscillation alive with the same period

9.105 T. Osada, N. Miura, G. Saito: Physica B **143**, 403-405 (1986)

9.106 V.M. Yakovenko: Phys. Rev. Lett. **68**, 3607-3610 (1992)

9.107 A.G. Lebed: Phys. Rev. Lett. **74**, 4903-4906 (1995)

9.108 A.G. Lebed: I.F. Schegolev memorial issue: J. Phys. I (France) **6**, 1819-1836 (1996)

9.109 A.G. Lebed: Synth. Met. **85**, 1615-1616 (1997)

9.110 L.P. Gor'kov, A.G. Lebed: Phys. Rev. B **51**, 1361-1365 (1995)

9.111 K. Yamaji: J. Phys. Soc. Jpn. **56**, 1101-1110 (1987)

9.112 X. Yan, M.J. Naughton, R.V. Chamberlin, L.Y. Chiang, S.Y. Hsu, P.M. Chaikin: Synth. Met. **27**, B145-B150 (1988)

9.113 X. Yan, M.J. Naughton, R.V. Chamberlin, S.Y. Hsu, L.Y. Chiang, J.S. Brooks, P.M. Chaikin: Phys. Rev. B **36**, 1799-1802 (1987)

9.114 R.W. Stark, C.B. Friedberg: J. Low Temp. Phys. **14**, 112-146 (1974)

9.115 M.Ya. Azbel, P.M. Chaikin: Phys. Rev. Lett. **59**, 582-585 (1987)

9.116 T. Osada, N. Miura: Solid State Commun. **69**, 1169-1172 (1989)

9.117 J.S. Brooks, R.G. Clark, R.H. McKenzie, R. Newbury, R.P. Starrett, A.V. Skougarevsky, M. Tokumoto, S. Takasaki, J. Yamada, H. Anzai, S. Uji: Phys. Rev. B **53**, 14406-14410 (1996)

9.118 S. Uji, J.S. Brooks, M. Chaparala, S. Takasaki, J. Yamada, H. Anzai: Phys. Rev. B **55**, 14387-14391 (1997)

9.119 K. Kishigi, K. Machida: Phys. Rev. B **53**, 5461-5464 (1996)

9.120 S. Uji, J.S. Brooks, M. Chaparala, S. Takasaki, J. Yamada, H. Anzai: Phys. Rev. B **55**, 12446-12453 (1997)

9.121 A. G. Lebed, P. Bak: Phys. Rev. Lett. **63**, 1351-1317 (1989)

9.122 T. Osada, A. Kawasumi, S. Kagoshima, N. Miura, G. Saito: Phys. Rev. Lett. **66**, 1525-1528 (1991)

9.123 T. Osada, A. Kawasumi, S. Kagoshima, N. Miura, G. Saito: Physica C **185-189**, 2697-2698 (1991)

9.124 K. Maki: Phys. Rev. B **45**, 5111-5113 (1992)

9.125 A.G. Lebed: J. Phys. I (France) **4**, 351-355 (1994)

9.126 G.M. Danner, W. Kang, P.M. Chaikin: Phys. Rev. Lett. **72**, 3714-3717 (1994)

9.127 G. M. Danner, P.M. Chaikin: Phys. Rev. Lett. **75**, 4690-4693 (1995)

9.128 S.P. Strong, D.G. Clarke, P.W. Anderson: Phys. Rev. Lett. **73**, 1007-1010 (1994)

9.129 T. Osada, S. Kagoshima, N. Miura: Phys. Rev. Lett. **77**, 5261-5264 (1996)

9.130 H. Yoshino, K. Saito, K. Kikuchi, H. Nishikawa, K. Kobayashi, I. Ikemoto: J. Phys. Soc. Jpn. **64**, 2307-2310 (1995)

9.131 A.G. Lebed', N.N. Bagmet: Phys. Rev. B **55**, R8654-R8657 (1997)

9.132 J. S. Brooks, X. Chen, S.J. Klepper, S. Valfells, G.J. Athas, Y. Tanaka, T. Kinoshita, N. Kinoshita, M. Tokumoto, H. Anzai, C.C. Agosta: Phys. Rev. B **52**, 14457-14478 (1995)

9.133 H. Mori, S. Tanaka, M. Oshima, G. Saito, T. Mori, Y. Murayama, H. Inokuchi: Bull. Chem. Soc. Jpn. **63**, 2183-2190 (1990)

9.134 L. Duccasse, A. Fritsch: Solid State Commun. **91**, 201-204 (1994)

9.135 T. Sasaki, H. Sato, N. Toyota: Synth. Met. **41-43**, 2211-2214 (1991)

9.136 F.L. Pratt, T. Sasaki, N. Toyota: Phys. Rev. Lett. **74**, 3892-3895 (1995)

9.137 T. Osada, R. Yagi, A. Kawasumi, S. Kagoshima, N. Miura, M. Oshima, H. Mori, T. Nakamura, G. Saito: Synth Metals **41-43**, 2171-2174 (1991)

9.138 M.V. Kartsovnik, A.E. Kovalev, V.N. Laukhin, S.I. Pesotskii: J. Phys. I (France) **2**, 223-228 (1992)

9.139 Y. Iye, R.Yagi, N. Hanasaki, S. Kagoshima, H. Mori, H. Fujimoto, G. Saito: J. Phys. Soc. Jpn. **63**, 674-684 (1994)

9.140 T. Osada, R. Yagi, A. Kawasumi, S. Kagoshima, N. Miura, M. Oshima, G. Saito: Phys. Rev. B **41**, 5428-5431 (1990)

9.141 T. Osada, S. Kagoshima, N. Miura: Synth. Met. **70**, 931-934 (1995)

9.142 K. Kishigi, K. Machida: J. Phys. Soc. Jpn. **64**, 3853-3859 (1995)

9.143 D. Yoshioka: J. Phys. Soc. Jpn. **64**, 3168-3171 (1995)

9.144 P. Christ, W. Biberacher, H. Müller, K. Andres, E. Steep, A.G.M. Jansen: Physica B **204**, 153 (1995)

9.145 P. Christ, W. Biberacher, H. Müller, K. Andres: Solid State Commun. **91**, 451 (1994)

9.146 T. Sasaki, A. Lebed, T. Fukase, N. Toyota: Phys. Rev. B **54**, 12969-12978 (1996)

9.147   T. Sasaki, A. Lebed, T. Fukase, N. Toyota: Synth. Met. **86**, 2063-2064 (1997)

9.148   M.V. Kartsovnik, W. Biberacher, P. Christ, A.E. Kovalev, K. Andres, E. Steep, A.G.M. Jansen: Synth. Met. **86**, 1933-1936 (1997)

Chapter 10

10.1   *The Fullerenes*, ed. by H.W. Kroto, J.E. Fisher, D.E. Cox (Pergamon, Oxford 1993)

10.2   M.S. Dresselhaus, G. Dresselhaus, P.C. Eklund: *Science of Fullerenes and Carbon Nanotubes* (Academic, San Diego, CA 1996)

10.3   R.C. Haddon, L.E. Brus, K. Raghavaachari: Chem. Phys. Lett. **125**, 459 (1986)

10.4   J. Wu, Z.-X. Shen, D.S. Dessau, R. Cao, D.S. Marshall, P. Pianetta, I. Lindau, X. Yang, J. Terry, D.M. King, B.O. Wells, D. Elloway, H.R. Went, C.A. Brown, H. Hunziker, M.S. de Vries: Physica C **197**, 251 (1992)

10.5   P.A. Heiney: J. Phys. Chem. Solids **53**, 1333 (1992)

10.6   R. Tycko, G. Dabbagh, R.M. Fleming, R.C. Haddon, A.V. Makhija, S.M. Zahurak: Phys. Rev. Lett. **67**, 1886 (1991)

10.7   W.I.F. David, R.M. Ibberson, T.J.S. Dennis, J.P. Hare, K. Prassides: Europhys. Lett. **18**, 219 (1992)

10.8   S. Saito, A. Oshiyama: Phys. Rev. Lett. **66**, 2637 (1991)

10.9   J.H. Weaver, J.L. Martins, T. Komeda, Y. Chen, T.R. Ohno, G.H. Kroll, T. Troullier: Phys. Rev. Lett. **66**, 1741 (1991)

10.10   A. Oshiyama, S. Saito, N. Hamada, Y. Miyamoto: In [Ref. 10.1, pp. 287-302]

10.11   H. Ajie, M.M. Alvarez, S.J. Anz, R.D. Beck, F. Diederich, K. Fostiropoulous, D.R. Huffman, W. Krätschmer, Y. Rubin, K.E. Schriver, D. Sensharma, R.L. Whetten: J. Phys. Chem. **94**, 8630 (1990)

10.12   B.-L. Gu, Y. Maruyama, J.-Z. Yu, K. Ohno, Y. Kawazoe: Phys. Rev. B **49**, 16202 (1994)

10.13   S. Matsuura, T. Tsuzuki, T. Ishiguro, H. Endo, K. Kikuchi, Y. Achiba, I. Ikemoto: J. Phys. Chem. Solids **55**, 835 (1994)

10.14   T.N. Thomas, R.A. Taylor, J.F. Ryan, D. Mihailovic, R. Zamboni: Europhys. Lett. **25**, 403 (1994)

10.15   T. Tsubo, K. Nasu: J. Phys. Soc. Jpn. **63**, 2401 (1994)
        K. Harigaya, S. Abe: Phys. Rev. B **49**, 16746 (1994)

10.16   Y. Wang, J.M. Holden, A.M. Rao, W.-T. Lee, X.X. Bi, S.L. Ren, W.G. Lehman, G.T. Hager, P.C. Eklund: Phys. Rev. B **45**, 14396 (1992)

10.17   A. Skumanich: Chem. Phys. Lett. **182**, 486 (1991)

10.18   R.W. Lof, M.A. van Veenendaal, B. Koopmans, H.T. Jonkman, G.A. Sawatzky: Phys. Rev. Lett. **68**, 3924 (1992)

10.19   T. Arai, Y. Murakami, H. Suematsu, K. Kikuchi, Y. Achiba, I. Ikemoto: Solid State Commun. **81**, 827 (1992)

10.20   D.W. Murphy. M.J. Rosseinsky, R.M. Fleming, R. Tycko, A.P. Ramirez, R.C. Haddon, T. Siegrist, G. Dabbagh, J.C. Tully, R.E. Walstedt: J. Phys. Chem. Solids **53**, 1321 (1992)
        O. Zhou, D. E. Cox: In [Ref. 10.1, p. 203]

10.21 N. Hamada, S. Saito, Y. Miyamoto, A. Oshiyama: Jpn. J. Appl. Phys. **30**, L2036 (1991)

10.22 P. J. Benning, F. Stepniak, D. M. Poirier, J. L. Martins, J. H. Weaver, L. P. F. Chbante, R. E. Smalley: Phys. Rev. **B 47**, 13843 (1993)

10.23 J. L. Martins, N. Troullier: Phys. Rev. **B 46**, 1766 (1992)

10.24 J. H. Weaver, P.J. Benning, F. Stepniak, D.M. Poirier: J. Phys. Chem. Solids **53**, 1707 (1992)

10.25 Y. Iwasa, K. Tanaka, T. Yasuda, K. Koda, S. Koda: Phys. Rev. Lett. **69**, 2284 (1992)

10.26 K. Tanigaki, M. Kosaka, T. Manako, Y. Kubo, I. Hirosawa, K. Ueda, K. Prassides: Chem. Phys. Lett. **240**, 627 (1995)

10.27 J. Schulte, M. C. Böhm: Solid State Commun. **93**, 249 (1995)

10.28 R. C. Haddon A. F. Hebard, M. J. Rosseinsky, D. W. Murphy, S. J. Duclos, K. B. Lyons, B. Miller, J. M. Rosamilia, R. M. Fleming, A. R. Kortan, S. H. Glarum, A. V. Makhija, A. J. Muller, R. H. Elick, S. M. Zahurak, R. Tycko, G. Dabbagh, F. A. Thiel: Nature **350**, 321 (1991)

10.29 M. J. Rosseinsky, A. P. Ramirez, S. H. Glarum, D. W. Murphy, R. C. Haddon, A. F. Hebard, T. T. M. Plastra, A. R. Kortan, S. M. Zahurak, A. V. Makhija: Phys. Rev. Lett. **66**, 2830 (1991)
A. F. Hebard, M. J. Rosseinsky, R. C. Haddon, D. W. Murphy, S. H. Glarum, T. T. M. Palstra, A. P. Ramirez, A. R. Kortan: Nature **350**, 600 (1991)
K. Holczer, O. Klein, S. -M. Huang, R. B. Kaner, K. -J. Fu, R. L. Whetten, F. Diederich: Science **252** 1154 (1991)

10.30 K. Tanigaki, T. W. Ebbsen, S. Saito, J. Mizuki, J. S. Tsai, Y. Kubo, S. Kuroshima: Nature **352**, 222 (1991)

10.31 T. T. M. Palstra, O. Zhou, Y. Iwasa, P. E. Sulewski, R. M. Fleming, B. R. Zegarski: Solid State Commun. **93**, 327 (1995)

10.32 S. H. Irons, J. Z. Liu, P. Klavins, R. N. Shelton: Phys. Rev. **B 52**, 15517 (1995)

10.33 G. Sparn, J. D. Thompson, R. L. Whetten, S. -M. Huang, R. B. Kaner, F. Diederich, G. Grüner, K. Holczer: Phys. Rev. Lett. **68**, 1228 (1992)

10.34 C. E. Johnson, H. W. Jiang, K. Holczer, R. B. Kaner, R. L. Whetten, F. Diederich: Phys. Rev. **B 46**, 5880 (1992)

10.35 G. S. Boebinger, T. T. M. Palstra, A. Passner, M. J. Rosseinsky, W. D. Murphy, I. I Mazin: Phys. Rev. **B 46**, 5876 (1992)

10.36 R. Ikeda, T. Tsuneto: J. Phys. Soc. Jpn. **60**, 1337 (1991)

10.37 A. Otsuka, T. Ban, G. Saito, H. Ito, T. Ishiguro, N. Hosoito, T. Shinjo: Synth. Met. **55-57**, 3148 (1993)

10.38 J. G. Hou, V. H. Crespi, X. -D. Xiang, W. A. Vareka, G. Bricen, A. Zettl, M. L. Cohen: Solid State Commun. **86**, 643 (1993)

10.39 N. R. Wertermer, E. Helfand, P. C. Hohenberg: Phys. Rev. **147**, 295 (1969)

10.40 X. -D. Xiang, J. G. Hou, V. H. Crespi, A. Zettl, M. L. Cohen: Nature **361**, 54 (1993)

10.41 M. Dressel, L. Degiorgi, O. Klein, G. Grüner: J. Phys. Chem. Solids **54**, 1411 (1993)

10.42 L. Degiorgi, P. Wacher, G. Grüner, S.-M. Huang, J. Wiley, R. B. Kaner: Phys. Rev. Lett. **69**, 2987 (1992)

10.43 Z. Zhang, C. -C. Chen, S. P. Kelty, H. Dai, C. M. Lieber: Nature **353**, 333 (1991), Z. Zhang, C. -C. Chen, M. Lieber: Science **254**, 1619 (1991)

10.44 Y. J. Uemura, A. Keren, L. P. Lee, G. M. Luke, B. J. Sternlib, W. D. Wu, J. H. Brewer, R. L. Whetten, S. M. Huang, S. Lin, R. B. Kaner, F. Diederich, S. Donovans, G. Grüner, K. Holczer: Nature **352**, 605 (1991)

10.45 R. Tycko, G. Dabbagh, M. J. Rosseinsky, D. W. Murphey, R. M. Fleming, A. P. Ramirez, J. C. Tulley: Science **253**, 884 (1991)

10.46 R. Tycko, G. Dabbagh, M. J. Rosseinsky, D. W. Murphy, A. P. Ramirez, R. M. Fleming: Phys. Rev. Lett. **68**, 1912 (1992)

10.47 M. Baenitz, M. Heinze, k. Lüders, H. Werner, R. Schlögl: Solid State Commun. **91**, 337 (1994)

10.48 K. Tanigaki, I. Hirosawa, T. W. Ebbesen, J. Mizuki, J. -S. Tsai: J. Phys. Chem. Solids **54**, 1645 (1993)

10.49 M. J. Rosseinsky, D. W. Murphy, R. M. Fleming, R. Tycko, A. P. Ramirez, T. Siegrist, G. Dabbagh, S. E. Barrett: Nature **356**, 416 (1992)

10.50 A. Oshiyama, S. Saito: Solid State Commun. **82**, 41 (1992)

10.51 J. Fisher, P. A. Heiney: J. Phys. Chem. Solids **54**, 1725 (1993)

10.52 K. Parassides, C. Christides. L. M. Thomas, J. Mizuki, K. Tanigaki, I. Hirosawa, T. W. Ebbesen: Science **263**, 950 (1995), and references therein

10.53 C. M. Verma, J. Zaanen, K. Raghavachari: Science **254**, 989 (1991)

10.54 R. A. Jishi, M. S. Dresselhaus: Phys. Rev. B **45**, 2597 (1992)

10.55 M. Schluter, M. Lannoo, M. Needles, G. A. Baraff, D. Tomanek: Phys. Rev. Lett. **68**, 526 (1992)

10.56 V. H. Crespi, J. G. Hou, Z. -D. Xiang, M. L. Cohen, A. Zettl: Phys. Rev. B **46**, 12064 (1992)

10.57 I. I. Mazin, O. V. Dolgove, A. Golubov, S. V. Shulga: Phys. Rev. B **47**, 538 (1993)

10.58 F. C. Zhang, M. Ogata, T. M. Rice: Phys. Rev. Lett. **67**, 3452 (1991)

10.59 G. H. Chen, Y. J. Guo, N. Karasawa, W. A. Goddara III: Phys. Rev. B **48**, 13959 (1993)

10.60 S. Chakravaty, S. A. Kivelson, M. K. Salkola, S. Tewari: Science **254**, 970 (1991)

10.61 Y. Takada: Physica C **185-189**, 419 (1991)

10.62 T. W. Ebbesen, J. S. Tsai, K. Tanigaki, H. Hiura, Y. Shimakawa, Y. Kubo, I. Hirosawa, J. Mizuki: Physica C **203**, 163 (1992)

10.63 B. Burk, V. H. Crespi, A. Zettl, M. L. Cohen: Phys. Rev. Lett. **72**, 3706 (1994)

10.64 A. P. Ramairez, A. R. Kortan, M. J. Rossinsky, Sa. J. Duclos, A. M. Mujsche, R. C. Haddon, D. W. Murphy, A. V. Makhija, S. M. Zahurak, K. B. Lyons: Phys. Rev. Lett. **68**, 1058 (1992)

10.65 T. W. Ebbesen, J. S. Tsai, K. Tanigaki, J. Tabuchi, Y. Shimakawa, Y. Kubo, I. Hirosawa, J. Mizuki: Nature **355**, 620 (1992)

10.66 A. A. Zakhidov, K. Imaeda, D. M. Petty, K. Yakushi, H. Inokuchi, K. Kikuchi, I. Kikemoto, S. Suzuki, Y. Achiba: Phys. Lett. A **164**, 355 (1992)

10.67 C. -C. Chen, C. M. Lieber: Science **259**, 655 (1993), P. Auban-Senzier, G. Quirion, D. Jerme, P. Bernier, S. Della-Negra, C. Farbe, A. Rassat: Synth. Met. **56**, 3027 (1993)

10.68 K. Prassides, C. Christides, M. J. Rosseinsky, J. Tomkinson, D. W. Murphy, R. C. Haddon: Europhys. Lett. **19**, 629 (1992)

10.69 Y. Maniwa, T. Saito, A. Ohi, K. Mizoguchi, K. Kume, K. Kikuchi, I. Ikemoto, S. Suzuki, Y. Achiba, M. Kosaka, K. Tanigaki, T. W. Ebbesen: J. Phys. Soc. Jpn. **63**, 1139 (1994)

10.70 R. Tycko, G. Dabbagh, M. J. Rosseinsky, D. W. Murphy, R. M. Fleming, A. P. Ramirez, J. C. Tully: Science **253**, 884 (1991)

10.71 M. Kosaka, K. Tanigaki, T. W. Ebbesen, Y. Nakahara, K. Takeishi: Appl. Phys. Lett. **63**, 2561 (1993)

10.72 S. Saito, A. Oshiyama: Phys. Rev. **B 44**, 11536 (1991)

10.73 W. H. Wong, M. E. Hanson, W. G. Clark, G. Grüner, J. D. Thompson, R. L. Whenten, S. -M. Huang, R. B. Kaner, F. Diederich, P. Petit, J.-J. Andre, K. Holczer: Europhys. Lett. **18**, 79 (1992)
A. P. Ramirez, M. J. Rosseinsky, D. W. Murphy, R. C. Haddon: Phys. Rev. Lett. **69**, 1687 (1992)

10.74 T. Inabe, H. Ogata, Y. Maruyama, Y. Achiba, S. Suzuki, K. Kikuchi, I. Ikemoto: Phys. Rev. Lett. **69**, 3797 (1992)

10.75 C. T. Chen, L. H. Tjeng, P. Rudolf, G. Meigs, J. S. Rowe, J. Chen, J. P. McCauley Jr., A. B. Smith II, A. R. McGhie, W. J. Romanow, E. W. Plummer: Nature **352**, 603 (1991)

10.76 V.A. Stenger, C.H. Pennington, D.R. Buffinger, R.P. Ziebarth: Phys. Rev. Lett. **74**, 1649 (1995)

10.77 M. Cryot: J. Phys. **27**, 283 (1966)

10.78 R.F. Kiefl, W.A. MacFarlane, K.H. Chow, S. Dunsiger, T.L. Duty, T.M.S. Johnston, J.W. Schneider, J. Sonier, L. Brard, R.M. Stronger, J.E. Fisher, A.B. Smith III: Phys. Rev. Lett. **70**, 3987 (1993)

10.79 S. Chakravarty, S. A. Kivelson, M. I. Salkola, S. Tewari: Science **256**, 1306 (1992)

10.80 O. Zhou, R. M. Fleming, D. W. Murphy, M. J. Rosseinsky, A. P. Ramirez, R. B. van Dover, R. C. Haddon: Nature **362**, 433 (1993)

10.81 H. Shimoda, Y. Iwasa, Y. Miyamoto, Y. Maniwa, T. Mitani: Phys. Rev. B **54**, R15653 (1996)

10.82 A. R. Kortan, N. Kophylov, S. Glarum, E. M. Gyorgy, A. P. Ramirez, R. M. Fleming, F. A. Thiel, R. C. Haddon: Nature **355**, 529 (1992)

10.83 A. R. Kortan, N. Kophylov, S. Glarum, E. M. Gyorgy, A. P. Ramirez, R. M. Fleming, O. Zhou, F. A .Thiel, P. L. Trevor, R. C. Haddon: Nature **360**, 566 (1992). Later work termed $Ba_6 C_{60}$ as $Ba_4 C_{60}$

10.84 Y. Iwasa, H. Hayashi, T. Furudate, T. Mitani: Phys. Rev. B **54**, 14960 (1996)

10.85 E. Özdas, A. R. Kortan, N. Kopylov, A. P. Ramirez, T. Siegrist, K. M. Rabe, H. E. Bair, S. Schuppler, P. H. Critrin: Nature **375**, 126 (1995)

10.86 X. H. Chen, G. Roth: Phys. Rev. B **52**, 15534 (1995)

10.87 M. Kraus, S. Gärtner, M. Baenitz, M. Kanowski, H. M. Veith, C. T. Simmons, W. Krätschmer, V. Thommen, H. P. Lang, H.-J. Güntherodt, K. Lüders: Europhys. Lett. **15**, 419 (1992)

10.88 S. P. Kelty, C. -C. Chen, C. M. Lieber: Nature **352**, 223 (1991)

10.89 Z. Iqval: Science **254** , 826 (1991)

Chapter 11

11.1   H. Akamatu, H. Inokuchi, Y. Matsunaga: Nature **173**, 168-169 (1954)

11.2   K. Brass, E. Clar: Chem. Ber. **65**, 1660-1662 (1932)

11.3   E. Zinke, A. Pongratz: Chem. Ber. **69**, 1591-1593 (1936)

11.4   E. Müller, W. Wieseman: Chem. Ber. **69**, 2173-2174 (1936)

11.5   H. Akamatu, Y. Matsunaga: Bull Chem. Soc. Jpn. **26**, 364-372 (1953)

11.6   F.R.M. McDonnell, R.C. Pink, A.R. Ubbelohde: J. Chem. Soc. **1951**, 191-199 (1951)

11.7   R.S. Mulliken: J. Am. Chem. Soc. **74**, 811-824 (1952)

11.8   C. Krohnke, V. Enkelmann, G. Wegner: Angew. Chem. Int'l Edn. Engl. **19**, 918-919 (1980)

11.9   H.J. Keller, D. Nothe, H. Pritzkow, D. Wehe, M. Werner, P. Koch, D. Schweitzer: Mol. Cryst. Liq. Cryst. **62**, 181-200 (1980)

11.10   C.V. Ristagno, H.J. Shine: J. Org. Chem. **36**, 4050-4055 (1971)

11.11   L. Alcacer, A.H. Maki: J. Phys. Chem. **78**, 215-217 (1974)

11.12   D.S. Acker, R.J. Harder, W.R. Hertler, W. Mahler, L.R. Melby, R.E. Benson, W.E. Mochel: J. Am. Chem. Soc. **82**, 6408-6409 (1960)

11.13   D.B. Chesnut, P. Arthur Jr.: J. Chem. Phys. **36**, 2969-2975 (1962)

11.14   W.J. Siemons, P.E. Bierstedt, R.G. Kepler: J. Chem. Phys. **39**, 3523-3528 (1963)

11.15   R.G. Kepler: J. Chem. Phys. **39**, 3528-3532 (1963)

11.16   L.R. Melby: Can. J. Chem. **43**, 1448-1453 (1965)

11.17   I.F. Shchegolev: Phys. Status. Solidi (a) **12**, 9-45 (1972)

11.18   A.J. Epstein, S. Etemad, A.F. Garito, A.J. Heeger: Phys. Rev. B **5**, 952-977 (1972)

11.19   A.F. Garito, A.J. Heeger: Acc. Chem. Res. **7**, 232-240 (1974)

11.20   B.A. Scott, S.J. La Placa, J.B. Torrance, B.D. Silverman, B. Welber: J. Am. Chem. Soc. **99**, 6631-6639 (1977)

11.21   J.B. Torrance: Acc. Chem. Res. **12**, 79-86 (1979)

11.22   W.A. Little: Phys. Rev. A **134**, 1416-1424 (1964)

11.23   O.H. LeBlanc Jr.: J. Chem. Phys. **42**, 4307-4308 (1965)

11.24   B.H. Klanderman, D.C. Hoesterey: J. Chem. Phys. **51**, 377-382 (1969)

11.25   J.H. Perlstein, J.A. van Allan, L.C. Isett, G.A. Reynolds: Ann. Acad. Sci. (New York) **313**, 61-78 (1978)

11.26   J.H. Perlstein: Angew. Chem. Int'l Edn. Engl. **16**, 519-534 (1977)

11.27   D.L. Coffen: Tetrhedron Lett. **1970**, 2633-2636 (1970)

11.28   F. Wudl, G.M. Smith, E.J. Hufnagel: J. Chem. Soc., Chem. Commun. **1970**, 1453-1454 (1970)

11.29   N. Sato, G. Saito, H. Inokuchi: Chem. Phys. **76**, 79-88 (1983)

11.30   K. Seki: Mol. Cryst. Liq. Cryst. **171**, 255-270 (1989)

11.31   D.L. Lichtenberger, R.L. Johnston, K. Hinkelmann, T. Suzuki, F. Wudl: J. Am. Chem. Soc. **112**, 3302-3307 (1990)

11.32   J. de Vries, H. Steger, B. Kamke, C. Menzel, B. Weisser, W. Kamke, I.V. Hertel: Chem. Phys. Lett. **188**, 159-162 (1992)

11.33   L.-S. Wang, J. Conceicao, C. Jin, R.E. Smalley: Chem. Phys. Lett. **182**, 5-11 (1991)

11.34   T. Akutagawa, G. Saito: Bull. Chem. Soc. Jpn. **68**, 1753-1773 (1995)

11.35 R.C. Wheland: J. Am. Chem. Soc. **98**, 3926-3930 (1976)

11.36 G. Saito, J.P. Ferraris: Bull. Chem. Soc. Jpn. **53**, 2141-2145 (1980)

11.37 V. Khodorkovsky, J.Y. Becker: In *Organic Conductors, Fundamentals and Applications*, ed. by J.-P. Farges (Dekker, New York 1994) pp.75-114

11.38 S. Horiuchi, H. Yamochi, G. Saito, K. Sakaguchi, M. Kusunoki: J. Am. Chem. Soc. **118**, 8604-8622 (1996)

11.39 Y. Misaki, H. Fujiwara, T. Yamabe, T. Mori, H. Mori, S. Tanaka: Chem. Lett. **1994**, 1653-1656 (1994)

11.40 C.J. Schramm, R.P. Scaringe, D.R. Stojakovic, B.M. Hoffman, J.A. Ibers, T.J. Marks: J. Am. Chem. Soc. **102**, 6702-6713 (1980)

11.41 R.P. Shibaeva, V.F. Kaminskii, E.B. Yagubskii: Mol. Cryst. Liq. Cryst. **119**, 361-373 (1985)

11.42 Y. Misaki, H. Nishikawa, K. Kawakami, S. Koyanagi, T. Yamabe, M. Shiro: Chem. Lett. **1992**, 2321-2324 (1992)

11.43 T. Mori, H. Inokuchi, Y. Misaki, T. Yamabe, H. Mori, S. Tanaka: Bull. Chem. Soc. Jpn. **67**, 661-667 (1994)

11.44 K. Takimiya, A. Ohnishi, Y. Aso, T. Otsubo, F. Ogura, K. Kawabata, K. Tanaka, M. Mizutani: Bull. Chem. Soc. Jpn. **67**, 766-772 (1994). However, the metallic nature of these 1:1 materials [11.41-44] can also be interpreted by a very small amount of anion defficiency or a very narrow gap; but this has not been confirmed yet

11.45 R. Foster: *Organic Charge Transfer Complexes* (Academic, London 1969)

11.46 F.H. Herbstein: In *Perspectives in Structural Chemistry*, ed. by J.D. Dunitz, J.A. Ibers (Wiley, New York 1971) Vol.IV, p.169

11.47 F. Wudl: In *Chemistry and Physics of One-Dimensional Metals*, ed. by H.J. Keller (Plenum, New York 1976) pp.233-256; Acc. Chem. Res. **17**, 227-232 (1984)

11.48 D.O. Cowan, J.A. Fortkort, R.M. Metzger: In *Low-Dimensional Systems and Molecular Electronics*, ed. by R.M. Metzger, P. Day, G.C. Papavassiliou (Plenum, New York 1990) pp.1-22

11.49 P. Delhaes: *ibid*, pp.43-65

11.50 D.O. Cowan: In *Proc. 4th. Int'l Kyoto Conf. on New Aspects of Organic Chemistry*, ed. by Z. Yoshida, T. Shiba, Y. Oshiro (Prentice Hall, Englewood Cliffs, NJ 1992)

11.51 J.M. Williams, J.R. Ferraro, R.J. Thorn, K.D. Carlson, U. Geiser, H.H. Wang, A.M. Kini, H.-H. Whangbo: *Organic Superconductors (Including Fullerenes): Synthesis, Structure, Properties, and Theory* (Prentice Hall, Englewoods Cliffs, NJ 1992)

11.52 G. Saito: In *Metal Insulator Transitions Revisited*, ed. by P.P. Edwards, C.N.R. Rao (Taylor & Francis, London 1995) pp.231-267; Phos. Sulfur and Silicon **67**, 345-360 (1992)

11.53 C.J. Jacobsen, K. Mortensen, J.R. Andersen, K. Bechgaard: Phys. Rev. B **18**, 905-921 (1978)

11.54 N. Thorup, G. Rindolf, C.S. Jacobsen, K. Bechgaard, I. Johannsen, K. Mortensen: Mol. Cryst. Liq. Cryst. **120**, 349-352 (1985)

11.55 B. Hilti, C.W. Mayer, G. Rihs, H. Loeliger, P. Baltzer: Mol. Cryst. Liq. Cryst. **120**, 267-271 (1985)

11.56 C.J. Schramm, D.R. Stojakovic, B.M. Hoffman, T.J. Marks: Science **200**, 47-48 (1978)

11.57 B.N. Diel, T. Inabe, J.W. Lyding, K.F. Schoch Jr., C.R. Kannewurf, T.J. Marks: J. Am. Chem. Soc. **105**, 1551-1567 (1983)

11.58 T. Inabe, T.J. Marks, J.W. Lyding, R. Burton, C.R. Kannewurf: Mol. Cryst. Liq. Cryst. **118**, 353-356 (1985)

11.59 M.Y. Ogawa, B.M. Hoffman, S. Lee, M. Yudkowsky, W.P. Halperin: Phys. Rev. Lett. **57**, 1177-1180 (1986)

11.60 G.J. Ashwell, D.D. Eley, M.R. Willis: Nature **259**, 201-202 (1976)

11.61 G.J. Ashwell: Nature **290**, 686-688 (1981)

11.62 G.J. Ashwell: Mol. Cryst. Liq. Cryst. **86**, 147-154 (1982)

11.63 G.J. Ashwell, S.C. Wallwork, P.J. Rizkallah: Mol. Cryst. Liq. Cryst. **91**, 359-369 (1983)

11.64 T.J. Marks, D.W. Kalina: In *Extended Linear Chain Compounds*, ed. by J.S. Miller (Plenum, New York 1983) Vol. 1, pp. 197-331

11.65 B.M. Hoffman, J. Martinsen, L.J. Pace, J.A. Ibers: In *Extended Linear Chain Compounds*, ed. by J.S. Miller (Plenum, New York 1983) Vol. 3, pp. 459-549

11.66 M. Kobayashi, T. Enoki, K. Imaeda, H. Inokuchi, G. Saito: Phys. Rev. B **36**, 1457-1462 (1987)

11.67 L. Forro, A. Janossy, L. Zuppiroli, K. Bechgaard: J. Physique **43**, 977-981 (1982)

11.68 L. Zuppiroli: In *Semiconductor and Semimetals*, **27**, 437-481 (Academic, New York 1988)

11.69 E.M. Engler, V.V. Patel: J. Am. Chem. Soc. **96**, 7376-7378 (1974)

11.70 D.O. Cowan, R. McCullough, A. Bailey, K. Lerstrup, D. Talham, D. Herr, M. Mays: Phos. Sulfur and Silicon **67**, 277-294 (1992)

11.71 G. Saito, T. Enoki, K. Toriumi, H. Inokuchi: Solid State Commun. **42**, 557-560 (1982)

11.72 G. Saito, T. Enoki, H. Inokuchi, H. Kobayashi: J. Physique **44**, C3-1215-1218 (1983)

11.73 R.H. Friend, D. Jérome, J.M. Fabre, L. Giral, K. Bechgaard: J. Phys. C **11**, 263-275 (1978)

11.74 A. Bondi: J. Phys. Chem. **86**, 441-451 (1964)

11.75 I.F. Shchegolev, E.B. Yagubskii: In *Extended Linear Chain Compounds*, ed. by J.S. Miller (Plenum, New York 1982) Vol. 2, pp. 385-434

11.76 R.P. Shibaeva: *ibid* pp. 435-467

11.77 C. Weyl, L. Brossard, S. Tomic, D. Mailly, D. Jérome: Mol. Cryst. Liq. Cryst. **120**, 263-266 (1985)

11.78 B. Hilti, C.W. Mayer, E. Minder, K. Hauenstein, J. Pfeiffer, M. Rudin: Chimia **40**, 56-57 (1986)

11.79 A.N. Bloch, D.O. Cowan, K. Bechgaard, R.E. Pyle, R.H. Banks, T.O. Poehler: Phys. Rev. Lett. **34**, 1561-1564 (1975)

11.80 J.R. Cooper, M. Weger, D. Jérome, D. Lefur, K. Bechgaard, A.N. Bloch, D.O. Cowan: Solid State Commun. **19**, 749-754 (1976)

11.81 C.S. Jacobsen, K. Mortensen, J.R. Andersen, K. Bechgaard: Phys. Rev. B **18**, 905-921 (1978)

11.82 T.E. Phillips, T.J. Kistenmacher, A.N. Bloch, D.O. Cowan: J. Chem. Soc. Chem. Commun. **1976**, 334-335 (1976)

11.83 A. Andrieux, C. Duroure, D. Jérome, K. Bechgaard: J. Physique Lett. **40**, L381-L383 (1979)

11.84 D. Jérome, A. Mazaud, M. Ribault, K. Bechgaard: J. Phys. Lett. **41**, L95-L98 (1980)

11.85 D. Jérome, K. Bechgaard: Sci. Am. **247**, 50-59 (1982)

11.86 K. Shimizu, N. Tamitani, N. Takeshita, M. Ishizuka, K. Amaya, S. Endo: J. Phys. Soc. Jpn. **61**, 2353-2355 (1992)

11.87 A. Onodera, I. Shirotani, H. Inokuchi, N. Kawai: Chem. Phys. Lett. **25**, 296-298 (1974)

11.88 I. Shirotani, A. Onodera, Y. Kamura, H. Inokuchi, N. Kawai: J. Solid State Chem. **18**, 238-239 (1976)

11.89 T. Yokota, N. Takeshita, K. Shimizu, K. Amaya, A. Onodera, I. Shirotani, S. Endo: Czech. J. Phys. **46**, 817-818 (1996)

11.90 T. Imakubo, H. Sawa, R. Kato: Synth. Met. **73**, 117-122 (1995)

11.91 T. Imakubo, H. Sawa, R. Kato: J. Chem. Soc., Chem. Commun. **1995**, 1097-1098 and 1667-1668 (1995)

11.92 R. Kato, A. Kobayashi, A. Miyamoto, H. Kobayashi: Chem. Lett. **1991**, 1045-1048 (1991)

11.94 A. Kobayashi, T. Udagawa, H. Tomita, T. Naito, H. Kobayashi: Chem. Lett. **1993**, 2179-2182 (1993)

11.95 L.K. Montgomery, T. Burgin, J.C. Huffman, K.D. Carlson, J.D. Dudek, G.A. Yaconi, L.A. Megna, P.R. Mobley, W.K. Kwok, J.M. Williams, J.E. Schirber, D.L. Overmyer, J. Ren, C. Rovira, M.-H. Whangbo: Synth. Met. **55-57**, 2090-2095 (1993)

11.96 H. Kobayashi, H. Tomita, T. Naito, A. Kobayashi, F. Sasaki, T. Watanabe, P. Cassoux: J. Am. Chem. Soc. **118**, 368-377 (1996)

11.97 Only the preparation of dibenzo-TOF has been reported; O.G. Safiev, D.V. Nazarov, V.V. Zorin, D.L. Rukhmankulov: Khim. Geterotsikl. Soedin. **1988**, 852 (1988); through Chem. Abst. 110: 212656h (1989)

11.98 M.A. Beno, H.H. Wang, A.M. Kini, K.D. Carlson, U. Geiser, W.K. Kwok, J.E. Thompson, J.M. Williams, J. Ren, M.-H. Whangbo: Inorg. Chem. **29**, 1599-1601 (1990)

11.99 S. Kahlich, D. Schweitzer, I. Heinen, Song En Lan, B. Nuber, H.J. Keller, K. Winzer, H.W. Helberg: Solid State Commun. **80**, 191-195 (1991)

11.100 K. Bechgaard, D.O. Cowan, A.N. Bloch: J. Chem. Soc., Chem. Commun. **1974**, 937-938 (1974)

11.101 A. Moradpour, V. Peyrussan, I. Johansen, K. Bechgaard: J. Org. Chem. **48**, 388-389 (1983)

11.102 J.S. Chappell, A.N. Bloch, W.A. Bryden, M. Maxfield, T.O. Poehler, D.O.Cowan: J. Am. Chem. Soc. **103**, 2442-2443 (1981)

11.103 T.J. Kistenmacher, T.J. Emge, A.N. Bloch, D.O. Cowan: Acta Cryst. B **38**, 1193-1199 (1982)

11.104 J.M. Williams, M.A. Beno, J.C. Sullivan, L.M. Banovetz, J.M. Braam, G.S. Blackman, C.D. Carlson, D.L. Greer, D.M. Loesing, K. Carneiro: Phys. Rev. B **28**, 2873-2876 (1983)

11.105 J.M. Williams, M.A. Beno, J.C. Sullivan, L.M. Banovetz, J.M. Braam, G.S. Blackman, C.D. Carlson, D.L. Greer, D.M. Loesing: J. Am. Chem. Soc. **105**, 643-645 (1983)

11.106 T.J. Kistenmacher: Mol. Cryst. Liq. Cryst. **136**, 361-382 (1986)

11.107 F. Wudl, E. Aharon-Shalom: J. Am. Chem. Soc. **104**, 1154-1156 (1982)

11.108 G. Saito, T. Enoki, H. Inokuchi, H. Kumagai, J. Tanaka: Chem. Lett. **1983**, 503-506 (1983)

11.109 D.O. Cowan, M. Mays, M. Lee, R. McCullough, A. Baily, K. Lerstrup, T. Kistenmacher, T. Poehler, L-Y. Chiang: Mol. Cryst. Liq. Cryst. **125**, 191-204 (1985)

11.110 G. Saito, H. Kumagai, J. Tanaka, T. Enoki, H. Inokuchi: Mol. Cryst. Liq. Cryst. **120**, 337-340 (1985)

11.111 I.A. Howard: In *Semiconductors and Semimetals* **27**, 22-85 (Academic, New York 1988) Chap. 2

11.112 M. Mizuno, A.F. Garito, M.P. Cava: J. Chem. Soc., Chem. Commnun. **1978**, 18-19 (1978)

11.113 G. Steimecke, H.-J. Sieler, R. Kirmse, E. Hoyer: Phosphorus and Sulfur **7**, 49-55 (1979)

11.114 K. Hartke, T. Kissel, J. Quante, R. Matusch: Chem. Ber. **113**, 1898-1906 (1980)

11.115 T.K. Hansen, J. Becher, T. Jorgensen, K.S. Varma, R. Khedekar, M.P. Cava: Org. Synth. **73**, in press

11.116 J. Larsen, C. Lenoir: Synthesis **21**, 134-134 (1989)

11.117 T. Mori, H. Inokuchi: Solid State Commun. **59**, 355-359 (1986); Bull. Chem. Soc. Jpn. **60**, 402-404 (1987)

11.118 P. Pfeiffer: *Organische Molekülverbindungen*, 2nd edn. (Verlag von F.Enke, Stuttgart 1927) pp.341-346

11.119 H.M. McConnell, H.M. Hoffman, R.M. Metzger: Proc. Nat. Acad.Sci. (USA) **53**, 46-50 (1965)

11.120 J. Kommandeur: Am. Chem. Soc. Meeting, St. Louis; US Govt. Res. Develop. Rep. **68** (1), 170 (1968)

11.121 Z.G. Soos: In *Organic and Inorganic Low-Dimensional Crystalline Materials*, ed. by P. Delhaes, M. Drillon (Plenum, New York 1987) pp.47-61

11.122 J.B. Torrance, J.E. Vazquez, J.J. Mayerle, V.Y. Lee: Phys. Rev. Lett. **46**, 253-257 (1981)

11.123 S.S. Shaik: J. Am. Chem. Soc. **104**, 5328-5334 (1982)

11.124 T. Akutagawa, G. Saito, M. Kusunoki, K. Sakaguchi: Bull. Chem. Soc. Jpn. **69**, 2487-2511 (1996)

11.125 H. Kobayashi, A. Kobayashi, Y. Sasaki, G. Saito, H. Inokuchi: Bull. Chem. Soc. Jpn. **59**, 301-302 (1986)

11.126 P.C.W. Leung, T.J. Emge, M.A. Beno, H.H. Wang, J.M. Williams, V. Petricek, P. Coppens: J. Am. Chem. Soc. **107**, 6184-6191 (1985)

11.127 T. Mori, H. Inokuchi: Chem. Lett. **1987**, 1657-1660 (1987)

11.128 M.J. Rosseinsky, M. Kurmoo, D.R. Talham, P. Day, D. Chasseau, D. Watkin: J. Chem. Soc., Chem. Commun. **1988**, 88-90 (1988)

11.129 Y. Misaki, N. Higuchi, H. Fujiwara, T. Yamabe, T. Mori, H. Mori, S. Tanaka: Angew. Chem. Int'l Edn. Engl. **34**, 1222-1225 (1995)

11.130 R.N. Lyubovskaya, R.B. Lyubovskii, R.P. Shibaeva, M.Z. Aldoshina, L.M. Gol'denberg, L.P. Rozenberg, M.L. Khidekel, Yu.F. Shul'pyakov: Pis'ma Zh. Eksp. Teor. Fiz. **42**, 380-383 (1985)

11.131 R.N. Lyubobskaya, E.A. Zhilyaeva, A.V. Zvarykina, V.N. Lakhin, R.B. Lyubovskii, S.I. Pesotskii: Pis'ma Zh. Eksp. Teor. Fiz. **45**, 416-418 (1987)

11.132 H. Mori, I. Hirabayashi, S. Tanaka, T. Mori, Y. Maruyama, H. Inokuchi: Solid State Commun. **80**, 411-415 (1991)

11.133 T. Mori, K. Kato, Y. Maruyama, H. Inokuchi, H. Mori, I. Hirabayashi, S. Tanaka: Solid State Commun. **82**, 177-181 (1992)

11.134 M. Kurmoo, A.W. Graham, P. Day, S.L. Coles, M.B. Hursthouse, J.L. Caufield, J. Singleton, F.L. Pratt, W. Hayes, L. Ducasse, P. Guionneau: J. Am. Chem. Soc. **117**, 12209-12217 (1995)

11.135 P. Cassoux, L. Valade, H. Kobayashi, A. Kobayashi, R.A. Clark, A.E. Underhill: Coord. Chem. Rev. **110**, 115-160 (1991)

11.136 R.-M. Olk, B. Olk, W. Dietzsch, R. Kirmse, E. Hoyer: Coord. Chem. Rev. **117**, 99-131 (1992)

11.137 V. Petricek, K. Maly, P. Coppens, X. Bu, I. Cisarova, A.F. Jensen: Acta. Cryst. A **47**, 210-216 (1991)

11.138 K. Deuchert, S. Hünig: Angew. Chem. Int'l Edn. Engl. **17**, 875-886 (1978)

11.139 K. Kimura, S. Nakajima, K. Niki, H. Inokuchi: Bull. Chem. Soc. Jpn. **58**, 1010-1012 (1985)

11.140 R.M. Metzger, C.A. Panetta: In *Lower-Dimensional Systems and Molecular Electronics*, ed. by R.M. Metzger, P. Day, G.C. Papavassiliou (Plenum, New York 1990) pp. 611-625

11.141 K.A. Abboud, M.B. Clevenger, G.F. de Oliveira, D.R. Talham: J. Chem. Soc., Chem. Commun. **1993**, 1560-1562 (1993)

11.142 L.-K. Chou, M.A. Quijada, M.B. Clevenger, G.F. de Oliveira, K.A. Abboud, D.B. Tanner, D.R. Talham: Chem. Mater. **7**, 530-???? (1995)

11.143 R.P. Shibaeva, R.M. Robkovskaya, V.E. Korotkov, N.D. Kushch, E.B. Yagubskii, M.K. Makova: Synth. Met. **27**, A457-463 (1988)

11.144 T. Mori, H. Inokuchi: Bull. Chem. Soc. Jpn. **61**, 591-593 (1988)

11.145 P. Batail, K. Boubekeur, A. Davidson, M. Fourmigue, C. Lenoir, C. Livage, A. Penicaud: *The Physics and Chemistry of Organic Superconductors*, ed. by G. Saito, S. Kagoshima. Springer Proc. Phys., Vol. 51 (Springer, Berlin, Heidelberg 1990) pp. 353-356

11.146 J.-P. Legros, L. Valade, P. Cassoux: Synth. Met. **27**, 347 (1988)

11.147 R.A. Clark, A.E. Underhill, I.D. Parker, R.H. Friend: J. Chem. Soc., Chem. Commun. **1989**, 228-229 (1989)

11.148 S. Mazumdar, A.N. Bloch: Phys. Rev. Lett. **50**, 207-211 (1983)

11.149 A.J. Epstein, J.S. Miller: In *The Physics and Chemistry of Low Dimensional Solids*, ed. by L. Alcacer (Reidel, Dordrecht 1980) pp. 339-351

11.150 J.C. Scott, S. Etemad, E.M. Engler: Phys. Rev. B **17**, 2269-2275 (1978)

11.151 T. Hasegawa, S. Kagoshima, T. Mochida, S. Sugiura, Y. Iwasa. Solid State Commun. **103**, 489-493 (1997)

11.152 R. Kumai, A. Asamitsu, Y. Tokura: Priv. Commun (1997)

11.153 C. Coulon, P. Delhaes, J. Amiell, J.P. Manceau, J.M. Fabre, L. Giral: J. Physique **43**, 1721-1729 (1982)

11.154 T. Naito, A. Miyamoto, H. Kobayashi, R. Kato, A. Kobayashi: Chem. Lett. **1992**, 119-122 (1992)

11.155 M. Tokumoto, H. Anzai, K. Murata, K. Kajimura, T. Ishiguro: Jpn. J. Appl. Phys. **26**, Suppl. 26-3, 1977-1982 (1987)

11.156 N.C. Kushch, S.I. Pesotskii, V.N. Topnikov: JETP Lett. **56**, 506-509 (1992)

11.157 N.D. Kushch, L.I. Buravov, A.G. Khomenko, E.B. Yagubskii, L.P. Rosenberg, R.P. Shibaeva: Synth. Met. **53**, 155-160 (1993)

11.158 H. Tanaka, A. Kobayashi, T. Saito, K. Kawano, T. Naito, H. Kobayashi: Adv. Mater. **8**, 812-815 (1996)

11.159 T. Komatsu, N. Matsukawa, T. Inoue, G. Saito: J. Phys. Soc. Jpn. **65**, 1340-1354 (1996)

11.160 U. Geiser, H.H. Wang, K.D. Carlson, J.M. Williams, H.A. Charlier Jr., J.E. Heindl, G.A. Yaconi, B.H. Love, M.W. Lathrop, J.E. Schirber, D.L. Overmyer, J. Ren, M.-H. Whangbo: Inorg. Chem. **30**, 2586-2588 (1991)

11.161 X. Bu, A. Frost-Jensen, R. Allendoerfer, P. Coppens, B. Lederle, M.J. Naughton: Solid State Commun. **79**, 1053-1057 (1991)

11.162 T.J. Emge, H.H. Wang, P.C.W. Leung, P.R. Rust, J.D. Cook, P.L. Jackson, K.D. Carlson, J.M. Williams, M.-H. Whangbo, E.L. Venturini, J.E. Schirber, L.J. Azevedo, J.R. Ferraro: J. Am. Chem. Soc. **106**, 695-702 (1986)

11.163 T.J. Emge, H.H. Wang, U. Geiser, M.A. Beno, K.S. Webb, J.M. Williams: J. Am. Chem. Soc. **108**, 3849-3850 (1986)

11.164 M.-H. Whangbo, D. Jung, J. Ren, M. Evain, J.J. Novoa, F. Mota, S. Alvarez, J.M. Williams, M.A. Beno, A.M. Kini, H.H. Wang, J.R. Ferraro: In *The Physics and Chemistry of Organic Superconductors*, ed. by G.Saito, S. Kagoshima. Springer Proc. Phys., Vol.51 (Springer, Berlin, Heidelberg 1990) pp.262-266 and the referneces cited therein

11.165 J.J. Novoa, M.-H. Whangbo, J.M. Williams: Mol. Cryst. Liq. Cryst. **181**, 25-42 (1990)

11.166 T. Nakamura, G. Yunome, R. Azumi, M. Tanaka, H. Tachibana, M. Matsumoto, S. Horiuchi, H. Yamochi, G. Saito: J. Phys. Chem. **98**, 1882-1887 (1994)

11.167 K. Ogasawara, T. Ishiguro, S. Horiuchi, H. Yamochi, G. Saito: Jpn. J. Appl. Phys. **35**, L571-L573 (1996)

11.168 S. Horiuchi, H. Yamochi, G. Saito, J.K. Jeszka, A. Tracz, A. Sroczynska, J. Ulanski: Mol. Cryst. Liq. Cryst. **296** 365-382 (1997)

11.169 S. Kahlich, D. Schweitzer, C. Rovira, J.A. Paradis, M.-H. Whangbo, I. Heinen, H.J. Keller, B. Nuber, P. Bele, H. Brunner, R.P. Shibaeva: Z. Physik B **94**, 39-47 (1994)

11.170 H. Yamochi, T. Komatsu, N. Matsukawa, G. Saito, T. Mori, M. Kusunoki, K. Sakaguchi: J. Am. Chem. Soc. **115**, 11319-11327 (1993)

11.171 M. Tokumoto, H. Bando, K. Murata, H. Anzai, N. Kinoshita, K. Kajimura, T. Ishiguro, G. Saito: Synth. Met. **13**, 9-20 (1986)

11.172 J.M. Williams, M.A. Beno, H.H. Wang, U.W. Geiser, T.J. Emge, P.C.W. Leung, G.W. Crabtree, K.D. Carlson, L.J. Azevedo, E.L. Venturini, J.E. Schirber, J.F. Kwak, M.-H. Whangbo: Physica B **136**, 371-375 (1986)

11.173 T.J. Kistenmacher: Solid State Commun. **63**, 977-981 (1987)

11.174 H. Urayama, H. Yamochi, G. Saito, K. Nozawa, T. Sugano, M. Kinoshita, S. Sato, K. Oshima, A. Kawamoto, J. Tanaka: Chem. Lett. **1988**, 55-58 (1988)

11.175 G. Saito, H. Urayama, H. Yamochi, K. Oshima: Synth. Met. **27**, A331-A340 (1988)

11.176 M. Oshima, H. Mori, G. Saito, K. Oshima: Chem. Lett. **1988**, 1159-1162 (1988); and in *The Physics and Chemistry of Organic Superconductors*, ed. by G. Saito, S.

Kagoshima, Springer Proc. Phys., Vol.51 (Springer, Berlin, Heidelberg 1990) pp.257-261

11.177 H.H. Wang, K.D. Carlson, U. Geiser, W.K. Kwok, M.D. Vashon, J.E. Thompson, N.F. Larsen, G.D. McCabe, R.S. Hulsher, J.M. Williams: Physica C **166**, 57-61 (1990)

11.178 G. Saito: In *Lower Dimensional Systems and Molecular Electronics*, ed. by R.M. Metzger, P. Day, G.C. Papavassiliou (Plenum, New York 1991) pp.67-84

11.179 J.M. Williams, A.M. Kini, H.H. Wang, K.D. Carlson, U. Geiser, L.K. Montgomery, G.J. Pyrka, D.M. Watkins, J.M. Kommers, S.J. Boryschuk, A.V. Streiby Crouch, W.K. Kwok, J.E. Schirber, D.L. Overmyer, D. Jung, M.-H. Whangbo: Inorg. Chem. **29**, 3272-3274 (1990)

11.180 A.M. Kini, U. Geiser, H.H. Wang, K.D. Carlson, J.M. Williams, W.K. Kwok, K.G. Vandervoort, J.E. Thompson, D.L. Stupka, D. Jung, M.-H. Whangbo: Inorg. Chem. **29**, 2555-2557 (1990)

11.181 T. Komatsu, T. Nakamura, N. Matsukawa, H. Yamochi, G. Saito, H. Ito, T. Ishiguro, M. Kusunoki, K. Sakaguchi: Solid State Commun. **80**, 843-847 (1991)

11.182 H. Ito, M. Watanabe, Y. Nogami, T. Ishiguro, T. Komatsu, G. Saito, N. Hosoito: J. Phys. Soc. Jpn. **60**, 3230-3233 (1991)

11.183 M.-H. Whangbo, J.M. Williams, A.J. Schultz, T.J. Emge, M.A. Beno: J. Am. Chem. Soc. **109**, 90-94 (1987)

11.184 M.-H. Whangbo, J.J. Novoa, D. Jung, J.M. Williams, A.M. Kini, H.H. Wang, U. Geiser: In *Organic Superconductivity*, ed. by V.Z. Kresin, W.A. Little (Plenum, New York 1990) pp.243-266

11.185 J.M. Williams, A.J. Schultz, U. Geiser, K.D. Carlson, A.M. Kini, H.H. Wang, W.-K. Kwok, M.-H. Whangbo, J.E. Schirber: Science **252**, 1501-1508 (1991)

11.186 Y. Watanabe, T. Shimazu, T. Sasaki, N. Toyota: Synth. Met. **86**, 1917-1918 (1994)

11.187 Hard and Soft Acids and Bases, ed. by R.G. Pearson, Dowden, Hutchingson, Ross (Stroudsburg, PA 1973)

11.188 J. Caulfield, W. Lubczynski, F.L. Pratt, J. Singleton, D.Y.K. Ko, W. Hayes, M. Kurmoo, P. Day: J. Phys. Condensed Matter **6**, 2911-2924 (1994)

11.189 N. Toyota, T. Sasaki: Solid State Commun. **72**, 859-862 (1989)

11.190 N. Toyota, T. Sasaki: Solid State Commun. **74**, 361-365 (1990)

11.191 J.P. Pouget: Mol. Cryst. Liq. Cryst. **230**, 101-131 (1993)

11.192 Y. Watanabe, H. Sato, T. Sasaki, N. Toyota: J. Phys. Soc. Jpn. **60**, 3608-3611 (1991)

11.193 R.C. Yu, J.M. Williams, H.H. Wang, J.E. Thompson, A.M. Kini, K.D. Carlson, J. Ren, M.-H. Whangbo, P.M. Chaikin: Phys. Rev. B **44**, 6932-6936 (1991)

11.194 A.B. Harris, R.V. Lange: Phys. Rev. **157**, 295-314 (67)

11.195 G. Saito, A. Otsuka, A.A. Zakhidov: Mol. Cryst. Liq. Cryst. **284**, 3-14 (1996)

11.196 W. Krätchmer, L.D. Lamb, K. Fostiropoulos, D.R. Huffman: Nature **347**, 354-358 (1990)

11.197 J.L. Atwood, G.A. Koutsantonis, C.L. Raston: Nature **368**, 229-231 (1994)

11.198 T. Suzuki, K. Nakashima, S. Shinkai: Chem. Lett. **1994**, 699-702 (1994)

11.199 B. Nie, V.M. Rotello: J. Org. Chem. **61**, 1870-1871 (1996)

11.200 S.H. Yang, C.L. Pettiette, J. Conceicao, O. Cheshnovsky, R.E. Smalley: Chem. Phys. Lett. **139**, 233-238 (1987)

11.201 G. Saito, T. Teramoto, A. Otsuka, Y. Sugita, T. Ban, M. Kusunoki, K. Sakaguchi: Synth. Met. **64**, 359-368 (1994)

11.202 C. Bossard, S. Rigaut, D. Astruc, M-H. Delville, G. Felix, A.F-Bouvier, J. Amiell, S. Flandrois, P. Delhaes: J. Chem. Soc., Chem. Commun. **1993**, 333-334 (1993)

11.203 J. Stinchcombe, A. Penicaud, P. Bhyrappa, P.D.W. Boyd, C.A. Reed: J. Am. Chem. Soc. **115**, 5212-5217 (1993)

11.204 P.-M. Allemand, K.C. Khemani, A. Koch, F. Wudl, K. Holczer, S.Donovan, G. Grüner, J.D. Thompson: Science **253**, 301-303 (1991)

11.205 A.P. Ramirez: Condensed Matter News **3**, 9-23 (1994)

11.206 K. Khazeni, V.H. Crespi, J. Hone, A. Zettl, M.L. Cohen: Phys. Rev. B **56**, 6627-6630 (1997)

11.207 K. Tanigaki: Optoelectronics - Devices and Technologies: **10**, 231-246 (1995)

11.208 R.M. Fleming, A.P. Ramirez, M.J. Rosseinsky, D.W. Murphy, R.C. Haddon, S.M. Zahurak, A.V. Makhija: Nature **352**, 787-788 (1991)

11.209 K. Tanigaki, I. Hirosawa, T.W. Ebbesen, J. Mizuki, Y. Shimakawa, Y. Kubo, J.S. Tsai, S. Kuroshima: Nature **356**, 419-421 (1992)

11.210 G. Saito, T. Komatsu, T. Nakamura, H. Yamochi: MRS. Proc. **247**, 483-494 (1992)

11.211 H. Shimoda, Y. Iwasa, Y. Miyamoto, Y. Maniwa, T. Mitani: Phys. Rev. B **54**, R15653-R15656 (1996)

11.212 A. Otsuka, G. Saito, A.A. Zakhidov, K. Yakushi, M. Kusunoki, K. Sakaguchi: Mol. Cryst. Liq. Cryst. **285**, 187-192 (1996)

11.213 D.D. Perrin, W.L.F. Armarego, D.R. Perrin: *Purification of Laboratory Chemicals*, 2nd edn. (Pergamon, Oxford 1980)

11.214 A.R. McGhie, A.F. Garito, J. Heeger: J. Cryst. Growth **22**, 295-297 (1984)

11.215 G. Saito, T. Inukai: J. Jpn. Assoc. Cryst. Growth **16**, 2-16 (1989)

11.216 H. Anzai, M. Tokumoto, G. Saito: Mol. Cryst. Liq. Cryst. **125**, 385-392 (1985)

11.217 H. Anzai, M. Tokumoto, K. Takahashi, T. Ishiguro: J. Cryst. Growth **91**, 225 (1988)

11.218 T.C. Chiang, A.H. Reddoch, D.W. Williams: J. Chem. Phys. **54**, 2051-2055 (1971)

11.219 F.B. Kaufman, E.M. Engler, D.C. Green, J.Q. Chamber: J. Am. Chem. Soc. **98**, 1596-1597 (1976)

11.220 B.A. Scott, S.J. La Placa, J.B. Torrance, B.D. Silvermann, B. Welber: J. Am. Chem. Soc. **99**, 6631-6639 (1977)

11.221 E.E. Laukhina: Priv. Commun. (year 1997)

11.222 G. Saito, T. Sugano, H. Yamochi, M. Kinoshita, K. Oshima, M. Suzuki, C. Katayama, J. Tanaka: Chem. Lett. **1985**, 1037-1040 (1985)

11.223 M. Bousseau, L. Valade, J.P. Legros, P. Cassoux, M. Garbauskas, L.V. Interrante: J. Am. Chem. Soc. **108**, 1908-1916 (1986)

11.224 A.F. Hebard, M.J. Rosseinsky, R.C. Haddon, D.W. Murphy, S.H. Glarum, T.T.M. Palstra, A.P. Ramirez, A.R. Kortan: Nature **350**, 600-601 (1991)

11.225 H.H. Wang, A.M. Kini, B.M. Savall, K.D. Carlson, J.M. Williams, K.R. Lykke, P. Wurz, D.H. Parker, M.J. Pellin, D.M. Gruen, U. Welp, W.K. Kwok, S. Fleshler, G.W. Crabtree: Inorg. Chem. **30**, 2838-2839 (1991)

11.226 F. Bensebaa, B. Xiang, L. Kevan: J. Phys. Chem. **96**, 6118 (1992)

11.227 J.P. McCauley Jr., Q. Zhu, N. Coustel, O. Zhou, G. Vaughan, S.H. Idziak, J.E. Fischer, S.W. Tozer, D.M. Groski, N. Bykovetz, C.L. Lin, A.R. McGhie, B.H. Allen, W.J. Romanow, A.M. Denenstein, A.B. Smith III: J. Am. Chem. Soc. 113, 8537-8538 (1991)

11.228 X.-D. Xiang, J.G. Hou, G. Briceno, W.A. Vareka, R. Mostovoy, A. Zettl, V.H. Crespi, M.L. Cohen: Science 256, 1190-1191 (1992)

11.229 J.W. Dykes, P. Klavins, M.D. Lan, J.Z. Liu, R.N. Shelton: J. Superconductivity 7, 635-637 (1994)

11.230 K. Bechgaard, C.S. Jacobsen, K. Mortensen, H.J. Pedersen, N. Thorup: Solid State Commun. 33, 1119-1125 (1980)

11.231 I.F. Schegolev, E.B. Yagubskii, V.N. Laukhin: Mol. Cryst. Liq. Cryst. 126, 365-377 (1985)

11.232 R.P. Shibaeva, R.M. Lobkovskaya, E.B. Yagubskii, E.E. Kostyuchenko: Sov. Phys. - Crystallogr. 31, 267-271 (1986)

11.233 R.P. Shibaeva, R.M. Lobkovskaya, E.B. Yagubskii, E.E. Kostyuchenko: Sov. Phys. - Crystallogr. 31, 657-659 (1986)

11.234 D. Schweitzer, E. Gogu, I. Hennig, T. Klutz, H.J. Keller: Ber. Bunsenges. Phys. Chem. 91, 890-896 (1987)

11.235 H. Kobayashi, R. Kato, A. Kobayashi, Y. Nishio, K. Kajita, W. Sasaki: Chem. Lett. 1986, 833-836 (1986)

11.236 E.I. Zhilyaeva, R.N. Lyubovskaya, O.A. Dyachenko, T.G. Takhirov, V.V. Gritsenko, S.V. Konovalikhin: Synth. Met. 41-43, 2247-2250 (1991)

11.237 A. Ugawa, G. Ojima, K. Yakushi, H. Kuroda: Phys. Rev. B 38, 5122-5125 (1988)

11.238 H. Mori: Int'l J. Mod. Phys. B 8, 1-45 (1994)

11.239 H.H. Wang, A.M. Kini, L.K. Montgomery, U. Geiser, K.D. Carlson, J.M. Williams, J.E. Thompson, D.M. Watkins, W.K. Kwok, U. Welp, K.G. Vandervoort: Chem. Mater. 2, 482-484 (1990)

11.240 T. Komatsu, N. Matsukawa, T. Nakamura, H. Yamochi, G. Saito, H. Ito, T. Ishiguro: Phosphorus, Sulfur, and Silicon 67, 295-300 (1992)

11.241 K. Kanoda: Solid State Phys. 30, 240-254 (1995)

11.242 J.M. Williams, A.M. Kini, H.H. Wang, K.D. Carlson, U. Geiser, L.K. Montgomery, G.J. Pyrka, D.M. Watkins, J.M. Kommers, S.J. Boryschuk, A.V.S. Crouch, W.K. Kwok, J.E. Schirber, D.L. Overmyer, D. Jung, M.-H. Whangbo: Inorg. Chem. 29, 3272-3274 (1990)

11.243 H. Kim, A. Kobayashi, Y. Sasaki, R. Kato, H. Kobayashi: Chem. Lett. 1987, 1799-1802 (1987)

Appendix

A.1 K. Yamaji: J. Phys. Soc. Jpn. 58, 1520 (1989)

A.2 R. Yagi, Y. Iye, T. Osada, S. Kagoshima: J. Phys. Soc. Jpn. 59, 3069-3072 (1990)

A.3 J.M. Ziman: Principles of the Theory of Solids (Cambridge Univ. Press, 1965) Chapt. 9

A.4 V.G. Peschansky, J.A.R. Lopez, T.G. Yao: J. Phys. I (France) 1, 1469-1479 (1991)

A.5 M.V. Kartsovnik, V.N. Laukhin, S.I. Pesotskii, I. F. Schegolev, V.M. Yakovenko: J. Phys. I (France) **2**, 89-99 (1991)

A.6 Y. Kurihara: J. Phys. Soc. Jpn. **61**, 975-982 (1992)

B.1 P.G. de Gennes: *Superconductivity in Metals and Alloys* (Benjamin, New York 1966)

B.2 M. Tinkham: Physica C **235-240**, 3-8 (1994)

  W.J. Skocpol, M. Tinkham: Rep. Prog. Phys. **38**, 1049 (1975)

B.3 G. Blatter, M.V. Feigel'man, V.B. Geshkenbein, A.I. Larkin, V.M. Vinokur: Rev. Mod. Phys. **66**, 1127-1388 (1994)

B.4 W.E. Lawrence, S. Doniach: In *Proc. 12th Int'l Conf. Low Temperature Physics* (Kyoto 1970), ed. by E. Kanda (Keigaku, Tokyo 1971) p.361

B.5 R. Ikeda, T. Ohmi, T. Tsuneto: J. Phys. Soc. Jpn. **60**, 1051-1069 and 1337-1346 (1991)

B.6 H. Ito, M. Watanabe, Y. Nogami, T. Ishiguro, T. Komatsu, G. Saito, N. Hosoito: J. Phys. Soc. Jpn. **60**, 3230-3233 (1991)

B.7 H. Ito, Y. Nogami, T. Ishiguro, T. Komatsu, G. Saito, N. Hosoito: Jpn. J. Appl. Phys. **7**, 419-425 (1992)

B.8 S. Ullah, A.T. Dorsey: Phys. Rev. B **44**, 262-273 (1991)

B.9 U. Welp, S. Fleshler, W.K. Kwok, R.A. Klemm, V.M. Vinokur, J. Downey, B. Veal, G.W. Crabtree: Phys. Rev. Lett. **67**, 3180 (1991)

B.10 M. Lang, F. Steglich, N. Toyota, T. Sasaki: Phys. Rev. B **49**, 15227 (1994)

B.11 S. Friemel, C. Pasquier, Y. Loirat, D. Jérome: Physica C **259**, 181-186 (1996)

Subject Index

Umklapp scattering 87-89, 368, 370
uniaxial strain 192
unusual pairing, unconventional pairing 188, 197
upper critical magnetic field, see H_{c_2}

valence state 422
van der Waals radius 91
variational formulation 111, 328-343
vertex correction 284, 293
vibronic interaction 311
vortex lock-in 192

vortex motion, dynamics 191, 198
vortex pinning 191

warping 37
weak electron correlation 242
weak ferromagnetic hysteresis 153
Werthermer-Helfand-Hohenberg formula 384
Wigner crystal 41

X-ray study 84, 122

ζ-parameter 95

Springer Series in Solid-State Sciences

Editors: M. Cardona P. Fulde K. von Klitzing H.-J. Queisser

Springer Series in Solid-State Sciences

Editors: M. Cardona   P. Fulde   K. von Klitzing   H.-J. Queisser

Springer
and the
environment

At Springer we firmly believe that an
international science publisher has a
special obligation to the environment,
and our corporate policies consistently
reflect this conviction.
We also expect our business partners –
paper mills, printers, packaging
manufacturers, etc. – to commit
themselves to using materials and
production processes that do not harm
the environment. The paper in this
book is made from low- or no-chlorine
pulp and is acid free, in conformance
with international standards for paper
permanency.

Printing: Mercedesdruck, Berlin
Binding: Buchbinderei Lüderitz & Bauer, Berlin